Advanced Hydroinformatic Techniques for the Simulation and Analysis of Water Supply and Distribution Systems

Special Issue Editors

Manuel Herrera
Silvia Meniconi
Stefano Alvisi
Joaquín Izquierdo

MDPI • Basel • Beijing • Wuhan • Barcelona • Belgrade

MDPI

Special Issue Editors

Manuel Herrera
University of Bath
UK

Silvia Meniconi
University of Perugia
Italy

Stefano Alvisi
Università degli Studi di Ferrara
Italy

Joaquín Izquierdo
Universitat Politècnica de València
Spain

Editorial Office
MDPI
St. Alban-Anlage 66
Basel, Switzerland

This edition is a reprint of the Special Issue published online in the open access journal *Water* (ISSN 2073-4441) from 2017–2018 (available at: http://www.mdpi.com/journal/water/special_issues/Hydroinformatic_water_supply_distribution).

For citation purposes, cite each article independently as indicated on the article page online and as indicated below:

Lastname, F.M.; Lastname, F.M. Article title. *Journal Name* **Year**, *Article number*, page range.

First Edition 2018

ISBN 978-3-03842-953-1 (Pbk)
ISBN 978-3-03842-954-8 (PDF)

Table of Contents

About the Special Issue Editors

Manuel Herrera, Ph.D. in Hydraulic Engineering and Environmental Studies. Manuel's research is mainly based on Data Science and Urban Informatics—with an emphasis on urban water distribution systems, as well as challenges related to the built environment. After obtaining his Ph.D. from the Universitat Politècnica de València (Spain), Manuel did postdoctoral work at the Université Libre de Bruxelles (Belgium) and Imperial College London (UK), focusing on Multidisciplinary Design Optimization and Complex Networks approaches to Engineering applications. He is working now as Research Associate at the University of Bath (UK). All this work has led him to become co-author of dozens of JCR indexed journal papers, several book chapters in international publishers, and more than 100 conference papers in international meetings. Manuel has also been a tutor of an Electrical Engineering MSc thesis and two Doctoral Dissertations in Water Engineering; he is currently tutoring another two on-going Ph.D. theses.

Silvia Meniconi, Ph.D. in Civil Engineering, Marie Curie fellow at the Eindhoven University of Technology, and Associate Professor at the University of Perugia. Her main research interests are transient test-based techniques for the analysis of pipe system performance, modelling of transients, leak-law for leakage evaluation, pressure control by PRVs, and monitoring of microbiological quality in WDS. Silvia is the author of more than 100 JCR indexed journal contributions, refereed conferences papers, book chapters, books, and the guest editor of seven Special Issues in JCR indexed journals, as well as the principal investigator of seven research projects and a member of the research unit of several international, national, and local projects. She has been a member of the organizing and/or scientific committee of numerous conferences (e.g., HIC2018, WDSA2014, and CCWI2013), and chairperson at some international conferences (e.g., IAHR2017). She received the Sustainability Award at the International Water Exhibition, H2O 2016 for the Portable Pressure Wave Maker device for the diagnosis of pressure pipe systems.

Stefano Alvisi, Ph.D. in Engineering and Associate Professor at the University of Ferrara. His main research interests are water demand modeling and forecasting, pipe network rehabilitation and leakage management, and data-driven techniques for real time flood forecasting models. He is the author of more than one hundred papers, most of them in JCR indexed journals. Tutor of 4 Ph.D. students. He has been involved in several European, national, and local research projects and is member of the editorial boards of some international journals. He was granted with outstanding reviewer recognition by *Environmental Modelling and Software Journal* (Elsevier) (2014) and *Journal of Water Resources Planning and Management* (ASCE) (2016), and with Certificates of Excellence to recognize outstanding achievements as the best paper nominees in the "Battle of Water Calibration Network (BWCN)" (2010), in the "Battle of Background leakage assessment for water networks (BBLAWN) (2014) and in the ICLP2011 conference.

Joaquín Izquierdo, Ph.D. in Mathematics and full professor of Applied Mathematics at the Universitat Politècnica de València (UPV, Spain). Educational career developed in the UPV, lecturing in algebra, calculus, differential equations, numerical methods applied to engineering, etc. Joaquin's research activity includes being the Director of the research group FluIng (fluing.upv.es) within the Institute for Multidisciplinary Mathematics (IMM) devoted to mathematical modelling, knowledge-based systems, and DSSs in Engineering (mainly Urban Hydraulics). He is the author and editor of several books, and the author or co-author of more than 300 research papers, book chapters, and contributions to international events (https://www.researchgate.net/profile/Joaquin_Izquierdo/contributions). He is also tutor of 12 Doctoral Dissertations and, currently, six on-going theses, as has participated in consulting in water resources in the context of Urban Hydraulics. He has extensive experience in continuous professional development for Spanish and Latin-American professionals, and is the author of commercial computer simulation packages. He has periodically organized international scientific events and participated in R&D projects at various levels: local, regional, national, European, and international.

Preface to "Advanced Hydroinformatic Techniques for the Simulation and Analysis of Water Supply and Distribution Systems"

More than half of the world's population live in cities—with an increase of 1.5 billion city dwellers in the last 20 years. The United Nations predicts that this trend will continue; it is expected that up to 6.5 billion people will be living in cities by 2050. One of the main challenges for managers and authorities is to provide an adequate quality and quantity of water to people living and working in cities. This book encompasses 18 articles that comprise the special issue of the MDPI journal WATER, entitled: 'Advanced Hydroinformatic Techniques for the Simulation and Analysis of Water Supply and Distribution Systems'. A wide range of topics is presented by world-class researchers and academics that covers water distribution system design, optimisation of network performance assessment, monitoring and diagnosis of pressure pipe systems, optimal water quality management, and modelling and forecasting water demand. The book also explores new research avenues in urban hydraulics and hydroinformatics to aid the decision-making processes of water utility managers and practitioners.

Manuel Herrera, Silvia Meniconi , Stefano Alvisi and Joaquín Izquierdo

Special Issue Editors

water · MDPI

Editorial

Advanced Hydroinformatic Techniques for the Simulation and Analysis of Water Supply and Distribution Systems

Manuel Herrera [1,*] [iD], **Silvia Meniconi** [2] [iD], **Stefano Alvisi** [3] [iD] and **Joaquín Izquierdo** [4] [iD]

1   Department of Architecture & Civil Engineering, University of Bath, Claverton Down, Bath BA2 7AZ, UK
2   Department of Civil and Environmental Engineering, University of Perugia, via G. Duranti 93,
    06125 Perugia, Italy; silvia.meniconi@unipg.it
3   Department of Engineering, University of Ferrara, via Saragat, 1, 44122 Ferrara, Italy; stefano.alvisi@unife.it
4   Institute for Multidisciplinary Mathematics, Universitat Politècnica de València, Cno. de Vera, s/n,
    46022 Valencia, Spain; jizquier@upv.es
*   Correspondence: A.M.Herrera.Fernandez@bath.ac.uk; Tel.: +44-1225-38-4440

Received: 21 March 2018; Accepted: 4 April 2018; Published: 8 April 2018

Abstract: This document is intended to be a presentation of the Special Issue "Advanced Hydroinformatic Techniques for the Simulation and Analysis of Water Supply and Distribution Systems". The final aim of this Special Issue is to propose a suitable framework supporting insightful hydraulic mechanisms to aid the decision-making processes of water utility managers and practitioners. Its 18 peer-reviewed articles present as varied topics as: water distribution system design, optimization of network performance assessment, monitoring and diagnosis of pressure pipe systems, optimal water quality management, and modelling and forecasting water demand. Overall, these articles explore new research avenues on urban hydraulics and hydroinformatics, showing to be of great value for both Academia and those water utility stakeholders.

Keywords: water distribution design; water network performance; pressure pipe system; water quality; water demand

1. Introduction

One of the most complex structures that an intelligent city has to manage is its water distribution system (WDS), which must provide water to citizens in adequate quantity and quality. This complexity is twofold. On the one hand, it is well known that the classical hydraulic models that describe the phenomena that take place in a WDS are of a complex nature, given the characteristics of the equations that describe these phenomena and their eminently distributed nature. On the other hand, the galloping recent need to handle large amounts of data obtained from the monitoring of systems has brought classical complexity to new paradigms that need new ways of addressing the problems of Urban Hydraulics. In this regard, special attention should be given to the achievement of an adequate digital connection related to the availability of data in real time, which allows effective solutions for demand prediction and other water utilities operations.

In essence, WDSs must be adequately designed (in the case of new systems) and adequately rehabilitated (extensions, renovation, restoration, etc., in later stages) so that they supply the user at all times and places under given satisfactory conditions. WDSs must be adequately monitored in order to obtain quality data in real time that allows an efficient control of the system. In addition, suitable (optimal) operation for the quality service to be provided continuously, without interruption, is essential. Finally, intelligent management that is capable of reconciling conflicting objectives such as economic benefit and social satisfaction, among others, must be an inescapable condition for ruling a WDS.

To achieve these objectives, efficient techniques are necessary to overcome the complexity of the associated problems. For example, in the tasks of design and rehabilitation, optimization algorithms are needed that are capable of manipulating nonlinearity, the coexistence of variables of different types, the discrete nature of some processes, etc.; these requirements impose the need to transcend classical optimization and to use modern techniques of evolutionary optimization. The real-time monitoring of the quality of the service will be greatly benefited by efficient time series processing techniques and several other forms of intelligent data manipulation. The operation of a system can be defined in terms of certain operators and variables of Boolean type, which must be optimally defined and integrated into the models and data structures in an appropriate manner; again, efficient techniques of optimization and fusion of methodologies that are able to work with data in real time will be necessary. Finally, the management of WDSs is currently carried out through a wide range of elements, such as demand prediction, network sectorization, leak detection, system maintenance through appropriate policies, control of transients, evaluation of user satisfaction, etc. Moreover, some of the elements that intervene in decision-making are quantifiable, while others must be classified as intangible; therefore, it is crucial to have adequate techniques for handling the information to be manipulated, which will frequently be affected by uncertainty and subjectivity.

In the water supply industry, as in other fields, any improvement that can occur in the treatment and handling of big data will produce considerable and obvious benefits. For example, through the installation of advanced measurement infrastructure (AMI) and the more efficient treatment of the data obtained, it will be possible to reduce more effectively the unaccounted-for water in the short term. More generally, in the long term, a more efficient operation that is expected from such improvements will contribute to the excellence of the urban water cycle, one of the objectives of an intelligent city. The idea is to promote the implementation of the smart city concept from the perspective of water supply.

2. Overview of This Special Issue

This issue contains 18 papers which focus on some of the mentioned problems of water distribution system management. The key points are: (i) design of water system [1–4]; (ii) optimization of network performance assessment [5–8]; (iii) monitoring and diagnosis of pressure pipe system [9–11]; (iv) optimal water quality management [12–14]; and (v) modelling and forecasting of water demand [15–18].

2.1. Design of Water System

Four papers of this issue examine the first key point that is design of water system. Firstly, Mala-Jetmarova et al. present a systematic literature review of optimization of WDS design since the end of the 1980s [1]. The review classifies the examined papers by the following issues: the type of design problems (i.e., static or dynamic), the application area (i.e., new or existing systems, with the optional inclusions of system operations), the optimization model (i.e., objective functions, constraints, and decision variables), and the analysed networks. It pinpoints trends and limits and suggests further research directions. Specifically, it reveals that there is not a consensus about the best WDS design optimization model, and consequently, researchers would force themselves to compare and validate different methods on real case studies.

Secondly, different multiobjective evolutionary algorithms (MOEAs) are compared in [2] on four well-known benchmark networks (i.e., two loop, Hanoi, Fossolo, and Balerma irrigation networks), by taking into account two objective functions: cost minimization and resiliency index maximization. A new hybrid algorithm that combines differential evolution and harmony search algorithm has been proposed for WDS design and compared with five MOEAs (i.e., NSGA2, AMALGAM, Borg, "ε-MOEA", and "ε-NSGA2"): the comparison shows that the new approach outperforms the previous MOEAs.

Thirdly, in [3] a new approach has been developed: to reduce the search space it bounds the pipe diameter values by analysing two opposite extreme flow distribution scenarios (i.e., uniform and maximum flow distribution) and applying velocity constraints. This model has been coupled to a genetic algorithm (GA) to improve its performance. The approach is applied to two benchmark networks (i.e., again two-loop network and Hanoi networks) by taking into account the cost minimization as objective function. By means of the new approach, the search space is reduced to less than 3% of the total search space for both the analysed networks. The results are compared to the classical GA: the comparison shows that the new approach is much faster and more accurate.

Finally, Zheng et al. describe experimental tests aimed at the optimal design of circular drop manholes in urban drainage networks [4]. Particularly, free, pressurized, and constrained outflow conditions have been tested for different manhole heights. Tests show that the local head loss coefficient of the manhole strictly depends on the outflow conditions. As a concluding remark, some empirical equations have been proposed to evaluate this coefficient.

2.2. Optimization of Network Performance Assessment

Four papers refer to the second key point that is optimization of network performance assessment. First, Sadatiyan and Miller [5] introduce a multiobjective version of the Pollution Emission Pump Station Optimization tool (PEPSO). It can be used to find a pump schedule of a WDS to reduce both the electricity cost and pollution emissions, by measuring the Undesirability Index in a nondominated sorting genetic algorithm. Tests carried out on the WDS of Monroe City (MI, USA) and Richmond (UK) show that PEPSO can optimize and provide useful information in a very limited amount of time.

Second, the main aim of the paper by Zischg et al. [6] is to assist decision makers in testing various planning options and design strategies during long-term city transitions. The procedure consists of the automatic creation, simulation, and analysis of different WDS scenarios. The pressure head, water age, and pressure surplus have been taken into account. Moreover, if data are not available, the approach uses alternative systems with strong similarity to WDSs. The proposed methodology is applied to the Swedish town Kiruna, in which it allows understanding the lack of the sole design at the final-stage WDS for most of the future scenarios and planning options.

Third, a procedure has been developed by Ilaya-Ayza et al. [7] to define district metered areas (DMAs) in WDSs with intermittent supply. The chosen objective function is the water supply equity. The approach uses soft computing tools from graph theory and cluster analysis and both the company expert opinions and adequate supply times for each DMA have been taken into account. The considered case study is the water supply network of Oruro (Bolivia), for which the proposed sectorization allows a clear improvement of the resilience index of the entire network.

Fourth, di Nardo et al. [8] propose the application of graph spectral theory (GST) for the optimal network sectorization. The approach is applied to two case studies (i.e., the well-known C-Town network and a real small WDS of Parete, Italy), and GST allows ranking WDS nodes and selecting the most important nodes for monitoring water quality, flow, or pressure, and for defining the DMAs. The main advantage of such an approach is that this is based only on topological and geometric information and no hydraulic data—often not available—are required.

2.3. Monitoring and Diagnosis of Pressure Pipe System

Three papers are part of the third key point that is monitoring and diagnosis of pressure pipe systems. First, Duan [9] investigates analytically and numerically the impact of nonuniformities of pipe diameter on transient wave behavior. Specifically, it demonstrates the dependence of wave scattering on the relationship between the incident wave frequency and nonuniform pipe diameter frequency, and nine numerical tests have been carried out by varying the pipe diameter nonuniformities (i.e., regular or random) and this relationship. As a result, the wave scattering has a nonnegligible effect

3

on wave reflections and attenuation, and, consequently, it has to be taken into account in transient modelling—along with both the unsteady friction and viscoelasticity—and in the application of Transient Test-Based Techniques (TTBTs) for the diagnosis of pressure pipe systems.

Second, Lin [10] presents a hybrid heuristic optimization approach called leak detection ordinal symbiotic organism search (LDOSOS) for locating and sizing leaks in a WDS. This approach combines the ordinal optimization algorithm (OOA) and the symbiotic organism search (SOS) in an inverse transient analysis (ITA). Moreover, the problem of generation of pressure waves is discussed and SOS is used to determine the optimal transient generation point. The procedure is tested on two numerical case studies: a two-loop network with a constant head supply reservoir and two very closed leaks, and a more complex network with a supply node with a constant inflow rate, a larger range of pipe diameters and lengths, and two distant leaks. Tests show that the LDOSOS has the ability to detect leak number, location, and size, by speeding up the ITA convergence and improving the reliability of the results.

Third, in Meniconi et al. [11] TTBTs are used to detect system defects and characteristics by monitoring the pressure waves at key points. The transmission main of the city of Trento (I) was analysed and transient tests were executed by pump shutdown. By means of the comparison of the numerical model and the acquired pressure signal, the relevance of the topology, pipe material characteristics, transient energy dissipation, and defects has been explored. Specifically, two malfunctioning valves have been detected and a preliminary criterion for the skeletonization of the transmission mains has been proposed.

2.4. Optimal Management of Water Quality

Three papers of this issue analyse the optimal management of water quality. Specifically, in Meyers et al. [12] a long-term continuous study of discolouration mobilisation is presented along with a methodology to determine the approximate amount and origin of hydraulically mobilised turbidity in trunk mains. The methodology is validated on three UK trunk main networks, observed over a period of about three years. The results show that the mobilisation of discolouration material is mainly determined by hydraulic forces, and consequently can be modelled and predicted, and its origin can be approximately determined.

Furthermore, in de Melo et al. [13], the factors that influence the water quality of the Jucazinho reservoir in northeastern Brazil have been pointed out by a data base of nine years of water quality reservoir monitoring and a multivariate statistical technique (i.e., the Principal Component Analysis, PCA). The study points out the connection between water quality parameters and the rainfall that has an annual or seasonal pattern. Precisely, two principal components of the water quality of this reservoir have been selected by PCA. The first, ranging from an annual basis, explains the increase in the concentration of dissolved solids and the cyanobacteria proliferation as a function of the drought period, during which the turbidity and the levels of total phosphorus decrease. The second, ranging from a monthly basis, indicates the connection between the process of photosynthesis performed by cyanobacteria with the percentage of the volume of the dam.

Finally, Reynoso-Meza et al. [14] have incorporated two decision-making methodologies (i.e., Technique for Order of Preference by Similarity to Ideal Solution—TOPSIS, and Preference Ranking Organisation Method for Enrichment of Evaluations—PROMETHEE) in a Multiobjective Evolutionary Algorithm (MOEA). The analysed multiobjective problems are two typical water quality problems: the dissolved oxygen problem for the activated sludge wastewater treatment process and the water quality of a river polluted by a cannery industry (Pierce-Hall Cannery) and the effluent from three treatment plants. The case studies have validated the reliability of such approaches for the degree of flexibility to capture designers' preferences.

2.5. Modelling and Forecasting of Water Demand

Four papers of this issue examine the last key point, modelling and forecasting of water demand. Firstly, Letting et al. [15] present a water demand calibration approach. The approach is aimed at estimating the water demand multiplier at each node of a water distribution system model by minimizing the error between observed and simulated nodal head and pipe flow rates. An optimization approach based on Particle Swarm algorithm is used. Application to a simple case study (Epanet example Net1) and a medium-sized real network highlights that the approach can provide an accurate water demand multiplier estimation by using data observed in a limited number of properly placed sensors.

Secondly, Pastor-Jabaloyes at al. [16] present an automatic tool for smart metered water demand time series disaggregation into single-use events. The tool is based on a filter automatically calibrated by using NSGA-II algorithm, and on a cropping algorithm. Furthermore, a semiautomatic classification is subsequently performed in order to categorize the obtained single-use events into different water end uses in a household such as shower, toilet, etc. The tool is applied to water demand time series collected from 20 households featuring very different characteristics in terms of geographical location, number of inhabitants, and average daily consumptions.

Thirdly, Anele et al. [17] provide an overview of some methods for short-term water demand forecast pointing out their pro and cons. The methods considered are univariate time series, time series regression, artificial neural network, and hybrid methods (i.e., a combination of two or more of the previous methods). The methods are applied to a case study highlighting that univariate time series, time series regression, and hybrid models may be accurate and appropriate for short-term water demand forecast. However, these methods are not applicable in more general decision problem frameworks. Indeed, these methods cannot be used to understand and analyse the overall level of uncertainty in future demand forecasts and thus much more attention needs to be given to probabilistic forecasting methods for short-term water demand forecast.

Finally, and strictly related to the previous considerations, a Markov-chain-based approach for probabilistic short-term water demand forecasting is presented by Gagliardi et al. [18]. In particular, two models based on homogeneous and nonhomogeneous Markov chains are proposed. The models are capable of providing both a deterministic forecast of the future values of water demand, and a characterization of the stochastic behaviour of the forecasted values. The models are applied to water demand time series of three district metered areas in the UK, and the deterministic forecast compared with those provided by neural network-based and naïve forecasting models, highlighting that the homogeneous Markov chain model provides both an accurate deterministic forecast and useful information regarding the probability distribution of the forecast itself.

3. Conclusions

In addition to the complexity inherent to WDS management, there often is a need for online actions to accomplish decision-making processes in real time. Another challenge that water companies should face nowadays is handling the huge amount of data generated by supervisory control and data acquisition (SCADA) systems, smart water meters, and other cyber–physical systems. This Special Issue on "Advanced Hydroinformatic Techniques for the Simulation and Analysis of Water Supply and Distribution Systems" presents a number of powerful techniques able to cope with such a complexity associated with: the nature of the hydraulic models, real-time requirements, and large scale databases. Bio-inspired and evolutionary algorithms play an important role in dealing with these issues. This is the reason why various contributions presented herein are based on these techniques. Overall, the Special Issue encompasses a collection of proposals that can be classified as follows:

- optimization, both classical and evolutionary;
- definition of structures and tools for big data;
- neural networks, support vector machines, and other Machine Learning techniques;

- graph theory and methods for complex networks;
- efficient treatment of time series;
- agent-based systems;
- multi-attribute decision-making techniques;
- transient test-based techniques for the diagnosis of pressure pipe systems; and
- other mathematical and computational tools and techniques

These techniques have been developed within the field of the so-called Hydroinformatics in Urban Hydraulics, that is, with application to problems such as:

- smart water networks (intelligent measurement, intelligent analysis of measurement data, ...);
- online analysis of WDSs (prediction of online demand, estimation of states, ...);
- water quality aspects (water quality characterization, prediction of discolouration, ...);
- reduction of unaccounted-for water and optimization of operation (sectorization, leak detection, operation indicators, water balance, and benchmarking, ...);
- optimal operation (of pumping stations, scheduling, transient control, ...);
- efficient utilization of real-time monitoring signals through smart treatment of online raw data using suitable learning approaches

Contributions to this Special Issue, exploring those new research avenues on urban hydraulics and hydroinformatics, are expected to be of great value for both Academia and all water utility stakeholders. On top of this, important social benefits are expected from a number of research objectives that ultimately aim to guarantee a regular supply of clean water at the pressure and quality required at the network consumption points.

Author Contributions: Each of the authors contributed to the design, analysis and writing of the manuscript.

Conflicts of Interest: The authors declare no conflict of interest.

References

1. Mala-Jetmarova, H.; Sultanova, N.; Savic, D. Lost in Optimisation of Water Distribution Systems? A Literature Review of System Design. *Water* **2018**, *10*, 307. [CrossRef]
2. Yazdi, J.; Choi, Y.H.; Kim, J.H. Non-Dominated Sorting Harmony Search Differential Evolution (NS-HS-DE): A Hybrid Algorithm for Multi-Objective Design of Water Distribution Networks. *Water* **2017**, *9*, 587. [CrossRef]
3. Reca, J.; Martínez, J.; López, R. A Hybrid Water Distribution Networks Design Optimization Method Based on a Search Space Reduction Approach and a Genetic Algorithm. *Water* **2017**, *9*, 845. [CrossRef]
4. Zheng, F.; Li, Y.; Zhao, J.; An, J. Energy Dissipation in Circular Drop Manholes under Different Outflow Conditions. *Water* **2017**, *9*, 752. [CrossRef]
5. Sadatiyan A., S.M.; Miller, C.J. PEPSO: Reducing Electricity Usage and Associated Pollution Emissions of Water Pumps. *Water* **2017**, *9*, 640. [CrossRef]
6. Zischg, J.; Mair, M.; Rauch, W.; Sitzenfrei, R. Enabling Efficient and Sustainable Transitions of Water Distribution Systems under Network Structure Uncertainty. *Water* **2017**, *9*, 715. [CrossRef]
7. Ilaya-Ayza, A.E.; Martins, C.; Campbell, E.; Izquierdo, J. Implementation of DMAs in Intermittent Water Supply Networks Based on Equity Criteria. *Water* **2017**, *9*, 851. [CrossRef]
8. Di Nardo, A.; Giudicianni, C.; Greco, R.; Herrera, M.; Santonastaso, G.F. Applications of Graph Spectral Techniques to Water Distribution Network Management. *Water* **2018**, *10*, 45. [CrossRef]
9. Duan, H.-F. Transient Wave Scattering and Its Influence on Transient Analysis and Leak Detection in Urban Water Supply Systems: Theoretical Analysis and Numerical Validation. *Water* **2017**, *9*, 789. [CrossRef]
10. Lin, C.-C. A Hybrid Heuristic Optimization Approach for Leak Detection in Pipe Networks Using Ordinal Optimization Approach and the Symbiotic Organism Search. *Water* **2017**, *9*, 812. [CrossRef]
11. Meniconi, S.; Brunone, B.; Frisinghelli, M. On the Role of Minor Branches, Energy Dissipation, and Small Defects in the Transient Response of Transmission Mains. *Water* **2018**, *10*, 187. [CrossRef]

12. Meyers, G.; Kapelan, Z.; Keedwell, E. Data-Driven Study of Discolouration Material Mobilisation in Trunk Mains. *Water* **2017**, *9*, 811. [CrossRef]
13. De Melo, R.R.C.; Rameh Barbosa, I.M.B.; Ferreira, A.A.; Lee Barbosa Firmo, A.; da Silva, S.R.; Cirilo, J.A.; de Aquino, R.R.B. Influence of Extreme Strength in Water Quality of the Jucazinho Reservoir, Northeastern Brazil, PE. *Water* **2017**, *9*, 955. [CrossRef]
14. Reynoso-Meza, G.; Alves Ribeiro, V.H.; Carreño-Alvarado, E.P. A Comparison of Preference Handling Techniques in Multi-Objective Optimisation for Water Distribution Systems. *Water* **2017**, *9*, 996. [CrossRef]
15. Letting, L.K.; Hamam, Y.; Abu-Mahfouz, A.M. Estimation of Water Demand in Water Distribution Systems Using Particle Swarm Optimization. *Water* **2017**, *9*, 593. [CrossRef]
16. Pastor-Jabaloyes, L.; Arregui, F.J.; Cobacho, R. Water End Use Disaggregation Based on Soft Computing Techniques. *Water* **2018**, *10*, 46. [CrossRef]
17. Anele, A.O.; Hamam, Y.; Abu-Mahfouz, A.M.; Todini, E. Overview, Comparative Assessment and Recommendations of Forecasting Models for Short-Term Water Demand Prediction. *Water* **2017**, *9*, 887. [CrossRef]
18. Gagliardi, F.; Alvisi, S.; Kapelan, Z.; Franchini, M. A Probabilistic Short-Term Water Demand Forecasting Model Based on the Markov Chain. *Water* **2017**, *9*, 507. [CrossRef]

water

MDPI

Review

Lost in Optimisation of Water Distribution Systems? A Literature Review of System Design

Helena Mala-Jetmarova [1], Nargiz Sultanova [2] and Dragan Savic [1,*]

[1] College of Engineering, Mathematics and Physical Sciences, University of Exeter, Streatham Campus, North Park Road, Exeter, Devon EX4 4QF, UK; h.malajetmarova@exeter.ac.uk

[2] Faculty of Science and Technology, Federation University Australia, Mt Helen Campus, University Drive, Ballarat, Victoria 3350, Australia; n.sultanova@federation.edu.au

* Correspondence: d.savic@exeter.ac.uk; Tel.: +44-139-272-3637

Received: 16 January 2018; Accepted: 22 February 2018; Published: 13 March 2018

Abstract: Optimisation of water distribution system design is a well-established research field, which has been extremely productive since the end of the 1980s. Its primary focus is to minimise the cost of a proposed pipe network infrastructure. This paper reviews in a systematic manner articles published over the past three decades, which are relevant to the design of new water distribution systems, and the strengthening, expansion and rehabilitation of existing water distribution systems, inclusive of design timing, parameter uncertainty, water quality, and operational considerations. It identifies trends and limits in the field, and provides future research directions. Exclusively, this review paper also contains comprehensive information from over one hundred and twenty publications in a tabular form, including optimisation model formulations, solution methodologies used, and other important details.

Keywords: water distribution systems; optimisation; literature review; design; rehabilitation; algorithms

1. Introduction

Water distribution systems (WDSs) are one of the major infrastructure assets of the society, with new systems being continually developed reflecting the population growth, and existing systems being upgraded and extended due to raising water demands. Designing economically effective WDSs is a complex task, which involves solving a large number of simultaneous nonlinear network equations, and at the same time, optimising sizes, locations, and operational statuses of network components such as pipes, pumps, tanks and valves [1]. This task becomes even more complex when the optimisation problem involves a larger number of requirements for the designed system to comply with (e.g., water quality), includes additional objectives beside a least-cost economic measure (e.g., potential fire damage) and incorporates more real-life aspects (e.g., uncertainty, staging of construction).

The early research related to the design optimisation of WDSs can be dated from the 1890s to 1950s. It was based on the principle of economic velocity [2–4], which was gradually reviewed and replaced by establishing the minimum (annual) costs of the system (i.e., least-cost design) [5–7]. Due to lack of computational technology in that period, those previous studies involved manual calculations combined with graphical methods, often resulting in practical charts to derive economic pipe diameters. The development of the optimisation of WDS design, therefore, had been an incremental process over time and may have appeared to be "only too true that the design of the transmission and distribution system receives [at that period] little attention in spite of the great sums of money invested in such installations" [8].

A successive period from the 1960s to 1980s displays a more rapid progression, which was initiated by the introduction of digital computers to network analysis in 1957 [9]. The introduction of computers was subsequently followed by the development of iterative methods [10,11] and simulation packages [12,13] to solve simultaneous nonlinear network equations, and eventuated in the application of mathematical deterministic methods to solve WDS design optimisation problems. These methods, including linear programming (LP) [14], nonlinear programming (NLP) [15,16], and others [17], typically minimised the design or capital (and operational) costs of the system, which were combined into one economic measure.

Another significant advancement in the optimisation of WDSs represented an introduction of stochastic methods using principles of biological evolution [18] and natural genetics [19]. Nonetheless, it was not until the 1990s when these methods became more popular [20] due to their ability to solve complex, real-world problems for which deterministic methods incured difficulty or failed to tackle them at all [21,22], and to also control multiple objectives. The popularity of metaheuristics has resulted in a dramatic increase in the application [21,23] to optimal design of WDSs, with "the several hundred research papers written on the subject" by 2001 [24]. Optimisation of WDS design has also progressed from a cost-driven single-objective framework to multi-objective models, when various objectives that continually gain importance (e.g., environmental objectives, community objectives reflecting the level of service provided to customers) can be evaluated on more equal basis [25]. Some of the most recent developments include the use of an engineered (as opposed to a random) initial population to improve the algorithm convergence [26], application of online artificial neural networks (ANNs) to replace network simulations [27], analysis of the algorithm search behaviour [28] in relation to the WDS design problem features [29], and reduction of the search space [30] to increase computational efficiency.

2. Aim, Scope and Structure of the Paper

This paper aims to provide a comprehensive and systematic review of publications since the end of the 1980s to nowadays, which are relevant to the optimisation of WDS design, strengthening (i.e., pipe paralleling), expansion and rehabilitation. The purpose of the review is to enable one's speedy familiarisation with the scope of the field, insight in the overwhelming amount of publications available and realisation of the future research directions. This paper contributes to and goes beyond the existing review literature for the optimisation of WDS design and rehabilitation [20,21,31–39] by not only identifying trends and limitations in the field, but also by providing comprehensive information from over one hundred and twenty publications in a tabular form, including optimisation model formulations, solution methodologies used, and other important details.

The paper consists of two parts: (i) the main review and (ii) an appendix in a tabular form (further referred to as the table), each having a different structure and purpose. The main review is structured according to publications' design problems and general classification. The design problems cover application areas, such as new system design, existing system strengthening, expansion and rehabilitation, and time, uncertainty and performance considerations. The general classification captures all the main aspects of a design optimisation problem answering the questions: what is optimised (Section 4.1), how is the problem defined (Section 4.2), how is the problem solved (Section 4.3) and what is the application (Section 4.4)? The purpose of the main review is to provide the current status, analysis and synthesis of the current literature, and to suggest future research directions.

A significant portion of this review paper is represented by the table, which refers to over one hundred and twenty publications in a chronological order. Each paper is classified according to an optimisation model (i.e., objective functions, constraints, decision variables), water quality parameter(s), network analysis, optimisation method and test network(s) used. Obtained results as well as other relevant information are also included. The purpose of the table is to provide a representative list of publications on the topic detailing comprehensive information, so that it could be used as a primary reference point to identify one's papers of interest in a timely manner. Hence, it presents a unique and integral contribution of this review.

The structure of the paper is as follows:

- The main review: Design problems (Section 3), General classification of reviewed publications (Section 4), Future research (Section 5), Summary and conclusion (Section 6), List of terms (Section 7), List of abbreviations.
- The table: Appendix A.

3. Design Problems

Two types of a design problem have been identified based on the field progression as follows: (i) a traditional design (i.e., theoretical or static design) of a WDS with a single construction phase for an entire expected life cycle of the system usually considering fixed loading conditions reflecting maximum (and other) future demands (Section 3.1); (ii) an advanced design (i.e., real-life or dynamic design) of a WDS capturing the system modifications and growth (due to the development of the populated area) over multiple construction phases, including future uncertainties (e.g., in demands, pipe deterioration) and other performance considerations (Section 3.2).

3.1. Application Areas

3.1.1. New Systems: Design

Critical infrastructure, including water, energy and transport systems, is essential in ensuring the survival and wellbeing of populations worldwide. Since the ancient Greek civilisations, WDSs have been an important part of making human settlements sustainable, thus optimising these systems to meet various requirements has over time gained interest of researchers and practitioners alike. Generally, optimisation of WDS design involves determining sizes, locations and operational statuses of network components such as pipes, pumps, tanks and valves, while keeping the system design or capital (and operational) costs at their minimum. The problem scope is primarily dependent on a type of a WDS under consideration, which is either a branched or looped and gravity or pumped system.

A network topology, branched or looped, represents a fundamental distinction in the problem complexity at the network analysis stage due to a way of determining flows in pipes. In branched networks there is a unique flow distribution calculated directly using nodal demands, while in looped systems flows can undertake multiple and alternative paths from a source to a customer [40]. This possible variability results in iterative methods being required to solve pipe flows in looped networks, such as that described in [41].

Regarding gravity WDSs, a basic optimisation model minimises the design cost of the network subject to the nodal pressure requirements, with pipe sizes or diameters being the only decision variables [42–48]. Popular test networks used to solve such a problem are the two-loop network [14], Hanoi network [49] and Balerma irrigation network [50]. As far as pumped WDSs are concerned, the optimisation problem becomes more complex than in the case of gravity WDSs, because of the presence of pumps and tanks (see Section 3.1.3), which require selecting not only their sizes and locations [14,26,51,52], but also their operational statuses [14,29,53,54], as well as often running an extended period simulation (EPS) for multiple loading conditions. Unlike for gravity WDSs, there does not seem to be any test network that is frequently used by multiple authors for pumped WDSs.

Regarding test networks, nevertheless, study [26] comments that they are limited, in general, to simple transmission networks, so-called benchmark systems, excluding local distribution lines. This exclusion is mainly due to a dramatic increase in the problem dimension, thus computational time, if local pipes were included. A problem of excluding smaller distribution pipes from the optimisation is in oversizing the transmission mains, as local distribution networks provide alternative pathways and display significant capacity to carry when the transmission lines are out of service [26]. The lack of large and complex test networks has recently been addressed by a number of researchers [55–57] who developed methodologies for generating synthetic networks of varying sizes and complexity levels. Furthermore, several real-world networks have been used for the design competitions by international research teams working in the area of WDS design, including those that are described by [58,59].

The problem complexity further increases by considering multiple simultaneous objectives. Initially, single-objective optimisation models were used to formulate WDS design problems, in which all objectives are combined into one economic (i.e., least-cost) measure (see, for example, [14,51,60–62]). A multi-objective optimisation approach was possibly first applied in the late 1990s (Figure 1), maximising the network benefit on one hand and minimising the system cost (of network rehabilitation) on the other hand [63]. In studies of newly designed WDSs, in addition to the economic measure, the other objectives considered were the pressure deficit [30,62,64–67] or excess [68,69] at network nodes, the penalty cost for violating the pressure constraint [70], greenhouse gas (GHG) emissions [71–76] or emission cost [77], water discolouration risk [68] and water quality [78]. A multi-objective optimisation approach is considered "very appealing for engineers as it provides a tool to investigate interesting trade-offs", for example, a marginal pressure deficit can be outweighed by a considerable cost reduction [67].

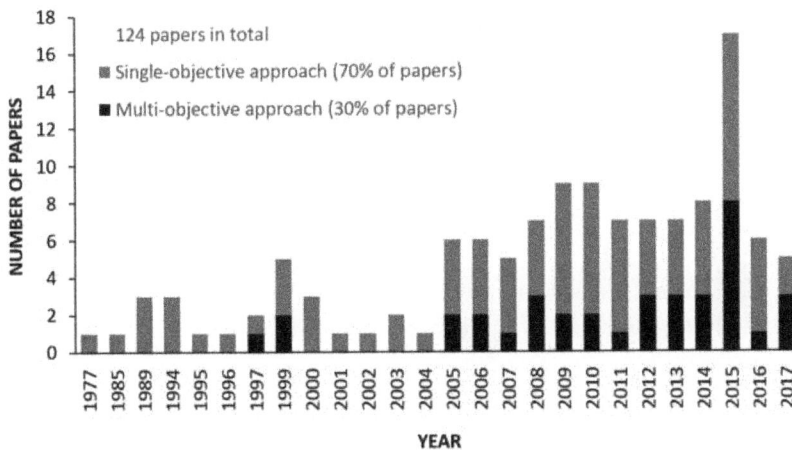

Figure 1. Papers (from Appendix A Table A1) by year and optimisation approach.

The single-objective approach benefits from being able to identify one best solution, which is then easy to analyse and implement. Multi-objective methods, on the other hand, result in a set of tradeoff (Pareto, non-dominated) solutions, which requires an additional step to select only one or a limited number of the promising solutions. Choosing such a reduced number of solutions from a potentially large (or even infinite) non-dominated set is likely to be difficult for any decision maker. This task makes the multi-objective approach less desirable as there is often a requirement to make a clear decision to be implemented. The research question resulting from this challenge is how to select the best solution(s) from the Pareto set, which may involve providing the decision makers with a practical and representative subset of the non-dominated set that is sufficiently small to be tractable [22]. For example, study [79] introduced game-theoretic bargaining models to take into account conflicting requirements and managed to reduce the solution sets to a reasonable size. Further investigation of the methodologies for identifying a handful of useful solutions, such as those where a small improvement in one objective would lead to a large deterioration in at least one other objective, is thus warranted. In addition to game-theoretic models, the approaches that are based on identifying 'knees' of the Pareto front or expected marginal utility, maximum convex bulge/distance from hyperplane, hypervolume contribution and local curvature [80] are all promising methods that require a thorough analysis on WDS problems.

3.1.2. Existing Systems

As a consequence of the development/growth and population density increase within urban areas, existing WDSs require to be upgraded to satisfy raising water demands. These upgrades involve system strengthening (i.e., pipe paralleling), rehabilitation (e.g., pipe cleaning and relining) and expansion. Even though these processes often take place within one WDS thus some of the research articles fall under all system strengthening, rehabilitation and expansion, they are divided into separate subsections in order to provide a systematic overview.

Strengthening

System strengthening represents a reinforcement of an existing WDS to meet future demands, through lying duplicated pipes in parallel to the existing water mains. It is also sometimes referred to as parallel network expansion [42] or pipe paralleling. The main objective and decision variables are, similar to the design of new WDSs, the minimisation of the design (or capital) cost and pipe diameters of duplicated pipes, respectively. Publically available test networks involving purely system strengthening include the New York City tunnels [81] and EXNET [82]. In addition, there are test networks considering system strengthening together with other design strategies (e.g., system expansion, rehabilitation), which include the 14-pipe network with two supply sources [20,83] and Anytown network [84]. Of those publically available test networks, the most frequently applied is the New York City tunnels, which was often the only network used to test the proposed methodology. These studies used genetic algorithm (GA) [85,86], combined with ANNs [87], fast messy GA (fmGA) [88] and non-dominated sorting genetic algorithm II (NSGA-II) [89] as a solution algorithm.

The complexity of an optimisation problem involving exclusively system strengthening as a design strategy can be substantially increased by incorporating water quality considerations. Such applications include, apart from pipe sizes as decision variables, also water quality decision variables that can be in a form of disinfectant (i.e., chlorine) dosage rates [27,87]. In order to reduce computational effort of those problems, ANNs were implemented to replace network simulations to a large extent. Further increase in the complexity presents the use of a multi-objective approach, with additional objectives being system robustness [89] (uncertainty and system robustness are contained in Section 3.2.3), the pressure deficit at network nodes [62,65], and the number of demand nodes with pressure deficit [65,90]. In those studies, a conflicting relationship was identified between the economic (i.e., least-cost) objective and pressure deficit/the number of nodes with pressure deficit. Based on such information, the decision maker is able to "quantitatively evaluate the cost of pressure constraints attenuation which implies a reduction in the system service to its consumers." Optimisation methods used in those studies were NSGA-II [65,89,90], strength Pareto evolutionary algorithm 2 (SPEA2) [65] and cross entropy (CE) [62].

Rehabilitation

Due to aging water infrastructure, which causes a decreased level of service in terms of water quantity as well as quality for customers, increased operation costs and leakage, pipe breaks and other issues, existing WDSs require rehabilitation in a timely manner. Large investments are and will be needed in the future to rehabilitate ever deteriorating pipe networks [91] reaching the end of their lifecycle. Network rehabilitation consists of the replacement of pipes with the same or larger diameter, cleaning, or cleaning and lining of existing pipes; with the main objective to minimise the pipe rehabilitation cost. Within an optimisation model, pipe replacement options can be represented by binary [17] or integer [92] decision variables to identify the pipes selected for replacement, and continuous [17] or integer [92] diameters, respectively, of the replaced pipes. Pipe rehabilitation options are often binary decision variables (i.e., 1 = cleaning/lining, 0 = no action) [17,93]. If a pipe is not scheduled for rehabilitation, it is expected to be subject of break repair over a longer planning horizon. Hence, study [17] added the expected pipe repair costs to the rehabilitation cost of the network. Because a network rehabilitation strategy also has a direct impact on pump operating costs and GHG emissions due to pumping (i.e., they are reduced with an increased quantity of rehabilitated pipes) [94], pump energy costs have been added to the total least-cost objective [17,95].

Some studies consider only a single economic objective to formulate a network rehabilitation problem [17], while other investigations apply a multi-objective optimisation framework in order to incorporate measures affecting the level of service provided to customers (i.e., 'community objectives'). Accordingly, additional objectives considered, beside the economic measure, include the network benefit [63], pressure violations at network nodes [68,95], velocity violations in pipes [95] causing potential sedimentation problems and subsequent water discolouration, water quality (i.e., disinfectant) deficiencies at network nodes [92], and potential fire damage expressed as lack of available fire flows [92]. To generate multi-objective optimal solutions, those studies use mainly metaheuristics or hyperheuristics, such as structured messy GA (SMGA) [63], NSGA-II [95], non-dominated sorting evolution strategy (NSES) [92], and evolution strategy (ES)/SPEA2 in a hyperheuristic framework with evolved mutation operators [68]. The resulting Pareto fronts can then serve decision makers in selecting a rehabilitation strategy that balances community objectives with a capital expenditure.

Note that publications included in this section belong to the category of static design, which involves a single network rehabilitation intervention for a near planning period, designed based on the current network status. Publications, which are concerned with staged rehabilitation interventions involving their timing over an extended planning horizon, are reviewed in Section 3.2.1.

Expansion

An expansion of a WDS means developing or expanding the existing system beyond its current boundary, with the main objective to minimise the total design (or capital) and operation cost. System expansion can be thought of as the following two interdependent design problems: (i) developing a new network that is connected to the existing one, and simultaneously (ii) strengthening, rehabilitating and upgrading the existing system in order to convey increased water demands. Hence, system expansion is the most complex WDS design problem as it can ultimately contain all aspects of designing new as well as existing systems. A typical example of the optimal network expansion is the Anytown network problem [84]. Essentially, the objective is to determine least-cost design and operation, using locations and sizes of new pipes (including duplicated pipes), pumps and tanks, as well as pipe rehabilitation options (i.e., cleaning and lining) as decision variables. Such extensive problems are often solved by combining a power of optimisation algorithms with "manual calculations and a good deal of engineering judgement" [84].

Although some studies solved the Anytown network problem as initially formulated [84], for example, study [83] by enumeration and [96] using GA, others included new aspects to the (original or modified) problem. Those aspects represent, for example, water quality [97] inclusive of the construction and operation costs of treatment facilities [53], new tank sizing approach (further discussed in Section 3.1.3) [93,98], and additional objectives, such as the network benefit incorporating multiple system performance criteria [93,99] or the hydraulic failure, fire flow deficit, leakage and water age with visual analytics used to explore the tradeoffs between numerous objectives [97]. These studies used SMGA [99], GA [53,93], and ε-NSGA-II [97] to solve the problem. Study [93] combined GA with fuzzy reasoning, where system performance criteria are individually assessed by fuzzy membership functions and combined using fuzzy aggregation operators.

An example of large system expansion represents the battle of the water networks II (BWN-II) optimisation problem, which involves the addition of new and parallel pipes, storage, operational controls for pumps and valves, and sizing of backup power supply, and includes the capital and operational costs, water quality, reliability and environmental considerations as performance measures [58]. This problem was solved by multiple authors within the Water Distribution Systems Analysis (WDSA) conference series [58]. Another example of large and real-world system expansion is presented in [100]. Apart from the decision variables for the BWN-II, it also includes selections of pipe routes, expansions of water treatment plants (WTPs) and configurations of pressure zones. The common approach that is applied to solve both of those optimisation problems was the use of engineering judgement, which led to a reduction in the number and type of decision variables. In the case of the study of [100], some eliminated variables were included in separate optimisation problems. Study [58] demonstrates that "different combinations of engineering experience, computational power and problem formulation can give similar results".

Despite the advances in optimisation methods developed for new system design, rehabilitation and/or expansion of WDS, most notably over the last three decades, the large, complex systems still represent a significant challenge to solve using a fully automated optimisation procedure. There are several reasons for that, including: (i) complexity resulting from a mixed-discrete, nonlinear optimisation problem with often conflicting and difficult to assess objectives and performance measures; (ii) the large network sizes normally encountered in practice, which translates into large search spaces where a global optimum is almost impossible to find; (iii) the so called No-Free-Lunch theorem [101], which says that not all of the optimisers are well suited to solving all problems, in other words, slow convergence of general population-based optimisation methodologies that do not utilise some form of traditional engineering experience/heuristics; and (iv) the lack of computational efficiency of network simulators required by modern population-based optimisation methods. A number of approaches have been developed to deal with these challenges, mainly aimed at increasing the computational efficiency of the optimisation process. Those improvements often include the division of a design problem into multiple phases [58] that can be solved separately, the involvement of engineering expertise and manual interventions [59] to reduce the search space, or the use of surrogate and meta-modelling to speed up the simulation process [27]. The work that is still needed in the WDS design optimisation area is to understand the link between the performance of an algorithm (and its operators) and certain topological features of a WDS (e.g., existence of pumps/tanks, loops), as indicated in [29].

3.1.3. Problem Elements

Pipes

Unlike other network elements (e.g., pumps, tanks, valves), pipes are always included in the optimisation of WDS design, as the basic model is to determine such pipe sizes (or diameters) for which the design cost of the network is minimal, subject to the nodal pressure requirement. Even though pipe decision variables are incorporated in every optimisation model, they do not seem to have been unified. Assuming a given layout of the pipe network, there are two types of a decision variable, pipe sizes/diameters, and pipe segment lengths of a constant (known) diameter. Pipe sizes/diameters are discrete by nature of the problem, because they are to be selected from a set of commercially available sizes, however both discrete and continuous values are used mainly depending on the optimisation method. Discrete sizes are used mostly for stochastic algorithms (i.e., metaheuristics) [42,70,85,88,102–109], whereas continuous sizes for deterministic methods [16,110,111]. In regards to continuous sizes, the final solution can be modified by splitting a link into two pipes of closest upper- and lower-sized commercially available discrete diameter [16].

WDS design optimisation problems, which use pipe sizes/diameters as decision variables, can be referred to as a single-pipe design [112,113], while problems with pipe segment lengths of a constant (known) diameter as a split-pipe design [112,113]. Pipe segment lengths of a constant (known) diameter are predominantly used in conjunction with deterministic algorithms [14,114,115] or hybrid methods (i.e., combined deterministic and stochastic methods) [113,116,117]. Single-pipe design with discrete decision variables can provide, compared to split-pipe design and continuous diameters solutions, high quality [102], or good quality results without unnecessary restrictions imposed by split-pipe design [42]. Even if only pipe diameters are optimised, the design of WDS is a complex problem that requires a careful selection of decision variables as to minimise the search space. The choice between direct representation of discrete pipe diameters and split-pipe solutions has largely been resolved in favour of the former, but further improvements in decision variable coding might be possible.

In cases of an unspecified network layout (e.g., when designing a new or extending an existing WDS), additional decision variables are required in order to determine or select pipe routes [52,100]. These variables can be formulated, for example, as binary selecting a link which should be included into the pipe route [52]. Pipe routes can also be considered when strengthening an existing WDS, as "parallel pipes do not necessarily have to be laid in the same street", they "may be laid in a parallel street or right-of-way that may not have existed at previous construction times" [118]. Another possible type of a pipe decision variable are pipe closures/openings to adjust a pressure zone boundary within a WDS [100].

Pumps

There are two main aspects of including pumps into the optimisation of WDS design. First, the pump design or capital cost and second, the pump operating cost due to electricity consumption. Typically, electricity consumption is one of the largest marginal costs for water utilities, with the price of electricity rising globally making it a dominant cost in managing WDSs. Therefore, "the presence of pumps requires that both the design and the operation of the network should be considered in the optimisation" [99]. Accordingly, the minimisation of the pump design or capital cost as well as the pump operating cost to achieve minimal amount of electricity consumed by pumps ought to be included in an optimisation model. Pump operating cost is usually calculated on annual basis using the typical daily demand patterns (i.e., EPS), but a longer period can be considered depending on the planning horizon of a case study, for example, 20 years [17,119], 100 years [72,76,77]. Because this cost occurs at different times in the future, its present value is required to be included in the objective function. This conversion of future economic effects into the current time is undertaken via a present value analysis (PVA), described in detail in [71,72,77], using zero, constant or time varying discount rates.

In the model, pumps are controlled by three types of a decision variable. Firstly, a pump location, which are used when designing a new or extending and upgrading an existing WDS. Possible options to consider are, for example, to predetermine a limited number of potential pump locations [93,120], to evaluate network nodes as potential pump locations (yes/no) via binary variables [52] or to upgrade the current pump stations where new pumps are to be installed in parallel to existing ones [99]. Secondly, a pump size, which can be included as a pump capacity [14,121], pump type [75,76], pumping power [17], pump head/height [52,122], pump operation curve/head-flow [93] or pump size in a combination with the number of pumps [26]. Thirdly, a pump schedule, which describes when the pump is on and off during a scheduling period (e.g., 24 h). It can be specified by a pumping power [53,54] or pump head [123] at each time step, the number of pumps in operation during 24 h [97], binary pump statuses [29], continuous options representing on/off times with a limit imposed on the number of pump switches [76], discrete options representing the time at which a pump is turned on/off using a predefined time step (e.g., 30 min) [75]. All of these decisions impact on the size of the search space and eventually on the computational efficiency of the optimisation algorithm used. A comparative study of various approaches would be useful to help determine what their advantages and disadvantages are and which one to use for a particular situation.

In terms of the model objectives, the pump design or capital and/or operating costs were mostly incorporated together with the costs of other network elements (e.g., pipes, tanks, valves) into one economic function (see, for example, [17,26,51,60,93,95,96,119]). Although a few studies, which considered the design and operating costs as part of separate objectives (e.g., [124]), reported on their conflicting tradeoff, this relationship was not confirmed for a higher-dimensional space when required to balance numerous objectives [97]. Additionally, the pump maintenance cost (see, for example, [61,62,121]) as well as the pump replacement and refurbishment cost [71,72,77] were accounted for. More recently, GHG emission cost or GHG emissions due to the electricity that is consumed by pumps [71–77] were introduced as an environmental objective. Similar to the pump operating cost, a PVA can be used for the pump maintenance, replacement and refurbishment costs, as well as GHG emissions/cost. Even though there is a significant tradeoff between economic and environmental objectives (i.e., GHG emissions decrease with the increasing costs and vice versa), GHG emissions can be considerably reduced by a reasonable increase in the costs [71,72]. Additional results indicate that the price of carbon has no effect on the tradeoff [77], whereas the discount rates do [72], the use of variable speed pumps (VSPs) (rather than fixed speed pumps (FSPs)) leads to significant savings in both total costs and GHG emissions [74].

The mixed-integer nature of pumps as decision variables and their often significant impact in terms of hydraulic behaviour of the entire system, makes them a difficult element to include and control its impact during an optimisation run. Furthermore, the increased complexity of modelling VSPs and their incorporation into the optimisation problem pose another difficulty that has to be tackled by modern optimisation algorithms. Finally, the formulation of various objectives, including maintenance requirements (i.e., often surrogated by the number of times a pump is switched on during the optimisation period), represents another challenge for including pumps into overall WDS design studies.

Tanks

In spite of having a valuable role in WDSs contributing to their reliability and efficiency [125], storage tanks (further in the text referred simply to as tanks) are not often included in WDS design optimisation problems. Several types of a decision variable have been used in the literature to control tanks in the model, and a few objectives (or objective functions) have been developed to mainly evaluate tank performance. However, the use of those variables as well as objectives seems to vary across studies with no general framework on how to model tanks available. As far as decision variables are concerned, they include tank locations [71,72,96–99,120], tank volumes [16,53,93,96,98,99], minimum (and maximum) operational levels [93,96,98,99], tank heads [78], tank elevations [14], ratio between diameter and height [98], ratio between emergency volume and total volume [98]. Study [99] compared two approaches to model tanks in terms of operational levels, first of which calculates tank levels analytically during the network analysis, and second of which includes tank levels as independent variables. Although they yielded similar results, the former approach obtained more robust solutions.

In regard to objectives, the most frequently used account for the tank design or capital cost, which is normally part of the total system costs (i.e., pipes, pumps, etc.) [16,53,76,93,96–99,120]. Furthermore, additional objectives have been introduced evaluating, along with others, the tank performance. These objectives are the network benefit, including storage capacity difference [99], safety and operational volume capacities, and the filling capacity of the tank [93], and system hydraulic failure including tank failure index [97]. A positive relationship was identified between the total cost of the system and network benefit [93,99], whereas a negative relationship exists between the cost and failure index [97]. The effect of changing the tank balancing volume, so called tank reserve size (TRS), on the minimisation of system cost and GHG emissions was also investigated [76]. It was identified that a larger TRS could assist in reducing GHG emissions with no additional cost by modifying pumping schedules.

In addition to pumps, the presence or absence of a tank can also play a significant role in changing hydraulic behaviour of a WDS. This presents a large challenge for any optimisation approach as it creates a discontinuity (i.e., a large change in behaviour with or without a tank at a particular location), which has to be properly managed by the algorithm. Additionally, the setup of the tank (i.e., the link to the system, overflow valve operation, consideration of upper/lower level limits) within a simulation model can also play a significant role in the efficiency of the optimisation run.

Valves

The inclusion of valves in WDS design optimisation problems appears to be rather sporadic and descriptions related to their implementation are often very brief with not many details provided. Studies [14,26] accounted for valves in the overall costs of the system, based on optimal valve locations. The optimisation of a real-life scale WDS incorporating not only transmission pipelines, but also local distribution pipelines, concluded that optimal valve locations are to be affected by the presence of local lines which "provide alternative pathways when the main lines are out of service" [26]. As shutdown of valves used to isolate a portion of the WDS during an emergency (e.g., pipe break or a water quality incident) creates a change in hydraulic behaviour, the valve numbers and locations play part in the overall system design, particularly when the reliability or resilience of the system is considered. For example, study [126] presented a methodology for optimal system design accounting for valve shutdowns. Another application of valves is using their settings to influence the pressure distribution in the network (via pressure reducing valves (PRVs)) [16], or to determine timing of flows and flow rate values (either via flow control valves (FCVs) or PRVs) [127].

Valves were also used to incorporate a simpler model of VSPs into the multi-objective optimisation of WDS design including total economic cost of the system (i.e., design and operation) and GHG emissions [74]. In such an application, a pump power estimation method uses a FCV combined with an upstream reservoir to represent a pump in the system, with the aim to maintain the flows via the FCV into the downstream tanks as close as possible to the required flows. Hence, the determination of the most appropriate FCV setting for calculating pump power is formulated as a single-objective minimisation problem that is subject to multiple flow constraints, which is implemented within a multi-objective GA (MOGA) framework [74].

A combined design of the isolating valve system and the pipe network presents a considerable challenge to optimisation methods. Not only that the number of decisions increases exponentially with the addition of valves, but also the consequences of various valve system designs can only be evaluated by investigating a large number of (probabilistic) scenarios making the whole process computationally inefficient. Furthermore, the location and status of isolating valves can form decision variables also when a WDS is to be divided into manageable subsystems. This is the case with the so-called district metering areas (DMAs), which have been first implemented in the UK primarily for leakage management purposes [128]. Due to the fact that the DMA optimal design is normally performed after a system has been constructed, this problem was deemed beyond the scope of this review paper.

3.2. Time, Uncertainty and Performance Considerations

3.2.1. Staged Design

A staged design represents an optimisation of a WDS over a long planning horizon divided into several construction phases, without considering future uncertainties (e.g., in demands, pipe deterioration). In other words, it is a deterministic dynamic design either over several prefixed time intervals or with timing decisions (i.e., year of action execution). The planning horizon can spread across a number of years to an expected life cycle of the system.

Initial work in the staged design is related to the development of multiquality water resources systems using a single-objective approach, which minimised the costs of water allocation, facilities expansion, water transportation, and losses caused by insufficient supply [129]. It was formulated as a LP optimisation problem, into which nonlinear water quality equations were incorporated using a successive linear approximation iterative scheme. An advantage of using a staged design was demonstrated by realising linkages between certain management processes and variables, and a particular planning period (prefixed time interval).

Concerning WDSs, the staged design is often applied to rehabilitate an existing system as this problem inherently involves the timing of ongoing works over an extended planning horizon. Both single- and multi-objective optimisation models were proposed to solve such problems. Single-objective models included beside the network rehabilitation [130], also network strengthening [131] and expansion [124,132] combined into one least-cost objective, while multi-objective models incorporated the network benefit [131] or the system operating costs [124,132] as additional objectives. Optimisation methods used were GA [130], SMGA [131] and NSGA-II [124,132]. As opposed to the study of [129], these models do not define prefixed time intervals, but include timing decision variables to schedule works, also referred to as event-based coding [124,132]. This coding dramatically reduces the search space, thus the computing and memory requirements, because it eliminates unnecessary zero values of a traditional coding based on a time-interval (e.g., yearly) basis [124]. A further search space reduction can be achieved by so called limited pipe representation introduced by [130], which involves placing an upper bound on the number of pipes considered for rehabilitation. These reductions in the search space and computing requirements are especially important for large size WDSs.

Moreover, the staged design was applied to extend and strengthen existing wastewater, recycled and drinking water systems applying an integrated optimisation scheme within a single-objective framework using GA [127], and to plan a new WDS considering two objectives, the construction costs and network reliability, using NSGA-II [118]. Both of these studies used prefixed time intervals to schedule the construction. In addition, study [118] analysed the effect of the scheduled construction on the network design using a set of scenarios reflecting different lengths of planning horizons (25–100 years), time intervals (25–100 years) and the number of construction phases (1–4). Both studies [118,127] confirmed that for long planning horizons, the staged design is cost effective. The system upgrades guarantee a predefined reliability and there is always opportunity to modify or redesign subsequent upgrades at the later stage, based on new up-to-date predictions of potential future development [118].

By introducing staged design to WDSs, it is obvious that the search space increases almost exponentially to accommodate decisions at various times in the planning horizon. This is one of the key challenges for deterministic staged design, as computational efficiency of optimisation algorithms plays even more significant role than with static design. Another difficulty for achieving the optimised staged design is that even if an optimal solution can be found for each of the intermediate time steps, the algorithm has to ensure that contiguity among the staged decision is maintained, i.e., that the decisions selected in the previous stages are retained in the subsequent ones. An approach by [133] presents one way of obtaining that contiguity of decisions, starting from the solution at one extreme of the Pareto front. However, this issue is still an under-researched area, which requires more investigation. All of the above challenges apply even when the future is assumed to be perfectly known, which is unfortunately not the case.

3.2.2. Flexible Design

A flexible design represents one of the most recent developments in the design optimisation of WDSs. Similar to a staged design, a flexible design represents an optimisation of a WDS over a long planning horizon divided into several construction phases, but with the consideration of future uncertainties (e.g., in demands, pipe deterioration, urban expansion scenarios). Specifically, it is a probabilistic dynamic design over several prefixed time intervals and with the planning horizon ranging from a number of years to an expected life cycle of the system. Such a design allows for flexible and adaptive planning, which is favoured by water organisations that are often encouraged to include risk and uncertainty in their long term plans.

Uncertainties included in the flexible design are related to future demands [122,134–136] and future network expansions [137]. They are implemented using either a probabilistic demand assessment [135] or scenario-based approach via demand/expansion scenarios [122,134,136,137]. A decision tree has been introduced to combine the uncertain demands and intervention measures into optional decision paths [135]. Analogously, studies [122,137] have proposed the use of real options

(ROs) approach, which is also based on decision trees that reflect future uncertainties. In ROs approach, a decision tree is formed by independent decision paths with assigned probabilities to each of the scenarios. This approach enables flexibility to be incorporated into the decision making process and to subsequently change the investment plan based on new circumstances [122].

The majority of the above studies apply multi-objective optimisation approach, including, besides an economic (least-cost) objective, the system resilience [135], reliability [136] or total pressure violations [137] as another objective. Stochastic optimisation algorithms, such as NSGA-II [135,136], simulated annealing (SA), and multi-objective SA [122,137] have been employed to solve flexible design problems, except for [134] who applied integer LP (ILP) combined with preprocessing methods to reduce the dimensionality of the problem. These preprocessing methods included separating the (branched) network into subnetworks, reducing the number of decision variables (e.g., velocity constraints were used to eliminate unsuitable pipe diameters) and solving each subnetwork separately. As a consequence, the quality of the solution was improved and the proposed methodology [134] can be applied to large size WDSs.

While comparing to a traditional deterministic design, the results indicate that a flexible design has a higher initial cost (i.e., in the first construction phases) [122,136], which enables the system to adapt to various future conditions. However, it outperforms a traditional design in terms of the total cost over the entire planning horizon [122,135].

The application of flexible optimisation methodologies in WDS design that consider long-term uncertainty and management options, is yet to be explored to a larger extent in the literature. One of the key reasons is that it is not clear how various types of uncertainties, i.e., stochastic vs. deep uncertainty or aleatoric vs. epistemic uncertainty, are best represented in the optimisation process. The other possible reason is that the flexible design incurs additional computational costs that affects the overall computational efficiency of the optimisation algorithm. However, as the planning and design exercises are done sporadically, the additional computational burden and costs are often justified. Future uncertainties that might have an impact on WDS design, including climate change, population movements and economic development, make flexible design probably the most promising area of research over the next few decades.

3.2.3. Resilient, Reliable and Robust Design

System resilience, reliability and robustness present performance characteristics of a WDS in relation to current and most importantly future uncertain conditions. Although there is no universally agreed definition of any of these measures, the resilience can be defined in broadest terms as the ability of a WDS to adapt to or recover from a significant disturbance, which can be internal (e.g., pipe failure) or external (e.g., natural disaster) (adapted from [138]). Similarly, the reliability can be defined as the ability of a WDS to provide expected service, and can be expressed as the probability that the system will be in service over a specific period of time (adapted from [139]). The robustness represents the ability of a WDS to maintain its functionality under all circumstances (adapted from [138]), or over everyday fluctuations that have the potential to cause low to moderate (i.e., not catastrophic) loss of performance [89].

A robust design problem of a WDS is primarily concerned with uncertainties in model parameters. These uncertainties are related mainly to future demands [89,110,121,123,140,141], but can also consider pipe roughnesses [89,110,140,141], minimum nodal pressure requirements [110], network expansions [137] and others [142]. Study [89] states that "neglecting uncertainty in the design process may lead to serious underdesign of water distribution networks".

Several approaches have been proposed in the literature to formulate a robust design problem for WDSs. Firstly, a redundant design approach which adds redundancy to the system to account for the uncertain parameters by assuming that those parameters are larger than expected [140]. Secondly, an integration approach where uncertainties are incorporated into the model formulation via either objective function [89] or constraints [140] sometimes referred to as a chance-constrained model [110]. Both of those approaches assume a probabilistic distribution of uncertain parameters and convert an original stochastic optimisation problem into a deterministic one. Thirdly, a two-phase optimisation approach that initially solves an optimisation problem with deterministic parameters (i.e., no uncertainties), and subsequently uses those obtained solutions as an initial population for a stochastic problem where future demands and pipe roughnesses are considered uncertain variables following a probability density function [141]. Fourthly, a scenario-based approach where the uncertainty is implemented via a set of scenarios, each assigned a probability [121]. Lastly and most recently, a robust counterpart (RC) approach which is non-probabilistic and incorporates the uncertainty through an ellipsoidal uncertainty set constructed according to the user-defined protection level [123].

Despite a number of approaches to incorporate robustness into the design of WDSs, the measure has been defined fairly well and consistently in the literature, and consequently it has been used in optimisation studies. This may be due to the advances in robust optimisation in other fields and/or due to the focus on non-catastrophic loss of performance that is associated with robust operation. However, the other two measures, reliability and most notably resilience, have not been defined consistently in the WDS literature or have been considered seriously only fairly recently. Therefore, this section focuses on robust design of WDSs, with resilience and reliability being outside of the scope of this review paper. This also indicates that future research efforts could be directed toward a consistent and agreed definition of reliability and resilience, with optimisation methods being capable of solving WDS design considering them as objectives/performance measures.

Robust design optimisation problems are mainly solved using stochastic methods, such as GA [140], NSGA-II [89,121], optimised multi-objective GA (OPTIMOGA) [141], PSO [142] and CE [123], except for [110], who solves it as a NLP problem.

3.2.4. Design for Water Quality

In the literature, water quality is incorporated into the WDS design optimisation problems in various ways concerning both an optimisation model and water quality measure used. In terms of optimisation models, single-objective as well as multi-objective exist which include water quality considerations. In the former, water quality related expenditures, such as the cost of disinfection [27,120], cost of water treatment [53] or cost of losses incurred by insufficient quality [129], are combined with the system design/capital (and operation) costs into one objective. Alternatively, water quality is included as a constraint to a single-objective model in a form of minimum (and maximum) disinfectant concentrations at the network nodes [87,143]. In the latter, water quality presents a sole objective, which is either water quality benefit (being maximised) [63,131], water quality deficiencies (being minimised) [92,97,144] or water quality reliability (being maximised) [78]. Regardless of an optimisation model used, study [120] highlighted an importance of incorporating water quality considerations with system design and operation in one optimisation framework, which enables promoting water quality in the design stage, rather than leaving potential water quality issues to be resolved during the system's operational phase. Indeed, study [78] reports a significant tradeoff between water quality objective based on disinfectant residual and the system capital costs (i.e., the best quality solutions correspond to higher costs and vice versa), and demonstrates the sensitivity of the obtained solutions to a disinfectant dosage rate. Interestingly, there was not tradeoff found between water quality objective based on water age and the cost.

Regarding the water quality measure, it is dependent on the system specifics, its requirements, and also the optimisation model advancements progressively implementing water quality objectives useful to system operators. Basic water quality parameters that are used in optimisation models of drinking WDSs are chlorine [27,87,120,143] and chloramine [120], modelled as non-conservative applying first order decay kinetics, adjusted by a higher decay rate in parts of the system when needed [120]. In contrast, conservative water quality parameters are typically important for regional multiquality WDSs. These parameters are either specified within an optimisation model, such as salinity [129] or unspecified being modelled in conjunction with the operation of treatment facilities [53]. In multi-objective optimisation models, both specific parameters and surrogate measures are used to quantify water quality objectives. Water quality benefit is expressed as a function of the length of renewed and/or lined old pipes, as aged pipes are considered to cause the development of microorganisms and water discolouration [63,131]. Water quality deficiencies can be represented by a performance function on disinfectant residual reflecting governmental regulations [92], water age [97], or the risk of water discolouration due to the potential material after daily conditioning shear stress [144]. Water quality reliability is based on disinfectant residual [145] and/or water age [78].

Optimisation methods used to solve WDS design problems including water quality considerations were LP [129], GA [53,87,100,120,143] and differential evolution (DE) [27] for single-objective models, and SMGA [63,131], NSES [92], ε-NSGA-II [97], NSGA-II and SPEA2 integrated with a heuristic Markov-chain hyper-heuristic (MCHH) [144] and ant colony optimisation (ACO) [78] for multi-objective models. These algorithms were mainly linked with a network simulator EPANET to solve network equations. Because these EPANET simulations, in particular water quality analyses, are very computationally demanding, they were replaced by ANNs [27,87,143] to reduce computational effort.

Unsurprisingly, introduction of water quality considerations increases the complexity of the quest for the optimal design considerably. This increased complexity is caused not only by the more complex simulations required to predict the temporal and spatial distribution of a variety of constituents within a distribution system, but also by the requirement to run shorter time step water quality computations [22]. Furthermore, computational efficiency is affected by the ability to model multiple constituents throughout the WDS via the EPANET Multi-Species Extension, EPANET-MSX [146].

4. General Classification of Reviewed Publications

Based on the selected literature analysis, the following are the four main criteria for the classification of design optimisation for WDSs: application area (Section 4.1), optimisation model (Section 4.2), solution methodology (Section 4.3) and test network (Section 4.4).

4.1. Application Area

As outlined in Section 3, there are four application areas: design of new systems (Section 3.1.1), strengthening, expansion and rehabilitation of existing systems (Section 3.1.2). Numerous publications do not deal only with those design optimisation problems, but also with the operational optimisation (see, for example, [14,26,53,71,120,135]), which is an equally important area if the total cost (i.e., including capital and operation expenditure) is considered. Hence, the system operation has been added to the following analysis. It represents papers optimising (mainly) the pump operation together with the system design, strengthening, expansion and/or rehabilitation. Figure 2 displays distribution of the application areas across the papers analysed and listed in Appendix A Table A1 as follows:

- Design of new systems is an application area with the highest representation (41%). Interestingly, an almost identical percentage (43%) totals applications for existing systems (i.e., strengthening, expansion and rehabilitation).
- An application area with the second highest representation (25%) is strengthening of existing systems.
- Expansion and rehabilitation of existing systems are both represented evenly by 9% of applications each.
- Optimisation of the system operation is represented by 16% of applications.

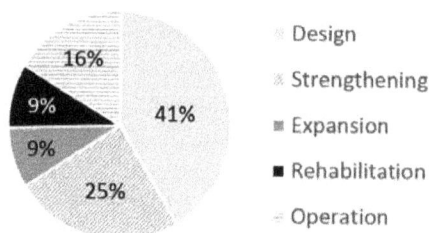

Figure 2. Application areas (of papers from Appendix A Table A1).

It is not surprising that design (and mostly using pipe diameters as decision variables) dominates the literature, which occurs mostly due to historical reasons. Namely, the sizing of pipes was addressed first in the literature, even before WDS simulation was possible. Other design variants, such as strengthening, expansion and rehabilitation, followed on, but use the same or quite similar performance measures and optimisation tools. The introduction of other network elements, such as pumps, tanks and valves, as well as various performance criteria, including water quality and operational considerations, appeared much later in the literature. Lately, more emphasis was put on robustness, reliability and resilience assessment of WDS design and operation, which seems to be the trend for the future.

4.2. Optimisation Model

An optimisation model is mathematically defined by three key components: objectives, constraints and decision variables. Figure 3 shows how many of these components are included in the optimisation models (of papers analysed in Appendix A Table A1), which indicates the degree of complexity of the formulation. Note that not all of the reviewed papers include mathematical formulations of an optimisation model used. Therefore, our assessment is limited to our interpretation of the provided information in the publications, where explicit formulation was partially presented or missing altogether.

- The number of objectives included in optimisation models ranges from one to six. The majority of models (69%) are single-objective, determining the least-cost design. The second largest proportion (27%) represents two-objective optimisation models. Multi-objective models including more than two objectives (i.e., 3–6 objectives) are very sparse as they represent only 4% of all formulations.
- The number of constraints incorporated in optimisation models ranges from zero to ten. Hydraulic constraints (such as conservation of mass of flow, conservation of energy and conservation of mass of constituent) are normally included as implicit constraints and are forced to be satisfied by a WDS modelling tool, such as EPANET, and thus are not included in these statistics. There are 5% of models with no constraints, which are mainly multi-objective optimisation models where the pressure requirement is defined as an objective rather than a constraint. Almost half models (48%) include only one constraint (mostly the minimum pressure requirement). A quarter of models (25%) incorporate two constraints. The proportion of optimisation models with exactly three or more (i.e., 4–10) constraints is 13% and 9%, respectively.
- The number of types of a decision (i.e., control) variable included in optimisation models ranges from one to 13. The majority of optimisation models (60%) uses one type of a decision variable, being a pipe diameter/size or pipe segment length of a constant (known) diameter. The use of more than one type of a decision variable is considerably less frequent and is represented by 16%, 11% and 13%, respectively, for two, three and more (i.e., 4–13) types of a decision variable.

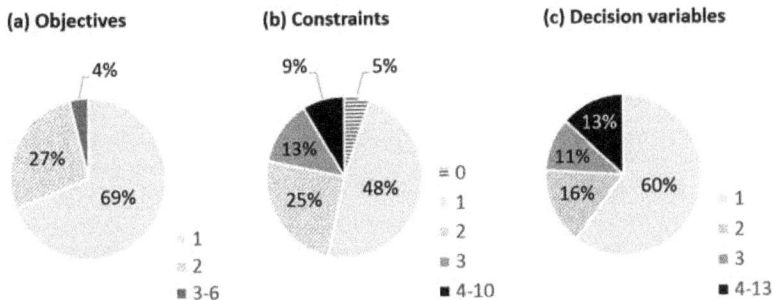

Figure 3. Optimisation models (of papers from Appendix A Table A1) by: (a) number of objectives, (b) number of constraints, (c) number of types of a decision variable, in an optimisation model.

Inspecting Figure 3, the question arises as to how many optimisation models there are, which include only one objective, one constraint and one type of a decision variable? There are, in total, 129 optimisation models formulated and solved in 124 papers listed in Appendix A Table A1. From those optimisation models, 30% (i.e., 39 models) consist of one objective (mostly design costs), one constraint (mostly the minimum pressure at nodes) and one type of a decision variable (mostly pipe diameters).

As indicated, the prevailing use of single-objective optimisation is probably caused by the preference to arrive at a single solution, which can be implemented by decision makers. On the other hand, the preference for one constraint seems surprising as the number of constraints of the problem depends on the complexity of the system and the number of criteria expressed as constraints rather than objectives. Finally, the number and types of decision variables appearing in the literature is a function of historical developments in the field and the increasing trend is expected in the future. Research questions still remain as how to best formulate the optimisation model for a particular case, and what effect the model formulation has on obtained solution(s) [22,23].

4.2.1. General Optimisation Model

A general multi-objective optimisation model for optimal design of a WDS can be formulated as:

$$\text{Minimise/maximise } (f_1(x), f_2(x), \ldots, f_n(x)) \tag{1}$$

subject to:

$$a_i(x) = 0, \quad i \in I = \{1, \ldots, m\}, \quad m \geq 0 \tag{2}$$

$$b_j(x) \leq 0, \quad j \in J = \{1, \ldots, n\}, \quad n \geq 0 \tag{3}$$

$$c_k(x) \leq 0, \quad k \in K = \{1, \ldots, p\}, \quad p \geq 0 \tag{4}$$

where Equation (1) represents objective functions to be minimised (e.g., system capital costs) or maximised (e.g., system reliability), Equations (2)–(4) present three types of a constraint, with x representing decision variables.

Objectives

Objectives of a general optimisation model of WDS design are listed in Table 1. They can be divided into four distinct groups according to their type. The first group represents *economic objectives* such as capital and rehabilitation costs, and expected operation and maintenance costs of the system. The second group are *community objectives*, which report on the level of service provided to WDS customers, and which, if inadequate, could eventuate in water supply related issues for those customers. Those objectives include, for example, a benefit function (using various performance criteria), water quality deficiencies, pressure deficit at demand nodes, hydraulic failure of the system and potential fire damages. The third group presents *performance objectives*, reflecting the operation of a WDS, specifically system robustness, reliability and resilience. These objectives, although ultimately indicating the level of service for WDS customers, have separate classification, due to their primary purpose to report on the performance in relation to a WDS rather than to customers. The fourth group represents *environmental objectives*, namely GHG emissions, consisting of capital emissions due to manufacturing and installation of network components applicable at the WDS construction phase, and operating emissions due to electricity consumption occurring during the WDS life cycle.

Table 1. Objectives of a general optimisation model.

Objective Type	Objectives	Reference (An Example)
Economic	*Capital costs* of the system, including purchase, installation and construction of network components (pipes, pumps, tanks, treatment plants, valves, etc.)	[53,74,121]
	Rehabilitation costs of the system, including pipe/pump replacement, pipe cleaning/lining, pipe break repair	[17,124] (pipes), [77] (pumps)
	Expected operation costs of the system, including pump stations, treatment plants and disinfectant dosage	[53] (pump stations and treatment plants), [27] (disinfectant dosage)
	Expected maintenance costs of the system	[121]
Community	*Benefit/benefit of the solution* (i.e., rehabilitation, expansion and strengthening) using various performance criteria by authors	[131] (welfare index to place greater importance on early improvements), [99] (quantity shortfalls as criteria), [93] (e.g., safety volumes and operational capacities as criteria)
	Water quality (e.g., disinfectant, sedimentation, discolouration) deficiencies or water age at customer demand nodes, water discolouration risk, velocity violations (causing sedimentation/discolouration)	[92,120] (water quality deficiencies), [97] (water age), [144] (water discolouration), [95] (velocity violations)
	Pressure deficit at customer demand nodes (maximum individual or total), or the number of demand nodes with the pressure deficit	[65] (maximum individual deficit), [66,68] (total deficit), [90] (the number of demand nodes)
	Hydraulic failure of the system expressed by the failure index	[97]
	Potential fire damages using either expected conditional fire damages or fire flow deficit	[142] (expected conditional fire damages), [92] (fire flow deficit)
Performance	*System robustness* using either a redundant design approach, integration approach (via objective function or constraints), two-phase optimisation approach, scenario-based approach or RC approach	[140] (redundant design), [89] (integration via objective function), [110,140] (integration via constraints), [141] (two-phase optimisation), [121] (scenario-based), [123] (RC)
	System reliability	[118]
	System resilience	[135]
Environmental	*GHG emissions* or emission costs consisting of capital emissions (due to manufacturing and installation of network components) and operating emissions (due to electricity consumption)	[77] (capital and operating GHG emission costs), [73,75] (capital and operating GHG emissions), [132] (operating GHG emission cost)

Constraints

Constraints of a general optimisation model of WDS design are described in Table 2 and divided into three groups as follows. *Hydraulic constraints* are given by physical laws governing the fluid flow in a pipe network. These constraints are incorporated in an optimisation model either explicitly often in conjunction with deterministic [147] and hybrid optimisation techniques [116,117], or implicitly by a

WDS modelling tool (e.g., EPANET) [26] and/or ANNs [27,87] normally in combination with stochastic optimisation algorithms. *System constraints* arise from limitations and operational requirements of a WDS, and include tank water level bounds, pressure/water quality requirements at demand nodes, etc. The ways to manage these constraints include an integration of EPANET (e.g., tank water levels), the augmented Lagrangian penalty method [17], a penalty function [26], a penalty function with a self-adaptive penalty multiplier [45,88], or a (modified) constraint tournament selection [148–150]. *Constraints on decision variables*, such as pipe diameters being limited to commercially available sizes and others, are handled explicitly by an optimisation algorithm.

Table 2. Constraints of a general optimisation model.

Constraints	Reference (An Example)
Hydraulic constraints given by physical laws of fluid flow in a pipe network: (i) conservation of mass of flow, (ii) conservation of energy, (iii) conservation of mass of constituent	[41]
System constraints given by limitations and operational requirements of a WDS, for example, minimum/maximum pressure at (demand) nodes and flow velocity in pipes, water deficit/surplus at storage tanks at the end of the simulation period, maximum water withdrawals from sources	[54] (limits on nodal pressure, storage tank deficit and water withdrawals from sources), [127] (limits on pipe velocity)
Constraints on decision variables x, for example, limits on pipe diameters, pipe segment lengths (so called split-pipe design), pump station capacities	[92] (limits on pipe diameters), [117] (limits on pipe segments), [121] (limits on pump stations)

Decision Variables

Decision variables of a general optimisation model of WDS design are listed in Table 3. They are grouped according to an element or aspect that drives the optimisation (i.e., pipes, pumps, tanks, valves, nodes, water quality and timing). In general, a pipe diameter/size is often the main (or the only) decision variable used in design optimisation of WDSs. Accordingly, a total of 60% optimisation models (of papers listed in Appendix A Table A1) use only one type of a decision variable (see Figure 3c), which is either a pipe diameter/size or the pipe segment length of a constant (known) diameter. As the complexity of an optimisation model increases, so does the number of types of a decision variable. An example of such an optimisation model could be an expansion and rehabilitation of an existing WDS with pumps, tanks and a treatment plant to meet future demands and water quality requirements.

Table 3. Decision variables of a general optimisation model.

Decision Variables	Reference (An Example)
Pipes: pipe diameters/sizes, pipe duplications, pipe rehabilitation options (pipe replacement, pipe cleaning/lining), pipe break repair, pipe segment lengths (so called split-pipe design), future pipe roughnesses, pipe routes, pipe closures/openings (to adjust a pressure zone boundary)	[75] (diameters), [132] (duplications, replacement, lining and break repair), [117] (segments), [141] (roughnesses), [52] (routes), [100] (routes and closures/openings)
Pumps: pump locations, pump sizes (pump capacities, pump types, pumping power, pump head/height or head-flow), the number of pumps, pump schedules (pumping power or pump head at each time step, the number of pumps in operation during 24 h, binary statuses at time steps, on/off times)	[52,99] (locations), [14] (locations and capacities), [75] (types), [17] (power), [52,122] (head/height), [93] (head-flow), [26] (sizes and the number of pumps), [53,123] (power or head at each time step), [97] (the number of pumps in operation), [29] (binary statuses), [75] (on/off times)
Tanks: tank locations, tank sizes/volumes, minimum operational level, ratio between diameter and height, ratio between emergency volume and total volume, tank heads	[98] (locations, sizes/volumes, minimum operational level, ratios), [78] (heads)
Valves: valve locations, valve settings (headlosses or flows)	[14] (locations), [16] (headlosses via a roughness coefficient), [127] (headlosses and flows)

Table 3. *Cont.*

Decision Variables	Reference (An Example)
Nodes: flowrates from sources, future nodal demands, threshold demands, hydraulic heads at junctions	[127] (flowrates), [135,141] (demands), [147] (heads)
Water quality: disinfectant dosage rates (at the sources, at the treatment plants, in the tanks), treatment removal ratios, treatment plant capacities	[143] (dosage at the sources), [27] (dosage at the treatment plants), [78] (dosage in the tanks), [53] (removal ratios), [121] (capacities)
Timing: year of action (e.g., network expansion, rehabilitation, pipe replacements) execution	[131] (network expansion and rehabilitation), [130] (pipe replacements)

Tables 1–3 provide a generic set of components used for formulating an optimisation problem involving initial design with subsequent operational management of a WDS. Particular circumstances being considered in different case studies may warrant only a portion of those components to be used.

4.3. Solution Methodology

An enormous effort has been dedicated to the application and development of optimisation methods to solve WDS design optimisation problems since the 1970s. Initially, deterministic methods namely LP [14,114,129], NLP [16,110] and mixed-integer NLP (MINLP) [17,115] were used. In the mid 1990s, after the first popular applications of a GA [20,151], there was a swing towards stochastic methods and they dominate the field since (see Figure 4). A great range of those methods has been applied to optimise design of WDSs to date, inclusive of (but not limited to) a GA [42,45,50,85,86,152–154], fmGA [88], non-crossover dither creeping mutation-based GA (CMBGA) [149], adaptive locally constrained GA (ALCO-GA) [155], SA [60], shuffled frog leaping algorithm (SFLA) [103], ACO [104,156], shuffled complex evolution (SCE) [157], harmony search (HS) [105,158,159], particle swarm HS (PSHS) [160], parameter setting free HS (PSF HS) [161], combined cuckoo-HS algorithm (CSHS) [162], particle swarm optimisation (PSO) [106,153,154], improved PSO (IPSO) [163], accelerated momentum PSO (AMPSO) [164], integer discrete PSO (IDPSO) [165], newly developed swarm-based optimisation (DSO) algorithm [150], scatter search (SS) [166], CE [61,62], immune algorithm (IA) [167], heuristic-based algorithm (HBA) [168], memetic algorithm (MA) [107], genetic heritage evolution by stochastic transmission (GHEST) [169], honey bee mating optimisation (HBMO) [170], DE [46,153,154,171], combined PSO and DE method (PSO-DE) [172], self-adaptive DE method (SADE) [173], NSGA-II [70], ES [68], NSES [92], cost gradient-based heuristic method [119], improved mine blast algorithm (IMBA) [174], discrete state transition algorithm (STA) [175], evolutionary algorithm (EA) [109], and convergence-trajectory controlled ACO (ACO_{CTC}) [176]. The vast majority of those studies solely solve a basic single-objective least-cost design problem (i.e., pipe cost minimisation constrained by the nodal pressure requirement) and use a small number of available benchmark networks (e.g., Hanoi network [49], New York City tunnels [81], two-loop network [14]) to test the proposed optimisation method. The usual result obtained was a better or comparable optimal solution reached more efficiently than by algorithms previously used in the literature, without providing an explanation as to why the selected algorithm performed better for a particular test network. It seems, therefore, that research have been trapped, to some extent, in applying new metaheuristic optimisation methods to relatively simple (from an engineering perspective) design problems, without understanding the underlying principles behind algorithm performance. Moreover, study [177] stresses that there has been "little focus on understanding why certain algorithm variants perform better for certain case studies than others".

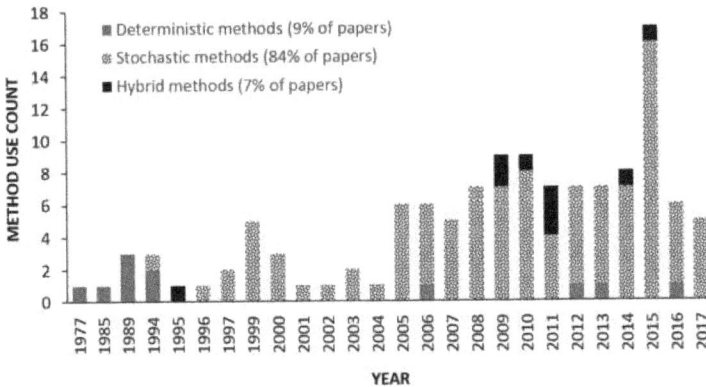

Figure 4. Optimisation methods (of papers from Appendix A Table A1) by year.

Over the past decade, an increase in the use of deterministic and hybrid methods (i.e., a combined deterministic and stochastic method) can be observed from Figure 4. These methods, which are computationally more efficient when comparing to stochastic methods, thus more suitable for large real-world applications, include ILP [51,134], MINLP [147], a combined GA and LP method (GA-LP/GALP) [113,117], combined GA and ILP method (GA-ILP) [178], combined binary LP and DE method (BLP-DE) [179], combined NLP and DE method (NLP-DE) [111], hybrid discrete dynamically dimensioned search (HD-DDS) [180], decomposition-based heuristic [52], optimal power use surface (OPUS) method paired with metaheuristic algorithms [47], and modified central force optimisation algorithm (CFOnet) [181]. However, WDS simulations may still be computationally prohibitive even with more efficient deterministic or hybrid optimisation methods, especially as the fidelity of the model and the number of decision variables increase [22].

The choice of the solution methodology depends on the type of problem being considered, the level of expertise of the analyst and the familiarity with the particular method/tool. Nonetheless, there is often no clear justification provided as to why a particular methodology has been selected over another and/or why an alternative methodology has not been tested. Quite often, this choice is based on the analyst's preference, level of familiarity, and software availability, rather than on a comparison of the tests performed using two or more solution methodologies. This practice makes it difficult to progress towards the development of meaningful guidelines for the application of different optimisation methods [177]. An interesting research question for further studies would be how to characterise and select the best optimisation method for a particular WDS design problem.

However, that being stated, several attempts have been made to compare or evaluate algorithm performance for both single- and multi-objective WDS design problems, but with an absence of a universally adopted method to date. A methodology for comparing the performance of various single-objective algorithms involves assessing the best solution obtained (which is straightforward contrary to multi-objective optimisation), the convergence speed, and the spread and consistency of the solutions using a number of random starting seeds and evaluations [153,154]. A methodology has also been developed to evaluate the performance of a particular algorithm by assessing the effectiveness of its parameters (such as crossover and mutation) applying their different values [182]. In multi-objective optimisation, in general, performance metrics were proposed and are commonly used to compare performance of various algorithms in terms of the quality of the Pareto fronts obtained (see, for example, [183–185]). A comparison of solutions is substantially more complex than in single-objective optimisation as there is no single performance metric both compatible and complete [186]. Possibly for that reason, some WDS design studies have limited their analysis to a visual comparison of solutions only (i.e., two-objective Pareto fronts), which was criticised by [187].

Most recent research, progressively, evaluates the performance and search behaviour of multi-objective algorithms in relation to their parameters and/or WDS features [28] (more such studies are listed in Section 4.3.2).

4.3.1. Computational Efficiency

Numerous advancements have been reported in the literature to improve the computational efficiency of both optimisation algorithms and network simulators. These developments include methods for search space reduction [45,63,88,95,99,120,131,188,189], parallel programming techniques [109], hybridisation of the evolutionary search with machine learning techniques to limit the number of function evaluations [67], surrogate models (metamodels) to replace network simulations [27,43,67,87,143], approximation of the objective function by shortening the EPS [119], and enhanced methods for speedy network simulations for large size WDSs [190].

Various techniques for search space reduction have been proposed, which can be broadly classified as algorithm-based and network-based methods. The algorithm-based techniques include the method for more efficient encoding of decision variables [63,99,131], a self-adaptive boundary search strategy for selection of the penalty factor within the optimisation algorithm to guide the search towards constraint boundaries [88], and the application of an artificial inducement mutation (AIM) to acceleratingly direct the search to the feasible region [95]. The network-based techniques analyse either the network as a whole or individual pipes. The former include a network stratification into upper, middle and lower diameter sets using engineering judgment [188], and the critical path method [45,191]. The latter involve the elimination of certain pipes from the optimisation based on their preliminary capacity assessment [120], application of a pipe index vector (PIV), a measure of the relative importance of pipes regarding their hydraulic performance in the network, which assists in exclusion of impractical and infeasible regions from the search space [189], and introduction of upper/lower bounds on pipe diameters based on the initial analysis [30].

In terms of replacing time consuming network simulations with faster means, three types of a surrogate model have been applied to the design optimisation of WDSs to date. These models include a linear transfer function (LTF) [43], Kriging [67] and ANNs [27,87,143], which are used more frequently than two previous ones. The purpose of a surrogate model is to approximate network simulations (hydraulics and/or water quality), hence reduce the calls of the simulation model during the optimisation. Kriging uses solutions visited during the search to model the search landscape [192]. ANNs can be divided into two groups, offline ANNs, which are trained only once at the beginning of the optimisation, and recently proposed online ANNs, which are "retrained periodically during the optimisation in order to improve their approximation to the appropriate portion of the search space" [27]. ANN metamodels are often used in conjunction with water quality simulations [27,87,143].

All of those methods have shown promise on a limited number of test cases or a specific case study. It would be interesting to conduct a thorough comparison of all of those on a selection of benchmark cases of various sizes and complexity.

4.3.2. Recent Developments

More recently, a number of advancements, such as improving and understanding the algorithm performance and others, have been proposed in the literature indicating potential directions for future research. Some of those developments are a consequence of an appeal of [23,177] "to counteract potential repetition and stagnation in this field" to continually produce too many papers using "an ever increasing number of EA variants" and "theoretical or very simplistic case studies".

Firstly, to improve the algorithm performance regarding the solution quality, an engineered initial population has been suggested [26,30,44,66,108]. Traditionally, a random (or naïve) initial population of solutions (expressed as pipe sizes) is used as a starting point for algorithms. An engineered initial population, in contrast, is created by taking into account the rules and hydraulics principles of water flow in a pipe network, or by performing pre-optimisation runs under various parameter scenarios (e.g., [30]). Another way to achieve better algorithm convergence, particularly for design problems incorporating water quality, is to run the algorithm with hydraulic analysis only for several first generations, and subsequently add water quality analysis [120]. Furthermore, a postoptimisation technique can be used to refine the solutions that are found by an optimisation algorithm to get closer to the global optimum [193]. Secondly, a range of strategies have been introduced to eliminate the tedious and time demanding process of calibrating algorithm parameters (i.e., selecting the best performing combination of parameter values) for a particular test problem. These strategies involve the use of a statistical analysis [158], evolved heuristics (i.e., hyper-heuristic approach) [68,144,194], and convergence trajectories [176]. Thirdly, several studies focused on analysing algorithm performance [195] and search behaviour [28,48,196] in relation to the WDS design problem features [29]. Lastly, methodological improvements using existing methods have been proposed rather than applying/developing new algorithms, with the aim to improve computational efficiency. Those improvements represent multiple-phase optimisation concepts [30], which can be combined with graph decomposition [46,69] or clustering [90] techniques.

4.4. Test Network

An enormous diversity exists among test networks used in optimisation of WDS design. These networks vary in size, complexity, and the types of network components that they contain (i.e., nodes, pipes, pumps, tanks, reservoirs/sources and/or valves). The simplest networks are small gravity WDSs with one source, a few nodes and pipes (see, for example, [14,60]), or simplified pumped WDSs including only one source, one pump, one pipe and one tank (see, for example, [71]). An example of a large network represents EXNET [82], which is a realistic WDS comprising two sources, control valves and almost 2500 pipes. Figure 5 categorises test networks that are used (in the papers listed in Appendix A Table A1) by network size. In order to be consistent with the previous review [22], network size is expressed by the number of nodes within a network. Networks, for which the number of nodes cannot be identified from the reviewed paper or the references provided, are excluded from the analysis. Figure 5 reveals that nearly a half of the networks (49%) is limited in size to 20 nodes and the majority of the test networks (84%) contains up to 100 nodes. This finding is analogous to operational optimisation of WDSs, where networks with up to 100 nodes represent 80% of applications [22].

Figure 5 illustrates that in the large body of literature, various WDS design formulations and optimisation methods have usually been tested using small, computationally cheap networks. This prevalent usage of small networks is in contrast to the requirement to optimise design of real-world systems that contain hundreds of thousand elements (including pumping stations, tanks and valves) causing a single EPS simulation to take minutes or even hours to execute even on powerful desktop computers. Consequently, large networks are not often considered by optimal design studies. This situation can be remedied by using latest developments in methods capable of generating realistic WDS networks by [55–57], who have each developed their own automatic network generation software.

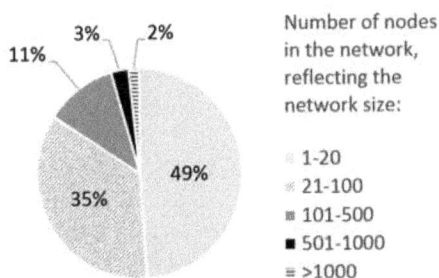

Figure 5. Test networks (of papers from Appendix A Table A1) by network size.

Real-world WDS design optimisation problems normally involve large size, complex-topology networks, comprising a number of elements of various types. Such a problem is often solved by combining a sophisticated simulation model (to potentially analyse both hydraulics and water quality) with a non-trivial optimisation method. The approach ought to satisfy the requirements of a water utility and other stakeholders for objectives, constraints, decision variables, as well as model assumptions. Although studies exist that report on successful solutions to such problems [100,127,197–199], they are limited possibly due to the complexity associated with mathematically formulating objectives and constraints and/or finding the best solution. Study [200] even speculates that the real-world considerations need to be explicitly quantified, "if it is possible to do so at all", otherwise the water industry will apply engineering judgment instead of any optimisation method to design WDSs.

Similar to network size, the frequency of use of test networks varies considerably, as some networks have been used only once, while others have been used repeatedly and by multiple authors. In particular, the prevalence of some networks attributes to their use as benchmark problems to test optimisation algorithms. These benchmark networks, all of which have been used (in the papers listed in Appendix A Table A1) 10 or more times, are listed in Table 4 in order of their usage count. They are, except for the Anytown network, gravity-fed WDSs with the common objective to determine the most economical pipe design. The popularity of those benchmark networks contributed to high percentages of the first two categories in Figure 5, because the majority of them are limited in size to 20 and 100 nodes, respectively.

Table 4. Frequently used test networks.

Test Network Name	No. of Nodes	Network Description	Optimisation Problem	Network Modified Versions	Network Usage Count *
Hanoi network [++] [49]	32	Network organised in three loops supplied by gravity from a single source	New system design (pipes)	Double Hanoi network, triple Hanoi network (both [113])	55
New York City tunnels [++] [81]	20	Tunnel system supplied by gravity from a single source, constituting the primary WDS of the New York city	Existing system strengthening (i.e., pipe paralleling) to meet projected demands	Double New York City tunnels [201]	42
Two-loop network [++] [14]	7	Small network with two loops supplied by gravity from a single source	New system design (pipes)	Adapted to system strengthening and expansion over a planning horizon [118]	40
Balerma irrigation network [++] [50]	447	Large looped network supplied by gravity from four sources, adapted from the existing irrigation network in Balerma, Spain	New system design (pipes)	N/A	20
Anytown network [84]	19	Hypothetical looped system supplied by three parallel pumps from a single source	Existing system strengthening, expansion and rehabilitation (pipes, pumps, tanks) to meet projected demands	** With additional source and tank [53], with additional tank [119] proposed by [83]	15

5. Future Research

Future research challenges for the optimisation of WDS design are illustrated in Figure 6 and divided into the following four groups: (i) model inputs, (ii) algorithm and solution methodology, (iii) search space and computational efficiency, and (iv) solution postprocessing. As far as model inputs are concerned, there is a requirement to explore how to best represent various types of uncertainties, i.e., stochastic vs. deep uncertainty or aleatoric vs. epistemic uncertainty, in the optimisation process. Additional future uncertainties, for example, climate change, population movements and economic development, might affect planning for optimal WDSs, and make flexible design one of the promising research areas over the next few decades. Another research challenge in regards to model inputs is to compare various approaches to pump decision variables, including VSPs and their coding, in order to determine their advantages, disadvantages and suitability for a particular case. Furthermore and overall, a research question remains how to best formulate the optimisation model for a particular case [22,23].

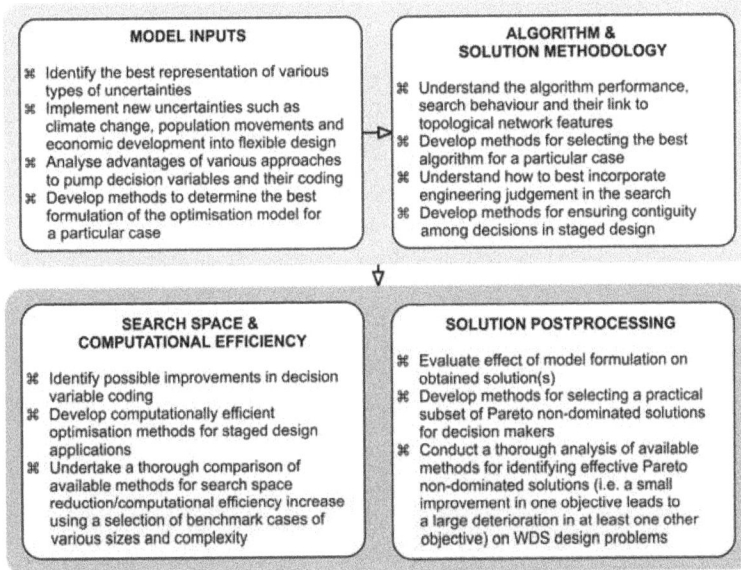

MODEL INPUTS

⌘ Identify the best representation of various types of uncertainties
⌘ Implement new uncertainties such as climate change, population movements and economic development into flexible design
⌘ Analyse advantages of various approaches to pump decision variables and their coding
⌘ Develop methods to determine the best formulation of the optimisation model for a particular case

ALGORITHM & SOLUTION METHODOLOGY

⌘ Understand the algorithm performance, search behaviour and their link to topological network features
⌘ Develop methods for selecting the best algorithm for a particular case
⌘ Understand how to best incorporate engineering judgement in the search
⌘ Develop methods for ensuring contiguity among decisions in staged design

SEARCH SPACE & COMPUTATIONAL EFFICIENCY

⌘ Identify possible improvements in decision variable coding
⌘ Develop computationally efficient optimisation methods for staged design applications
⌘ Undertake a thorough comparison of available methods for search space reduction/computational efficiency increase using a selection of benchmark cases of various sizes and complexity

SOLUTION POSTPROCESSING

⌘ Evaluate effect of model formulation on obtained solution(s)
⌘ Develop methods for selecting a practical subset of Pareto non-dominated solutions for decision makers
⌘ Conduct a thorough analysis of available methods for identifying effective Pareto non-dominated solutions (i.e. a small improvement in one objective leads to a large deterioration in at least one other objective) on WDS design problems

Figure 6. Future research challenges.

Concerning algorithm and solution methodology, a vast research area represents a progression towards better understanding of algorithm performance and its search behaviour. These aspects need to be further linked to the WDS design problem features including system topology (e.g., existence of pumps/tanks, loops) and initial population used. A related challenge is to eliminate a time consuming process of calibrating algorithm parameters to achieve a satisfactory performance, hence there is a question how to select the best performing combination of parameter values. Moreover, it is important to develop understanding related to the suitability of various optimisation methods for particular design problems and the incorporation of engineering judgement in the search. In relation to staged design, methods for ensuring contiguity among decisions, i.e., that the decisions selected in the previous stages are retained in the subsequent ones, are required.

Recently, there has been an observed increased interest in aspects of the search space and computational efficiency. Indeed, the reduction of the search space and an increase in the computational efficiency are significant particularly for real-world WDS optimisation problems as well as dynamic (i.e., staged and flexible) design, so they are expected to remain important and promising research areas into the future. The research community would benefit from a thorough comparison of existing methods for search space reduction and computational efficiency increase, which could use a selection of benchmark cases of various sizes and complexity. In addition to currently available methods for search space reduction, it might be possible to further improve decision variable coding.

Regarding solution postprocessing, an open question is how sensitive the obtained solution(s) is to the optimisation model used [22,23]. When multi-objective optimisation approach is used, a remaining challenge is to select a practical and representative subset of the non-dominated solutions, which could be useful for the decision makers. Accordingly, there is a need for methods to identify a handful of effective solutions, such as those where a small improvement in one objective leads to a large deterioration in at least one other objective. The existing approaches, including maximum convex bulge/distance from hyperplane, hypervolume contribution, and local curvature [80] are all promising and require a thorough analysis on WDS design problems.

6. Summary and Conclusion

A systematic literature review of optimisation of water distribution system (WDS) design since the end of the 1980s to nowadays has been presented. The publications included in this review are relevant to the design of new WDSs, strengthening, expansion and rehabilitation of existing WDSs, and also consider design timing, parameter uncertainty, water quality and operational aspects. The value of this review paper is that it brings together a large number of publications for design optimisation of WDSs, just under three hundred in total, which have been published over the past three decades. Therefore, it may enable researchers to identify one's articles of interest in a timely manner. The review analyses the current status, identifies trends and limits in the field, describes a general optimisation model, suggests future research directions. Exclusively, this review paper also contains comprehensive information for over one hundred and twenty publications in a tabular form, including optimisation model formulations (i.e., objectives, constraints, decision variables), solution methodologies used and other important details.

This review has identified the following main limits in the field and future research directions. It was demonstrated that there is still no agreement among researchers and practitioners on how to best formulate a WDS design optimisation model, how to include all relevant objectives and constraints, and whether and how to take into account various sources of uncertainty, while still allowing for an efficient search for the best solution to be achieved. Although a plethora of generic and problem-specific optimisation methods have been developed and applied over the years, there is no consensus on what optimisation method is best for a particular design problem, whether a single or multiple-phase optimisation concept is to be used, and how engineering judgement can best be incorporated in the search. Therefore, a concerted effort by the research community is required to develop methods for objective comparison and validation of various optimisation algorithms and concepts on large, real-world problems. In addition, an analysis of available methods for reducing the search space, increasing computational efficiency, as well as selecting effective Pareto non-dominated solutions representing a practical subset for decision makers, is needed using WDS design problems of various sizes and complexity. In spite of the overwhelming amount of literature that has been published over the past three decades, design optimisation of WDSs faces considerable research challenges in the years to come.

7. List of Terms

- Deterministic dynamic design = staged design over a long planning horizon divided into several construction phases, without considering future uncertainties.
- Deterministic static design = traditional design with a single construction phase for an entire expected life cycle of the system, without considering future uncertainties.
- Dynamic design = staged (i.e., real-life) design capturing the system modifications/growth over a long planning horizon divided into several construction phases (adopted from [118]).
- Hydraulic constraints = constraints arising from physical laws of fluid flow in a pipe network, such as conservation of mass of flow, conservation of energy, conservation of mass of constituent.
- Optimisation approach = single-objective approach or multi-objective approach.
- Optimisation method = method, either deterministic or stochastic, used to solve an optimisation problem.

- Optimisation model = mathematical formulation of an optimisation problem inclusive of objective functions, constraints and decision variables.
- Probabilistic dynamic design = flexible design over a long planning horizon divided into several construction phases, with considering future uncertainties.
- Probabilistic static design = traditional design with a single construction phase for an entire expected life cycle of the system, with considering future uncertainties.
- Simulation model = mathematical model or software used to solve hydraulics and water quality network equations.
- Single pipe design = design which uses pipe sizes/diameters as decision variables (either discrete or continuous).
- Solution = result of optimisation, either from feasible or infeasible domain, so we refer to a 'feasible solution' or 'infeasible solution', respectively. In mathematical terms though an 'infeasible solution' is not classified as a solution.
- Split-pipe design = design which uses pipe segment lengths of a constant (known) diameter as decision variables.
- Static design = traditional (i.e., theoretical) design with a single construction phase for an entire expected life cycle of the system (adopted from [118]).
- System constraints = constraints arising from the limitations of a WDS or its operational requirements, such as water level limits at storage tanks, limits for nodal pressures or constituent concentrations, tank volume deficit etc.

Conflicts of Interest: The authors declare no conflict of interest.

Abbreviations

ACO	ant colony optimisation
ACO_{CTC}	convergence-trajectory controlled ant colony optimisation
ACS	ant colony system
AEF	average emissions factor
AIM	artificial inducement mutation
ALCO-GA	adaptive locally constrained genetic algorithm
AMPSO	accelerated momentum particle swarm optimisation
ANN	artificial neural network
AS	ant system
AS_{elite}	elitist ant system
AS_{rank}	elitist rank ant system
BB	branch and bound
BB-BC	big bang-big crunch
BLIP	binary linear integer programming
BLP-DE	combined binary linear programming and differential evolution
BWN-II	battle of the water networks II (optimisation problem)
CA	cellular automaton
CAMOGA	cellular automaton and genetic approach to multi-objective optimisation
CANDA	cellular automaton for network design algorithm
CC	chance constraints
CDGA	crossover dither creeping mutation genetic algorithm
CE	cross entropy
CFO	central force optimisation
CGA	crossover-based genetic algorithm with creeping mutation
CMBGA	non-crossover dither creeping mutation-based genetic algorithm
CR	crossover probability (parameter)
CS	cuckoo search

CSHS	combined cuckoo-harmony search
CTM	cohesive transport model
D	design
dDE	dither differential evolution
DDSM	demand-driven simulation method
DE	differential evolution
DMA	district metering area
DPM	discoloration propensity model
DSO	newly developed swarm-based optimisation algorithm
EA	evolutionary algorithm
EA-WDND	evolutionary algorithm for solving water distribution network design
EEA	embodied energy analysis
EEF	estimated (24-h) emissions factor (curve)
EF	emissions factor
EPANETpdd	pressure-driven demand extension of EPANET
EPS	extended period simulation
ES	evolution strategy
F	mutation weighting factor (parameter)
FCV	flow control valve
fmGA	fast messy genetic algorithm
FSP	fixed speed pump
GA	genetic algorithm
GA-ILP	combined genetic algorithm and integer linear programming
GA-LP/GALP	combined genetic algorithm and linear programming
GANEO	genetic algorithm network optimisation (program)
GENOME	genetic algorithm pipe network optimisation model
GHEST	genetic heritage evolution by stochastic transmission
GHG	greenhouse gas (emissions)
GOF	gradient of the objective function
GP	genetic programming
GRG2	generalised reduced gradient (solver)
GUI	graphical user interface
HBA	heuristic-based algorithm
HBMO	honey bee mating optimisation
HD-DDS	hybrid discrete dynamically dimensioned search
HDSM	head-driven simulation method
HMCR	harmony memory considering rate (parameter)
HMS	harmony memory size (parameter)
HS	harmony search
IA	immune algorithm
IDPSO	integer discrete particle swarm optimisation
ILP	integer linear programming
IMBA	improved mine blast algorithm
IPSO	improved particle swarm optimisation
KLSM	Kang and Lansey's sampling method [26]
LCA	life cycle analysis
LHS	Latin hypercube sampling
LINDO	linear interactive discrete optimiser
LM	Lagrange's method
LP	linear programming
LTF	linear transfer function
MA	memetic algorithm
MBA	mine blast algorithm

MBLP	mixed binary linear problem
MCHH	Markov-chain hyper-heuristic
MdDE	modified dither differential evolution
MENOME	metaheuristic pipe network optimisation model
mIA	modified immune algorithm
MILP	mixed integer linear programming
MINLP	mixed integer nonlinear programming
MMAS	max-min ant system
MO	multi-objective
MODE	multi-objective differential evolution
MOEA	multi-objective evolutionary algorithm
MOGA	multi-objective genetic algorithm
MSATS	mixed simulated annealing and tabu search
NBGA	non-crossover genetic algorithm with traditional bitwise mutation
NFF	needed fire flow
NLP	nonlinear programming
NLP-DE	combined nonlinear programming and differential evolution
NSES	non-dominated sorting evolution strategy
NSGA-II	non-dominated sorting genetic algorithm II
OP	operation
OPTIMOGA	optimised multi-objective genetic algorithm
OPUS	optimal power use surface
PAR	pitch adjustment rate (parameter)
PESA-II	Pareto envelope-based selection algorithm II
PHSM	prescreened heuristic sampling method
PIV	pipe index vector
PRV	pressure reducing valve
PSF HS	parameter setting free harmony search
PSHS	particle swarm harmony search
PSO	particle swarm optimisation
PSO-DE	combined particle swarm optimisation and differential evolution
PVA	present value analysis
RC	robust counterpart (approach)
ROs	real options (approach)
RS	random sampling
RST	random search technique
SA	simulated annealing
SADE	self-adaptive differential evolution
SAMODE	self-adaptive multi-objective differential evolution
SCA	shuffled complex algorithm
SCE	shuffled complex evolution
SDE	standard differential evolution
SE	search enforcement
SFLA	shuffled frog leaping algorithm
SGA	crossover-based genetic algorithm with bitwise mutation
SMGA	structured messy genetic algorithm
SMODE	standard multi-objective differential evolution (i.e., optimising the whole network directly without decomposition into subnetworks)
SMORO	scenario-based multi-objective robust optimisation
SO	single-objective
SPEA2	strength Pareto evolutionary algorithm 2

SS	scatter search
SSSA	scatter search using simulated annealing as a local searcher
STA	state transition algorithm
TC	time cycle
TRS	tank reserve size
TS	tabu search
VSP	variable speed pump
WCEN	water distribution cost-emission nexus
WDS	water distribution system
WDSA	water distribution systems analysis (conference)
WPP	water purification plant
WSMGA	water system multi-objective genetic algorithm
WTP	water treatment plant

Appendix A

Table A1. Papers reviewed in a chronological order.

ID, Authors (Year) [Ref] SO/MO * Brief Description	Optimisation Model (Objective Functions +, Constraints **, Decision Variables ++)	Water Quality Network Analysis Optimisation Method	Notes
1. Alperovits and Shamir (1977) [14] SO Optimal water distribution system (WDS) design and operation with split pipes considering multiple loading conditions using linear programming (LP) with a two-phase procedure.	Objective (1): Minimise (a) the overall capital cost of the network including pipes, pumps, reservoirs and valves, (b) present value of operating costs (pumps, penalties on operating the dummy valves). Constraints: (1) Min/max pressure limits, (2) sum of lengths of pipe segments within an arc equals to the length of this arc, (3) non-negativity requirement for the length of pipe segments. Decision variables: (1) Flows in pipes as primary variables, (2) length of pipe segments of constant pipe diameter (so called split-pipe decision variables), (3) dummy valve variables to represent multiple loadings (demands), (4) pump locations and capacities, (5) valve locations, (6) reservoir elevations, (7) pump operation statuses, (8) valve settings.	Water quality: N/A. Network analysis: Initial flow distribution is to be specified, flows are then redistributed using a gradient method within an optimisation process. Optimisation method: LP gradient method.	• A looped network is used. • A nonlinear problem is replaced by a linear problem. Hierarchical decomposition is iteratively applied as follows. In the first phase, LP solves the problem for the given flow distribution. In the second phase, flows in the network are updated and so on. • The method considers multiple loading conditions (i.e., peak and low demands) simultaneously, and is applicable for real complex systems. • The method gives only a local optimum. • Results: The optimal solutions represent a decrease in the total cost of 3–9% for the test networks, comparing to the costs for the initial flow distributions. • Test networks: (1) Two-loop network supplied by gravity (incl. 7 nodes), (2) two-loop network with a pump and balancing reservoir (incl. 8 nodes), (3) real network with 65 pipes and 2 pumps (incl. 52 nodes).
2. Schwarz et al. (1985) [129] SO Optimal development of a regional multiquality water resources system over a planning horizon (e.g., several years) using LP.	Objective (1): Minimise the costs of (a) water supply (water), (b) temporary curtailment of water supply, (c) network expansion, (d) conveying water, (e) excess salination. Major constraints: (1) Water quantity bounds, (2) water quality bounds, (3) regional water balance (quantity), (4) capacity expansion of the network, (5) annual source water balance (quantity), (6) annual source mass balance (salinity), (7) node mass balance (salinity). Decision variables: (1) Target water supply (m³/year), (2) temporary curtailment of water supply (m³/year), (3) capacity expansion (m³/day), (4) conveyance of water (m³/day), (5) amount of water used from storage (m³/day), (6) salinity (mg/L).	Water quality: Salinity. Network analysis: TEKUMA model [202,203]. Optimisation method: TEKUMA model [202,203] using LP.	• Seasonal variations and probabilities of climatic states are included. • Constituent (i.e., salinity) mass balance equations make the model nonlinear. These nonlinear equations are incorporated into the LP model by using a successive linear approximation iterative scheme. • The TEKUMA model was developed to determine "the plan of allocation, capacity expansion, production, transportation and operation that maximises the sum of all benefit - the sum of all water-related values minus the sum of all investment and operating costs and losses incurred by insufficient supply". • Results: Some of the typical quality management processes are demonstrated, such as that the source salinity increases steadily as a result of saline return flows, desalination is economically justified after the third period, etc. • Test networks: (1) Simplified system with one source and one customer (incl. 3 nodes), (2) real-world regional water supply system in Southern Arava, Israel, consisting of 59 consumer groups in 9 regions, 31 water sources, 77 links, considering 3 planning horizons, 3 climatic zones and 3 seasons.
3. Kessler and Shamir (1989) [114] SO Optimal WDS design with split pipes using LP with a two-phase procedure.	Objective (1): Minimise (a) the design cost of the network (pipes). Constraints: (1) Pressure limitations at selected nodes, (2) sum of lengths of pipe segments within an arc equals to the length of this arc. Decision variables: (1) Lengths of pipe segments of constant (all available) pipe diameters (so called split-pipe decision variables), (2) flows in pipes.	Water quality: N/A. Network analysis: Flow in pipes is calculated using projected gradient method within an optimisation process. Optimisation method: LP gradient method.	• A looped network is used, assuming that flow distribution is known. In the first phase, the pipeline cost for a known flow distribution is minimised using LP. In the second phase, flows are redistributed based on the gradient of the objective function (GOF). These steps repeat iteratively converging to a local optimum. • In contrast to [204], it is proved that the mathematical expression of the GOF is independent of the initial choice of the sets of loops and paths, which are used for formulation of the head constraints (conservation of energy). • Results: The optimal solution obtained is comparable to the best known solution [205] with the flows distributed more evenly. • Test networks: (1) Two-loop network supplied by gravity (incl. 7 nodes) [14].

Table A1. *Cont.*

ID. Authors (Year) [Ref] SO/MO * Brief Description	Optimisation Model (Objective Functions *, Constraints **, Decision Variables ††)	Water Quality Network Analysis Optimisation Method	Notes
4. Lansey and Mays (1989) [16] SO Optimal WDS design, rehabilitation and operation considering multiple loading conditions using nonlinear programming (NLP) with a two-phase procedure.	Objective (1): Minimise (a) the design cost of the network including pipes, pumps and tanks, (b) penalty cost for violating nodal pressure heads. Constraints: (1) Lower and upper pressure bounds at nodes, (2) design constraints (i.e., storage requirements), (3) general constraints. Decision variables: (1) Pipe diameters (continuous), (2) pump sizes (horsepower or head-flow), (3) valve settings, (4) tank volumes.	Water quality: N/A. Network analysis: KYPIPE [12]. Optimisation method: NLP solver generalised reduced gradient (GRG2) [206].	• The solution algorithm consists of an inner and outer loop, where the inner loop links KYPIPE with GRG2 and the outer loop updates penalty parameters. The augmented Lagrangian penalty method is used to incorporate nodal pressure head constraints in the objective function. • The final solution is modified, so that a pipe within a link is split into two pipes of upper- and lower-sized commercially available (discrete) diameters closest to the obtained optimal (continuous) diameter. • Multiple demand loads are analysed (i.e., combination of instantaneous peak, daily peak and fire demands at the selected nodes). • Results: The method determines optimal sizes/settings of all network components with the limitation of continues (rather than discrete) values for pipes and pumps. • Test networks: (1) Anytown network [84] with modifications (incl. 16 nodes), (2) network example 5A (incl. 13 nodes) from KYPIPE [12].
5. Lansey et al. (1989) [110] SO Optimal WDS design including uncertainties in demands, minimum pressure requirements and pipe roughnesses using NLP.	Objective (1): Minimise (a) the design cost of the network (pipes). Constraints: (1) Conservation of mass of flow and energy, (2) min pressure at the nodes, (3) pipe diameter bigger than or equal to zero. Decision variables: (1) Pipe diameters (continuous), (2) pressure head at nodes.	Water quality: N/A. Network analysis: Network hydraulics is included as a constraint to the optimisation model. Optimisation method: NLP solver GRG2 [206].	• The model includes uncertainties in (i) demands, (ii) minimum pressure head requirements and (iii) pipe roughnesses, they are included as chance constraints (CC). • Constraints (1) and (2), initially expressed as probabilities, are transformed into a deterministic form using the concept of the cumulative probability distribution, where model uncertainties are assumed to be normal random variables. The final chance-constrained optimisation model represents a NLP problem. • Results: A more reliable WDS design is obtained when including uncertainties. • Test networks: (1) Two-loop network supplied by gravity (incl. 7 nodes), (2) more realistic size network with 33 pipes (incl. 16 nodes).
6. Eiger et al. (1994) [115] SO Optimal WDS design with split pipes using mixed-integer NLP (MINLP) with a two-phase procedure.	Objective (1): Minimise (a) the design cost of the network (pipes). Constraints: (1) Pressure limitations at selected nodes, (2) sum of lengths of pipe segments within an arc equals to the length of this arc. Decision variables: (1) Lengths of pipe segments of constant (all available) pipe diameters (so called split-pipe decision variables), (2) flows in pipes. Note: Same formulation as in Kessler and Shamir (1989).	Water quality: N/A. Network analysis: Flow in pipes is calculated using projected gradient method within an optimisation process. Optimisation method: Branch and bound (BB) method.	• The optimisation model is decomposed into two models, inner (linear) and outer (nonsmooth and nonconvex) problems, which are solved by the LP solver CPLEX [207] and bundle trust region method, respectively. The dimension of the outer problem is significantly reduced by an affine transformation. This process is further referred to as primal. • Using the duality theory, a dual problem paired with the original problem is formulated and solved by CPLEX to estimate a global lower bound of the solution. This process is further referred to as dual. • Both of these processes, primal and dual, are combined in a BB type algorithm. • Results: The proposed method produces better (and feasible) solutions than previously used methods [14,49]. • Test networks: (1) Two-loop network supplied by gravity (incl. 7 nodes) [14], (2) Hanoi network (incl. 32 nodes) [49], (3) complex two-loop network (incl. 8 nodes) [14], (4) real network (incl. 52 nodes) [14].

Table A1. *Cont.*

ID. Authors (Year) [Ref] SO/MO * Brief Description	Optimisation Model (Objective Functions +, Constraints **, Decision Variables ++)	Water Quality Network Analysis Optimisation Method	Notes
7. Kim and Mays (1994) [17] SO Optimal WDS rehabilitation and operation over a planning horizon (e.g., 20 years) using MINLP with a two-phase procedure.	Objective (1): Minimise the sum of the present value of the (a) pipe replacement cost, (b) pipe rehabilitation cost, (c) expected pipe repair (i.e., break repair) cost, (d) pump energy cost. Constraints: (1) Demand supplied to each node should be greater or equal to the required demand, (2) min/max pressure at demand nodes, (3) constraints on binary decision variables representing pipe replacement and rehabilitation options, (4) constraints on continuous decision variables representing the diameter of the replaced pipe and pump horsepower. Decision variables: (1) Pipe replacement option (binary), (2) pipe rehabilitation option (binary), (3) pipe diameters of the replaced pipes (continuous), (4) pump horsepower (continuous).	Water quality: N/A. Network analysis: KYPIPE [12]. Optimisation method: BB method combined with GRG2 [206].	• The optimisation problem is formulated as a MINLP problem. This problem is divided into the following two phases within an optimisation procedure. • The NLP subproblem, which involves continuous decision variables, such as pipe diameters and pumping powers, is solved by GRG2 linked with KYPIPE. Nodal pressure head constraints are implemented using the augmented Lagrangian penalty method. • The master problem, which involves binary decision variables, such as pipe replacement and rehabilitation options, is solved by a BB implicit enumeration procedure. • The global optimum cannot be guaranteed. • <u>Results</u>: The method is able to find optimal solutions, which is supported by the comparison with the minimum cost obtained from the 1000 random system configurations. • <u>Test networks</u>: (1) Simple network with 4 pipes and 1 pump (incl. 3 nodes), (2) network with 17 pipes and 1 pump (incl. 12 nodes), (3) network with 43 pipes and 1 pump (incl. 27 nodes).
8. Murphy et al. (1994) [96] SO Optimal WDS strengthening, expansion, rehabilitation and operation considering multiple loading conditions using genetic algorithm (GA).	Objective (1): Minimise the design cost of the network including (a) pipes, (b) pumps, (c) tanks, and (d) the pump energy costs. Constraints: (1) Limits for nodal pressure heads, (2) limits for tank water levels. Decision variables: Options for (1) new pipes, (2) duplicated pipes, (3) cleaned/lined pipes, (4) pumps, (5) tanks.	Water quality: N/A. Network analysis: Unspecified steady state hydraulic solver. Optimisation method: GA.	• Pipe costs are calculated for the lengths of new pipes (i.e., network expansion), pipes laid in parallel to the existing pipes as duplications (i.e., network strengthening), and existing pipes cleaned and lined (i.e., network rehabilitation). • Four demand loadings are considered, these include instantaneous peak flow and three fire flow conditions at various nodes around the network. • <u>Results</u>: The obtained solution compares favourably with the previous designs presented in [94]. • <u>Test networks</u>: (1) Anytown network (incl. 19 nodes) [84].
9. Loganathan et al. (1995) [116] SO Optimal WDS design and strengthening with split pipes using a combination of LP, multistart local search and simulated annealing (SA) in a two-phase procedure.	Objective (1): Minimise (a) the design cost of the network (pipes). Constraints: (1) Min pressure at the nodes, (2) sum of pipe segment lengths must be equal to the link length, (3) nonnegativity of segment lengths. Decision variables: (1) Lengths of pipe segments of known diameters (so called split-pipe decision variables).	Water quality: N/A. Network analysis: Explicit mathematical formulation (steady state). Optimisation method: Combined LP, multistart local search and SA.	• The problem is solved by using an inner-outer optimisation procedure as follows. • Inner: for a fixed set of flows, a LP problem is solved to obtain (least-cost) pipe diameters and heads. • Outer: the flows are altered (optimised) using two global optimisation techniques, multistart local search and SA. Initially, a set of flows corresponding to a near optimal spanning tree of the network is found. The flows in the looped network are then taken as the perturbed tree link flows. • <u>Results</u>: The proposed optimisation method yields better least-cost designs than those previously reported in the literature. • <u>Test networks</u>: (1) Two-loop network supplied by gravity (incl. 7 nodes) [14], (2) New York City tunnels (incl. 20 nodes) [81].

Table A1. Cont.

ID. Authors (Year) [Ref] SO/MO * Brief Description	Optimisation Model (Objective Functions +, Constraints **, Decision Variables ++)	Water Quality Network Analysis Optimisation Method	Notes
10. Dandy et al. (1996) [85] SO Optimal WDS strengthening using GA.	Objective (1): Minimise (a) the sum of material and construction costs of pipes, (b) the penalty cost for violating the pressure constraints. Constraints: (1) Min/max pressure limits at certain network nodes, (2) min diameters for certain pipes in the network. Decision variables: (1) Pipe diameters (discrete diameters are coded using binary substrings).	Water quality: N/A. Network analysis: KYPIPE [12] and another hydraulic solver developed for the paper. Optimisation method: GA.	• Improved GA is used incorporating: (i) variable power scaling of the fitness function using a new variable exponent, which is initially kept low to preserve population diversity and allow global exploration, and gradually increases to emphasise fitter strings; (ii) adjacency or creeping mutation operator, which allows local exploration; (iii) Gray codes instead of binary codes representing decision variables to ensure that nearby designs are coded similarly. • Results: A solution found by the improved GA for the New York tunnels problem is the lowest-cost feasible discrete solution yet published. Test networks: (1) New York City tunnels (incl. 20 nodes) [81].
11. Halhal et al. (1997) [63] MO Optimal WDS rehabilitation and strengthening over a planning horizon (e.g., several years) using structured messy GA (SMGA).	Objective (1): Maximise the weighted sum of the following benefits of the network: (a) hydraulic performance, (b) physical integrity of the pipes, (c) system flexibility, (d) water quality. Objective (2): Minimise the cost (supply and installation) of the network including (a) new parallel pipes (i.e., duplication), (b) cleaning and lining existing pipes, (c) replacing existing pipes. Constraints: (1) Costs cannot exceed the available budget. Decision variables: String comprising 2 substrings: (1) substring consisting of pipe numbers, (2) substring consisting of decisions associated with those pipes (8 possible decisions). Note: One MO model including both objectives.	Water quality: A general water quality consideration. Network analysis: EPANET. Optimisation method: SMGA.	• Hydraulic performance benefit is quantified as the difference between the pressure deficiencies in the initial network and in the solution found. Physical integrity benefit is quantified using break repair costs for the renewed pipes. System flexibility and water quality benefits are quantified using the total diameter for the parallel pipes and the total length of renewed or relined pipes. Regarding water quality, old pipes usually create sites for the development of microorganisms and/or discoloured water. • A SMGA is introduced. It starts by evaluating all possible single variable decisions, the best of which are kept for the initial population. As the algorithm progresses, the short strings are concatenated to form longer strings. This enables to start with cheaper solutions which stay under budget from the very beginning. The SMGA encodes only those decision variables which are active thereby reducing the search space. • Results: SMGA displays outstandingly superior performance over the standard GA for the real network. Test networks: (1) Small looped network with 15 pipes (incl. 9 nodes), (2) real network with 167 pipes and 1 reservoir for a town of 50000 inhabitants in Morocco (incl. 115 nodes).
12. Savic and Walters (1997) [42] SO Optimal WDS design and strengthening using GA.	Objective (1): Minimise (a) the design cost of the network (pipes). Constraints: (1) Min pressure at the nodes. Decision variables: (1) Pipe diameters (discrete).	Water quality: N/A. Network analysis: Network solver based on the EPANET. Optimisation method: GANET [208] using GA.	• A program GANET for least-cost pipe network design is developed, implementing a modified GA. The modifications include, for example, the use of Gray codes instead of binary codes, allowing some infeasible solutions to join the population and help guide the search. • Discrete diameters solutions (obtained by GANET) are compared to split-pipes and continuous diameters solutions, previously published in the literature by [14,114,115,204]. • Results: GANET produced good designs without unnecessary restrictions imposed by split-pipe or linearising assumptions. Test networks: (1) Two-loop network supplied by gravity (incl. 7 nodes) [14], (2) Hanoi network (incl. 32 nodes) [49], (3) New York City tunnels (incl. 20 nodes) [81].

Table A1. Cont.

ID. Authors (Year) [Ref] SO/MO*, Brief Description	Optimisation Model (Objective Functions+, Constraints**, Decision Variables**)	Water Quality, Network Analysis, Optimisation Method	Notes
13. Cunha and Sousa (1999) [102] SO Optimal WDS design using SA.	Objective (1): Minimise (a) the design cost of the network (pipes). Constraints: (1) Min pressure at the nodes, (2) min pipe diameter. Decision variables: (1) Pipe diameters (discrete).	Water quality: N/A. Network analysis: Newton method to solve hydraulic equations to obtain flows and heads. Optimisation method: SA.	• The optimisation problem is solved as follows: initial set of pipe diameters is selected, hydraulic equations are solved using a Newton method, constraints are checked, SA is performed and the process is repeated until the optimal solution is found. • Discrete diameters solutions (obtained by SA) are compared to split-pipes, continuous diameters, as well as discrete diameter solutions, previously published in the literature by [14,42,49,114,115,204,209–211]. • Results: SA can provide high quality solutions for network design problems. • Test networks: (1) Two-loop network supplied by gravity (incl. 7 nodes) [14], (2) Hanoi network (incl. 32 nodes) [49].
14. Gupta et al. (1999) [188] SO Optimal WDS strengthening and expansion using GA with search space reduction.	Objective (1): Minimise (a) the design cost of the network (pipes), (b) penalty for violating minimum residual head. Constraints: (1) Min residual head, (2) min desirable velocity in a pipe. Decision variables: (1) Pipe diameters (discrete).	Water quality: N/A. Network analysis: ANALIS [212]. Optimisation method: GA.	• In GA, the solution is represented by a chromosome to avoid the conversion of binary coding to discrete pipe sizes. • The test networks are stratified into upper, middle and lower diameter sets using engineering judgment, which helps reduce the search space and facilitate faster convergence to the optimum. • Results: GA provides a better solution in general while compared with the NLP technique. Additionally, the GA convergence considerably improved by providing initial information on network stratification. • Test networks: (1) Network with 38 pipes (incl. 23 nodes), (2) same as network (1) with a significantly different demand pattern, (3) network with 52 pipes (incl. 31 nodes), (4) same as network (3) with a different demand pattern, (5) network with 28 pipes (incl. 18 nodes), (6) network with 13 pipes (incl. 11 nodes).
15. Halhal et al. (1999) [131] MO Optimal WDS rehabilitation and strengthening over a planning horizon (i.e., 10 years) using SMGA.	Objective (1): Maximise the present value of the benefit of the network rehabilitation over the planning period (incorporating the welfare index), using the following performance criteria: (a) carrying capacity, (b) physical integrity of the pipes, (c) system flexibility, (d) water quality. Objective (2): Minimise (a) the present value of the rehabilitation costs over the planning period. Constraints: (1) Rehabilitation costs less than or equal to the budget. Decision variables: String comprising 3 substrings (1) location substring: pipe numbers of pipes scheduled for rehabilitation (integer), (2) decision substring: rehabilitation option (integer), (3) timing substring: year of rehabilitation execution (integer). Note: One MO model including both objectives.	Water quality: A general water quality consideration. Network analysis: Unspecified solver (steady state). Optimisation method: SMGA.	• Carrying capacity is represented by the hydraulic performance, which is calculated as the sum of nodal pressure excesses and shortfalls weighted by the demand flows. Physical integrity is included as a function of the breakage repair costs, with new pipes considered break-free. System flexibility is determined as a function of the number of new parallel pipes. Water quality is included as a function of the length of renewed and/or lined pipes having Hazen-Williams coefficient below a specified limit. Old corroded pipes are considered to cause the development of microorganism and discoloured water. • SMGA has flexible coding and variable string length. Its difference from a conventional GA is that it uses, besides common GA operators, a process of concatenation. Basically, it starts with a population of one-element strings corresponding to a single decision variable (e.g., rehabilitation option for one pipe only) and gradually increases the length of the strings as populations evolve. The advantage of the SMGA is in reducing the space searched, while considering only the pipes which need alteration as opposed to all pipes in a conventional GA. • Results: The impact of varying parameters (interest and inflation rates, welfare index, pipe roughness) on the optimal solutions is presented. For example, higher welfare index enables greater initial investment and benefit. • Test networks: (1) Simple system with 15 pipes and 1 reservoir (incl. 9 nodes).

Table A1. *Cont.*

ID. Authors (Year) [Ref] SO/MO * Brief Description	Optimisation Model (Objective Functions *, Constraints **, Decision Variables ++)	Water Quality Network Analysis Optimisation Method	Notes
16. Montesinos et al. (1999) [86] SO Optimal WDS strengthening using GA.	Objective (1): Minimise (a) the design cost of the network (pipes). Constraints: (1) Min pressure at the nodes, (2) max velocity in the pipes. Decision variables: (1) Pipe diameters (discrete).	Water quality: N/A. Network analysis: Newton–Raphson method [10]. Optimisation method: GA.	• A modified GA with several changes to selection and mutation is introduced. "In each generation a constant number of solutions is eliminated, the selected ones are ranked for crossover and the new solutions are allowed to undergo at most one mutation". The GA convergence significantly increases as a result of these modifications. • A penalty factor is defined as a function of a number of constraint violations (not taking into account the degree of violation). • Results: The modified GA found the best-known solution for the test network in fewer evaluations than previous GA algorithms. • Test networks: (1) New York City tunnels (incl. 20 nodes) [81].
17. Walters et al. (1999) [99] MO Optimal WDS strengthening, expansion, rehabilitation and operation with multiple loading conditions and two approaches to model tanks using SMGA. Note: Discussion: [213], Erratum to Discussion: [214]	Objective (1): Maximise the weighted sum of the benefits of the network rehabilitation, using the following performance criteria: (a) nodal pressure shortfall, (b) storage capacity difference, (c) tank operating level difference or tank flow difference. Objective (2): Minimise (a) the capital cost of the network including pipes, pumps, tanks, (b) present value of the energy consumed during a specified period. Constraints: (1) Pressure constraints for different loading patterns, (2) flow constraints into and out of the tanks. Decision variables: String comprising 2 substrings (1) location substring: pipes, pumps, tanks (integer of 1 or 2 digits), (2) decision substring (expansion/rehabilitation options): pipes, pumps, tanks (integer of 1, 2 or 5 digits). Note: One MO model including both objectives.	Water quality: N/A. Network analysis: Unspecified solver (steady state). Optimisation method: SMGA.	• Two approaches to model tanks are tested, which differ in a way they determine the operating levels for new tanks. The first approach computes tank levels analytically during the network analysis, the second approach includes tank levels as independent variables. For the test network, both approaches yielded similar results, with the first approach obtaining more robust solutions in slightly increased computational time. • The previously published SMGA [63,131] is expanded to include not only pipe rehabilitation, but also pump and tank installations as decision variables. Variable mutation rate as a function of the string length and the nature of the decision variable is used. For more information about SMGA, see [131]. • Results: Two solutions are presented, the cheapest feasible solution and the most operationally satisfactory solution (preferred by the authors). These solutions are 4–5% cheaper than any previously published solutions to the Anytown problem. • Test networks: (1) Anytown network (incl. 19 nodes) [84].
18. Costa et al. (2000) [60] SO Optimal WDS design and operation using SA.	Objective (1): Minimise the capital cost of the network including (a) pipes, (b) pumps, (c) present value of pump energy costs. Constraints: (1) Min head bound on demand nodes. Decision variables: (1) Pipe diameters (discrete), (2) pump sizes (discrete).	Water quality: N/A. Network analysis: Newton–Raphson method [10]. Optimisation method: SA.	• Operating costs of pumps are calculated in terms of operating hours per year. The network model presents a realistic representation of the pump behaviour, including the head characteristic curve. • Results: The algorithm reaches optimal solutions with the average number of 11817–13454 simulations for the test networks. • Test networks: (1) Gravity network with one reservoir (incl. 9 nodes), (2) network with one pump and one reservoir (incl. 10 nodes), (3) network with one pump and 2 reservoirs (incl. 11 nodes).

Table A1. *Cont.*

ID, Authors (Year) [Ref] SO/MO* Brief Description	Optimisation Model (Objective Functions *, Constraints **, Decision Variables **)	Water Quality Network Analysis Optimisation Method	Notes
19. Dandy and Hewitson (2000) [120] SO Optimal WDS design, strengthening and operation including water quality considerations using GA with search space reduction.	Objective (1): Minimise (a) the capital cost of new pipes, pumps and tanks, present value of (b) pump energy costs, (c) likely cost to the community due to waterborne diseases, (d) likely community cost due to disinfection by-products, (e) community cost of chlorine levels that exceed acceptable limits, (f) cost of disinfection, (g) penalty cost for violating constraints. Constraints: (1) Min pressure at the demand nodes, (2) tanks must refill at the end of the cycle. Decision variables: (1) Sizes of new and duplicate pipes, (2) sizes of new pumps and tanks, (3) locations of new pumps and tanks, (4) decision rules for operating the system, (5) dosing rates of chloramine/chlorine at selected points.	Water quality: Chloramine, chlorine. Network analysis: EPANET (extended period simulation (EPS)). Optimisation method: GA.	• A total of 6 different demand patterns are used, ranging from peak instantaneous to 38 days of winter demand. • The periods of simulation for each season, which reflect the residence times for a season, were found to be necessary in order to reach a pseudo steady state for that season. • The problem is very complex (the GA string consists of 222 integer variables) with long run times. To reduce the size of the search space, a run with peak instantaneous demand was undertaken, then a run with peak daily demand. Out of 206 only 40 pipes that were duplicated in either of these runs were included as options in the total system analysis. It was found that better overall convergence occurred if the GA was run with hydraulic analysis only for several first generations, water quality analysis was subsequently added. • Results: The advantages of including design, operations and water quality in a single framework are demonstrated. In the design phase, allowance can be made for reducing residence times, thus improving water quality. • Test networks: (1) Yorke Peninsula, a rural area west of Adelaide, Australia.
20. Vairavamoorthy and Ali (2000) [43] SO Optimal WDS design and strengthening incorporating a linear transfer function (LTF) model to approximate network hydraulics using GA.	Objective (1): Minimise (a) the capital cost of the network (pipes), (b) penalty for violating the pressure constraints. Constraints: (1) Min/max pressure at the nodes. Decision variables: (1) Pipe diameters (discrete).	Water quality: N/A. Network analysis: Steady state hydraulic solver based on the gradient method [215]. Optimisation method: GA.	• Real coding is applied instead of binary coding (binary and Gray coding often generate redundant states which do not represent any of the design variables). • A variable penalty coefficient is introduced that depends on the degree of constraint violation. • LTF model is proposed which approximates the hydraulic behaviour of the system, so there is no need for each population member to be evaluated by a hydraulic solver. Instead, the LTF is used to estimate pressures for each string generated by the GA. • Results: Obtained solutions are favourable while compared to the results of previous studies [42,85,216]. • Test networks: (1) Hanoi network (incl. 32 nodes) [49], (2) New York City tunnels (incl. 20 nodes) [81].
21. Dandy and Engelhardt (2001) [130] SO Optimal WDS rehabilitation (considering only pipe replacement) over a planning horizon (i.e., 20 years) using GA.	Objective (1): Minimise (a) the system cost of the rehabilitated network (pipes)—present values of pipe failure costs (i.e., repair costs of existing and new pipes) and pipe replacement costs are considered. Constraints (case 1): N/A. Constraints (case 2): (1) Allowable budget for each time step (i.e., 5-year block). Constraints (case 3): (1) As above in the case 2, (2) min pressure at the nodes, (3) max velocity in the pipes. Decision variables (case 1): (1) Replacement decision (0 = no replacement, 1 = replace). Decision variables (case 2): (1) Timing of the replacement ("all pipe representation"); or (1) pipe to be replaced (integer) ("limited pipe representation"). Decision variables (case 3): (1)–(2) as above in the case 2, (3) diameter of the new pipe (integer).	Water quality: N/A. Network analysis: EPANET (Case 3 only). Optimisation method: GA.	• The economic analysis of the system is undertaken in three stages as follows. • Case 1 "single time-step case" is to decide if the pipes in the network need immediate replacement or should be left in operation. • Case 2 "multiple time-step case" is to schedule pipe replacements for the next 20 years, in 5-year steps. • Case 3 "multiple time-step case with changing diameters" is to determine the diameter of the replaced pipes, which is included as a decision variable. • Failure prediction equations were developed for the test network based on recorded failure data. • Two ways to represent chromosomes in GA are considered. One is "all pipe representation", which includes the decision bit for all the pipes in the network; the other is "limited pipe representation", where an upper limit to the number of pipes to be replaced is considered. The size of the search space for the latter representation is smaller than the first one, therefore it is considered a more preferable representation. • Results: The GA demonstrated ability to schedule future works. • Test networks: (1) The EL103N pressure zone, Adelaide, Australia.

Table A1. Cont.

ID. Authors (Year) [Ref] SO/MO *, Brief Description	Optimisation Model (Objective Functions +, Constraints **, Decision Variables ++)	Water Quality, Network Analysis, Optimisation Method	Notes
22. Wu and Simpson (2002) [88] SO Optimal WDS strengthening using fast messy GA (fmGA).	Objective (1): Minimise (a) the design cost of the network (pipes), (b) penalty for violating the pressure constraint. Constraints: (1) Min pressure at the nodes. Decision variables: (1) Pipe diameters (discrete).	Water quality: N/A. Network analysis: EPANET. Optimisation method: fmGA.	• A self-adaptive boundary search strategy is proposed for selection of the penalty factor within the GA. It evolves and adapts the penalty factor, so the search is guided to the boundary of the feasible and infeasible spaces. The penalty factor is treated as another decision variable (part of the solution string). In addition, a heuristic rule is developed to adjust the lower and upper boundaries of the penalty factor. • Results: The proposed algorithm finds the least-cost solution in the case study more effectively than a GA without the boundary search strategy. • Test networks: (1) New York City tunnels (incl. 20 nodes) [81].
23. Eusuff and Lansey (2003) [103] SO Optimal WDS design and strengthening using shuffled frog leaping algorithm (SFLA).	Objective (1): Minimise (a) the design cost of the network (pipes), (b) penalty cost for pressure head violations. Constraints: (1) Min pressure at the nodes. Decision variables: (1) Pipe diameters (discrete).	Water quality: N/A. Network analysis: EPANET. Optimisation method: SFLA.	• SFLA is a hybrid between particle swarm optimisation (PSO) (which provides local search tool) and shuffled complex evolution (SCE) algorithm (which helps move towards global solution). • The difference between the SFLA and GA is that in the SFLA an improved idea can be passed between all individuals of the population versus parent-child only interaction in GA. • Results: When compared to GA and SA in regards to the efficacy, SFLA is more efficient as it found the best-known optimal solutions in fewer iterations. • Test networks: (1) Two-loop network supplied by gravity (incl. 7 nodes) [14], (2) Hanoi network (incl. 32 nodes) [49], (3) New York City tunnels (incl. 20 nodes) [81].
24. Maier et al. (2003) [104] SO Optimal WDS strengthening, expansion and rehabilitation using ant colony optimisation (ACO).	Objective (1): Minimise (a) the design cost of the network (pipes), (b) penalty cost for violating the pressure constraint. Constraints: (1) Min pressure at the nodes. Decision variables: (1) Pipe diameters (discrete), (2) pipe rehabilitation options (binary).	Water quality: N/A. Network analysis: WADISO [217], final solutions checked by EPANET. Optimisation method: ACO.	• The main difference between GA and ACO is in generating the trial solutions. In GAs, trial solutions are represented as strings of genetic material, new solutions are obtained by modifying previous solutions, so the system memory is embedded in the actual trial solutions. In ACO, the system memory is contained in the environment, rather than the trial solutions, hence ACO may be more advantageous in certain types of applications. • A modification is made to the way pheromone concentration is changed, which ensures that the method does not get trapped in a local optimum. • Results: The comparison of GA and ACO shows that ACO is a good alternative to GA, having found the same solution in a similar number of iterations for the 14-pipe network, and a better (lower cost) solution with a significantly higher computational efficiency for the New York City tunnels. • Test networks: (1) 14-pipe network with two supply sources (incl. 10 nodes) [20], (2) New York City tunnels (incl. 20 nodes) [81].
25. Liong and Atiquzzaman (2004) [157] SO Optimal WDS design using SCE.	Objective (1): Minimise (a) the design cost of the network (pipes), (b) penalty cost for violating the pressure head bound. Constraints: (1) Min nodal pressure head bound, (2) min/max bound on pipe sizes. Decision variables: (1) Pipe sizes (converted to commercially available diameters).	Water quality: N/A. Network analysis: EPANET. Optimisation method: SCE [218].	• SCE is an evolutionary algorithm (EA) combined with a simplex algorithm [219]. The original SCE algorithm is modified to accommodate high number of decision variables. • The SCE algorithm is compared to the GA, SA, GLOBE [220] and SFLA. • Results: For the two-loop network, SCE converged after a significantly lower number of function evaluations, and for both test networks, SCE found an optimal solution notably faster, than other optimisation techniques. • Test networks: (1) Two-loop network supplied by gravity (incl. 7 nodes) [14], (2) Hanoi network (incl. 32 nodes) [49].

Table A1. Cont.

ID. Authors (Year) [Ref] SO/MO* Brief Description	Optimisation Model (Objective Functions +, Constraints **, Decision Variables **)	Water Quality Network Analysis Optimisation Method	Notes
26. Broad et al. (2005) [87] SO Optimal WDS strengthening including water quality considerations using offline artificial neural networks (ANNs) and GA.	Objective (1): Minimise (a) the design cost of the network (pipes), (b) penalty cost for violating pressure head, (c) penalty cost for violating chlorine residual. Constraints: (1) Min/max pressure at the nodes, (2) min/max chlorine residual at the nodes. Decision variables: (1) Pipe diameters, (2) chlorine dosing rates.	Water quality: Chlorine. Network analysis: Offline ANN. Optimisation method: GA.	• The methodology uses ANN as a substitute for a simulation model in order to reduce the computational time. • Because it is unlikely that ANN is able to perfectly represent the simulation model, two techniques are used to combat ANN inaccuracies as follows. • The first technique is to ensure feasibility, so solutions found by the ANN-GA are evaluated by EPANET in 3 stages: (i) each new best solution found by the ANN-GA is evaluated by EPANET; (ii) several top solutions are evaluated by EPANET when GA converges; (iii) local search is conducted after GA convergence. • The second technique is to adjust the constraints to cater for ANN underestimating or overestimating pressure and chlorine residuals. • Results: While optimising with ANN, the most time is spent on training the ANN. If the training time is included, the overall time saving for ANN-GA is 21% compared to EPANET-GA. Otherwise, ignoring the training time, the ANN-GA is 700 faster than EPANET-GA. • Test networks: (1) New York City tunnels (incl. 20 nodes) [81].
27. Farmani et al. (2005) [65] MO Optimal WDS design and strengthening using non-dominated sorting genetic algorithm II (NSGA-II) and strength Pareto evolutionary algorithm 2 (SPEA2).	Objective (1): Minimise (a) the design cost of the network (pipes). Objective (2): Minimise (a) the maximum individual head deficiency at the network nodes. Objective (3) (only for the EXNET test network): Minimise (a) the number of demand nodes with head deficiency. Constraints: N/A. Decision variables: (1) Pipe diameters (discrete). Note: Two MO models, the first including objectives (1) and (2) (applied to the New York City tunnels and Hanoi network); the second objectives (1), (2) and (3) (applied to the EXNET network).	Water quality: N/A. Network analysis: EPANET. Optimisation method: NSGA-II and SPEA2 are compared.	• NSGA-II and SPEA2 are compared in terms of non-dominated fronts obtained by these algorithms, using (i) graphical presentation, (ii) binary ε-indicator, (iii) binary coverage indicator, (iv) (only for EXNET test network) volume-based indicator. • Results: NSGA-II and SPEA2 are comparable and have the potential to find Pareto optimal solutions for WDS design problems. The results further show that SPEA2 outperformed NSGA-II in both MO optimisation problems, which is illustrated by graphical presentation as well as all indicators. • Test networks: (1) New York City tunnels (incl. 20 nodes) [81], (2) Hanoi network (incl. 32 nodes) [49], (3) simplified EXNET water network (serves a population of approximately 400000) [82].
28. Keedwell and Khu (2005) [44] SO Optimal WDS design using a combined cellular automaton for network design algorithm (CANDA) and GA (CANDA-GA) including an engineered initial population.	Objective (1): Minimise (a) the design cost of the network (pipes), (b) penalty for violating the pressure constraint. Constraints: (1) Min/max pressure at the nodes. Decision variables: (1) Pipe diameters (discrete).	Water quality: N/A. Network analysis: EPANET. Optimisation method: CANDA-GA.	• A heuristic-based cellular automaton (CA) [221] approach is introduced to provide a good initial population of solutions for GA runs. • A CA consists of an interconnected set of nodes that use a number of rules to update the state of every node according to the states of neighbouring nodes. The rules here are based on the intuitive knowledge of how the WDSs operate, so they are similar to engineering judgment. An important feature of the CA is that updates for every node are performed in parallel. CA does not produce the best solutions but it is capable of producing a good approximate solution in much less network simulations than GA. • Results: For the two-loop network, the results of GA and CANDA-GA are similar, with CANDA-GA producing a slightly better solution. For both real networks, CANDA-GA finds feasible solutions whereas GA fails to do so. • Test networks: (1) Two-loop network supplied by gravity (incl. 7 nodes) [14], (2) network A: real network with a single reservoir and 632 pipes, UK (incl. 535 demand nodes) [222], (3) network B: real network with a single reservoir and 1277 pipes, UK (incl. 1106 nodes).

Table A1. Cont.

ID, Authors (Year) [Ref] SO/MO *, Brief Description	Optimisation Model (Objective Functions *, Constraints **, Decision Variables **)	Water Quality Network Analysis Optimisation Method	Notes
29. Ostfeld (2005) [53] SO Optimal design and operation of multiquality WDSs including multiple loading conditions and water quality considerations using GA.	Objective (1): Minimise (a-D⁷) the construction costs of pipes, tanks, pump stations and treatment facilities, (b-OP⁷⁷) annual operation costs of pump stations and treatment facilities. Constraints: (1) Min/max heads at consumer nodes, (2) max permitted amounts of water withdrawals at sources, (3) tank volume deficit at the end of the simulation period, (4) min/max concentrations at consumer nodes, (5) removal ratio constraints. Decision variables: D: (1) Pipe diameters, (2) tank max storage, (3) max pumping unit power, (4) max removal ratios at treatment facilities, OP: (5) scheduling of pumping units, (6) treatment removal ratios.	Water quality: Unspecified conservative parameters. Network analysis: EPANET (EPS). Optimisation method: GA.	• Time horizon is 24 h, with a varied energy tariff and unsteady water flow conditions. Similar to [223], cyclic water quality behaviour is not accomplished, so the results depend on the initial settings of the concentrations at the nodes. • Multiple loading conditions (demands) are used. • Sensitivity analysis is performed with the following modifications to the data or constraints. The two-loop network: increased minimum pressure constraint at one consumer node, increased maximum concentration limit for all consumer nodes, increased operational unit treatment cost coefficient. The Anytown network: reduced unit power cost of pump construction and energy tariffs, altered pressure and concentration constraints at one consumer node, decreased elevation at one consumer node. • Results: The model explicitly addresses the conjunctive design and operation problem of quantity, pressure and quality simultaneously under unsteady hydraulics, but is expensive in terms of the computational resources. • Test networks: (1) Two-loop network with 3 sources (incl. 6 demand nodes) [223], (2) Anytown network [84] with modifications (incl. 16 nodes).
30. Vairavamoorthy and Ali (2005) [189] SO Optimal WDS design using GA with a pipe index vector (PIV) and search space reduction in a three-phase procedure.	Objective (1): Minimise (a) the design cost of the network (pipes). Constraints: (1) Min/max pressure at the nodes. Decision variables: (1) Pipe diameters (discrete).	Water quality: N/A. Network analysis: Explicit mathematical formulation (steady state for peak demands). Optimisation method: GA with PIV.	• PIV, a measure of the relative importance of pipes regarding their hydraulic performance in the network, is introduced. Using PIV, impractical and infeasible regions can be excluded from the search space, enabling quicker generation of feasible solutions resulting in substantial computational time savings. • The proposed method involves the following three steps: (i) establishing tighter bound constraints on all pipes using simple heuristics before the GA starts; (ii) calculating a pipe index, ranking the pipes and dividing them into groups (i.e., constructing PIV), and generating the initial population using PIV; (iii) reducing the search space during the GA itself. • It is found that calculating pipe indices is computationally expensive, therefore a surrogate measure is proposed to compute them. • Results: The proposed method outperforms the standard GA in both convergence and computational time. • Test networks: (1) Alandur network, Madras, India (incl. 82 nodes) [224], (2) Hanoi network (incl. 32 nodes) [49].
31. Vamvakeridou-Lyroudia et al. (2005) [93] MO Optimal WDS strengthening, expansion, rehabilitation and operation considering multiple loading conditions using GA with fuzzy reasoning.	Objective (1): Minimise (a) the design cost of the network including pipes, pumps and tanks. Objective (2): Maximise the benefit/quality of the solution, using the following system performance criteria (constraints): (a) min pressure at the nodes, (b) max velocity in the pipes, (c) safety volume capacity for tanks, (d) safety volume capacity for the network as a whole, (e) pump operational capacity, (f) operational volume capacity for tanks, (g) filling capacity for tanks, (h) operational volume capacity for the network as a whole, (i) filling capacity for the network as a whole. Constraints: N/A. Decision variables: (1) Commercially available pipe diameters (integer), (2) cleaning/lining of existing pipes (binary: 0 = no action, 1 = cleaning/lining), (3) the number of new pumps (integer) with pre-defined operation curve, (4) volume of a new tank (integer, 0 = no tank), (5) min operational level of this tank (integer). Note: One MO model including both objectives.	Water quality: N/A. Network analysis: EPANET. Optimisation method: GA combined with fuzzy reasoning.	• Fuzzy reasoning is introduced. System performance criteria are individually assessed by fuzzy membership functions and combined using fuzzy aggregation operators. A fuzzy set and fuzzy membership functions are defined for each performance criterion/each loading/each network element, based on previous experience [225]. Membership functions are provided with linguistic tags (e.g., "tolerant", "strict", "very strict") to enable implementation of decision maker requirements for specific network elements. • Fuzzy aggregation operators used are weighted means and classic fuzzy intersection, which are ANDlike aggregators covering a wide range and varying in strength. The model is flexible: if a decision maker wishes to omit one or more criteria, the weight assigned to it can be set to zero. On the contrary, should more criteria be added (e.g., resilience), the modular approach allows for additions and modifications, without affecting the structure of the multiobjective model and algorithm. • A novel approach for the inclusion of tanks within the GA is proposed, taking into account the tank shape. • Results: A better solution in terms of cost is obtained than any other previously published, despite the multiple criteria applied for the extensive and stricter benefit function. • Test networks: (1) Anytown network (incl. 19 nodes) [84].

Table A1. Cont.

ID, Authors (Year) [Ref] SO/MO * Brief Description	Optimisation Model (Objective Functions +, Constraints **, Decision Variables ++)	Water Quality Network Analysis Optimisation Method	Notes
32. Atiquzzaman et al. (2006) [64] MO Optimal WDS design using NSGA-II.	Objective (1): Minimise (a) the design cost of the network (pipes). Objective (2): Minimise (a) the total pressure deficit at the network nodes. Constraints: (1) Pipe diameters limited to commercially available sizes, (2) min pressure at the nodes, (3) lower and upper limit of total pressure deficit, (4) lower and upper limit of total network cost. Decision variables: (1) Commercially available pipe diameters (integer). Note: One MO model including both objectives.	Water quality: N/A. Network analysis: EPANET. Optimisation method: NSGA-II.	• The aim is to yield "alternative" solutions, which are particularly useful when the associated network cost of the optimal solution is beyond the available budget. Hence, solutions provided are within the (i) available budget and (ii) tolerated total nodal pressure deficit. The total pressure deficit is accompanied with the list of nodes at which pressure deficit occurs and a value of their individual nodal pressure deficit. This information assists in deciding whether the magnitude of the pressure violation may be tolerated. • Results: There is more than one solution with the same network cost and yet different total pressure deficits. Additionally, there are several solutions with about the same total pressure deficit for the same network cost. • Test networks: (1) Two-loop network supplied by gravity (incl. 7 nodes) [14].
33. Geem (2006) [105] SO Optimal WDS design and strengthening using harmony search (HS).	Objective (1): Minimise (a) the design cost of the network (pipes), (b) the penalty cost for violating the pressure constraint. Constraints: (1) Min pressure at the nodes. Decision variables: (1) Pipe diameters (discrete).	Water quality: N/A. Network analysis: EPANET. Optimisation method: HS.	• An improved HS algorithm, which adopts both memory consideration and pitch adjustment operations, is proposed. The algorithm is compared to the methods previously used in the literature, including LP, GA, SA and tabu search (TS). • Results: For all test networks, the HS obtained either the same or 0.28–10.26% cheaper solution than other algorithms. The HS also required fewer function evaluations than other meta-heuristic algorithms. • Test networks: (1) Two-loop network supplied by gravity (incl. 7 nodes) [14], (2) Hanoi network (incl. 32 nodes) [49], (3) New York City tunnels (incl. 20 nodes) [81], (4) GoYang network, South Korea (incl. 22 nodes) [226], (5) BakRyun network, South Korea (incl. 35 nodes) [227].
34. Keedwell and Khu (2006) [66] MO Optimal WDS design using cellular automation and genetic approach to multi-objective optimisation (CAMOGA) and NSGA-II including an engineered initial population.	Objective (1): Minimise (a) the design cost of the network (pipes). Objective (2): Minimise (b) the total head deficit at the network nodes. Constraints: (1) Max total head deficit. Decision variables: (1) Pipe diameters (discrete). Note: One MO model including both objectives.	Water quality: N/A. Network analysis: EPANET. Optimisation method: CAMOGA and NSGA-II are compared.	• An extension of the paper by [44] including a novel hybrid CAMOGA. CAMOGA consists of the following two phases: (i) CANDA [44] to generate good 'near' Pareto-optimal solutions with only a small number of iterations; (ii) NSGA-II to enhance and expand the solutions found in the previous step. • The paper also compares the performance of CAMOGA and NSGA-II using a visual comparison of obtained Pareto fronts and the S-metric [228]. • Results: CAMOGA can provide good solutions with very few network simulations, and that it outperforms NSGA-II in the efficiency of obtaining similar Pareto fronts. • Test networks: (1) Network A: real network with a single reservoir and 632 pipes, UK (incl. 535 demand nodes) [222], (2) network B: real network with a single reservoir and 1277 pipes, UK (incl. 1106 nodes).

Table A1. Cont.

ID, Authors (Year) [Ref] SO/MO *, Brief Description	Optimisation Model (Objective Functions +, Constraints **, Decision Variables ++)	Water Quality, Network Analysis, Optimisation Method	Notes
35. Reca and Martínez (2006) [50] SO Optimal WDS and irrigation network design using GA.	Objective (1): Minimise (a) the design cost of the network (pipes), (b) penalty for violating the pressure constraint. Constraints: (1) Min pressure at the nodes, (2) min/max flow velocities. Decision variables: (1) Pipe diameters (discrete).	Water quality: N/A. Network analysis: EPANET. Optimisation method: Genetic algorithm optimisation model (GENOME) using GA.	• GENOME is developed particularly for optimisation of looped irrigation networks. It is based on a GA with modifications and improvements to adapt for this specific problem. • An integer coding scheme is used to code the chromosomes. A stochastic sampling mechanism based on the "roulette wheel algorithm" is used for selection. Three crossover strategies are used: one-point, two-point and uniform. • Results: The results are compared to 20 previous studies (for two-loop network) and 17 previous studies (for Hanoi network) from the literature. They indicate that GENOME is able to obtain the best published results in a reasonable computational time. However, some adjustments would be required to improve its performance for complex networks. • Test networks: (1) Two-loop network supplied by gravity (incl. 7 nodes) [14], (2) Hanoi network (incl. 32 nodes) [49], (3) Balerma irrigation network, Almería, Spain (incl. 447 nodes).
36. Samani and Mottaghi (2006) [51] SO Optimal WDS design, operation and maintenance using integer LP (ILP). Note: Discussion: [229]	Objective (1): Minimise (a) the capital cost of the network (pipes), (b) capital, operation and maintenance costs of pumps and reservoirs. Constraints: (1) Only one pipe diameter per network branch, (2) only one pump or reservoir per network location, (3) min/max pressure at the nodes, (4) min/max velocity in the pipes. Decision variables: (1) Integer variables related to pipe diameters and pumps/reservoirs.	Water quality: N/A. Network analysis: Unspecified hydraulic solver (a single loading condition). Optimisation method: Linear interactive discrete optimiser (LINDO) program using BB method.	• Nonlinear objective function and constraints are linearised. • A procedure that iterates between a hydraulic solver and ILP solver is employed. • The test network (1) is used to demonstrate the validity of the procedure as it can be solved by enumeration. • An issue related to poorly selected initial decision variables is reported, when no feasible solution could be found and the program will stop. This issue can be overcome by setting wider limits for the pressure and velocity constraints to provide a better initial guess of decision variables. • Results: The proposed method can find good solutions; for the two-loop network, the solution obtained is comparable to previously published results in the literature. The proposed method converges very quickly. • Test networks: (1) Simple network with 3 pipes and one reservoir in a looped system (incl. 3 nodes), (2) two-loop network supplied by gravity (incl. 7 nodes) [14], (3) network with 15 pipes, 2 reservoirs, a pump and a check valve (incl. 15 nodes).
37. Suribabu and Neelakantan (2006) [106] SO Optimal WDS design using PSO.	Objective (1): Minimise (a) the design cost of the network (pipes), (b) penalty for violating the pressure constraint. Constraints: (1) Min pressure at the nodes. Decision variables: (1) Pipe diameters (discrete).	Water quality: N/A. Network analysis: EPANET. Optimisation method: PSONET program using PSO.	• PSONET program, which uses PSO algorithm, is developed and its performance compared to the previous studies from the literature having applied GA, SA, SFLA and shuffled complex algorithm (SCA). • Results: For both test networks, the PSO obtained competitive solutions, but in a lower number of function evaluations than GA, SA and SFLA. • Test networks: (1) Two-loop network supplied by gravity (incl. 7 nodes) [14], (2) Hanoi network (incl. 32 nodes) [49].

Table A1. *Cont.*

ID. Authors (Year) [Ref] SO/MO *, Brief Description	Optimisation Model (Objective Functions +, Constraints **, Decision Variables ++)	Water Quality, Network Analysis, Optimisation Method	Notes
38. Babayan et al. (2007) [89] MO Optimal robust WDS strengthening considering uncertainties in future demands and pipe roughnesses using NSGA-II.	Objective (1): Minimise (a) the design cost of the network/rehabilitation. Objective (2): Maximise (a) the level of network robustness. Constraints: (1) Design/rehabilitation options are limited to the discrete set of available options. Decision variables: (1) Design/rehabilitation option index (discrete). Note: One MO model including both objectives.	Water quality: N/A. Network analysis: EPANET. Optimisation method: NSGA-II.	• Network robustness is represented by the probability that the nodal pressure head is equal to or above the minimum requirement for that node, considering the uncertainties in (i) the future demands and (ii) pipe roughnesses. These uncertainties are assumed to be independent and random following some pre-specified probability density function. • To reduce computational complexity, the original stochastic formulation of robustness objective is replaced by the deterministic formulation. • The model is able to handle uncertainties in different types of parameters and with various probability distribution functions. • Results: When compared to deterministic solutions from the literature, the obtained results demonstrate that "neglecting uncertainty in the design process may lead to serious underdesign of water distribution networks". • Test networks: (1) New York City tunnels (incl. 20 nodes) [81].
39. Lin et al. (2007) [166] SO Optimal WDS design and strengthening using scatter search (SS).	Objective (1): Minimise (a) the design cost of the network (pipes). Constraints: (1) Min pressure at the nodes. Decision variables: (1) Pipe diameters (discrete).	Water quality: N/A. Network analysis: EPANET. Optimisation method: SS.	• SS, a population-based evolutionary method, is introduced and compared to the algorithms previously used in the literature, including GA, SA, SFLA, ACO and TS. • Results: The SS is able to obtain solutions as good as or better than the other methods both in the quality of solution and efficiency. • Test networks: (1) Two-loop network supplied by gravity (incl. 7 nodes) [14], (2) Hanoi network (incl. 32 nodes) [49], (3) New York City tunnels (incl. 20 nodes) [81].
40. Perelman and Ostfeld (2007) [61] SO Optimal WDS design, operation and maintenance using cross entropy (CE).	Objective (1): Minimise (a) (all test networks) the design cost of the network (pipes), (b) (test network (3) only) construction costs of pumps and tanks, (c) (test network (3) only) operation and maintenance costs of pumps. Constraints: (1) Min pressure at the nodes. Decision variables: (1) Pipe diameters (discrete).	Water quality: N/A. Network analysis: EPANET. Optimisation method: CE for combinatorial optimisation [230].	• An adaptive stochastic algorithm, based on the CE for combinatorial optimisation, is proposed. In this method flows, heads and pipe diameters are solved simultaneously. • Results: The CE found the best-known solution for the two-loop network, and improved the best-known solutions for the test networks (2) and (3). For all test networks, the solutions were obtained with a considerably lower number of function evaluations than previously reported in the literature [42,231]. • Test networks: (1) Two-loop network supplied by gravity (incl. 7 nodes) [14], (2) Hanoi network (incl. 32 nodes) [49], (3) two-loop network with pumping and storage (incl. 7 nodes) [231].
41. Tospornsampan et al. (2007) [112] SO Optimal WDS design and strengthening with split pipes using SA.	Objective (1): Minimise (a) the design cost of the network (pipes). Constraints: (1) Min/max pressure at the nodes, (2) min/max diameter for the pipes, (3) min discharge for the pipes, (4) the total length of pipe segments equal to the length of the corresponding link, (5) nonnegativity for pipe segment lengths. Decision variables: (1) Two pipe diameters for each link (discrete), (2) pipe segment lengths (continuous) for the first diameter.	Water quality: N/A. Network analysis: Not specified. Optimisation method: SA.	• Split-pipe design of looped WDSs is proposed. • The number of decision variables for split-pipe design is triple to the number of links. For each link, two pipe diameters and the segment length for the pipe of the first diameter need to be calculated. • A constraint of the minimum pipe segment length, which must be equal or more than 5% of its link length, is imposed to the Hanoi network. • Results: The obtained solutions are compared to the solutions from the literature for both split-pipe and single pipe designs. The proposed methodology found the lowest cost solutions yet published to date for all tested networks. • Test networks: (1) Two-loop network supplied by gravity (incl. 7 nodes) [14], (2) Hanoi network (incl. 32 nodes) [49], (3) New York City tunnels (incl. 20 nodes) [81].

Table A1. *Cont.*

ID, Authors (Year) [Ref] SO/MO * Brief Description	Optimisation Model (Objective Functions +, Constraints **, Decision Variables ++)	Water Quality Network Analysis Optimisation Method	Notes
42. Zecchin et al. (2007) [156] SO Optimal WDS design and strengthening using ACO.	Objective (1): Minimise (a) the design cost of the network (pipes), (b) the penalty cost for violating the pressure constraint. Constraints: (1) Min pressure at the nodes. Decision variables: (1) Pipe diameters (discrete). Note: Formulated in [201].	Water quality: N/A. Network analysis: EPANET. Optimisation method: ACO (5 algorithms).	• The paper compares 5 different formulations of ACO algorithms, namely the original one: ant system (AS) [232], and four variations: ant colony system (ACS) [233], elitist ant system (AS$_{elite}$) [232], elitist rank ant system (AS$_{rank}$) [234], max-min ant system (MMAS) [235]. • Results: "AS$_{rank}$ and MMAS stand out from the other ACO algorithms in terms of their consistently good performances". They also outperformed all other algorithms previously applied to same test networks in the literature. • Test networks: (1) Two reservoir network (incl. 10 nodes) [20], (2) New York City tunnels (incl. 20 nodes) [81], (3) Hanoi network (incl. 32 nodes) [49], (4) double New York City tunnels (incl. 39 nodes) [201].
43. Chu et al. (2008) [167] SO Optimal WDS strengthening using immune algorithm (IA).	Objective (1): Minimise (a) the design cost of the network (pipes), (b) penalty for violating the pressure constraint. Constraints: (1) Min pressure at the nodes. Decision variables: (1) Pipe diameters (discrete).	Water quality: N/A. Network analysis: Not specified. Optimisation method: IA and modified IA (mIA) are compared.	• IA, a heuristic algorithm which imitates the immune system defending against the invaders in a biological body, is introduced. The objective function and constraints are represented by antigens, the string of decision variables is represented by antibodies. Crossover and mutation operators from GA are used in producing the new antibodies to avoid the local minima. • Additionally, mIA is developed. Within the mIA optimisation procedure, GA is used (due to its good global search capability) to screen the initial repertoire (initial strings) of the IA. • Results: Both the IA and mIA found solutions as good as those obtained by GA and fmGA in other studies, in significantly fewer evaluations. Moreover, mIA exhibits far superior computational efficiency than GA or IA individually. • Test networks: (1) New York City tunnels (incl. 20 nodes) [81].
44. Jin et al. (2008) [95] MO Optimal WDS rehabilitation and operation using NSGA-II with artificial inducement mutation (AIM) to accelerate algorithm convergence.	Objective (1): Minimise (a) the rehabilitation cost of the network involving pipe replacement, (b) energy cost for pumping. Objective (2): Minimise (a) the sum of the velocity violations (shortfalls or excesses) weighted by the pipe flow. Objective (3): Minimise (a) the sum of pressure violations (excesses) weighted by the node demand. Constraints: (1) Pipe diameters limited to available standard diameter set. Decision variables: (1) Pipe diameters (real). Note: One MO model including all objectives.	Water quality: N/A. Network analysis: EPANET. Optimisation method: NSGA-II with AIM.	• A new mutation method called AIM is introduced, which acceleratingly directs the population convergence to the feasible region, and then uses normal mutation (i.e., one point random mutation) searching for the best solution within the feasible region. • To evaluate algorithm performance, the optimisation problem is solved by NSGA-II with and without AIM. • The test network to be optimised is an existing network displaying too high pipe velocities and too low nodal pressures in some areas due to an increase in water consumption. The optimisation aims to rehabilitate the network by replacing existing pipes with larger diameter pipes (no cleaning or lining of pipes is considered). • Results: NSGA-II with AIM outperforms NSGA-II without AIM in terms of convergence speed as well as the quality of the solutions obtained. • Test networks: (1) Network resembling the EPANET Example 3 (incl. 92 nodes) network [236].

Table A1. Cont.

ID. Authors (Year) [Ref] SO/MO * Brief Description	Optimisation Model (Objective Functions +, Constraints **, Decision Variables **)	Water Quality Network Analysis Optimisation Method	Notes
45. Kadu et al. (2008) [45] SO Optimal WDS design using GA with search space reduction.	Objective (1): Minimise (a) the design cost of the network (pipes), (b) penalty for violating the pressure constraint. Constraints: (1) Min pressure at the nodes. Decision variables: (1) Pipe diameters (discrete).	Water quality: N/A. Network analysis: GRA-NET, a hydraulic solver based on gradient method [215]. Optimisation method: GA-WAT program using GA.	• GRA-NET and GA-WAT are developed. • A modified GA is used with the following operators: the tournament selection, the multiparent, universal parent and basic crossover, a nonuniform and neighbour mutation [237–239]. The operators are selected randomly. • Self-adapting penalty multiplier to handle the constraints and scaled fitness function are used. • Real-coding scheme, in which discrete diameters are directly used to form solution strings, is adopted. • The solution space is substantially reduced by applying the critical path method [191] as follows. A tree is identified that approximates the original looped network, the links are classified as primary and secondary. Primary links are the pipes forming the shortest paths from the source to each demand node. Hydraulic gradient levels are obtained at the intermediate demand nodes, then flows and diameters for the links are obtained. Candidate diameters are obtained for each link based on the previous information and these are used in generating the initial population of GA. The modified GA with search space reduction is more effective, especially for large networks. • Results: The modified GA with search space reduction is more effective, especially for large networks. • Test networks: (1) Single source network with 7 links (incl. 5 demand nodes), (2) Hanoi network (incl. 32 nodes) [49], (3) two-reservoir network with 34 links (incl. 26 nodes).
46. Ostfeld and Tubaltzev (2008) [54] SO Optimal WDS design and operation considering multiple loading conditions using ACO.	Objective (1): Minimise (a) the pipe construction costs, (b) annual pump operation costs, (c) pump construction costs, (d) tank construction costs, (e) penalty function for violating pressure at nodes. Constraints: (1) Min/max pressure at consumer nodes, (2) max water withdrawals from sources, (3) tank volume deficit at the end of the simulation period. Decision variables: (1) Pipe diameters, (2) pump power at each time interval.	Water quality: N/A. Network analysis: EPANET (EPS). Optimisation method: ACO, compared to the previous study also using ACO [104].	• Time horizon is 24 h, with a varied energy tariff. • Multiple loading conditions (demands) are used. • Sensitivity analysis is performed for algorithm parameters, such as the maximum number of iterations, the discretisation number, quadratic and triple penalty functions, the initial number of ants, the number of ants subsequent to initialisation, the number of best ants solutions for pheromone updating. • Results: The proposed ACO produced better results than the ACO of [104]. However, it is difficult to anticipate which method is better in general as the performance always depends on model calibration for a specific problem. • Test networks: (1) Two-loop network with 3 sources (incl. 6 demand nodes) [223], (2) Anytown network [84] with modifications (incl. 16 nodes).
47. Perelman et al. (2008) [62] MO Optimal WDS design, strengthening, operation and maintenance using CE.	Objective (1): Minimise (a) (both test networks) the design cost of the network (pipes), (b) (test network (2) only) construction costs of pumps and tanks, (c) (test network (2) only) operation and maintenance costs of pumps. Objective (2): Minimise (a) the maximum pressure deficit of the network demand nodes. Constraints: N/A. Decision variables: (1) Pipe diameters (discrete). Note: One MO model including both objectives.	Water quality: N/A. Network analysis: EPANET. Optimisation method: CE for combinatorial optimisation [230].	• An extension of the paper by [61] to a multi-objective optimisation approach, particularly by using the rank of the generated elite solutions to update the CE probabilities instead of using fitness function values. • CE is compared to NSGA-II using the following performance metrics: (i) generational distance [240]; (ii) distance measure [241] for assessing the proximity of individual solutions of a Pareto front to the best approximated Pareto front; (iii) distribution measure [241] for evaluating the diversity of the solutions along the Pareto frontier. • Results: The CE method demonstrates a high potential of receiving good solutions with a relatively low number of function evaluations. It is robust and reliable, and provides improved results when compared to the NSGA-II. • Test networks: (1) New York City tunnels (incl. 20 nodes) [81], (2) two-loop network with pumping and storage (incl. 7 nodes) [231].

Table A1. *Cont.*

ID, Authors (Year) [Ref] SO/MO *, Brief Description	Optimisation Model (Objective Functions +, Constraints **, Decision Variables ++)	Water Quality, Network Analysis, Optimisation Method	Notes
48. Van Dijk et al. (2008) [152] SO Optimal WDS design and strengthening using GA with an improved convergence.	Objective (1): Minimise (a) the design cost of the network (pipes), (b) penalty for violating the pressure constraint. Constraints: (1) Min pressure at the nodes. Decision variables: (1) Pipe diameters (discrete).	Water quality: N/A. Network analysis: EPANET. Optimisation method: Genetic algorithm network optimisation (GANEO) program using GA.	• GANEO program, based on GA, is developed. Modifications are made to crossover and mutation to improve GA convergence. • A new approach of determining penalty for not meeting the pressure requirements at the nodes depending on the degree of failure and importance of the pipe (higher or lower flow) is developed. • Results: GANEO produced comparable results, in a limited number of generations in relation to other GA-based methods used in the literature. • Test networks: (1) New York City tunnels (incl. 20 nodes) [81], (2) Hanoi network (incl. 32 nodes) [49], (3) two-loop network supplied by gravity (incl. 7 nodes) [14].
49. Wu et al. (2008) [71] MO, SO Optimal WDS design and operation including greenhouse gas (GHG) emissions using multi-objective GA (MOGA).	Objective (1): Minimise (a) the capital cost of the network including pipes and pumps, (b) present value of pump replacement costs, (c) present value of pump operating costs (due to electricity consumption). Objective (2): Minimise GHG emissions including (a) capital GHG emissions (due to manufacturing), (b) present value of operating GHG emissions (due to electricity consumption). Constraints: (1) Min flowrate in pipes. Decision variables: (1) Pipe sizes (discrete), (2) pump sizes (discrete), (3) tank locations (discrete). Note: One MO model including both objectives, one SO model including objective (1).	Water quality: N/A. Network analysis: EPANET. Optimisation method: MOGA (based on NSGA-II).	• Present value analysis (PVA) using Gamma discounting is applied to evaluate operating costs and pump replacement costs during the life of the system. • Evaluation of GHG emissions is undertaken using life cycle analysis (LCA), where only pipes are considered as they account for most of the material usage. Two sources of emissions are considered: emissions during manufacturing of pipes and during operation of the system. Embodied energy analysis (EEA) is performed to evaluate the former, whereas PVA to evaluate the latter. • A single average energy tariff is used. • The constraints are handled by constrained tournament selection method. • For both test networks, multi-objective and single-objective optimisation is performed. For the first network, a full enumeration of solutions is also carried out to show that the MOGA has found all Pareto optimal solutions. • Results: There is a significant tradeoff between economic and environmental objectives. Considerable reduction in GHG emissions can be achieved by a reasonable increase in the cost. The discount rate values have significant impacts on the PVA results. • Test networks: (1) One-pipe pumping system (incl. 1 node), (2) multi-pump system (incl. 4 nodes).
50. Dandy et al. (2009) [127] SO Optimal expansion, strengthening and operation of wastewater, recycled and potable water systems for planning purposes using GA.	Objective (1): Minimise the total design cost of (a) wastewater, (b) recycled and potable networks. Constraints: Wastewater system: (1) Max surcharge in gravity sewers, (2) min/max velocity in rising mains, (3) treatment plant capacity. Potable/recycled systems: (4) Min pressure at the nodes. Decision variables: Wastewater system: Options for (1) trunk sewers upgrades, (2) new diversion sewers, (3) pump stations upgrades, (4) new pump stations, (5) new storage facilities, (6) new treatment plants. Potable/recycled systems: Options for (7) new/duplicate pipelines, (8) new/expanded pump stations, (9) new storages, (10) valve settings, (11) pump controls, (12) potable top-ups, (13) flowrates from sources.	Water quality: Not specified. Network analysis: Not specified. Optimisation method: GA.	• Optimisation of wastewater, recycled and potable water systems is performed simultaneously by linking together two optimisation models, one for wastewater and the other for recycled and potable water. The interface between those two models occurs at wastewater and recycled water treatment plants (WTPs). Three different combinations of locations of the plants are considered. Linking the wastewater solution with the potable/recycled water solutions involves pairing solutions from compatible source scenarios. • The optimisation of recycled/potable water systems is undertaken for a 24-h dry summer day demand for ultimate build out (year 2030) using 5-year increments. Possible future demands of a potential new development are considered. • Results: The feasibility of an integrated approach to the planning problem considered is demonstrated. This approach "is likely to make third pipe systems more attractive and to lead to significant savings in the use of limited water supplies". • Test networks: (1) Hume/Epping corridor, north Melbourne, Australia.

Table A1. Cont.

ID. Authors (Year) [Ref] SO/MO* Brief Description	Optimisation Model (Objective Functions+, Constraints**, Decision Variables++)	Water Quality Network Analysis Optimisation Method	Notes
51. di Pierro et al. (2009) [67] MO Optimal WDS design using ParEGO and LEMMO with a limited number of function evaluations.	Objective (1): Minimise (a) the total cost of the network (pipes). Objective (2): Minimise (a) the head deficit. Constraints: (1) Min head at the nodes. Decision variables: (1) Pipe diameters (discrete). Note: One MO model including both objectives.	Water quality: N/A. Network analysis: EPANET (EPS for the test network (2)). Optimisation method: Hybrid algorithms ParEGO [192] and LEMMO [242].	• The paper aims to use algorithms capable of satisfactory performance with a limited number of function evaluations. • ParEGO is based on surrogate modelling "Kriging" to model the search landscape from solutions visited during the search [192]. LEMMO is based on the hybridisation of the evolutionary search with machine learning techniques. These algorithms are tested against Pareto envelope-based selection algorithm II (PESA-II) [243] which can address simultaneously proximity and diversity (two success measures) of an approximation of the Pareto front and performed well on difficult problems. The best solutions for the problems have been obtained by PESA-II. • Results: For the network (1), LEMMO can achieve results similar to PESA-II with a significant (90%) reduction in hydraulic simulations. For the network (2), it performed well in identifying solutions interesting from an engineering perspective (i.e., solutions with small pressure deficit). ParEGO performed worse than LEMMO, but it can still be successfully applied to reduce the number of function evaluations for small to medium-size problems. • Test networks: (1) Medium-size network with 34 pipes, Apulia, Southern Italy (incl. 24 nodes) [244], (2) network A: real network with a single reservoir and 632 pipes, UK (incl. 535 demand nodes) [222].
52. Geem (2009) [160] SO Optimal WDS design and strengthening using particle swarm HS (PSHS).	Objective (1): Minimise (a) the design cost of the network (pipes). Constraints: (1) Min pressure at the nodes. Decision variables: (1) Pipe diameters (discrete).	Water quality: N/A. Network analysis: EPANET. Optimisation method: PSHS.	• An application of a particle swarm concept to the original HS algorithm to enhance its performance is presented. • The memory consideration operation in HS is replaced by the particle swarm operation where the new harmony (new vector) is formed with a certain probability (called particle swarm rate) using the best-known solution vector. • PSHS is compared with several other methods, such as GA, SA, SFLA, ACO, CE, HS, SS, and mixed SA and TS (MSATS). • Results: The PSHS algorithm performed well, especially for small-scale and medium-scale networks, for which it found the best solution in a lower number of evaluations than other methods. For the large networks, it was inferior only to HS. • Test networks: (1) Two-loop network supplied by gravity (incl. 7 nodes) [14], (2) Hanoi network (incl. 32 nodes) [49], (3) Balerma irrigation network, Almeria, Spain (incl. 447 nodes) [50], (4) New York City tunnels (incl. 20 nodes) [81].
53. Giustolisi et al. (2009) [141] SO, MO Optimal robust WDS design considering uncertainties in demands and pipe roughness using optimised multi-objective GA (OPTIMOGA) with a two-phase procedure.	Objective (1) (for a deterministic phase): Minimise (a) the design cost of the network (pipes), (b) pressure deficit at the critical node (i.e., the worst-performing node). Objective (2) (for a stochastic phase): Minimise (a) the design cost of the network (pipes). Objective (3) (for a stochastic phase): Maximise (a) the robustness of the network. Constraints: (1) Min pressure at the nodes. Decision variables: (1) Pipe diameters (discrete) (for both deterministic and stochastic problems), (2) future nodal demands (for stochastic problem only), (3) future pipe roughnesses (for stochastic problem only). Note: One SO model (i.e., deterministic) including objective (1); one MO model (i.e., stochastic) including objectives (2) and (3).	Water quality: N/A. Network analysis: Demand-driven analysis [11]. Optimisation method: OPTIMOGA [245].	• The network robustness is defined based on the worst-performing node (that is a constraint should be fulfilled at the most critical node). • The optimisation consists of two phases as follows: (i) the optimal design is found deterministically (a single-objective problem); (ii) using the obtained solutions as initial population, the robust design is found multi-objectively (cost minimisation and robustness maximisation) and stochastically considering future nodal demands and pipe roughnesses uncertain variables. This two-phase procedure is to reduce the computational time required by the stochastic phase. • Several probability density functions (mainly beta functions) are introduced and tested to model uncertain variables in different ways. • Results: The proposed two-phase optimisation procedure results in noticeable computational savings. "The entire procedure permits the simultaneous realisation of two major objectives: overall network robustness can be improved and the most important mains in terms of network reliability may be identified from the difference in the deterministic and stochastic solutions. Results illustrate the procedure's effectiveness in yielding information of practical engineering value". • Test networks: (1) Apulian network, Southern Italy (incl. 23 nodes).

Table A1. *Cont.*

ID, Authors (Year) [Ref] SO/MO * Brief Description	Optimisation Model (Objective Functions +, Constraints ++, Decision Variables ++)	Water Quality Network Analysis Optimisation Method	Notes
54. Krapivka and Ostfeld (2009) [117] SO Optimal WDS design with split pipes using a combination of GA and LP (GA-LP) in a two-phase procedure.	Objective (1): Minimise (a) the design cost of the network (pipes). Constraints: (1) Min pressure at the nodes, (2) sum of pipe segment lengths must be equal to the link length, (3) nonnegativity of segment lengths. Decision variables: (1) Lengths of pipe segments of known diameters (so called split-pipe decision variables). Note: Formulated in [116].	Water quality: N/A. Network analysis: Explicit mathematical formulation (steady state). Optimisation method: Combined GA-LP.	• An extension of the paper by [116] with the following modifications: ∘ To solve the outer problem, a GA is used instead of SA. ∘ The solution is constrained to the lowest cost spanning tree layout with the spanning tree chords (the missing pipes) kept at the minimum permissible diameters. (This solution is further improved by the GA). • Results: The results obtained are similar to the results presented in [116]. The proposed methodology is superior to the standard GA (without the refinement of using a spanning tree with minimal chord diameters). • Test networks: (1) Two-loop network supplied by gravity (incl. 7 nodes) [14].
55. Mohan and Babu (2009) [168] SO Optimal WDS design using heuristic-based algorithm (HBA).	Objective (1): Minimise (a) the design cost of the network (pipes). Constraints: (1) Min head at the nodes. Decision variables: (1) Pipe diameters (discrete).	Water quality: N/A. Network analysis: EPANET. Optimisation method: HBA.	• Heuristic optimisation method, which uses implicit information provided by the network, is proposed. Initially, all pipes are assigned the minimum available diameter size, then all diameters are increased until minimum head requirement at the nodes is met. Finally, certain diameters are decreased or increased based on the head loss information from the network. • HBA is compared to other heuristic optimisation methods (rule-based gradient approach, CANDA) as well as stochastic optimisation methods (GA, SA, SFLA). • Results: HBA finds a better solution than other heuristic methods. Compared to stochastic methods, the cost obtained by HBA is slightly higher, but the number of evaluations is significantly lower. • Test networks: (1) Two-loop network supplied by gravity (incl. 7 nodes) [14]. (2) Hanoi network (incl. 32 nodes) [49].
56. Mora et al. (2009) [158] SO Optimal WDS design using HS with optimised algorithm parameters.	Objective (1): Minimise (a) the design cost of the network (pipes), (b) penalty for violating the pressure constraint. Constraints: (1) Min pressure at the nodes. Decision variables: (1) Pipe diameters (discrete).	Water quality: N/A. Network analysis: Not specified. Optimisation method: HS.	• The aim of the paper is to find the most suitable combination of HS parameters, which would ensure not only a better solution to be obtained, but also a lower number of iterations to reach such a solution. Parameters considered are harmony memory size (HMS), memory considering rate (HMCR) and pitch adjustment rate (PAR). • HS parameters are optimised using a statistical analysis of the HS performance, for which 54,000 simulations is performed with varying values of HS parameters. The optimal cost of 6081 thousands, the smallest value ever published for the Hanoi network in the literature, was only found 4 times out of 54,000. • The concept of "good solutions" is introduced. It is the capacity of an algorithm to obtain a set of solutions, which exceed the minimum cost by no more than 3%. • Results: HMS has a key influence on obtaining good solutions and also on the number of iterations, while PAR does not have a great impact on the results. • Test networks: (1) Hanoi network (incl. 32 nodes) [49].

Table A1. *Cont.*

ID. Authors (Year) [Ref] SOMO * Brief Description	Optimisation Model (Objective Functions +, Constraints **, Decision Variables ++)	Water Quality Network Analysis Optimisation Method	Notes
57. Rogers et al. (2009) [100] SO Optimal WDS expansion, operation and maintenance planning with reliability and water quality considerations over a planning horizon (i.e., 25 years) using GA.	Objective (1): Minimise the life cycle cost of the network including (a) capital costs, (b) energy costs, (c) operation costs, (d) maintenance costs, (e) penalty cost for violating constraints. Constraints: (1) Min pressure at the nodes, (2) min/max storage facility levels, (3) min/max watermain velocities. Decision variables: Options for (1) watermains (pipe sizing and routes), (2) new pump stations, (3) pump station expansions, (4) elevated storage facilities, (5) reservoir expansions, (6) control valves, (7) expansions at the two existing water purification plants (WPPs), (8) pressure zone configurations (pressure zone boundaries).	Water quality: Water age (as a surrogate measure for water quality). Network analysis: EPANET. Optimisation method: GANET using GA, and a heuristic solver for postprocessing.	• The following optimisation strategy is adopted: Preliminary capacity-driven solutions are generated and evaluated by EPANET-GANET. Design criteria (e.g., the minimum sizing of specific infrastructure elements) are updated to ensure that the final solutions meet reliability and water quality requirements. This process is repeated to arrive at near optimal solutions. The review of near optimal solutions led to a reduction in the number and variety of the decision variable options. For example, pressure zone configuration options were eliminated from the optimisation model and were run as separate optimisation problems. • The optimisation results are evaluated using H2OMap Water, a GIS-enabled hydraulic simulation package. Operating scenarios involving critical infrastructure failures are developed and tested. • A heuristic solver is used to arrive at the final optimal solution from near optimal solutions generated by GA. Results: The results assisted in formulating practical conclusions and recommendations for large and complex WDS optimisation problems. Test networks: (1) City of Ottawa WDS, Canada.
58. Tolson et al. (2009) [180] SO Optimal WDS design and strengthening using hybrid discrete dynamically dimensioned search (HD-DDS).	Objective (1): Minimise (a) the design cost of the network (pipes), (b) penalty for violating the pressure constraint. Constraints: (1) Min pressure at the nodes. Decision variables: (1) Pipe diameter option numbers (integer).	Water quality: N/A. Network analysis: EPANET. Optimisation method: HD-DDS.	• An adaptation of the paper by [246] for continuous optimisation to discrete optimisation with a new termination criterion. • HD-DDS, which is not a population-based algorithm, combines global and two local search techniques (i.e., hybrid approach). Local search heuristics used are "one-pipe change" and "two-pipe change", which cycle through all possible ways to change the solution by modifying the diameter of one or two pipes at a time, respectively. No parameter tuning is required as there is only one parameter with a fixed value. Constraints are handled equivalently to Deb's tournament selection in GAs [148]. • A hydraulic simulator is only required for a fraction of the solutions evaluated. • The results are compared to the results from the literature obtained by other heuristics including GA, CE, PSO, MSATS and MMAS (ACO). Results: The ability of HD-DDS to find near global optimal solutions is the same or better than other heuristics while being more computationally efficient. Test networks: (1) New York City tunnels (incl. 20 nodes) [81], (2) double New York City tunnels (incl. 39 nodes) [201], (3) Hanoi network (incl. 32 nodes) [49], (4) GoYang network, South Korea (incl. 22 nodes) [226], (5) Balerma irrigation network, Almería, Spain (incl. 447 nodes) [50].

Table A1. *Cont.*

ID. Authors (Year) [Ref] SO/MO * Brief Description	Optimisation Model (Objective Functions +, Constraints **, Decision Variables ++)	Water Quality Network Analysis Optimisation Method	Notes
59. Banos et al. (2010) [107] SO Optimal WDS design using memetic algorithm (MA).	Objective (1): Minimise (a) the design cost of the network (pipes), (b) penalty for violating the pressure constraint. Constraints: (1) Min pressure at the nodes, (2) min/max flow velocities. Decision variables: (1) Pipe diameters (discrete).	Water quality: N/A. Network analysis: EPANET. Optimisation method: MA.	• MA, an extension of EAs which apply local search processes in the agents, is introduced. MA is compared to GA, SA, MSATS, SS using SA as a local searcher (SSSA) and binary linear integer programming (BLIP) method. • To compare the algorithms, the termination criterion used is the number of fitness function evaluations (except for BLIP that does not have a fitness function), which is a function of the number of links and possible pipe diameters. To avoid the randomness due to the use of different initial solutions, they are all obtained by taking the largest diameter pipes in the test networks. To achieve a good performance of each metaheuristic, a parametric analysis is performed. • The computer model called MENOME (metaheuristic pipe network optimisation model) [247] is used which integrates all algorithms, EPANET, a graphical user interface (GUI) and database management module. • Results: A dominance of MA over other algorithms is demonstrated, particularly for large-size problems. • Test networks: (1) Two-loop network supplied by gravity (incl. 7 nodes) [14], (2) Hanoi network (incl. 32 nodes) [49], (3) Balerma irrigation network, Almeria, Spain (incl. 447 nodes) [50].
60. Bolognesi et al. (2010) [169] SO Optimal WDS design and strengthening using genetic heritage evolution by stochastic transmission (GHEST).	Objective (1): Minimise (a) the design cost of the network (pipes), (b) penalty for violating the head constraint. Constraints: (1) Min head at the nodes. Decision variables: (1) Pipe diameters (discrete).	Water quality: N/A. Network analysis: EPANET. Optimisation method: GHEST.	• GHEST, a multi population evolutionary strategy method, is introduced. It uses two different complementary processes to search for the optimal solution. The first process synthesizes and transmits the genetic patrimony (heritage) of the parent solutions using their statistical indicators, while the second process called "shuffle" avoids local minima when the evolutionary potential of the population appears to be exhausted. • An extensive comparison of GHEST with previously used optimisation methods (such as ACO, GA, HS, LP, MSATS, PSHS, SA, SCE, SFLA) from the literature is presented. • Results: GHEST is able to find the same or better solution when compared to other algorithms. In particular, better results using a decreased number of evaluations are achieved for large-size problems. • Test networks: (1) Two-loop network supplied by gravity (incl. 7 nodes) [14], (2) Hanoi network (incl. 32 nodes) [49], (3) New York City tunnels (incl. 20 nodes) [81], (4) Balerma irrigation network, Almeria, Spain (incl. 447 nodes) [50].

Table A1. *Cont.*

ID. Authors (Year) [Ref] SO/MO * Brief Description	Optimisation Model (Objective Functions +, Constraints **, Decision Variables ++)	Water Quality Network Analysis Optimisation Method	Notes
61. Cisty (2010) [113] SO Optimal WDS design with split pipes using a combined GA and LP method (GALP) in a two-phase procedure.	Objective (1): Minimise (a) the design cost of the network (pipes). Constraints: (1) The sum of the unknown lengths of the individual diameters in each section has to be equal to its total length, (2) total pressure losses in a hydraulic path between a pump/tank and every critical node should be equal to or less than the known value (based on the minimum pressure requirements), (3) the lengths are positive (and greater than a nominated minimum value). Decision variables: (1) Lengths of selected pipe diameters for each section.	Water quality: N/A. Network analysis: Explicit mathematical formulation, EPANET used only for the computation of friction headlosses. Optimisation method: GALP.	• A split-pipe design is used. Hence, search space is smaller comparing to optimising pipe diameters, because the chromosomes correspond to the number of loops in the network rather than the number of pipes. • LP is more reliable to find the global optimum than heuristic methods, but is only suitable for branched networks. Therefore, GA is used to decompose the looped network into a group of branched networks, then LP is applied to optimise those branched networks. So, GA is used as an outer algorithm, LP as an inner algorithm, embedded into a GA fitness function. • It is suggested to refine the methodology by introducing a preprocessing stage with half the genes in the chromosomes. This stage is dealt with a suitable GA method, then the solutions are passed onto the main stage with full chromosomes. A postprocessing stage can be included, which also refines the solutions, again using only half the genes in the chromosomes, but a different half than in the preprocessing stage. • Fine tuning GA parameters is not necessary, as the algorithms performs consistently with different parameter values. • The extensions of the Hanoi network are introduced in order to test the method on greater problems. Those extensions are built so that the optimal solution can be evaluated. It is thus possible to compare the results produced by GALP with the global solutions for the problems. • Results: GALP consistently finds better solutions than those presented in the literature. • Test networks: (1) Hanoi network (incl. 32 nodes) [49], (2) double Hanoi network (incl. 62 nodes), (3) triple Hanoi network (incl. 92 nodes).
62. Filion and Jung (2010) [142] SO Optimal WDS design including fire flow protection using PSO.	Objective (1): Minimise (a) the design cost of the network (pipes), (b) cost of potential economic damages by the fire (expected conditional fire damages). Constraints: (1) Max velocity in the pipes. Decision variables: (1) Pipe diameters (discrete).	Water quality: N/A. Network analysis: EPANET. Optimisation method: PSO.	• Potential fire damages are incorporated directly into the objective function and are assigned a weight to reflect their importance relative to design costs. • New integration-based method to estimate the expected damages by the fire is developed. The uncertainty in needed fire flow (NFF) is included. Maximum day demands are considered to be known. • Minimum pressure constraint is excluded since corresponding violations are "accounted for in the damage component of objective function under the maximum day demand+fire condition". • Sensitivity analysis is performed to investigate the sensitivity of diameters, design costs, fire damages and total costs to changes in mean and standard deviation of fire flow. • Tradeoff curves for design costs and fire damage costs are generated. • Results: The uncertainty in fire flow has a little impact on pipe sizing and cost for the two-loop network. For the real-world network, 150 mm diameters provide adequate hydraulic capacity and make design costs and damages insensitive to fire damage weighting. • Test networks: (1) Two-loop network supplied by gravity (incl. 7 nodes) [14], (2) real-world network (incl. 29 nodes) [248].

Table A1. *Cont.*

ID, Authors (Year) [Ref] SO/MO * Brief Description	Optimisation Model (Objective Functions +, Constraints ++, Decision Variables ++)	Water Quality Network Analysis Optimisation Method	Notes
63. Mohan and Babu (2010) [170] SO Optimal WDS design using honey bee mating optimisation (HBMO).	Objective (1): Minimise (a) the design cost of the network (pipes), (b) penalty for violating the head constraint. Constraints: (1) Min head at the nodes. Decision variables: (1) Pipe diameters (discrete).	Water quality: N/A. Network analysis: EPANET. Optimisation method: HBMO [249].	• HBMO is introduced, different values of parameters tested and the sensitivity analysis presented. • Performance of HBMO in terms of the obtained solution and number of evaluations is compared to the other algorithms (GA, SA, SFLA) from the literature. Results: The multiple-queen colony is essential with the number of queen bees increasing with the increase in the number of pipes in the system. HBMO can obtain comparable results as other algorithms using a reduced number of evaluations. Test networks: (1) Two-loop network supplied by gravity (incl. 7 nodes) [14], (2) Hanoi network (incl. 32 nodes) [49].
64. Prasad (2010) [98] SO Optimal WDS strengthening, expansion, rehabilitation and operation with a new approach for tank sizing considering multiple loading conditions using GA.	Objective (1): Minimise the capital cost of the network including (a) pipes, (b) pumps, (c) tanks, (d) present value of the energy cost. Constraints: (1) Min pressure at the nodes, (2) max velocity in the pipes, (3) volume of water pumped greater than or equal to the system daily demand, (4) tanks recover their levels by the end of the simulation period, (5) total tank inflows greater than or equal to total tank outflows, (6) bounds on decision variables. Decision variables: For pipes: (1) New/duplicate diameters (integer), (2) options for existing pipes (0 = no change, 1 = clean and line). For pumps: (3) the number of pumps (integer). For tanks: (4) Location (integer), (5) total volume (real), (6) min operational level (real), (7) ratio between diameter and height (real), (8) ratio between emergency volume and total volume (real).	Water quality: N/A. Network analysis: EPANET (EPS). Optimisation method: GA.	• A new approach for tank sizing is proposed, which eliminates explicit consideration of some operational constraints. • EPS is conducted for each trial solution during the optimisation to enable accurate calculation of energy cost. • The pressure constraints are treated as hard constraints, so they are not to be violated. In contrast, all other constraints are treated as soft constraints, so the sum of normalized violation must be less than a specified value. Constraint handling is undertaken by ranking the solutions. • Two scenarios are analysed, the first considering all constraints except pressure constraints for normal day loading and the second considering all constraints. Results: Designs obtained are cheaper comparing to designs proposed by other researchers under similar performance conditions, but with different tank sizing methods. The solution for the first scenario violates pressure constraints for normal day loading as expected. The solution for the second scenario is superior in terms of both the cost and hydraulic performance. Test networks: (1) Anytown network (incl. 19 nodes) [84].
65. Suribabu (2010) [171] SO Optimal WDS design, strengthening, expansion and rehabilitation using differential evolution (DE).	Objective (1): Minimise (a) the design cost of the network (pipes), (b) penalty for violating the pressure constraint. Constraints: (1) Min pressure at the nodes. Decision variables: (1) Pipe diameters (discrete at the initialisation, converted to continuous in the DE process and back to discrete before the selection for the next generation), (2) pipe rehabilitation options.	Water quality: N/A. Network analysis: EPANET. Optimisation method: DE [250].	• DE resembles an EA and differs in an application of crossover and mutation. • The 14-pipe test network requires expansion and possibly rehabilitation (with pipes being cleaned, duplicated or left alone). • DE is compared to other optimisation methods (such as ACO, GA, HS, PSO, SA, SCE, SFLA) from the literature. Results: DE proves to be very effective as it finds optimal or near optimal solutions with a lower number of functions evaluations. Test networks: (1) Two-loop network supplied by gravity (incl. 7 nodes) [14], (2) Hanoi network (incl. 32 nodes) [49], (3) New York City tunnels (incl. 20 nodes) [81], (4) 14-pipe network with two supply sources (incl. 10 nodes) [20].

Table A1. Cont.

ID, Authors (Year) [Ref] SO/MO *, Brief Description	Optimisation Model (Objective Functions +, Constraints **, Decision Variables ++)	Water Quality Network Analysis Optimisation Method	Notes
66. Wu et al. (2010) [77] MO, SO Optimal WDS design and operation including GHG emissions over a planning horizon (i.e., 100 years) using water system multi-objective GA (WSMGA).	Objective (1): Minimise (a) the capital cost of the network including pipes and pumps (i.e., purchase and installation of pipes and pumps, and construction of pump stations), (b) present value of pump replacement/refurbishment costs, (c) present value of pump operating costs (i.e., electricity consumption). Objective (2): Minimise GHG emission cost including (a) capital GHG emissions (i.e., manufacturing and installation of pipes), (b) present value of operating GHG emissions (i.e., electricity consumption). Constraints: (1) System must be able to deliver at least the average flow(s) on the peak day to the tank(s). Decision variables: (1) Pipe sizes (discrete), (2) pump sizes (discrete). Note: One MO model including both objectives; one SO model summing up objectives (1) and (2).	Water quality: N/A. Network analysis: Not specified. Optimisation method: WSMGA (used for both single-objective and multi-objective problems, based on NSGA-II).	• An extension of the paper by [72] including carbon pricing while accounting for GHG emissions priced at a certain level (i.e., monetary value). • The question is raised "whether the introduction of carbon pricing under an emission trading scheme will make the use of a multi-objective optimisation approach obsolete or whether such an approach can provide additional insights that are useful in a decision-making context". A comparison between using single-objective and multi-objective approaches is presented. • A pipe network service life of 100 years and a pump service life of 20 years are assumed. • Because the test network (1) is very small with only 442 solutions, full enumeration and non-dominated sorting was used to optimise the system instead of GA. • Results: A multi-objective approach requires more computational effort and domain knowledge than a single-objective approach, but provides decision makers with more detailed information by showing the tradeoffs between the conflicting objectives. The authors note that the price of carbon has no effect on the tradeoff, hence it is recommended not to be used for the WDS optimisation of accounting for GHG emissions, resulting in the tradeoff between system costs in dollars and GHG emissions in tons. • Test networks: (1) Simple network with 1 tank and 1 pump station with 10 fixed speed pumps (FSP's) (incl. 1 node), (2) network with 1 pump, 8 pipes and 3 tanks (incl. 5 nodes).
67. Wu et al. (2010) [72] MO Optimal WDS design and operation including GHG emissions over a planning horizon (i.e., 100 years) using WSMGA.	Objective (1): Minimise (a) the capital cost of the network including pipes and pumps (i.e., purchase and installation of pipes and pumps, and construction of pump stations), (b) present value of pump replacement/refurbishment costs, (c) present value of pump operating costs (i.e., electricity consumption). Objective (2): Minimise GHG emissions including (a) capital GHG emissions (i.e., manufacturing and installation of pipes), (b) present value of operating GHG emissions (i.e., electricity consumption). Constraints: (1) Min pressure at the nodes. Decision variables: (1) Pipe sizes (discrete), (2) pump selection (discrete), (3) tank location selection (discrete). Note: One MO model including both objectives.	Water quality: N/A. Network analysis: EPANET. Optimisation method: WSMGA (based on NSGA-II with several modifications).	• PVA is used to account for future costs and emissions. A number of different discount rates is used in PVA for the evaluation of objective functions. • Two discount rate scenarios are used. In the first scenario, costs are discounted at different rates and GHG emissions are not discounted at all. In the second scenario, costs and GHG emissions are all discounted at the same rate. • A system design life of 100 years and a pump service life of 20 years are assumed. • Results: There is a significant tradeoff between the two objectives for both discount rate scenarios. This tradeoff notably improves the decision maker's understanding of the search space and shows which design is the most economical in reducing GHG emissions. It is found that the Pareto front is very sensitive to the discount rates, thus the selection of discount rates has a considerable impact on final decision making. • Test networks: (1) Simple network with one source, 9 pipes and one tank location (selected from two possible locations) (incl. 4 nodes).

Table A1. *Cont.*

ID. Authors (Year) [Ref] SO/MO * Brief Description	Optimisation Model (Objective Functions +, Constraints **, Decision Variables **)	Water Quality Network Analysis Optimisation Method	Notes
68. Geem and Cho (2011) [161] SO Optimal WDS design using parameter setting free HS (PSF HS).	Objective (1): Minimise (a) the design cost of the network (pipes), (b) penalty cost for violating the pressure constraint. Constraints: (1) Min pressure at the nodes. Decision variables: (1) Pipe diameters (discrete).	Water quality: N/A. Network analysis: EPANET. Optimisation method: PSF HS.	• The paper develops a new method for dynamic updating of the two major parameters in HS, HMCR and PAR, without resorting to trial and error approach to set their values. The authors argue that even metaheuristic algorithms have their advantages over the traditional algorithms, their disadvantage is a tedious and time consuming setting of parameters. • Basically, the parameters HMCR and PAR are set up automatically, but two other parameters are needed to do so: number of iterations with central parameter values and amount of noise effect. Proper values for these two amounts need to be investigated in the future. Results: PSF HS found the global solution 10 times out of 20 runs for the two-loop network, as opposed to the standard HS finding it only twice. This favourable result is believed to be due to automatic parameter settings in the iterations. Good results are obtained for the Hanoi network as well, reaching the global solution in fewer iterations than other algorithms (ACO, CE, GA, HS, SS). Test networks: (1) Two-loop network supplied by gravity (incl. 7 nodes) [14], (2) Hanoi network (incl. 32 nodes) [49].
69. Geem et al. (2011) [159] SO Optimal WDS design using HS.	Objective (1): Minimise (a) the design cost of the network (pipes). Constraints: (1) Min/max pressure at the nodes, (2) min/max velocity in the pipes. Decision variables: (1) Pipe diameters (discrete).	Water quality: N/A. Network analysis: EPANET. Optimisation method: HS.	• The velocity constraint is included to eliminate water hammer and sedimentation in pipes. • The methodology was intended to apply to three test networks, but only one test network is presented. Other test networks considered were the two-loop network [14] and the Hanoi network [49]. However, the methodology was not suitable for those test networks due to the velocity constraint for pipes. • A comparison of HS with LP, which was originally used to design the test network by [251], is presented. Results: HS obtains about 20% cheaper solution than LP. Test networks: (1) Yeosu network, South Korea (incl. 19 nodes) [251].
70. Goncalves et al. (2011) [52] SO Optimal WDS design and operation using a decomposition-based heuristic with a three-phase procedure.	Objective (1): Minimise (a) the investment cost of pipes, (b) investment cost of pumps and the power cost, (c) energy cost of the system. Constraints: (1) Each hydrant visited by exactly one path, (2) each junction/withdrawal visited at the most by one path, (3) a single diameter selected for an arc, (4) one pressure class selected for an arc, (5) min/max velocity in arcs, (6) max pressure in arcs, (7) min pressure at the hydrants, (8) min/max height for a pump at the nodes, (9) min/max land area to irrigate downstream the arcs, (10) binary and nonnegativity constraints. Decision variables: (1) Arc included into the route (0 = no, 1 = yes), (2) diameter assigned to the arc (0 = no, 1 = yes), (3) pressure class assigned to the arc (0 = no, 1 = yes), (4) pump installed at the node (0 = no, 1 = yes), (5) pumping height of installed pumps, (6) water flow in arcs, (7) land area to irrigate downstream the arcs.	Water quality: N/A. Network analysis: Explicit mathematical formulation. Optimisation method: Steiner tree constructive-based heuristic followed by improved local search heuristic (first subproblem), simple calculation of flows and irrigated areas (second subproblem), CPLEX [207] (third subproblem).	• The paper solves optimal design of a non-looped irrigation system. The problem considered is to find the routes from sources to consumer nodes, water flows in pipes and irrigated areas downstream of pipes, and diameters and thicknesses of pipes, and locations and powers of pumps. • Two new mixed binary nonlinear formulations of the problem are proposed: an initial model and a reformulated model to reduce nonlinearities of the initial model. • To solve the problem, it is sequentially decomposed into the following three subproblems: (i) building the network layout; (ii) computing the water flows and irrigated areas (a system of linear equations); (iii) dimensioning the network pipes and pumps, and locating the pumps (a mixed binary linear problem (MBLP)), defined for the network tree, flows and irrigated areas resulting from the previous two subproblems. • The computational experiments are undertaken using 12 randomly generated networks built from a real network in Portugal [252] to simulate different real case situations. This real network consists of three different zones and contains one source, 39 hydrants, 13 junctions and 279 pipes. Results: The proposed methodology is suitable for the problem at hand, with the average relative optimality gap calculated for all cases with known optimum 2.30%. Test networks: (1)–(12) Small test networks consisting of five different types (depending on the dimension of the network irrigated area), each possessing 10 nodes and a number of arcs (ranging between 20 and 40).

Table A1. *Cont.*

ID, Authors (Year) [Ref] SO/MO * Brief Description	Optimisation Model (Objective Functions +, Constraints **, Decision Variables ***)	Water Quality Network Analysis Optimisation Method	Notes
71. Haghighi et al. (2011) [178] SO Optimal WDS design using a combined GA and ILP method (GA-ILP) in a two-phase procedure.	Objective (1): Minimise (a) the design cost of the network (pipes). Constraints: (1) Min/max pressure limits, (2) min/max velocity in the pipes, (3) only one diameter for each pipe can be assigned. Decision variables: (1) Zero-unity variables related to the pipe diameters.	Water quality: N/A. Network analysis: EPANET. Optimisation method: GA-ILP.	• Using ILP, the search space thus the number of evaluations is considerably reduced. For ILP purposes, the looped network is transformed into a quasi-branched network by ignoring one pipe in each loop. These ignored pipes are not optimised in the ILP, but are assigned a fixed diameter for the objective function calculation. The quasi-branched network is optimised using a BB method. This process creates an inner loop. GA creates an outer loop, where the pipes which were ignored are optimised. The method iterates between ILP and GA. • Results: The GA-ILP method finds the optimal solution in a very fast and efficient manner, which is due to ILP preventing blind and time consuming searches in the GA and promoting each chromosome to a near optimal design. • Test networks: (1) Hanoi network (incl. 32 nodes) [49], (2) two-reservoir network with 34 links (incl. 26 nodes) [45].
72. Qiao et al. (2011) [163] SO Optimal WDS design using improved PSO (IPSO).	Objective (1): Minimise (a) the design cost of the network (pipes), (b) penalty cost for violating the pressure constraints. Constraints: (1) Min/(max) pressure at the nodes. Decision variables: (1) Pipe diameters (discrete).	Water quality: N/A. Network analysis: Not specified. Optimisation method: IPSO.	• The optimisation method combines PSO with disturbance (in order to escape local minima) with DE (in order to keep the population diversity). • IPSO performance is compared with several other methods such as HBMO, Lagrange's method (LM), PSO, random search technique (RST), SFLA and SS. • Results: IPSO performs well and reduces the possibility of trapping into a local optimum. • Test networks: (1) Serial network with 3 pipes (incl. 3 nodes) [253], (2) branched network with 3 pipes (incl. 3 nodes) [253], (3) two-loop network supplied by gravity (incl. 7 nodes) [14].
73. Wu et al. (2011) [73] MO Optimal WDS design and operation including GHG emissions over a planning horizon (i.e., 100 years), analysing sensitivity of tradeoffs between economic costs and GHG emissions, using WSMGA.	Objective (1): Minimise (a) the capital cost of the network including pipes and pumps (i.e., purchase and installation of pipes and pumps, and construction of pump stations), (b) present value of pump replacement/refurbishment costs, (c) present value of pump operating costs (i.e., electricity consumption). Objective (2): Minimise GHG emissions including (a) capital GHG emissions (i.e., manufacturing and installation of pipes), (b) present value of operating GHG emissions (i.e., electricity consumption). Constraints: (1) Min pressure at the nodes, (2) min flowrates within the system. Decision variables: (1) Pipe sizes (discrete). Note: One MO model including both objectives.	Water quality: N/A. Network analysis: EPANET. Optimisation method: WSMGA (based on NSGA-II with several modifications).	• An extension of the papers by [72,77] including sensitivity of tradeoffs between total economic costs and GHG emissions to electricity tariff and generation (i.e., emission factors). Three electricity tariff options and three emission factor options both over a time horizon of 100 years are considered. • The pump power estimation method [74] is used to estimate the maximum pump capacity and the annual electricity consumption for calculation of pump operating costs and operating GHG emissions. • To test the sensitivity of the optimisation results to the electricity tariff and emission factors, two optimisation scenarios (each for one factor) are considered. In each scenario, one factor is varied and the remaining factor is set at the moderate value of the three options considered, giving a total of 5 combinations of the two factors. • Results: Electricity tariffs impact significantly on the cost of the network, but little on GHG emissions. High electricity tariffs in the future can remove some networks from the Pareto front, indicating further possible reduction of GHG emissions by managing the water and energy industries jointly. In contrast, emission factors have no effect on the cost of the network. • Test networks: (1) Network with 1 pump, 8 pipes and 3 tanks (incl. 5 nodes) (adapted from [77]).

Table A1. *Cont.*

ID. Authors (Year) [Ref] SO/MO * Brief Description	Optimisation Model (Objective Functions +, Constraints **, Decision Variables ++)	Water Quality Network Analysis Optimisation Method	Notes
74. Zheng et al. (2011) [111] SO Optimal WDS design and strengthening using a combined NLP and DE method (NLP-DE) in a three-phase procedure.	Objective (1): Minimise (a) the design cost of the network (pipes). Constraints: (1) Min pressure at the nodes, (2) min/max diameter of pipes. Decision variables: (1) Pipe diameters (continuous for NLP, discrete for DE where continuous diameters are rounded to the nearest commercial pipe sizes after the mutation process).	Water quality: N/A. Network analysis: Explicit mathematical formulation for NLP, EPANET for DE. Optimisation method: NLP-DE.	• The methodology consists of three distinct steps as follows. • The shortest distance tree is determined for a looped network. This tree is part of the network graph, which contains only shortest paths from the sources to all demand nodes. It is assumed that the effective way to deliver demands is along the shortest path. The shortest distance tree is identified using a Dijkstra algorithm, which is modified in this paper to cover multisource WDSs. • A NLP solver is applied to the obtained shortest distance tree to optimise pipe diameters. The energy conservation constraint is not considered for NLP, because the shortest distance tree has no loops. The NLP solution with continuous diameters is an approximate solution to the original WDS. Missing pipes from the shortest distance tree are assigned the minimum allowable diameters. • A DE algorithm is applied to optimise the original looped network. The initial population for DE is seeded with diameters in the proximity of the continuous pipe sizes obtained by a NLP solver and with the minimum allowable diameters assigned to the missing pipes in the previous step. • Results: NLP-DE found optimal solutions with an extremely fast convergence speed. In addition, it found the new lowest cost solutions for the test networks (3) and (4). • Test networks: (1) New York City tunnels (incl. 20 nodes) [81], (2) Hanoi network (incl. 32 nodes) [49], (3) Zhi Jiang network, China (incl. 113 demand nodes), (4) Balerma irrigation network, Almería, Spain (incl. 447 nodes) [50].
75. Artina et al. (2012) [70] MO Optimal WDS design using parallel NSGA-II.	Objective (1): Minimise (a) the design cost of the network (pipes). Objective (2): Minimise (a) the penalty cost for violating the pressure constraint. Constraints: (1) Min pressure at the nodes. Decision variables: (1) Pipe diameters (discrete). Note: One MO model including both objectives.	Water quality: N/A. Network analysis: EPANET. Optimisation method: Parallel NSGA-II.	• Parallelisation of NSGA-II is implemented in order to reduce the computational time and improve the quality of solutions obtained. • Two parallel models, global and island, are used. In the global model, the selection and mating is performed globally, but "at each generation the fitness evaluation of solutions is distributed in a balanced way". In the island model, the population is divided into several subpopulations (i.e., islands), which evolve independently, but occasionally a migration between islands occurs. Additional parameters are necessary in the island model, being frequency and number of migrating solutions and the criterion for selecting the migrants. • Results: The global model reduces the computational time. On the other hand, the island model improves the quality of solutions due to an introduction of fundamental changes in the algorithm exploration method. Some parameter configurations (i.e., criteria for selecting the migrants) in the island model can find better solutions compared with the serial version of the algorithm. More observations are made in relation to the configuration of island model. • Test networks: (1) Hanoi network (incl. 32 nodes) [49], (2) Modena network, Italy (incl. 272 nodes) [254].

Table A1. Cont.

ID, Authors (Year) [Ref] SO/MO *, Brief Description	Optimisation Model (Objective Functions +, Constraints **, Decision Variables ++)	Water Quality Network Analysis Optimisation Method	Notes
76. Bragalli et al. (2012) [147] SO Optimal WDS design and strengthening using MINLP.	Objective (1): Minimise (a) the design cost of the network (pipes). Constraints: (1) Min/max pipe diameters/pipe cross sectional areas, (2) min/max hydraulic heads, (3) flow bounds. Decision variables: (1) Pipe flows, (2) pipe diameters/pipe cross sectional areas, (3) hydraulic heads at junctions.	Water quality: N/A. Network analysis: Explicit mathematical formulation. Optimisation method: BONMIN (an open source MINLP code) [255] using BB method.	• The methodology starts with preliminary smooth continuous NLP relaxation which accurately models the problem. In the model, the discrete objective function (due to discretised cost data) is transformed into a continuous polynomial, and headloss in pipes (Hazen-Williams) has a smooth relaxation. Subsequently, the diameters are discretised by introducing additional binary variables indicating when a specific diameter is selected for a pipe. They are further replaced by a cross sectional area (in the constraints), which removes the nonlinearities and nonconvexity from flow bound constraints. Finally, a MINLP solver can be applied. The MINLP code has been adapted to better suit the model formulation. The quality of the solutions obtained is checked by (i) comparing with the lower bounds on the solutions obtained using the global optimisation software Baron [256]; (ii) comparing with other results from the literature obtained mainly by metaheuristics; (iii) implementing and comparing with mixed-integer LP (MILP) technique. Results: Effective solutions are presented, both in terms of quality and accuracy, which are immediately usable in practice as diameters decrease from the sources towards the points further away from the sources (which is not the case for majority of the methods presented in the literature). Test networks: (1) Two-loop network supplied by gravity (incl. 7 nodes) [14], (2) Hanoi network (incl. 32 nodes) [49], (3) Blacksburg network (incl. 31 nodes) [257], (4) New York City tunnels (incl. 20 nodes) [81], (5) Foss_poly_0 network, Italy (incl. 37 nodes) [254], (6) Foss_iron network, Italy (incl. 37 nodes) [254], (7) Foss_poly_1, Italy (incl. 37 nodes) [254], (8) Pescara network, Italy (incl. 71 nodes) [254], (9) Modena network, Italy (incl. 272 nodes) [254]. Note: Test networks (5)–(9) are available from www.or.deis.unibo.it/research_pages/ORinstances/ORinstances.htm (accessed on 10 September 2017).
77. Kang and Lansey (2012) [26] SO Optimal WDS design and operation including the integrated transmission-distribution network considering multiple loading conditions using GA with an engineered initial population.	Objective (1): Minimise (a) the pipe construction (the sum of the base installation cost, trenching and excavation, embedment, backfill and compaction costs, and valve, fitting, and hydrant cost), (b) pump construction cost, (c) pump operation cost (energy consumed by pumps), (d) penalty for violating the pressure constraint. Constraints: (1) Min pressure at the nodes for three demand loading conditions (average, instantaneous peak and fire flows). Decision variables: (1) Pipe sizes, pump station capacity including (2) pump sizes and (3) the number of pumps.	Water quality: N/A. Network analysis: EPANET. Optimisation method: GA (for optimisation), a new heuristic (for generating an engineered initial population to improve the GA convergence).	• The optimisation of integrated transmission-distribution network is presented. The distribution network (a part of the system which delivers water to individual households) is usually not considered in the WDS design optimisation because of the large number of variables. • A new heuristic is proposed to generate initial population considering hydraulic behaviour of the system, so the velocities in the selected pipe sizes fall below the pre-defined flow velocity threshold. To maintain the diversity in the optimisation process, half of the initial population is generated by the new heuristic and the other half randomly. • The following main assumptions are made: no uncertainty in demand, one constant efficiency parameter to represent pumps, constant energy tariff, and one fire flow demand pattern. • There are 4 design scenarios considered: (i) the distribution network is excluded from the model; (ii) the distribution network is included in the model, but its pipe sizes are fixed at minimum values (i.e., are not optimised); (iii) and (iv) both transmission and distribution networks are included in the model, the initial population is generated by the proposed heuristic and randomly, respectively. Results: The comparison of scenarios (i) and (ii) shows that the pipes in the transmission network tend to be oversized if the distribution network is excluded from the model. The comparison of scenarios (iii) and (iv) shows that the new heuristic considerably improves the convergence of the GA in terms of speed as well as the quality of the solution. Test networks: (1) Real system with one source, one pump station and 1274 pipes (incl. 936 nodes).

Table A1. *Cont.*

ID, Authors (Year) [Ref] SO/MO *, Brief Description	Optimisation Model (Objective Functions *, Constraints **, Decision Variables **)	Water Quality Network Analysis Optimisation Method	Notes
78. Kanta et al. (2012) [92] MO Optimal WDS redesign/rehabilitation (pipe replacement) including fire damage and water quality objectives using non-dominated sorting evolution strategy (NSES).	Objective (1): Minimise (a) the potential fire damage, calculated as lack of available fire flows at selected hydrant nodes taking into account the importance of a hydrant location. Objective (2): Minimise (a) the water quality deficiencies, represented by a performance function on chlorine residual at selected monitoring nodes reflecting governmental regulations for drinking water quality. Objective (3): Minimise (a) the system redesign cost, expressed as a ratio of actual redesign cost over maximum expected redesign cost. Constraints: (1) Min pressure at the hydrant nodes, (2) pipe diameters limited to commercially available sizes, (3) max number of pipe decision variables (i.e., pipes to be replaced). Decision variables: (1) Pipes selected for replacement (integer), (2) diameters of replaced pipes (integer). Note: One MO model including all objectives.	Water quality: Disinfectant (i.e., chlorine). Network analysis: EPANET (demand-driven analysis to calculate the fire flows, using a hydrant lifting technique to satisfy the pressure constraint). Optimisation method: NSES.	• The method provides the flexibility to select a mitigation plan for urban fire events best suited for decision makers' needs. • NSES, a modification of NSGA-II for an evolution strategy (ES)-based implementation to address difficulties for heuristics posed by the WDS optimisation problems, is proposed. It differs from the standard NSGA-II in the application of specialised operators, such as representation, mutation and selection. • NSES is tested on three test problems of varying degrees of difficulty and compared to NSGA-II and PAES using a deviation metric [258]. Subsequently, it is applied to a WDS optimisation problem using two scenarios, fire flow at three and six hydrants, respectively. • EPANET simulations are executed as follows. Fire flow analysis is performed separately for each hydrant. Water quality analysis (incl. hydraulics) is simulated without a fire flow demand over 168 hours to reach dynamic equilibrium for chlorine residuals. • Results: NSES outperforms (for three test problems used) both NSGA-II and PAES in spreading solutions across the Pareto front and in maintaining solution diversity. NSES also demonstrated the capability to produce Pareto optimal solutions across several objectives. However, almost no solutions were found in the "high fire flow—low water quality—high cost' region of the objective domain, which is influenced by the disinfectant decay parameters and the characteristics of the particular WDS. • Test networks: (1) Virtual city of Micropolis (incl. 1262 nodes) [259,260].
79. McClymont et al. (2012) [194] SO Optimal WDS rehabilitation (pipe resizing) using ES with evolved mutation heuristics.	Objective (1): Minimise (a) the design cost of the network (pipes). Constraints: (1) Min/max pressure at the nodes, (2) max velocity in the pipes. Decision variables: (1) Pipe diameters (discrete).	Water quality: N/A. Network analysis: Not specified. Optimisation method: ES.	• A decision tree generative hyper-heuristic approach is presented which uses genetic programming (GP) to evolve novel mutation heuristics for the WDS design optimisation. The decision tree is based on domain knowledge in the form of node head conditions to inform the mutation to upstream pipes. For example, the upstream pipes may be too large or too small if a node has excessive head or head deficit, respectively. • Mutation heuristics evolve using NSGA-II and are evaluated on their ability to search for good solutions to the Hanoi test problem. The best 5 mutation heuristics are compared against a tuned Gaussian mutation using the Anytown network and three real networks. • Results: The importance of testing evolved heuristics for over-fitting is highlighted. Mutation heuristics display an improvement in performance over traditional heuristics such as Gaussian mutation. • Test networks: (1) Anytown network (incl. 19 nodes) [84], (2) real network with 7 pipes, (3) real network with 29 pipes, (4) real network with 81 pipes.
80. Sedki and Ouazar (2012) [172] SO Optimal WDS design and strengthening using a combined PSO and DE method (PSO-DE).	Objective (1): Minimise (a) the design cost of the network (pipes), (b) penalty cost for violating the pressure constraint. Constraints: (1) Min pressure at the nodes. Decision variables: (1) Pipe diameters (discrete).	Water quality: N/A. Network analysis: EPANET. Optimisation method: PSO-DE.	• A hybrid PSO-DE method is developed to overcome the problem of premature convergence in PSO. In this method, PSO finds the region of optimal solution, then combined PSO and DE find the optimal point. • PSO-DE is compared to the standard PSO as well as other methods from the literature (ACO, CE, GA, HS, SA, SFLA, SS). • Results: For the two-loop and Hanoi networks, PSO-DE found the best-known solution in fewer iterations than other algorithms. For the New York City tunnels, PSO-DE found a slightly better feasible solution in a lower number of evaluations than the solution reported in the literature to date. • Test networks: (1) Two-loop network supplied by gravity (incl. 7 nodes) [14], (2) Hanoi network (incl. 32 nodes) [49], (3) New York City tunnels (incl. 20 nodes) [81].

Table A1. *Cont.*

ID, Authors (Year) [Ref] SO/MO* Brief Description	Optimisation Model (Objective Functions +, Constraints **, Decision Variables **)	Water Quality Network Analysis Optimisation Method	Notes
81. Wu et al. (2012) [74] MO Optimal WDS design, operation and maintenance including GHG emissions, incorporating variable speed pumps (VSPs) using MOGA.	Objective (1): Minimise the total economic cost of the system including (a) capital cost (i.e., purchase, installation and construction of network components), (b) present value of operating costs (i.e., electricity consumption due to pumping), (c) present value of maintenance and end-of-life costs. Objective (2): Minimise the total GHG emissions of the system including (a) capital GHG emissions (i.e., manufacturing and installation of network components), (b) present value of operating GHG emissions (i.e., electricity consumption due to pumping), (c) present value of maintenance and end-of-life emissions. Constraints: (1) Min flowrates within the system. Decision variables: (1) Pipe sizes (discrete). Note: One MO model including both objectives.	Water quality: N/A. Network analysis: EPANET. Optimisation method: MOGA.	• The aim is to incorporate VSPs into an optimal design of WDSs. • A pump power estimation method is developed to incorporate VSPs. This method uses a flow control valve (FCV) combined with an upstream reservoir to represent a pump in the system, so that the flows (via FCV) into the downstream tanks are maintained as close as possible to the required flows. Therefore, the task of determining the most appropriate FCV setting for calculating pump power is formulated as a single-objective minimisation problem subject to multiple flow constraints. To solve this problem, the false position method [261] in conjunction with EPANET is used. • VSPs are compared to FSPs within the defined multi-objective optimisation problem. • In the case study, only capital and operating costs and emissions are considered (maintenance and end-of-life costs and emissions are omitted). Results: The use of VSPs leads to significant savings in total cost as well as GHG emissions. "The effectiveness of replacing FSPs with VSPs in reducing operating costs and emissions is more significant for a smaller pipe diameter system with higher dynamic heads (friction losses) relative to static heads". Test networks: Network with 1 pump, 8 pipes and 3 tanks (incl. 5 nodes) [77].
82. Fu et al. (2013) [97] MO Optimal WDS strengthening, expansion, rehabilitation and operation including multiple loading conditions and water quality objective applying many-objective visual analytics using ε-NSGA-II.	Objective (1): Minimise the capital cost for network expansion/rehabilitation including (a) pipes, (b) storage tanks, (c) pumps. Objective (2): Minimise (a) the operating cost of the system (i.e., energy cost for pump operation) during a design period. Objective (3): Minimise hydraulic failure of the system, expressed by the total system failure index (SFI) combining (a) nodal failure index and (b) tank failure index. Objective (4): Minimise (a) the fire flow deficit, representing the potential fire damage. Objective (5): Minimise (a) the total leakage of the system, considering background leakage from pipes only (calculated based on the pipe pressure). Objective (6): Minimise (a) the water age. Constraints: N/A. Decision variables: (1) Pipe diameters for new pipes (integer), (2) options for existing pipes including cleaning and lining or duplicating with a parallel pipe (integer), (3) tank locations (integer), (4) the number of pumps in operation during 24 hours (integer). Note: One MO model including all objectives.	Water quality: Water age (as a surrogate measure for water quality). Network analysis: Pressure-driven demand extension of EPANET (EPANETpdd) (EPS). Optimisation method: ε-NSGA-II.	• The optimisation model is formulated with no constraints, because the objective functions used meet all the criteria. • Nodal hydraulic failure is quantified as a fraction of time during which pressure at the node drops below the required pressure, the consequence of which is defined as water shortage at this node relative to the total demand of the entire WDS at that time. • Tank hydraulic failure is identified by the water level at the end of EPS being lower than at the beginning of simulation, which can cause potential problems for the following time period. • Five loading conditions are considered: average day flow, instantaneous flow, and three fire flow conditions. • The fire flow deficit objective is considered as the average deficit across the three fire flow conditions. • The leakage and water age are calculated for the average day flow condition. • ε-NSGA-II is chosen over NSGA-II as it has a better computational efficiency, which is important for many-objective optimisation due to a high computational burden. • Visual analytics are used to explore the tradeoffs between 6 objectives. The visualisation of the 6 objectives is achieved by placing three objectives (capital costs, system failure and leakage) on axes in a 3D chart, and representing the other 3 objectives through the colour, orientation and size of the cones which indicate the solutions. Also, lower-dimensional subproblem tradeoffs can be observed using convention Pareto fronts in 2D and 3D. Results: The results indicate relationships between individual objectives. For example, the capital and operating costs have a very different relationship with water age and leakage, which would not be revealed if the costs were aggregated into one objective. This paper highlights benefits therefore of many-objective optimisation approach in supporting more informed, transparent decision-making, in the WDS design process. Test networks: (1) Anytown network (incl. 19 nodes) [84].

Table A1. *Cont.*

ID. Authors (Year) [Ref] SO/MO *, Brief Description	Optimisation Model (Objective Functions +, Constraints **, Decision Variables ++)	Water Quality Network Analysis Optimisation Method	Notes
83. Kang and Lansey (2013) [121] MO Scenario-based robust optimal planning of an integrated water and wastewater system considering demand uncertainties using NSGA-II.	Objective (1): Minimise (a) the systems initial construction cost (pipes, pumps, tanks, wastewater plants), (b) expected operation and maintenance costs, (c) adaptive construction cost to expand the system if needed, (d) penalty cost for violating constraints. Objective (2): Minimise (a) the variability of actual costs across scenarios for the design solution, calculated as the standard deviation. Constraints: (1) Min pressure at the nodes, (2) min velocity in the sewer pipes, (3) max pump station capacities, (4) max storage tank sizes. Decision variables: (1) Pipe sizes (discrete), (2) pump station capacities (discrete), (3) wastewater treatment plant capacities (discrete). Note: One MO model including both objectives.	Water quality: N/A. Network analysis: N/A. Optimisation method: NSGA-II.	• Scenario-based multi-objective robust optimisation (SMORO) model for planning and designing a regional scale integrated water and wastewater system is proposed. SMORO solves deterministic problems in a scenario-based structure to effectively implement the stochastic factors inherent in the problem. • Uncertain parameters in the model are potable and reclaimed water demands, which are implemented through scenarios. A set of 5 scenarios (base condition, low growth, high growth, low reclamation, high reclamation) is developed, with the same probability assigned to each scenario. • Initially, the problem is solved individually for every scenario as regular single-objective optimisation problems. Subsequently, postoptimisation regret computation is performed. The regret cost is an overpayment or a supplementary cost due to overdesign or underdesign, respectively, owing to the implemented decision being made with imperfect information about the future. "In other words, the regret cost represents the risk that the implemented decision will be more costly than a decision made". Finally, the multi-objective problem with two objectives (costs and variability) is solved simultaneously for all scenarios. • Results: A single-objective solution is cost effective only for the design scenario; but in all other cases is inferior with possibly substantial regret cost. In contrast, SMORO provides a robust and flexible system design via a balanced solution in terms of initial investment and future risk. It is demonstrated that system demand is the most critical uncertainty in system design. • Test networks: (1) Water system planning (water supply and reuse water networks) in southeast Tucson, Arizona.
84. McClymont et al. (2013) [144] MO Optimal WDS design and rehabilitation including the water discolouration risk using NSGA-II and SPEA2 integrated with a new heuristic Markov–chain hyper-heuristic (MCHH).	Objective (1): Minimise (a) the cost of network infrastructure (pipes), (b) penalty for violating the pressure constraint, (c) penalty for violating the velocity constraint. Objective (2): Minimise (a) the water discolouration risk expressed as the sum of cumulative potential material after daily conditioning shear stress for all pipes in the network, (b) penalty for violating the pressure constraint, (c) penalty for violating the velocity constraint. Objective (3): Minimise (a) the sum of the cumulative head excess, (b) penalty for violating the pressure constraint, (c) penalty for violating the velocity constraint. Constraints: (1) Min head at the nodes, (2) max velocity in the pipes. Decision variables: (1) Pipe diameters. Note: One MO model including all objectives.	Water quality: Water discolouration. Network analysis: EPANET, discolouration propensity model (DPM). Optimisation method: NSGA-II and SPEA2 integrated with MCHH.	• This paper presents least-cost design of WDSs with a reduced risk of water discolouration (i.e., self-cleaning networks), thus reduced long-term maintenance and operational burdens of the system. • A new heuristic MCHH is proposed. It is applied after each generation of solutions having been attained and evaluated. Essentially, MCHH learns which simple heuristic within the algorithm (e.g., crossover, mutation) performs most effectively and adjusts the likelihood of their selection accordingly. • For the optimisation, NSGA-II and SPEA2 are integrated with MCHH. Four extra heuristics in addition to crossover and mutation are supplied to the algorithms with MCHH. Both the original algorithms NSGA-II and SPEA2 and the MCHH variants are run on the problem. • For comparison, NSGA-II and SPEA2 are also integrated with two other hyper-heuristics (Simple Random and TSRoulWheel). • To calculate the discolouration risk, DPM software which implements a cohesive transport model (CTM) [262,263] is used. So, the algorithms are linked with both EPANET and DPM. Results: An improvement in performance obtained by MCHH variants over the original algorithms is demonstrated. When compared with Simple Random and TSRoulWheel, it is shown that MCHH is able to find a wider range of solutions across the networks. • Test networks: (1) Two-loop network supplied by gravity (incl. 7 nodes) [14], (2) Anytown network (incl. 19 nodes) [84], (3) Hanoi network (incl. 32 nodes) [49], (4) small network with 68 pipes, South West of England (incl. 52 nodes), (5) medium-size network with 107 pipes, South West of England, (incl. 81 nodes), (6) large network with 213 pipes, South West of England (incl. 160 nodes).

Table A1. *Cont.*

ID. Authors (Year) [Ref] SO/MO * Brief Description	Optimisation Model (Objective Functions +, Constraints **, Decision Variables ++)	Water Quality Network Analysis Optimisation Method	Notes
85. Zhang et al. (2013) [134] SO Optimal design, strengthening, expansion and operation of a reclaimed WDS considering demand uncertainty with the time-staged construction over a planning horizon (i.e., 20 years) using ILP.	Objective (1): Minimise (a) the cost of installing pipes, (b) cost of constructing pump stations, (c) pump energy cost of operating the system, at the stage one (time horizon 0–10 years), (d) expected cost of installing additional pipes, pumps and operating the system, at the stage two (time horizon 10–20 years). Constraints: (1) Min pressure at the nodes for peak demands, (2) min pressure at the nodes for average demands, (3) only one pipe size selected for each link, (4) only one pump size selected for average demands, (5) only one pump size selected for peak demands, (6) ensuring that the existing pump station is either expanded or a new one constructed at the stage two, (7) binary constraints. Decision variables (stage 1): (1) Pipe of size j installed in link i, (2) pump size p installed at station s for peak demands, (3) same as (2) for average demands. Decision variables (stage 2): (4) Additional pipe of size k installed for link i, (5) if no pump installed at stage 1, pump size p installed at station s for peak demands, (6) if pump installed at stage 1, additional pump of size p installed at station s for peak demands, (7) pump size p installed at station s for average demands. Note: All decision variables are binary (0 = no, 1 = yes).	Water quality: N/A. Network analysis: Explicit mathematical formulation. Optimisation method: GAMS CPLEX solver [207] using branch and cut method.	• The paper presents a two-stage stochastic integer problem for a planning horizon of 20 years, so there are 2 stages of construction decisions: current decisions (for time horizon 0–10 years) and expansion decisions in 10 years' time (for time horizon 10–20 years). • The network structure is branched (due to reliability not being as important in a reclaimed water network), nonlinear hydraulic equations are linearised. • Preprocessing methods are developed to reduce the dimensionality of the problem (i.e., reducing the number of pipe and pump decisions). The network is separated into subnetworks, and pipe and pump size reduction is performed for each subnetwork. The set of permissible pipe diameters is reduced using velocity constraints. Each subnetwork is solved separately. • Uncertain future demands in expansion prospects (stage 2) of the system are considered. The uncertainties are handled by a discrete set of scenarios, with 81 scenarios used in the test problem. • Sensitivity analysis is performed to test changes in total cost, and pipe and pump decisions under varying demands, energy costs, annual discount rates and pipe material prices. • Results: Preprocessing considerably reduces problem dimension, improves solution quality, and enables to solve large problems. In regards to sensitivity analysis, mean demands are the most significant driving factor with respect to total costs. • Test networks: (1) Network with one source (wastewater treatment plant), 4 pump stations and 56 pipes (incl. 56 demand nodes).
86. Zheng et al. (2013) [46] SO Optimal design of a multisource WDS using network decomposition and DE in a two-phase procedure.	Objective (1): Minimise (a) the design cost of the network (pipes). Constraints: (1) Min/max pressure at the nodes. Decision variables: (1) Pipe diameters (discrete).	Water Quality: N/A. Network analysis: EPANET. Optimisation method: DE (the modification based on the approach of [171] to manage a discrete problem).	• The proposed method consists of the following two steps. • Network decomposition: A graph decomposition method is developed to divide the original network into subnetworks, so that the only one unique source supplies each subnetwork. • Multistage optimisation: Each subnetwork is optimised (i.e., first-stage optimisation) using DE. The combined optimal solutions for the subnetworks produce an approximate solution for the total network. However, this approximate optimal solution needs to be further improved because some of the pipes were not included in the optimisation due to network partitioning. Therefore, the entire original network is optimised (i.e., second-stage optimisation) using the initial population seeded from the optimal solutions of the subnetworks obtained from the first-stage optimisation. • Results: The final solution from the second-stage optimisation is close to the approximate solution found in the first-stage optimisation. Comparison with the standard DE (a whole of network optimisation) and other methods from the literature demonstrate that the proposed method exhibits better performance in terms of solution quality and convergence speed. • Test networks: (1) Two-reservoir network (incl. 4 nodes), (2) two-reservoir network with 34 links (incl. 26 nodes) [45], (3) real three-reservoir network, China (incl. 199 demand nodes), (4) Balerma irrigation network, Almeria, Spain (incl. 447 nodes) [50].

Table A1. *Cont.*

ID. Authors (Year) [Ref] SO/MO *, Brief Description	Optimisation Model (Objective Functions +, Constraints **, Decision Variables ++)	Water Quality, Network Analysis, Optimisation Method	Notes
87. Zheng et al. (2013) [173] SO Optimal WDS design and strengthening using a self-adaptive DE method (SADE).	Objective (1): Minimise (a) the design cost of the network (pipes). Constraints: (1) Min head at the nodes. Decision variables: (1) Pipe diameters (integer, with continuous values created during the mutation process which are then truncated to the nearest integer size).	Water quality: N/A. Network analysis: EPANET. Optimisation method: SADE.	• The paper introduces three new contributions as follows. • Mutation weighting factor (F) and crossover probability (CR) parameters of the SADE are encoded into a solution string and hence are adapted through evolution (i.e., are not pre-specified). • F and CR parameters are applied at the individual level rather than generational level like in the standard DE, so different parameters can be used for different individuals. • A new termination criterion for the SADE is proposed. The algorithm is terminated when all the individuals in the population have similar objective function values, which is checked using the coefficient of variation. • Constraint tournament selection [148] is used to handle constraints. • A sensitivity analysis is performed for different population sizes. • Results: The SADE displays good performance for both the solution quality and efficiency, with a reduced need to fine-tune algorithm parameter values. • Test networks: (1) New York City tunnels (incl. 20 nodes) [81], (2) Hanoi network (incl. 32 nodes) [49], (2) double New York City tunnels (incl. 39 nodes) [201], (4) Balerma irrigation network, Almería, Spain (incl. 447 nodes) [50].
88. Zheng et al. (2013) [149] SO Optimal WDS design and strengthening using non-crossover dither creeping mutation-based GA (CMBGA).	Objective (1): Minimise (a) the design cost of the network (pipes). Constraints: (1) Min pressure at the nodes. Decision variables: (1) Pipe diameters (discrete).	Water quality: N/A. Network analysis: EPANET. Optimisation method: CMBGA.	• Unlike the standard GA, the proposed CMBGA does not use crossover. It only uses selection and newly proposed dither creeping mutation replacing generally used bitwise mutation. The new parameter is randomly generated throughout the algorithm run rather than being preselected. It is also varies for each individual of the population. • To handle constraints, constraint tournament selection [148] is used. • CMBGA is compared with 4 other GA variants, including a crossover-based GA with bitwise mutation (SGA), a crossover-based GA with creeping mutation (CGA), a non-crossover GA with traditional bitwise mutation (NBGA), and a crossover dither creeping mutation GA (CDGA). • Results: CMBGA exhibits improvements in finding optimal solutions compared with the other GA variants and displays a comparable performance to the other EAs (MMAS and HD-DDS). • Test networks: (1) New York City tunnels (incl. 20 nodes) [81], (2) Hanoi network (incl. 32 nodes) [49].
89. Aghdam et al. (2014) [164] SO Optimal WDS design and strengthening using accelerated momentum PSO (AMPSO).	Objective (1): Minimise (a) the design cost of the network (pipes), (b) penalty for violating the pressure constraint. Constraints: (1) Min pressure at the nodes. Decision variables: (1) Pipe diameters (discrete).	Water quality: N/A. Network analysis: EPANET. Optimisation method: AMPSO.	• A new version of PSO called AMPSO is proposed in order to increase the convergence rate of the algorithm and avoid getting trapped in local optima. • The increased convergence rate is achieved by introducing new adaptive terms into the velocity update equation of the algorithm. These terms decrease or increase the movement step size relative to being close or far from the global optimum, respectively. The convergence rate can be thus enhanced by large or short steps proportional to the value of the cost function. • To avoid getting trapped in local minima, so-called momentum terms are introduced into the position updating formula of the algorithm. These momentum terms "determine the influence of the past position changes on the current direction of movement in the search space". • AMPSO is compared with three other heuristic methods from the literature (GA, ACO and PSO-DE). • Results: AMPSO exhibits the efficiency when compared with other heuristics. • Test networks: (1) Hanoi network (incl. 32 nodes) [49], (2) New York City tunnels (incl. 20 nodes) [81].

Table A1. Cont.

ID, Authors (Year) [Ref] SO/MO* Brief Description	Optimisation Model (Objective Functions+, Constraints**, Decision Variables++)	Water Quality Network Analysis Optimisation Method	Notes
90. Bi and Dandy (2014) [27] SO Optimal WDS design and strengthening including water quality considerations using online ANN and DE.	Objective (1): Minimise (a) the design cost of the network (pipes), (b) the net present value of chlorine cost over a planning horizon. Constraints: (1) Min head at the nodes, (2) min chlorine concentration at the nodes. Decision variables: (1) Pipe diameters (discrete), (2) chlorine dosage rates at the WTPs.	Water quality: Chlorine. Network analysis: Online ANN. Optimisation method: DE.	• ANN is proposed to replace a hydraulic and water quality simulator in order to reduce the computational effort of those simulations. Online DE-ANN method is designed, where ANN is retrained throughout the optimisation process (so called online ANN) to improve approximation of the portion of search space under consideration and DE performs the optimisation. A local search strategy is used to improve the final solution obtained by DE-ANN. • To reduce the run time, the ANN training is performed only for the selected critical nodes, which are determined before the optimisation using data from EPANET. There are 3 parameters used for training: the size of the training data, the number of generations between retrainings, and the number of retrainings. • To ensure the feasibility of generated solutions at each generation, the best solution is compared to the previous generations' best solution. If different, it is checked by EPANET for feasibility, if cheaper, it is noted as the current best solution. • The demands for the test network (1) are constant, whereas for the test networks (2) and (3) they vary within the 24 h cycle. • The performance of the proposed online DE-ANN method is compared with the DE-EPANET method and offline DE-ANN method where the ANN model is trained only at the beginning of the optimisation (see, for example, [87,264]. • Results: The online DE-ANN outperforms the offline DE-ANN in terms of efficiency and solution quality. In comparison to DE-EPANET, the online DE-ANN displays a substantial improvement in computational efficiency, while still producing good quality solutions. • Test networks: (1) New York City tunnels (incl. 20 nodes) [81], (2) modified New York City tunnels (incl. 20 nodes), (3) hypothetical Jilin network (incl. 28 nodes).
91. Creaco et al. (2014) [118] MO Optimal WDS design, strengthening and expansion accounting for construction phasing in prefixed time intervals (i.e., 25 years) over a planning horizon (i.e., 100 years) using NSCA-II.	Objective (1): Minimise (a) the total present worth construction cost of the network (pipes), calculated as the sum of the present worth costs of the n upgrades, (b) penalty for violating the pressure surplus constraint. Objective (2): Maximise (a) the network reliability, calculated as the minimum pressure surplus over the whole construction time. Constraints: (1) Pressure surplus bigger or equal to zero. Decision variables: (1) Pipe diameters (coded as integer numbers), with the genes consistently ordered (within each individual) according to the construction phases. Note: One MO model including both objectives.	Water quality: N/A. Network analysis: Demand-driven analysis [11]. Optimisation method: Modified NSCA-II.	• The aim is to optimise a phased WDS construction in prefixed time intervals over an expected life cycle, where nodal demands increase in time without uncertainty. • Modified NSCA-II used encodes genes with integer numbers instead of real numbers. • The solutions provide the pipe diameters which have to be laid in the various sites (inclusive of pipes laid in parallel to existing pipes) at the various time intervals. The following two scenarios are considered for network growth: (i) the network topology is constant in time, so no network expansion occurs over the planning horizon; (ii) the network topology changes in time, so network expansion occurs over the planning horizon. • Three different types of optimisation are performed for each network scenario as follows: (i) four construction phases with 25-year intervals over 100-year planning horizon; (ii) one construction phase over 25-year planning horizon; (iii) one construction phase over 100-year planning horizon. The objective is to assess how construction phasing affects network design. • Results: Optimisation of WDS design with construction phasing leads to better results than the traditional single construction phase approach. • Test networks: (1) Two-loop network supplied by gravity (incl. 7 nodes) [14].

Table A1. *Cont.*

ID, Authors (Year) [Ref] SO/MO * Brief Description	Optimisation Model (Objective Functions [+], Constraints [++], Decision Variables [++])	Water Quality Network Analysis Optimisation Method	Notes
92. Ezzeldin et al. (2014) [165] SO Optimal WDS design using integer discrete PSO (IDPSO).	Objective (1): Minimise (a) the design cost of the network (pipes), (b) penalty cost for violating the pressure constraint. Constraints: (1) Min pressure at the nodes, (2) min/max pipe diameters. Decision variables: (1) Pipe diameters (discrete).	Water quality: N/A. Network analysis: Newton-Raphson method [10]. Optimisation method: IDPSONET program using IDPSO.	• A new boundary condition and a new initialisation method are proposed for PSO. • The boundary condition is called billiard boundary condition. When a particle reaches the boundary, it is reflected back to the search space with its velocity remaining the same (only the sign changes). This technique gives the particle a bigger chance to find its global solution. Usually, a velocity clamping technique is used in PSO. The new boundary condition is tested against 5 other boundary conditions for the two-loop network. • In a new initialisation method, the initial position of the solution vector is set to one side of the boundary with the maximum available diameters. • Results: IDPSO reached the known optimal solution in a reduced number of evaluations for the two-loop network, and it improved the solutions previously found in the literature for the two-reservoir network. • Test networks: (1) Two-loop network supplied by gravity (incl. 7 nodes) [14], (2) two-reservoir network with 34 links (incl. 26 nodes) [45].
93. Johns et al. (2014) [155] SO Optimal WDS design, strengthening and operation using adaptive locally constrained GA (ALCO-GA).	Objective (1): Minimise (a) (all test networks) the design cost of the network (pipes), (b) (test network (4) only) cost of tanks, (c) (test network (4) only) pump energy cost. Constraints: (1) Min pressure at the nodes. Decision variables: (1) Pipe diameters (discrete), (2) (test network (4) only) tank locations (binary), (3) (test network (4) only) the number of pumps in operation during 24 h at every 1-h time step (binary).	Water quality: N/A. Network analysis: Not specified. Optimisation method: ALCO-GA.	• Heuristic-based mutation operator which utilises hydraulic head information and an elementary heuristic to allow earlier location of feasible solutions in the optimisation process are proposed. • Constraint handling is performed through the use of the modified mutation operator. • If only the heuristic-based mutation operator is applied (i.e., without random bitwise mutation) throughout the whole optimisation process, it causes premature convergence on a suboptimal solution. Therefore, the fitness gradient monitor is employed, which controls the probability that the heuristic-based mutation operator is used based on the rate of convergence of the best solution in the population. • Results: ALCO-GA displays faster convergence than the standard GA and often obtains better solutions than solutions from the literature obtained by the standard GA. • Test networks: (1) Two-loop network supplied by gravity (incl. 7 nodes) [14], (2) New York City tunnels (incl. 20 nodes) [81], (3) network B: real network with a single reservoir and 1277 pipes, UK (incl. 1106 nodes), (4) modified Anytown network (incl. 19 nodes) [84] (the options to duplicate/clean/line existing pipes are removed).
94. McClymont et al. (2014) [68] MO Optimal WDS rehabilitation (pipe resizing) using ES with evolved mutation operators in a three-phase procedure.	Objective (1): Minimise (a) the design cost of the network (pipes). Objective (2): Minimise (a) the total head deficit at the nodes. Constraints: N/A. Decision variables: (1) Pipe diameters (discrete). Note: One MO model including both objectives.	Water quality: N/A. Network analysis: Not specified. Optimisation method: ES.	• An extension of the paper by [194] developing a hyper-heuristic approach by using GP to evolve (optimise) mutation operators for the bi-objective WDS design optimisation. • A generative hyper-heuristic framework consists of the following three phases. • Initialisation phase, which generates random population of mutation operators and sample network designs (using the Hanoi training network) which are fixed. • Generation phase, which creates an optimisation loop, where the mutation operators are varied and evaluated using sample network designs. The best mutation operators are then selected to propagate into the next generation and the process repeats until a termination criterion is met. SPEA2 is used to optimise mutation operators. • Evaluation phase, which evaluates the best evolved mutation operators and applies them to a set of three test networks (the Anytown network and two real networks). • A comparison of the best 10 varied evolved mutation operators with each other and also with the standard Gaussian mutation operator is performed using the hypervolume indicator [265]. • Results: The method enables to classify the evolved mutation operators in terms of their robustness and impact on convergence characteristics. • Test networks: (1) Anytown network (incl. 19 nodes) [84], (2) real network with one source, (3) real network with two sources.

Table A1. Cont.

ID, Authors (Year) [Ref] SO/MO *, Brief Description	Optimisation Model (Objective Functions +, Constraints **, Decision Variables ++)	Water Quality, Network Analysis, Optimisation Method	Notes
95. Roshani and Filion (2014) [124] MO Optimal WDS rehabilitation, strengthening, expansion and operation with asset management strategies over a planning horizon (i.e., 20 years) using NSGA-II with event-based coding.	Objective (1): Minimise the present value of the capital costs of the network including (a) pipe replacement, (b) pipe duplication, (c) pipe lining, (d) installation of new pipes. Objective (2): Minimise the present value of the operating costs including (a) lost water to leakage, (b) break repair, (c) electricity to pump water. Constraints: (1) Max. yearly annual budget for the total of all costs (excluding leakage), (2) min pressure at the nodes, (3) max velocity in the pipes. Decision variables: (1) Time of rehabilitation, (2) place of rehabilitation, type of rehabilitation including (3) the diameter of a pipe being replaced/duplicated and (4) the diameter of a new pipe in an area slated for future growth, (5) the type of lining technology used. Note: One MO model including both objectives.	Water quality: N/A. Network analysis: EPANET. Optimisation method: NSGA-II.	• An event-based algorithm for optimal timing of water main rehabilitation is introduced. A new gene coding scheme, which reduces the chromosome length, thus saves computer memory and increases speed of convergence, is developed. The chromosome length is reduced by only coding the rehabilitation events, rather than coding all years of the planning horizon (20 years) with mostly zero entries where no rehabilitation occurs. • Savings achieved using asset management strategies by synchronising road reconstruction works with water main replacement/rehabilitation (called infrastructure adjacency discount) and obtaining discounts for purchasing large numbers of water main pipes (called quantity discount) are accounted for. • Four scenarios are used to investigate the impact of different asset management strategies on the optimisation process, where different variations of infrastructure adjacency discounts, quantity discounts and annual budget constraints are applied. Pipe leakage, pipe break and pipe roughness forecasting models are used. Sensitivity analysis is performed to examine the sensitivity of the capital and operation costs to uncertainties in water demands, initial break rate, break growth rate, initial leak rate, leak growth rate, and pipe roughness. Results: A budget constraint prohibits from investing early and heavily in pipe rehabilitation. This pipe rehabilitation postponement leads to an increase in operation costs linked to leakage, breaks and energy use in unimproved pipes. The capital and operation costs decrease when applying discounts, with pipe lining being favoured over pipe replacement and duplication. Test networks: (1) Fairfield network in Amherstview and Odessa, Ontario, Canada.
96. Zheng et al. (2014) [179] SO Optimal WDS design and strengthening using a combined binary LP and DE method (BLP-DE) in a three-phase procedure.	Objective (1): Minimise (a) the design cost of the network (pipes), (b) penalty cost for violating the nodal head requirement. Constraints: (1) Total head loss used by the pipes (from the source to a node) should be less than the value of the head at the source minus the head requirement at a node, (2) only one pipe diameter selected for each link. Decision variables: (1) Pipe diameters (binary for BLP, continuous for DE rounded to the nearest commercially available discrete diameters after the mutation process).	Water quality: N/A. Network analysis: EPANET Optimisation method: BLP-DE.	• The proposed BLP-DE method takes advantages of both BLP (being able to efficiently provide a global optimum for a tree network) and DE (being able to generate good quality solutions for a loop network with a reduced search space). However, this method is not appropriate for least-cost design of networks, which have only loops or only trees. • The proposed BLP-DE method involves the following three stages: (i) network decomposition into trees and the core using a graph algorithm; (ii) optimisation of the trees using BLP; (iii) optimisation of the core using DE while incorporating the optimal solutions for the trees. Results: For the New York City tunnels and Hanoi networks, BLP-DE found the best-known solutions with a significantly improved efficiency compared to numerous other algorithms from the literature. For the real network, BLP-DE found better quality solutions than standard DE (SDE) also with an improved efficiency. Test networks: (1) New York City tunnels (incl. 20 nodes) [81], (2) Hanoi network (incl. 32 nodes) [49], (3) real network with one source and 96 pipes, China (incl. 85 demand nodes).

Table A1. *Cont.*

ID. Authors (Year) [Ref] SO/MO * Brief Description	Optimisation Model (Objective Functions +, Constraints **, Decision Variables ++)	Water Quality Network Analysis Optimisation Method	Notes
97. Basupi and Kapelan (2015) [135] MO Optimal flexible WDS strengthening, expansion, rehabilitation and operation considering demand uncertainty and optimal intervention paths in prefixed time intervals (i.e., 25 years) over a planning horizon (i.e., 50 years) using NSGA-II.	Objective (1): Minimise the total intervention cost including (a) capital cost of rehabilitation intervention, (b) pump energy consumption cost. Objective (2): Maximise (a) the end-of-planning horizon system resilience, using a resilience index [266]. Constraints: (1) Min head requirement at the nodes. Decision variables: Intervention options (discrete) including (1) addition of new pipes, (2) duplication/cleaning/lining of existing pipes, (3) addition and (4) sizing of new tanks, (5) pump schedules, (6) threshold demands (discrete). Note: One MO model including both objectives.	Water quality: N/A. Network analysis: EPANET. Optimisation method: NSGA-II.	• Future demand uncertainty, following a probability density function, is considered. Simulations (Monte Carlo or Latin Hypercube) around the traditionally projected water demand are employed to reflect the possible scenarios of future demand realisation at certain decision points. • Decision trees are used to represent the uncertain demands and the respective flexible design intervention plans. The decision tree has optional intervention paths consisting of a set of intervention measures. There is path-dependence, which means that the extent of future design interventions depends on the previous intervention path undertaken. • Planning horizon of 50 years, divided into two design stages of 25 years, is used. • The proposed flexible design with optional intervention paths into the future is compared with the deterministic design with a single set of interventions for each design stage's future demand in the analysed planning horizon. • The sensitivity analyses of both the cost discount rate and the standard deviation scenarios across the planning horizon are investigated. • Results: The optimal flexible design under future demand uncertainty outperforms the corresponding optimal deterministic design in terms of the cost and resilience objectives, because it enables the system to adapt in addition to simply postpone interventions. The flexible design methodology is more sensitive to the cost discount rate than the level of demand uncertainty. • Test networks: (1) New York City tunnels (incl. 20 nodes) [81], (2) Anytown network (incl. 19 nodes) [84].
98. Bi et al. (2015) [108] SO Optimal WDS design using GA with an engineered initial population.	Objective (1): Minimise (a) the design cost of the network (pipes), (b) penalty cost for violating the pressure constraints. Constraints: (1) Min/(max) pressure at the nodes. Decision variables: (1) Pipe diameters (discrete).	Water quality: N/A. Network analysis: EPANET. Optimisation method: GA (for optimisation), a new heuristic sampling method (PHSM) (for generating an engineered initial population to improve the GA convergence).	• A new PHSM is proposed to determine the initial population of the EAs using engineering experience and domain knowledge to improve convergence of the algorithms. • The PHSM procedure is designed as follows: (i) assigning pipe sizes based on the knowledge that pipe diameters decrease with a greater distance from sources; (ii) adjusting pipe sizes based on the velocity threshold; (iii) ensuring diversity in the initial population by generating it from a distribution, so pipe diameters from step (ii) have the highest probability of being selected. • PHSM is compared to Kang and Lansey's sampling method (KLSM) [26] and two other sampling methods which do not use domain knowledge, such as random sampling (RS) and Latin hypercube sampling (LHS). • The number of decision variables of 7 test networks used varies from 34 to 1274. • Results: PHSM outperforms other sampling methods in terms of computational efficiency as well as the solution quality, and its advantage increases with network size. • Test networks: (1) Hanoi network (incl. 32 nodes) [49], (2) extended Hanoi network (incl. 32 nodes) (a number of diameter options is increased), (3) Zhi Jiang (ZJ) network, China (incl. 113 demand nodes) [111], (4) Balerma irrigation network, Almeria, Spain (incl. 447 nodes) [50], (5) rural network (incl. 379 nodes) [154], (6) Foss_poly_1, Italy (incl. 37 nodes) [147], (7) modified Kang and Lansey's network (KLmod) (incl. 936 nodes) [26].

Table A1. Cont.

ID. Authors (Year) [Ref] SO/MO* Brief Description	Optimisation Model (Objective Functions +, Constraints **, Decision Variables ++)	Water Quality Network Analysis Optimisation Method	Notes
99. Creaco et al. (2015) [136] MO Optimal WDS design, strengthening and expansion accounting for demand uncertainty and construction phasing in prefixed time intervals (i.e., 25 years) over a planning horizon (i.e., 100 years) using NSGA-II.	Objective (1): Minimise (a) the total present worth construction cost of the network including (a) the cost of installing pipes at new sites, (b) cost of installing pipes in parallel to existing pipes. Objective (2): Maximise (a) the network reliability, calculated as the minimum pressure surplus over the whole construction time. Constraints: (1) Min pressure at the nodes. Decision variables: (1) Pipe diameters (coded as integer numbers). Note: One MO model including both objectives.	Water quality: N/A. Network analysis: Demand-driven analysis [11]. Optimisation method: Modified NSGA-II.	• An extension of the paper by [118] taking into account uncertainty in demand growth. "The uncertainty in the water demand is obtained by expressing the parameters of the demand-growth model by means of a (discrete) random variable of given probability mass function". • A set of 81 demand-growth scenarios is developed, the first three of which have a constant demand-growth rate, whereas the others have a randomly variable demand-growth rate over the planning horizon. The reliability which is maximised in the second objective is in fact a discrete random variable, reflecting different demand-growth scenarios. • Four construction phases with 25-year intervals over 100-year planning horizon are considered. The number of pipes to be inserted at each phase is assumed to be known. • Different types of optimisation are performed: (i) probabilistic second objective optimisation using the entire set of 81 demand-growth scenarios; (ii) deterministic second objective optimisations, applying a constant demand-growth rate (i.e., the first three demand-growth scenarios), where the second objective function is the crisp minimum temporal surplus (instead of the discrete random variable) over the planning horizon. • Results: Optimisation of construction phasing, accounting for demand growth uncertainty, leads to the network being sized more conservatively (larger pipe diameters are evident mainly in the first construction phases), which makes the network more flexible to adapt itself to various conditions of demand growth over time. • Test networks: (1) Network of a town in northern Italy (incl. 26 nodes) [267], skeletonised from the original network [268].
100. Dziedzic and Karney (2015) [119] SO Optimal WDS design, strengthening and operation considering multiple loading conditions over a planning horizon (i.e., 20 years) using cost gradient-based heuristic method with computational time savings.	Objective (1): Minimise (a) the pump energy cost, (b) damage cost, (c) capital cost of the network (pipes). Constraints: (1) Min pressure at the nodes. Decision variables: (1) Pipe diameters (discrete).	Water quality: N/A. Network analysis: EPANET. Optimisation method: Cost gradient-based heuristic method.	• The aim is to reduce the computational time of the optimisation process of a WDS with multiple loading conditions. • A gradient search is applied and the objective function is approximated by shortening the extended period analysis. So, a shorter time period is used to estimate the hydraulic variations of the system and the costs for the full planning horizon. The demands within the shorter time cycle (TC) should match the demand probabilities in the full analysis period and their variation. • The ratio between the gradients of energy dissipation cost, damage cost and pipe cost is calculated at each iteration (i.e., one TC). The pipes with the minimum and maximum cost gradient ratios are identified, the pipe with a minimum (below 1) and maximum (above 1) cost ratio is downsized and upsized, respectively. • Hourly iterations were used initially to generate a rough solution, which was then optimised with the 100-day TC to represent demand variations. Significantly, these short TC results, when extrapolated, accurately depict the costs of the full 20-year planning horizon. The optimisation process took approximately 1 h. • The damage cost is computed according to the pressures (the probabilities are given) from EPANET. Three types of damage are considered: (i) the pressure falls below 14 m and fire erupts simultaneously; (ii) the pressure is between 14 and 26 m causing for example backup pumps to fail; (iii) the pressure is above 88 m potentially leading to a pipe burst. • Four additional scenarios were optimised: reduced demand, increased damage cost, increased energy cost, and varying roughness. • Results: Shorter TCs can be used to approximate full time horizon costs. The method is useful in cases where more computationally intensive methods are infeasible. • Test networks: (1) Anytown network (incl. 19 nodes) [84].

Table A1. Cont.

ID, Authors (Year) [Ref] SO/MO * Brief Description	Optimisation Model (Objective Functions +, Constraints **, Decision Variables ++)	Water Quality Network Analysis Optimisation Method	Notes
101. Marques et al. (2015) [122] SO Optimal WDS design, strengthening, expansion and operation with a real options (ROs) concept and demand uncertainty, accounting for construction phasing in prefixed time intervals (10–20 years) over a planning horizon (60 years) using SA.	Objective (1): Minimise (a) the cost of the initial solution to be implemented in year zero (for interval 0–20 years) incl. pipes, pumps and pump energy costs, (b) cost of the future conditions incl. pipes, pumps and pump energy costs (cost of all scenarios weighted by the corresponding probability of each scenario), (c) regret term incl. pipes, pumps and pump energy costs (squared differences between the cost of the solution to implement and the optimal cost for each scenario). Constraints: (1) Min/max pressure at the nodes, (2) min pipe diameter, (3) only one commercial diameter assigned to a pipe. Decision variables: (1) Pipe diameters (discrete), (2) pump heads.	Water quality: N/A. Network analysis: EPANET. Optimisation method: SA.	• The ROs concept is proposed, which allows flexibility to be included in the decision making process. The regret term introduced in the objective function captures a situation of making decisions without perfect information (i.e., an implemented solution can be suboptimal and the regret term represents the risk of such a decision). • Uncertainties in future demands are implemented. Three demand conditions are used, one of them considers instantaneous peak discharge and fire flow at one node. • Various network expansion options are considered to predict alternative future developments. • Combining all the different conditions and expansion options, a total of 8 scenarios are derived, which form a decision tree. • Planning horizon of 60 years divided into 4 intervals is used. It is assumed that interval 1 (T = 1, 20 years) requires no modifications and conditions will not change. T = 2 and T = 3 are 10-year intervals with potential network expansion. Pumps should be replaced in T = 2 and T = 4. Also in T = 4, the demand should be predicted, two scenarios here are demand increasing by 20% and demand remaining constant. For the first 40 years, the demand would increase at a constant rate of 10% per decade. • In order to understand the difference of using ROs in the flexible design of WDSs, the ROs concept and a traditional design are compared. • Results: Compared to a traditional design, the ROs solution enables saving resources if an extended and uncertain planning horizon is considered. Accordingly, the ROs solution has a higher initial cost (the first 20 years), yet the total cost over 60 years is lower. • Test networks: (1) Simple network supplied from a single reservoir (incl. 10 nodes), inspired by [269].
102. Marques et al. (2015) [137] MO Optimal WDS design, expansion and operation with a ROs concept and network expansion uncertainty, accounting for construction phasing in prefixed time intervals (20 years) over a planning horizon (60 years) using multi-objective SA.	Objective (1): Minimise (a) the cost of the initial solution to be implemented in year zero (for interval 0–20 years) incl. pipes, pumps, pump energy costs, carbon emissions cost for pipes and energy (b) cost of the future conditions incl. pipes, pumps, pump energy costs, carbon emissions cost for pipes and energy (cost of all scenarios weighted by the corresponding probability of each scenario). Objective (2): Minimise (a) total pressure violations for future scenarios (the sum of pressure violations for each scenario, each interval (starting from T = 2), each demand condition and each network node). Constraints: (1) Min pressure at the nodes, (2) min pipe diameter, (3) only one commercial diameter assigned to a pipe. Decision variables: (1) Pipe diameters (discrete). Note: One MO model including both objectives.	Water quality: N/A. Network analysis: EPANET. Optimisation method: Multi-objective SA.	• An extension of the papers by [122,270] considering a multi-objective approach with carbon emissions and uncertainties related to the future expansion scenarios of the network. • Similar to [122,270], ROs concept is applied, which uses a decision tree to reflect different scenarios (there is a total of 8 scenarios). • Planning horizon of 60 years divided into 3 intervals is used. Two kinds of minimum pressures are considered: desirable and admissible. In the first interval (T = 1, 20 years), the pressure cannot fall below the desirable minimum pressure. • The constraint of minimum pressure at the nodes aims to obtain higher values, thus fewer pressure violations, for scenarios with high occurrence probabilities. • The test network used can be expanded into four different areas, and also one area can be depopulated. • Results: The carbon emission costs have an insignificant influence on the objective function value. Energy and pipe costs are conflicting objectives. • Test networks: (1) Network supplied by three reservoirs (incl. 14 nodes) inspired by the study of [271].

Table A1. Cont.

ID. Authors (Year) [Ref] SO/MO * Brief Description	Optimisation Model (Objective Functions +, Constraints **, Decision Variables ++)	Water Quality Network Analysis Optimisation Method	Notes
103. McClymont et al. (2015) [29] SO Optimal WDS design and operation, investigating linkages between algorithm search operators and the WDS design problem features, using elitist EA.	Objective (1): Minimise (a) the design cost of the network (pipes), (b) the energy cost of running pumps. Constraints: (1) Min/max pressure at the nodes, (2) max velocity in the pipes. Decision variables: (1) Pipe diameters (discrete), (2) pump statuses (binary).	Water quality: N/A. Network analysis: EPANET. Optimisation method: Elitist EA.	• The aim is to bring insight into the interaction between an algorithm search operator and the WDS design problem. For that purpose, 60 artificial test networks are designed specifically, so they isolate individual features. These networks are then used to evaluate the impact of network features on operator performance. • "The method is as follows: (1) select operators, (2) select problems, (3) identify problem features, (4) synthesize artificial problems, (5) test on artificial problems, (6) analyse results and determine linkages, (7) select the most appropriate operators for selected problems, (8) test on actual problems, (9) analyse results". Such a systematic and quantitative approach provides detailed information (e.g., what linkages, if any, exist between the performance of an operator and certain WDS features) about an algorithm's suitability to optimise certain types of problem. • The following 6 operators are tested: mutation (random and 1 step size variation), crossover (uniform and n-point), and pipe smoothing and pipe expander (designed specifically for WDS problems). • Two types of experiments were conducted, one to test the effects of operators individually, the other to test the effect of the pairs of operators. • Results: Operator performance and problem search spaces are linked, which is verified using three well known benchmark problems. • Test networks: (1)–(60) Artificial networks based on 3 simple systems (looped, branched and hybrid), (61) two-loop network supplied by gravity (incl. 7 nodes) [14], (62) Hanoi network (incl. 32 nodes) [49], (63) Anytown network (incl. 19 nodes) [84].
104. Roshani and Filion (2015) [132] MO Optimal WDS rehabilitation, strengthening, expansion and operation with GHG emissions over a planning horizon (i.e., 20 years) using NSGA-II with event-based coding.	Objective (1): Minimise the present value of the capital costs of the network including (a) pipe replacement, (b) pipe duplication, (c) pipe lining, (d) installation of new pipes. Objective (2): Minimise the present value of the operating costs including (a) lost water to leakage, (b) break repair, (c) electricity to pump water, (d) carbon cost associated with electricity use. Constraints: (1) Min pressure at the nodes, (2) max velocity in the pipes. Decision variables: (1) Time of rehabilitation, (2) place of rehabilitation, type of rehabilitation including (3) the diameter of a pipe being replaced/duplicated and (4) the diameter of a new pipe in an area slated for future growth, (5) the type of lining technology used. Note: One MO model including both objectives.	Water quality: N/A. Network analysis: EPANET. Optimisation method: NSGA-II.	• An extension of the paper by [124] including energy use and GHG emissions linked to electricity consumption due to pumping, leakage, and increases in pipe wall roughness due to pipe aging. The paper also analysis impact of two carbon reduction strategies (carbon tax and discount rates) on WDS rehabilitation. • Event-based rehabilitation timing approach of [124] is used. • Six carbon-abatement scenarios are examined, involving different combinations of carbon tax and discount rates, for two different GHG emissions intensity factors (low and high yearly emissions). • Results: Adopting a low discount rate and levying a carbon tax has a small impact on energy use and GHG emissions reduction. A low discount rate and the application of a carbon tax has a modest impact on leakage and pipe breaks reduction, and encourages an early rehabilitation investment to reduce the ongoing costs of leakage, pipe repair, energy, and GHG emissions. • Test networks: (1) Fairfield network in Amherstview and Odessa, Ontario, Canada.
105. Sadollah et al. (2015) [174] SO Optimal WDS design and strengthening using improved mine blast algorithm (IMBA).	Objective (1): Minimise (a) the design cost of the network (pipes). Constraints: (1) Min pressure at the nodes. Decision variables: (1) Pipe diameters (discrete).	Water quality: N/A. Network analysis: EPANET. Optimisation method: IMBA.	• An improved algorithm based on a mine blast algorithm (MBA) is developed for least-cost design of WDSs. MBA is inspired by the process of mine explosions. Similar to other metaheuristics, it starts with an initial population (the number of shrapnel pieces), further followed by exploration and exploitation phases. • The modifications in the IMBA concern the exploitation phase and distance reduction of each shrapnel piece. In particular, the exploitation equations are modified to avoid problems with the dimension of the search space, where the perception of direction is replaced by moving to the best solutions. • IMBA is compared to a large number of other algorithms (14 to 17 for each test network) in terms of the solution quality and computational effort. • Results: IMBA reached a cheaper design than other algorithms for at least one test network. For the other two test networks, IMBA found the best-known design in fewer function evaluations. • Test networks: (1) Hanoi network (incl. 32 nodes) [49], (2) New York City tunnels (incl. 20 nodes) [81], (3) Balerma irrigation network, Almeria, Spain (incl. 447 nodes) [50].

Table A1. *Cont.*

ID. Authors (Year) [Ref] SO/MO * Brief Description	Optimisation Model (Objective Functions +, Constraints ++, Decision Variables ++)	Water Quality Network Analysis Optimisation Method	Notes
106. Saldarriaga et al. (2015) [47] SO Optimal WDS design using optimal power use surface (OPUS) method paired with metaheuristic algorithms.	Objective (1): Minimise (a) the design cost of the network (pipes), (b) penalty for violating the pressure constraint. Constraints: (1) Min pressure at the nodes. Decision variables: (1) Pipe diameters (discrete).	Water quality: N/A. Network analysis: Not specified. Optimisation method: OPUS combined with: GA in REDES, GA in GANETXL, GA in MATLAB, HS in REDES, SA in MATLAB, greedy algorithm in REDES.	• The OPUS algorithm is paired with metaheuristic methods whereby the solutions obtained through OPUS are used as hot start (i.e., initial population) for the metaheuristics applied subsequently. • The OPUS method uses deterministic hydraulic principles drawn from analysing energy use and flow distribution in the network. • Results: The proposed optimisation method consistently marginally reduces the costs obtained through the OPUS algorithm (up to 1%) and substantially increases the number of iterations in every case (around 3 orders of magnitude). Authors argue, therefore, that it is not worth "to refine a solution that is already very close to the optimum and required minimum computational and human effort to be reached" (through the OPUS algorithm). • Test networks: (1) Hanoi network (incl. 32 nodes) [49], (2) Balerma irrigation network. Almeria, Spain (incl. 447 nodes) [50], (3) Taichung network, Taiwan (incl. 20 nodes) [272], (4) hypothetical network R28 (incl. 39 nodes) created at the Water Distribution and Sewer Systems Research Centre (CIACUA) of the University of Los Andes in Bogota, Colombia.
107. Stokes et al. (2015) [76] MO Optimal WDS design and operation including GHG emissions over a planning horizon (i.e., 100 years), investigating the effect of changing tank reserve size (TRS), using Borg multi-objective EA (MOEA).	Objective (1): Minimise (a) the construction costs of the network (pipes, pumps, tanks), (b) operating costs (electricity consumed by pumps). Objective (2): Minimise GHG emissions associated with the system (a) construction, (b) operation (electricity consumed by pumps). Constraints: (1) Min pressure at the nodes, (2) the total volume pumped equal to or greater than the total demand during the EPS. Decision variables: (1) Pipe diameters (discrete), (2) pump types (discrete), (3) pump scheduling decision variable (continuous). Note: One MO model including both objectives. For the test network (1), both design and operation components are included; for the test network (2) (D-town), only operation components are included.	Water quality: N/A. Network analysis: EPANET (EPS). Optimisation method: Borg MOEA [273].	• The effect of changing (i) the storage tank balancing volume or TRS and (ii) time-varying emissions factors (EFs) on the minimisation of costs and GHG emissions in WDSs is investigated. • Four different TRS scenarios (for 3, 6, 12 and 24-h supply under average-day demand) and two different EF cases (an estimated 24-h time-varying EF (EEF) curve and an average EF (AEF)) are used. The TRS volumes are altered by changing the tank diameter, rather than lower and upper water levels which would impact on the system hydraulic. • Planning horizon of 100 years is considered and is used for calculating electricity costs, GHG emissions and pump replacement costs. • Peak and off-peak electricity tariffs are used. • Results: A larger TRS can help to reduce GHG emissions when the emissions intensity of electricity fluctuates during each day. This reduction in GHG emissions represents only 2–4% for a new WDS, but occurs with no additional cost as it allows pumping to be moved to the off-peak tariff period. However, when these fluctuations do not occur or are not considered when evaluating pumping operational GHG emissions (i.e., AEF is used), increasing the TRS results in no reduction of the cost or GHG emissions. • Test networks: (1) Two-pump network with 23 pipes (incl. 15 nodes), (2) modified D-town network (incl. 348 non-zero demand nodes) from the battle of the water networks II (BWN-II) [58,274].

Table A1. *Cont.*

ID. Authors (Year) [Ref] SO/MO * Brief Description	Optimisation Model (Objective Functions +, Constraints **, Decision Variables ++)	Water Quality Network Analysis Optimisation Method	Notes
108. Stokes et al. (2015) [75] MO Optimal WDS design and operation including GHG emissions considering varying emission factors, electricity tariffs and water demands using NSGA-II.	Objective (1): Minimise (a) the design costs of the network (pipes and pumps), (b) operating costs (electricity consumed by pumps). Objective (2): Minimise GHG emissions associated with the system (a) design (pipes), (b) operation (electricity consumed by pumps). Constraints: (1) Min pressure at the nodes, (2) the sum of the instantaneous pump supply equal to or greater than the sum of the instantaneous water demands. Decision variables: (1) Pipe diameters (discrete), (2) pump types (discrete), (3) pump schedules (discrete options representing the time at which a pump is turned on/off, using a time step of 30 minutes). Note: One MO model including both objectives.	Water quality: N/A. Network analysis: EPANET (EPS). Optimisation method: NSGA-II.	• Water distribution cost-emission nexus (WCEN) computational freeware framework is introduced for consolidating computational tools to solve WDS optimisation problems. A range of time-dependent operational conditions (e.g., EFs, electricity tariffs, water demands, pumping operational management options) can be considered. • For this study, hydraulic and pumping operational simulation, cost and GHG emissions calculation and MO heuristic optimisation are integrated. • Four operational scenarios are used: the first scenario reflects "standard" practices (i.e., steady state simulation with an average emission factor, electricity tariff and water demand), the other 3 scenarios use additional simulation complexity and flexibility (i.e., unsteady state simulation with varying emission factors, electricity tariffs and water demands). Results: Compared to standard simulation practices, considering both short-term (e.g., daily) and long-term (e.g., monthly and annual) variations can significantly affect the design, pumping operational management options as well as their costs and GHG emissions. Test networks: (1) Simple network with 23 pipes (incl. 15 nodes) [76].
109. Wang et al. (2015) [195] MO Optimal WDS design, strengthening and rehabilitation of well-known benchmark problems with the aim to obtain the best-known approximation of the true Pareto front using various MOEAs.	Objective (1): Minimise (a) the design costs of the network (pipes). Objective (2): Maximise (a) the network resilience [275]. Constraints: (1) Min/max pressure at the nodes (max pressure only for some test networks), (2) max velocity in the pipes (only for some test networks). Decision variables: (1) Diameters of new or duplicate pipes (integer) (duplicate pipes only for some test networks), (2) cleaning of existing pipes or do-nothing option (integer) (only for some test networks). Note: One MO model including both objectives.	Water quality: N/A. Network analysis: EPANET. Optimisation method: Five state-of-the-art MOEAs are used including AMALGAM [276], Borg [273], NSGA-II [258], ε-MOEA [277], ε-NSGA-II [278].	• The aim is to obtain the best-known approximation of the true Pareto front (PF) for a set of benchmark problems, in order to create a single point of reference. • MOEAs parameters are not fine-tuned, instead the recommended settings are used. • An innovative projection plot is applied to facilitate the MOEAs comparison in terms of convergence and diversity. Results: The true PFs for small problems and the best-known PFs for the other problems are obtained. No algorithm is completely superior to the others. Nevertheless, NSGA-II shows generally the best achievements across all the benchmark problems. Test networks: (1) Two-reservoir network [83], (2) two-loop network supplied by gravity (incl. 7 nodes) [14], (3) BakRyan network, South Korea (incl. 35 nodes) [227], (4) New York City tunnels (incl. 20 nodes) [81], (5) Blacksburg network (incl. 31 nodes) [257], (6) Hanoi network (incl. 32 nodes) [49], (7) GoYang network, South Korea (incl. 22 nodes) [226], (8) Fossolo network, Italy (incl. 37 nodes) [254], (9) Pescara network, Italy (incl. 71 nodes) [254], (10) Modena network, Italy (incl. 272 nodes) [254], (11) Balerma irrigation network, Almeria, Spain (incl. 447 nodes) [50], (12) Exeter network (serves a population of approximately 400,000) [82].

Table A1. *Cont.*

ID. Authors (Year) [Ref] SO/MO *, Brief Description	Optimisation Model (Objective Functions +, Constraints **, Decision Variables ++)	Water Quality Network Analysis Optimisation Method	Notes
110. Zheng (2015) [196] SO. Optimal WDS design and strengthening using four DE variants with a comparison of their searching behaviour.	Objective (1): Minimise (a) the design cost of the network (pipes), (b) penalty cost for violating the pressure constraint. Constraints: (1) Min pressure at the nodes. Decision variables: (1) Pipe diameters (discrete, with continuous values adjusted to the closest discrete sizes according to [111]).	Water quality: N/A. Network analysis: EPANET. Optimisation method: DE (4 variants).	• The aim is to investigate the impact of different parameterisation strategies on the DE's searching performance (exploration and exploitation) through the real-time behaviour analysis using a series of proposed metrics. • The following four variants of DE algorithm are used: (i) the SDE algorithm with fixed mutation (F) and crossover (CR) parameter values; (ii) the dither DE (dDE) variant [279] with the randomised F; (iii) the modified dDE (MdDE) variant with the randomised F and CR; (iv) the SADE variant [173] with the self-adapted F and CR along the searching process. The modified DE is proposed specifically for this study. • Six performance metrics, which measure search quality, search progress and convergence, are used to compare DE algorithms. • Results: The dDE, MdDE and SADE outperformed the SDE algorithm only in the middle to later searching periods. The SADE offered a larger number of improved solutions than the other DE variants in the exploitative periods. The MdDE has a greater exploratory ability than the SADE in the later searching period, hence found better solutions when a very large computational budget was available for the complex test network (3). • Test networks: (1) New York City tunnels (incl. 20 nodes) [81], (2) Balerma irrigation network, Almeria, Spain (incl. 447 nodes) [50], (3) large network with five reservoirs and 1278 pipes (incl. 936 nodes), originally introduced by [26], modified by [280].
111. Zheng et al. (2015) [69] MO. Optimal WDS design considering multiple loading conditions using multi-objective DE algorithm (MODE) with a graph decomposition technique.	Objective (1): Minimise (a) the design cost of the network (pipes), (b) penalty for violating the pressure head constraint. Objective (2): Maximise (a) the minimum head excess across the network of multiple demand loading cases, (b) penalty for violating the pressure head constraint. Constraints: (1) Min/max allowable pipe diameters. Decision variables: (1) Pipe diameters (discrete). Note: One MO model including both objectives.	Water quality: N/A. Network analysis: EPANET (EPS for the second objective). Optimisation method: MODE.	• The graph decomposition technique is proposed to improve the efficiency of MOEAs for WDS design optimisations. It allows to decompose the original network into a series of more manageable subnetworks (subproblems), which are optimised individually with significantly higher efficiency than the original network. • Subsequently, the propagation method is used to evolve Pareto fronts of the subnetworks towards the Pareto front of the original full network without the need to run the hydraulic simulation of the full network. • MODE, based on a single-objective DE algorithm [111], is developed. For comparison purposes, MODE is applied in conjunction with as well as without the graph decomposition technique when the whole network is optimised directly (referred to as SMODE). MODE is also compared with NSGA-II applied to the whole network optimisation. • Results: MODE exhibits significantly better performance than both conventional full-search methods SMODE and NSGA-II and its efficiency is more notable for larger networks. • Test networks: (1) Real-world network with 112 pipes and 24 demand loading cases, China (incl. 99 demand nodes), (2) BWN network with 433 pipes and 24 demand loading cases (incl. 387 demand nodes) [281].

Table A1. *Cont.*

ID. Authors (Year) [Ref] SO/MO * Brief Description	Optimisation Model (Objective Functions +, Constraints **, Decision Variables ++)	Water Quality Network Analysis Optimisation Method	Notes
112. Zheng et al. (2015) [48] SO Optimal WDS design using DE, analysing impact of algorithm parameters on its search behaviour.	Objective (1): Minimise (a) the design cost of the network (pipes), (b) penalty cost for violating the pressure constraint. Constraints: (1) Min pressure at the nodes. Decision variables: (1) Pipe diameters (discrete, with continuous values produced in the initialisation and mutation processes of DE converted to the nearest discrete pipe diameters).	Water quality: N/A. Network analysis: EPANET. Optimisation method: DE.	• The aim is to investigate search behaviour (exploration and exploitation) of DE as a function of the two control parameters: mutation weighting factor (F) and crossover probability (CR). The six metrics are developed to measure the population variance, search quality, convergence properties, the percentage of the time spent in feasible and infeasible regions, and the percentage of improved solutions within each generation. • The results are compared with prior theoretical results using WDS design problems. Test problems used (Hanoi, ZJ and Balerma networks) have different sizes and complexity (34, 164 and 454 decision variables, respectively). Results: An improved knowledge on search behaviour of DE via parameters F and CR is obtained. It was found that (i) there is excellent agreement between predicted and observed population variance as well as the lower bound of parameter F; (ii) DE performance is more dominated by parameter F; (iii) high CR value (CR > 0.8) often reduces DE's diversity with a rapid speed likely resulting in premature convergence. Test networks: (1) Hanoi network (incl. 32 nodes) [49], (2) Zhi Jiang (ZJ) network, China (incl. 113 demand nodes) [111], (3) Balerma irrigation network, Almería, Spain (incl. 447 nodes) [50].
113. Zhou et al. (2015) [175] SO Optimal WDS design and strengthening using discrete state transition algorithm (STA).	Objective (1): Minimise (a) the design cost of the network (pipes), (b) penalty for violating the pressure constraint. Constraints: (1) Min pressure at the nodes. Decision variables: (1) Pipe diameters (discrete).	Water quality: N/A. Network analysis: Newton-Raphson method [10]. Optimisation method: Discrete STA [282]. Note: Continuous STA [283].	• A reduction in the computational complexity of solving network continuity equations (linear equations) and energy equations (nonlinear equations) simultaneously is presented. Basically, some pipe flows are initially fixed as known to solve the linear equations and then substituted into the nonlinear equations. Consequently, the number of network linear and nonlinear equations is reduced to the number of closed simple loops. • For the two-loop network, the influence of penalty coefficient and one of the STA parameters called the search enforcement (SE) on the algorithm performance is studied. The knowledge gained is used in the optimisation of other test networks. Results: The penalty coefficient has a significant impact on the search ability and solution feasibility, whereas SE does not affect the STA performance explicitly. Discrete STA is able to find the best-known solutions with fewer function evaluations. Test networks: (1) Two-loop network supplied by gravity (incl. 7 nodes) [14], (2) Hanoi network (incl. 32 nodes) [49], (3) New York City tunnels (incl. 20 nodes) [81], (4) triple Hanoi network (incl. 92 nodes).

Table A1. *Cont.*

ID, Authors (Year) [Ref] SO/MO * Brief Description	Optimisation Model (Objective Functions +, Constraints **, Decision Variables ++)	Water Quality Network Analysis Optimisation Method	Notes
114. Andrade et al. (2016) [143] SO Optimal WDS design with improved offline ANNs to replace water quality simulations and the probabilistic approach to generate training data sets, using GA.	Objective (1): Minimise (a) the system cost of the network (the pipe and installation costs). Constraints: (1) Min pressure at the nodes, (2) min chlorine concentration at the nodes. Decision variables: (1) Pipe diameters (discrete), (2) chlorine dosages at the water source (discrete).	Water quality: Chlorine. Network analysis: EPANET, offline ANN (for water quality analyses). Optimisation method: GA.	• The aim is to improve the performance of an offline ANN applied to the WDS design problems in terms of their architecture and training data, which affect their speed and accuracy. • The probabilistic approach is introduced to generate a large set of networks (training data sets) resembling those analysed by an optimisation method after its initial iterations. ANNs trained with these networks are compared against ANNs trained with conventional random networks. • The conventional multi-ANN architecture versus two single ANN architectures are also compared. Regarding the multi-ANN architecture, there are multiple ANNs each individually forecasting concentration at a single node. Concerning the first single ANN architecture, concentrations at all network nodes are forecast. The second single ANN architecture has only one output neuron (for one node) to estimate the minimum concentration in a WDS regardless of its location. • Therefore, six types of ANNs, resulting from the combinations of the two training data sets (the new introduced one and conventional random) and the three ANN architectures are analysed with respect to speed and accuracy. Results: For a small WDS, there is no advantage in using multi-ANN architecture with a single output neuron over single ANN architectures; a probabilistic data set has no advantage over a conventional random data set. For a large WDS, multi-ANN architecture with a single output neuron outperforms the two other architectures analysed; a probabilistic data set is significantly superior to a conventional random data set. Test networks: (1) Hanoi network (incl. 32 nodes) [49], (2) modified Kang and Lansey's network (incl. 517 demand nodes) [26].
115. Jabbary et al. (2016) [181] SO Optimal WDS design using a modified central force optimisation algorithm (CFOnet).	Objective (1): Minimise (a) the design cost of the network (pipes), (b) penalty cost of violating the pressure constraint, (c) penalty cost of violating the velocity constraint. Constraints: (1) Min/max commercial pipe diameters, (2) min/max velocity in the pipes, (3) min/max pressure at the nodes. Decision variables: (1) Pipe diameters (discrete).	Water Quality: N/A. Network analysis: EPANET. Optimisation method: CFOnet.	• CFOnet, a deterministic metaheuristic method based on the rules of gravity, is applied to the WDS design optimisation. CFO uses a set of probes flying through space. The probes move under the influence of an accelerated force created by the gravitational attraction of masses in decision space. Due to the large computed acceleration values in the WDS problem, a normalisation operator is introduced to decelerate the probes so they remain inside of the decision space. Among the modifications, a new deterministic mutation operator is proposed, which prevents the algorithm to be locally trapped. • The method is compared with the original CFO method and other methods (GA, GA-ILP, PSO, LP) previously applied to the two test networks considered. Results: CFOnet shows significantly better results over CFO, 55% and 94% improvement for the Kadu and Khorramshahr networks, respectively. When compared to other methods (GA, GA-ILP and PSO) for the Kadu network and LP for Khorramshahr network, the improvement is 3–4%. Test networks: (1) Kadu network (incl. 26 nodes) [45], (2) Khorramshahr network (incl. 39 nodes) [284].

Table A1. *Cont.*

ID. Authors (Year) [Ref] SO/MO *, Brief Description	Optimisation Model (Objective Functions +, Constraints ++, Decision Variables ++)	Water Quality Network Analysis Optimisation Method	Notes
116. Schwartz et al. (2016) [123] SO Optimal robust WDS design and operation considering multiple loading conditions and demand uncertainty using the robust counterpart (RC) approach and CE.	Objective (1): Minimise the construction and operation costs of the network including (a) pipe capital costs, (b) tank capital costs, (c) pump station capital cost, (d) energy costs related to the operation of the system during a TC of operation. Constraints: (1) Min/max tank water volumes at the last time period of the cycle, (2) min desired nodal heads, (3) tank closure constraints defined by the difference between the tank water level at the start and end of the TC. Decision variables: (1) Pipe diameters (discrete), (2) pump station heads at all time periods reflecting the pump curve needed for the system.	Water quality: N/A. Network analysis: Explicit mathematical formulation (nonlinear equations are linearised). Optimisation method: CE for combinatorial optimisation [230],[285].	• The RC approach, which incorporates the uncertainty without the need for full stochastic information, is used. • The approach utilises characteristics of data distribution as opposed to assuming the entire probability density function. It uses simple statistical measures such as mean and covariance matrix to replace the original stochastic model with the deterministic model. Ellipsoidal uncertainty set, required by RC, is constructed using the mean value and the covariance matrix, according to the user-defined protection level. An obtained solution is robust and optimal to all possible scenarios in the uncertainty set. • Multiple time periods and multiload consumption patterns taking into account the temporal and spatial correlations simultaneously are used. • The system is tested under two different probability distributions, normal and uniform, on two test networks. • Results: The proposed method is robust under both normal and uniform distributions. Some of the tank volume obtained for a high protection level will not be utilised in reality and will perform as a safety factor withstanding the unexpected consumption unlike the deterministic solution. • Test networks: (1) Simple network (incl. 3 demand nodes) adopted from [286], (2) network with 2 sources and 65 pipes (incl. 48 demand nodes) adopted from [14].
117. Sheikholeslami and Talatahari (2016) [150] SO Optimal WDS design using a newly developed swarm-based optimisation (DSO) algorithm.	Objective (1): Minimise (a) the design cost of the network (pipes). Constraints: (1) Min pressure at the nodes. Decision variables: (1) Pipe diameters (discrete).	Water quality: N/A. Network analysis: EPANET. Optimisation method: DSO algorithm.	• A new DSO algorithm, which integrates accelerated PSO with big bang-big crunch (BB-BC) algorithm, is proposed for optimal design of WDSs. • To preserve the diversity of the swarm and avoid premature convergence to local optima, BB-BC concepts are introduced into the global and local search steps of accelerated PSO. In addition, a harmony memory concept from the HS algorithm is adopted to ensure that the particles do not leave the search space. • A modified constraint tournament selection is used for handling the constraints. • Another rule is added stating that infeasible solutions with slight violations are considered as feasible, which is to maintain the diversity of the population. • Results: While comparing with other methods from the literature, DSO found the best-known solutions in a lower number of evaluations for the GoYang and Hanoi networks, and exhibited comparable performance for the Balerma network. • Test networks: (1) GoYang network, South Korea (incl. 22 nodes) [226], (2) Hanoi network (incl. 32 nodes) [49], (3) Balerma irrigation network, Almeria, Spain (incl. 447 nodes) [50].

Table A1. *Cont.*

ID, Authors (Year) [Ref] SO/MO * Brief Description	Optimisation Model (Objective Functions +, Constraints **, Decision Variables ++)	Water Quality Network Analysis Optimisation Method	Notes
118. Sheikholeslami et al. (2016) [162] SO Optimal WDS design using a combined cuckoo-HS algorithm (CSHS) in a two-phase procedure.	Objective (1): Minimise (a) the design cost of the network (pipes), (b) penalty for violating the pressure constraint. Constraints: (1) Min pressure at the nodes. Decision variables: (1) Pipe diameters (discrete).	Water quality: N/A. Network analysis: EPANET. Optimisation method: CSHS algorithm.	• The proposed CSHS algorithm is a two-phase algorithm. It employs the cuckoo search (CS) algorithm in the first stage, and the HS algorithm in the second stage. To overcome the weaknesses of CS (i.e., slow convergence rate and no information exchange between the individuals of the population), some HS components are integrated with CS. HMCR enables CS to use a memory containing the search history, which assists in generating new solutions; PAR from HS serves as a mutation operator and speeds up the convergence. • A self-adaptive technique is used to adjust HMCR and PAR during the optimisation process to alter the performance of the algorithm. • Dynamic penalty factor which increases towards the end of the optimisation process is used. • Sensitivity analysis is performed for the main parameters of the algorithm (scaling factor, discovering probability of alien eggs/solutions) using the Hanoi network. • Results: CSHS outperformed the standard CS and the majority of other meta-heuristics previously applied to the test networks in terms of efficiency. • Test networks: (1) Hanoi network (incl. 32 nodes) [49], (2) double Hanoi network (incl. 62 nodes), (3) Balerma irrigation network, Almeria, Spain (incl. 447 nodes) [50], (4) network of a town in southeast China (incl. 192 demand nodes) [281].
119. Zheng et al. (2016) [28] MO Optimal WDS design and strengthening, analysis and comparison of the searching behaviour of NSGA-II, self-adaptive multi-objective DE (SAMODE) and Borg.	Objective (1): Minimise (a) the total network cost, including pipe material and construction costs. Objective (2): Maximise (a) the network resilience. Constraints: (1) Min/max pressure at the nodes, (2) min/max velocity in the pipes. Decision variables: (1) Pipe diameters (discrete). Note: One MO model including both objectives.	Water quality: N/A. Network analysis: EPANET. Optimisation method: NSGA-II, SAMODE, and Borg are compared.	• An extension of the paper by [195] analysing the run-time searching behaviour of MOEAs to understand how they arrive at their final performance. Six performance metrics, categorised as solution quality, spacing and convergence metrics, are used to measure algorithm's search effectiveness and convergence properties in both the objective and decision spaces. • The relationship between algorithm operators and behavioural properties is analysed. • Results: A fundamental understanding of the working mechanisms of MOEAs is developed. Guidance on the selection of appropriate algorithms (operators) for particular optimisation problems is provided. NSGA-II is good at obtaining solutions covering a large extent of the Pareto front, and Borg is a good choice when computational resources are limited. • Test networks: (1) New York City tunnels (incl. 20 nodes) [81], (2) Hanoi network (incl. 32 nodes) [49], (3) Fossolo network, Italy (incl. 37 nodes) [254], (4) Pescara network, Italy (incl. 71 nodes) [254], (5) Modena network, Italy (incl. 272 nodes) [254], (6) Balerma irrigation network, Almeria, Spain (incl. 447 nodes) [50].

Table A1. *Cont.*

ID, Authors (Year) [Ref] SO/MO *, Brief Description	Optimisation Model (Objective Functions +, Constraints **, Decision Variables ++)	Water Quality, Network Analysis, Optimisation Method	Notes
120. Avila-Melgar et al. (2017) [109] SO Optimal WDS design using EA in a grid computing environment.	Objective (1): Minimise (a) the design cost of the network (pipes). Constraints: (1) Min/max pressure at the nodes, (2) min/max velocity in the pipes. Decision variables: (1) Pipe diameters (discrete).	Water quality: N/A. Network analysis: EPANET. Optimisation method: EA.	• An evolutionary method is coupled with EPANET to create an EA for solving water distribution network design (EA-WDND) problems. • The method is implemented in a grid environment and uses parallel computing techniques. • Results: EA-WDND obtains the best-known solution for the two-loop network. The best solution found for the Balerma network is an improvement of 12.5% over the current best-known solution. • Test networks: (1) Two-loop network supplied by gravity (incl. 7 nodes) [14], (2) Hanoi network (incl. 32 nodes) [49], (3) Balerma irrigation network, Almeria, Spain (incl. 447 nodes) [50].
121. Cisty et al. (2017) [30] MO Optimal WDS design using NSGA-II with a two-phase procedure and search space reduction.	Objective (1): Minimise (a) the design cost of the network (pipes). Objective (2): Minimise (a) the total head deficit in the network. Constraints: N/A. Decision variables: (1) Pipe diameters (discrete). Note: One MO model including both objectives.	Water quality: N/A Network analysis: Not specified. Optimisation method: NSGA-II (for both phases of the optimisation procedure).	• A two-phase optimisation procedure is proposed as follows: in the first phase, suboptimal solutions are searched for; in the second phase, the optimisation problem is solved with a reduced search space based on these solutions. • The first phase consists of running NSGA-II several times with varying parameters (population size, number of generations, crossover and mutation). The aim is to obtain different suboptimal solutions. • The second phase has the following two alternatives: (i) diameters from the first phase's suboptimal solutions are used; (ii) flows in the pipes from suboptimal solutions are used. In both cases, the search space is reduced by introducing upper and lower bounds of diameters for all the pipes based on the diameters and flows obtained in the first phase. • The recommendations regarding the use of the proposed methodology are as follows. If a solution with the lowest cost possible is sought after, perform approximately 10 optimisation runs in the first phase and subsequently use the first alternative (with diameters) of the second phase. If a solution with the shortest computational time is required, perform only one optimisation run in the first phase and subsequently use the second alternative (based on the flows) of the second phase. • Results: Compared with previous results from the literature, the proposed methodology displays a slightly better performance in terms of the cost as well as the computational effort. The key finding from the computational experiments is that it is possible to obtain competitive results with simple, existing optimisation methods provided their adequate and methodological utilisation. • Test networks: (1) Balerma irrigation network, Almeria, Spain (incl. 447 nodes) [50].

Table A1. *Cont.*

ID. Authors (Year) [Ref] SO/MO* Brief Description	Optimisation Model (Objective Functions+, Constraints**, Decision Variables++)	Water Quality Network Analysis Optimisation Method	Notes
122. Muhammed et al. (2017) [90] MO Optimal WDS strengthening using a cluster-based technique and NSGA-II in a two-phase procedure.	Objective (1): Minimise (a) the total capital cost of duplicated pipes. Objective (2): Minimise (a) the total number of demand nodes with pressure below the minimum pressure requirement. Constraints: (1) The sum of the pressure deficiencies in all the nodes with negative pressure. Decision variables: (1) Pipe diameters (discrete). Note: One MO model including both objectives.	Water quality: N/A. Network analysis: EPANET. Optimisation method: GANETXL [208] using NSGA-II.	• The optimisation procedure consists of the following two phases: (i) the network is partitioned into a number of clusters (subsystems); (ii) the pipes which can have a direct impact on system performance are identified and considered as design variables in the optimisation. • The network is mapped into an undirected graph. For network clustering, the modularity-based method is applied to divide the graph into clusters with stronger internal than external connections. • The clustering method is implemented using an open source program Gephi [287], widely used for graph network visualisation. • The only rehabilitation option considered is pipe duplication. Candidate pipes for rehabilitation are selected based on three strategies: (i) rehabilitation of intercluster water transmission pipes with pressure deficiencies; (ii) rehabilitation of feed pipelines between the clusters with pressure deficiencies, or pipes in the path between sources and clusters; (iii) the combination of the previous two strategies. Results: Strategy (iii) generated a Pareto front which dominates the Pareto fronts obtained by the other two strategies. It also shows a better performance when compared with the whole search space (all pipes used as design variables) and engineering judgement-based optimisation strategies. • Test networks: (1) EXNET water network (incl. 1891 nodes) [82].
123. Shokoohi et al. (2017) [78] MO Optimal WDS design including water quality objective using ACO.	Objective (1): Minimise (a) the construction cost of the network (pipe cost, excavation, demolition etc.), (b) chlorine cost calculated as one-year chlorine usage (applied in the tanks). Objective (2A): Maximise (a) water quality reliability based on chlorine residual [145]. Objective (2B): Maximise (a) water quality reliability based on water age. Objective (2C): Maximise (a) combined water quality reliability based on both chlorine residual and water age. Constraints: (1) Min/max pressure at the nodes, (2) max velocity in the pipes. Decision variables: (1) Pipe diameters (discrete), (2) tank heads (discrete), (3) chlorine injection dosages in the tanks (discrete). Note: Three two-objective optimisation models, where the objective (1) is combined with either objective (2A), (2B) or (2C).	Water quality: Chlorine, water age. Network analysis: EPANET (EPS). Optimisation method: ACO.	• The aim is to investigate the effect of water quality on WDS design. A new water age penalty curve is developed. The existing chlorine residual penalty curve [288] is used. • Project lifetime considered is 22 years. • The following four scenarios are analysed, all of them using EPS: (i) Hydraulic analysis is based on demand-driven simulation method (DDSM), objectives (1) and (2A) are used. (ii) Hydraulic analysis is based on head-driven simulation method (HDSM), min pressure constraint is not considered, another constraint to secure at least 95% supply of water demand is applied, objectives (1) and (2A) are used. (iii) DDSM method is used, objectives (1) and (2B) are considered. (iv) DDSM method is used, objectives (1) and (2C) are considered, hence both chlorine residual and water age are used as the water quality parameters. Results: Scenario (i) offers cheaper solutions than the original design (i.e., already constructed in Jahrom). Scenario (ii) has cheaper solutions than scenario (i), but there is a risk of pressure deficit at some nodes. Scenario (iii) offers only marginal improvement in the reliability objective with a relatively significant increase in the construction costs. In scenario (iv), all the differences between solutions are in chlorine reliability, so water age reliability does not have any significant impact on solutions. • Test networks: (1) Jahrom WDS, zone 3, South of Iran (incl. 44 nodes).

Table A1. *Cont.*

ID. Authors (Year) [Ref] SO/MO * Brief Description	Optimisation Model (Objective Functions +, Constraints **, Decision Variables ++)	Water Quality Network Analysis Optimisation Method	Notes
124. Zheng et al. (2017) [176] SO Optimal WDS design and strengthening using convergence-trajectory controlled ACO (ACO$_{CTC}$) algorithm with parameter-adaptive strategy.	Objective (1): Minimise (a) the design cost of the network (pipes), (b) penalty for violating the pressure constraint. Constraints: (1) Min pressure at the nodes. Decision variables: (1) Pipe diameters (discrete).	Water quality: N/A. Network analysis: EPANET. Optimisation method: ACO$_{CTC}$.	• Parameter-adaptive strategy for ACO algorithms is developed, which enables pre-specified parameter trajectories to be followed and ensures the convergence to increasingly higher fitness subregions in decision space for a given computational budget. The algorithm parameters are automatically adjusted to balance search diversification and intensification (exploration and exploitation). • ACO$_{CTC}$ is based on AS$_{rank}$ [234]. • A total of eight different convergence trajectories (ranging from emphasis on high diversification to high intensification) and three computational budgets (low, moderate and high) are applied to six test networks. • Results: There is a strong relationship between the convergence trajectory in decision space and the searching quality in objective space. The convergence trajectories can significantly impact on the solution quality. The trajectory with a slight emphasis on intensification performed best overall, irrespective of the computational budget. New best-known solutions were found for the Pescara, and Kang and Lansey's test networks. • Test networks: (1) New York City tunnels (NYTP) (incl. 20 nodes) [81], (2) Hanoi network (HP) (incl. 32 nodes) [49], (3) Fossolo network (FOS), Italy (incl. 37 nodes) [254], (4) Pescara network (PES), Italy (incl. 71 nodes) [254], (5) Balerma irrigation network (BN), Almeria, Spain (incl. 447 nodes) [50], (6) Kang and Lansey's network (KL) (incl. 936 nodes) [26].

Notes: * SO = Single-objective (approach/model), MO = Multi-objective (approach/model). + Objective function is referred to as 'objective' in the column below due to space savings. ** Conservation of mass of flow, conservation of energy, and conservation of mass of constituent (for water quality network analysis) are not listed. ++ Control variables are listed, state variables resulting from network hydraulics are not necessarily listed. ? D = Design. ?? OP = Operation.

References

1. Mays, L.W. *Reliability Analysis of Water Distribution Systems*; American Society of Civil Engineers (ASCE): New York, NY, USA, 1989.
2. Tuttle, G.W. The economic velocity of transmission of water through pipes. *Eng. REC* **1895**, *XXXII*, 258.
3. True, A.O. Economical sizes for water force mains. *J. AWWA* **1937**, *29*, 536–547.
4. Braca, R.M.; Happel, J. New cost data bring economic pipe sizing up to date. *Chem. Eng.* **1953**, *60*, 180–187.
5. Genereaux, R.P. Fluid-flow design methods. *Ind. Eng. Chem.* **1937**, *29*, 385–388. [CrossRef]
6. Camp, T.R. Economic pipe sizes for water distribution systems. *Trans. Am. Soc. Civ. Eng.* **1939**, *104*, 190–213.
7. Lischer, V.C. Determination of economical pipe diameters in distribution systems. *J. AWWA* **1948**, *40*, 849–867.
8. Aldrich, E.H. Solution of transmission problems of a water system. *Trans. Am. Soc. Civ. Eng.* **1937**, *103*, 1579–1619.
9. Ormsbee, L.E. The history of water distribution network analysis: The computer age. In Proceedings of the 8th Annual Water Distribution Systems Analysis Symposium, Cincinnati, OH, USA, 27–30 August 2006.
10. Martin, D.W.; Peters, G. The application of newton's method to network analysis by digital computer. *J. Inst. Water Eng.* **1963**, *17*, 115–129.
11. Todini, E.; Pilati, S. A gradient algorithm for the analysis of pipe networks. In *Computer Applications in Water Supply: Vol. 1—System Analysis and Simulation*; Coulbeck, B., Orr, C.H., Eds.; Wiley: London, UK, 1988; pp. 1–20.
12. Wood, D.J. *Computer Analysis of Flow in Pipe Networks Including Extended Period Simulations: User's Manual*; Office of Continuing Education and Extension of the College of Engineering of the University of Kentucky: Lexington, KY, USA, 1980.
13. Rossman, L.A. *Epanet Users Manual*; Risk Reduction Engineering Laboratory, U.S. Environmental Protection Agency (EPA): Cincinnati, OH, USA, 1993.
14. Alperovits, E.; Shamir, U. Design of optimal water distribution systems. *Water Resour. Res.* **1977**, *13*, 885–900. [CrossRef]
15. Su, Y.C.; Mays, L.W.; Duan, N.; Lansey, K.E. Reliability-based optimization model for water distribution systems. *J. Hydraul. Eng.* **1987**, *113*, 1539–1556. [CrossRef]
16. Lansey, K.E.; Mays, L.W. Optimization model for water distribution system design. *J. Hydraul. Eng. ASCE* **1989**, *115*, 1401–1418. [CrossRef]
17. Kim, J.H.; Mays, L.W. Optimal rehabilitation model for water-distribution systems. *J. Water Resour. Plan. Manag. ASCE* **1994**, *120*, 674–692. [CrossRef]
18. Cembrowicz, R.G.; Krauter, G.W. Optimization of urban and regional water supply systems. In Proceedings of the Conference on Systems Approach for Development, IFAC, Cairo, Egypt, 26–29 November 1977.
19. Goldberg, D.E.; Kuo, C.H. Genetic algorithms in pipeline optimization. In Proceedings of the PSIG Annual Meeting, Albuquerque, NM, USA, 24–25 October 1985.
20. Simpson, A.R.; Dandy, G.C.; Murphy, L.J. Genetic algorithms compared to other techniques for pipe optimization. *J. Water Resour. Plan. Manag. ASCE* **1994**, *120*, 423–443. [CrossRef]
21. Nicklow, J.; Reed, P.; Savic, D.; Dessalegne, T.; Harrell, L.; Chan-Hilton, A.; Karamouz, M.; Minsker, B.; Ostfeld, A.; Singh, A.; et al. State of the art for genetic algorithms and beyond in water resources planning and management. *J. Water Resour. Plan. Manag. ASCE* **2010**, *136*, 412–432. [CrossRef]
22. Mala-Jetmarova, H.; Sultanova, N.; Savic, D. Lost in optimisation of water distribution systems? A literature review of system operation. *Environ. Model. Softw.* **2017**, *93*, 209–254. [CrossRef]
23. Maier, H.R.; Kapelan, Z.; Kasprzyk, J.; Kollat, J.; Matott, L.S.; Cunha, M.C.; Dandy, G.C.; Gibbs, M.S.; Keedwell, E.; Marchi, A.; et al. Evolutionary algorithms and other metaheuristics in water resources: Current status, research challenges and future directions. *Environ. Model. Softw.* **2014**, *62*, 271–299. [CrossRef]
24. Walski, T.M. The wrong paradigm—Why water distribution optimization doesn't work. *J. Water Resour. Plan. Manag. ASCE* **2001**, *127*, 203–205. [CrossRef]
25. Pareto, V. *Cours D'economie Politique*; F. Rouge: Lausanne, Switzerland, 1896.
26. Kang, D.S.; Lansey, K. Revisiting optimal water-distribution system design: Issues and a heuristic hierarchical approach. *J. Water Resour. Plan. Manag. ASCE* **2012**, *138*, 208–217. [CrossRef]
27. Bi, W.; Dandy, G. Optimization of water distribution systems using online retrained metamodels. *J. Water Resour. Plan. Manag. ASCE* **2014**, *140*. [CrossRef]

28. Zheng, F.; Zecchin, A.C.; Maier, H.R.; Simpson, A.R. Comparison of the searching behavior of nsga-ii, samode, and borg moeas applied to water distribution system design problems. *J. Water Resour. Plan. Manag.* **2016**, *142*. [CrossRef]

29. McClymont, K.; Keedwell, E.; Savic, D. An analysis of the interface between evolutionary algorithm operators and problem features for water resources problems. A case study in water distribution network design. *Environ. Model. Softw.* **2015**, *69*, 414–424. [CrossRef]

30. Cisty, M.; Bajtek, Z.; Celar, L. A two-stage evolutionary optimization approach for an irrigation system design. *J. Hydroinform.* **2017**, *19*, 115–122. [CrossRef]

31. Shamir, U. Optimization in water distribution systems engineering. *Eng. Optim.* **1979**, *11*, 65–84.

32. Walski, T.M. State-of-the-art pipe network optimization. In *Computer Applications in Water Resources*; ASCE: Buffalo, NY, USA, 1985; pp. 559–568.

33. Lansey, K.E.; Mays, L.W. Optimization models for design of water distribution systems. In *Reliability Analysis of Water Distribution Systems*; Mays, L.R., Ed.; ASCE: New York, NY, USA, 1989; pp. 37–84.

34. Goulter, I.C. Systems analysis in water-distribution network design: From theory to practice. *J. Water Resour. Plan. Manag. ASCE* **1992**, *118*, 238–248. [CrossRef]

35. Walters, G.A. A review of pipe network optimization techniques. In *Pipeline Systems*; Springer: Dordrecht, The Netherlands, 1992; pp. 3–13.

36. Dandy, G.C.; Simpson, A.R.; Murphy, L.J. A review of pipe network optimisation techniques. In Proceedings of the Watercomp 93: 2nd Australasian Conference on Computing for the Water Industry Today and Tomorrow, Melbourne, Australia, 30 March–1 April 1993.

37. Engelhardt, M.O.; Skipworth, P.J.; Savic, D.A.; Saul, A.J.; Walters, G.A. Rehabilitation strategies for water distribution networks: A literature review with a uk perspective. *Urban Water* **2000**, *2*, 153–170. [CrossRef]

38. Lansey, K.E. The evolution of optimizing water distribution system applications. In Proceedings of the 8th Annual Water Distribution Systems Analysis Symposium, Cincinnati, OH, USA, 27–30 August 2006.

39. De Corte, A.; Sörensen, K. Optimisation of gravity-fed water distribution network design: A critical review. *Eur. J. Oper. Res.* **2013**, *228*, 1–10. [CrossRef]

40. Walski, T.M.; Chase, D.V.; Savic, D.A.; Grayman, W.; Beckwith, S.; Koelle, E. *Advanced Water Distribution Modeling and Management*; Haestad Methods Press: Waterbury, CT, USA, 2003.

41. Rossman, L.A. *Epanet 2 Users Manual*; U.S. Environmental Protection Agency (EPA): Cincinnati, OH, USA, 2000.

42. Savic, D.A.; Walters, G.A. Genetic algorithms for least-cost design of water distribution networks. *J. Water Resour. Plan. Manag. ASCE* **1997**, *123*, 67–77. [CrossRef]

43. Vairavamoorthy, K.; Ali, M. Optimal design of water distribution systems using genetic algorithms. *Comput. Aided Civ. Infrastruct. Eng.* **2000**, *15*, 374–382. [CrossRef]

44. Keedwell, E.; Khu, S.-T. A hybrid genetic algorithm for the design of water distribution networks. *Eng. Appl. Artif. Intell.* **2005**, *18*, 461–472. [CrossRef]

45. Kadu, M.S.; Gupta, R.; Bhave, P.R. Optimal design of water networks using a modified genetic algorithm with reduction in search space. *J. Water Resour. Plan. Manag. ASCE* **2008**, *134*, 147–160. [CrossRef]

46. Zheng, F.; Simpson, A.R.; Zecchin, A.C. A decomposition and multistage optimization approach applied to the optimization of water distribution systems with multiple supply sources. *Water Resour. Res.* **2013**, *49*, 380–399. [CrossRef]

47. Saldarriaga, J.; Páez, D.; León, N.; López, L.; Cuero, P. Power use methods for optimal design of wds: History and their use as post-optimization warm starts. *J. Hydroinform.* **2015**, *17*, 404–421. [CrossRef]

48. Zheng, F.; Zecchin, A.C.; Simpson, A.R. Investigating the run-time searching behavior of the differential evolution algorithm applied to water distribution system optimization. *Environ. Model. Softw.* **2015**, *69*, 292–307. [CrossRef]

49. Fujiwara, O.; Khang, D.B. A two-phase decomposition method for optimal design of looped water distribution networks. *Water Resour. Res.* **1990**, *26*, 539–549. [CrossRef]

50. Reca, J.; Martínez, J. Genetic algorithms for the design of looped irrigation water distribution networks. *Water Resour. Res.* **2006**, *42*. [CrossRef]

51. Samani, H.M.V.; Mottaghi, A. Optimization of water distribution networks using integer linear programming. *J. Hydraul. Eng. ASCE* **2006**, *132*, 501–509. [CrossRef]

52. Goncalves, G.M.; Gouveia, L.; Pato, M.V. An improved decomposition-based heuristic to design a water distribution network for an irrigation system. *Ann. Oper. Res.* **2011**, *219*, 1–27. [CrossRef]

53. Ostfeld, A. Optimal design and operation of multiquality networks under unsteady conditions. *J. Water Resour. Plan. Manag. ASCE* **2005**, *131*, 116–124. [CrossRef]
54. Ostfeld, A.; Tubaltzev, A. Ant colony optimization for least-cost design and operation of pumping water distribution systems. *J. Water Resour. Plan. Manag. ASCE* **2008**, *134*, 107–118. [CrossRef]
55. De Corte, A.; Sörensen, K. Hydrogen: An artificial water distribution network generator. *Water Resour. Manag.* **2014**, *28*, 333–350. [CrossRef]
56. Möderl, M.; Sitzenfrei, R.; Fetz, T.; Fleischhacker, E.; Rauch, W. Systematic generation of virtual networks for water supply. *Water Resour. Res.* **2011**, *47*. [CrossRef]
57. Trifunović, N.; Maharjan, B.; Vairavamoorthy, K. Spatial network generation tool for water distribution network design and performance analysis. *Water Sci. Technol.* **2012**, *13*, 1–19. [CrossRef]
58. Marchi, A.; Salomons, E.; Ostfeld, A.; Kapelan, Z.; Simpson, A.; Zecchin, A.; Maier, H.; Wu, Z.; Elsayed, S.; Song, Y.; et al. The battle of the water networks ii (bwn-ii). *J. Water Resour. Plan. Manag. ASCE* **2014**, *140*. [CrossRef]
59. Giustolisi, O.; Berardi, L.; Laucelli, D.; Savic, D.; Kapelan, Z. Operational and tactical management of water and energy resources in pressurized systems: Competition at wdsa 2014. *J. Water Resour. Plan. Manag. ASCE* **2015**, *142*. [CrossRef]
60. Costa, A.L.H.; Medeiros, J.L.; Pessoa, F.L.P. Optimization of pipe networks including pumps by simulated annealing. *Br. J. Chem. Eng.* **2000**, *17*, 887–896. [CrossRef]
61. Perelman, L.; Ostfeld, A. An adaptive heuristic cross-entropy algorithm for optimal design of water distribution systems. *Eng. Optim.* **2007**, *39*, 413–428. [CrossRef]
62. Perelman, L.; Ostfeld, A.; Salomons, E. Cross entropy multiobjective otimization for water distribution systems design. *Water Resour. Res.* **2008**, *44*. [CrossRef]
63. Halhal, D.; Walters, G.A.; Ouazar, D.; Savic, D.A. Water network rehabilitation with structured messy genetic algorithm. *J. Water Resour. Plan. Manag. ASCE* **1997**, *123*, 137–146. [CrossRef]
64. Atiquzzaman, M.; Liong, S.-Y.; Yu, X. Alternative decision making in water distribution network with nsga-II. *J. Water Resour. Plan. Manag. ASCE* **2006**, *132*, 122–126. [CrossRef]
65. Farmani, R.; Savic, D.A.; Walters, G.A. Evolutionary multi-objective optimization in water distribution network design. *Eng. Optim.* **2005**, *37*, 167–183. [CrossRef]
66. Keedwell, E.; Khu, S.-T. A novel evolutionary meta-heuristic for the multi-objective optimization of real-world water distribution networks. *Eng. Optim.* **2006**, *38*, 319–333. [CrossRef]
67. Di Pierro, F.; Khu, S.-T.; Savic, D.; Berardi, L. Efficient multi-objective optimal design of water distribution networks on a budget of simulations using hybrid algorithms. *Environ. Model. Softw.* **2009**, *24*, 202–213. [CrossRef]
68. McClymont, K.; Keedwell, E.C.; Savić, D.; Randall-Smith, M. Automated construction of evolutionary algorithm operators for the bi-objective water distribution network design problem using a genetic programming based hyper-heuristic approach. *J. Hydroinform.* **2014**, *16*, 302–318. [CrossRef]
69. Zheng, F.; Simpson, A.; Zecchin, A. Improving the efficiency of multi-objective evolutionary algorithms through decomposition: An application to water distribution network design. *Environ. Model. Softw.* **2015**, *69*, 240–252. [CrossRef]
70. Artina, S.; Bragalli, C.; Erbacci, G.; Marchi, A.; Rivi, M. Contribution of parallel nsga-ii in optimal design of water distribution networks. *J. Hydroinform.* **2012**, *14*, 310–323. [CrossRef]
71. Wu, W.; Simpson, A.R.; Maier, H.R. Multi-objective genetic algorithm optimisation of water distribution systems accounting for sustainability. In *Proceedings of Water Down Under 2008*; Engineers Australia: Barton, Australia, 2008.
72. Wu, W.; Simpson, A.; Maier, H. Accounting for greenhouse gas emissions in multiobjective genetic algorithm optimization of water distribution systems. *J. Water Resour. Plan. Manag ASCE* **2010**, *136*, 146–155. [CrossRef]
73. Wu, W.; Simpson, A.R.; Maier, H.R. Sensitivity of optimal tradeoffs between cost and greenhouse gas emissions for water distribution systems to electricity tariff and generation. *J. Water Resour. Plan. Manag. ASCE* **2011**, *138*, 182–186. [CrossRef]
74. Wu, W.; Simpson, A.R.; Maier, H.R.; Marchi, A. Incorporation of variable-speed pumping in multiobjective genetic algorithm optimization of the design of water transmission systems. *J. Water Resour. Plan. Manag. ASCE* **2012**, *138*, 543–552. [CrossRef]

75. Stokes, C.S.; Simpson, A.R.; Maier, H.R. A computational software tool for the minimization of costs and greenhouse gas emissions associated with water distribution systems. *Environ. Model. Softw.* **2015**, *69*, 452–467. [CrossRef]

76. Stokes, C.S.; Maier, H.R.; Simpson, A.R. Effect of storage tank size on the minimization of water distribution system cost and greenhouse gas emissions while considering time-dependent emissions factors. *J. Water Resour. Plan. Manag. ASCE* **2015**, *142*. [CrossRef]

77. Wu, W.; Maier, H.R.; Simpson, A.R. Single-objective versus multiobjective optimization of water distribution systems accounting for greenhouse gas emissions by carbon pricing. *J. Water Resour. Plan. Manag. ASCE* **2010**, *136*, 555–565. [CrossRef]

78. Shokoohi, M.; Tabesh, M.; Nazif, S.; Dini, M. Water quality based multi-objective optimal design of water distribution systems. *Water Resour. Manag.* **2017**, *31*, 93–108. [CrossRef]

79. Beygi, S.; Bozorg Haddad, O.; Fallah-Mehdipour, E.; Marino, M.A. Bargaining models for optimal design of water distribution networks. *J. Water Resour. Plan. Manag. ASCE* **2014**, *140*, 92–99. [CrossRef]

80. Bhattacharjee, K.; Singh, H.; Ryan, M.; Ray, T. Bridging the gap: Many-objective optimization and informed decision-making. *IEEE Trans. Evol. Comput.* **2017**, *21*, 813–820. [CrossRef]

81. Schaake, J.C.; Lai, D. *Linear Programming and Dynamic Programming Application to Water Distribution Network Design*; Report No. 116; Hydrodynamics Laboratory, Department of Civil Engineering, Massachusetts Institute of Technology: Cambridge, MA, USA, 1969.

82. Farmani, R.; Savic, D.A.; Walters, G.A. "Exnet" benchmark problem for multi-objective optimization of large water systems. In Proceedings of the Modelling and Control for Participatory Planning and Managing Water Systems, IFAC Workshop, Venice, Italy, 29 September–1 October 2004.

83. Gessler, J. Pipe network optimization by enumeration. In *Computer Applications for Water Resources*; Torno, H.C., Ed.; ASCE: New York, NY, USA, 1985; pp. 572–581.

84. Walski, T.M.; Brill, E.D.J.; Gessler, J.; Goutler, I.C.; Jeppson, R.M.; Lansey, K.; Lee, H.-L.; Liebman, J.C.; Mays, L.; Morgan, D.R.; et al. Battle of network models: Epilogue. *J. Water Resour. Plan. Manag. ASCE* **1987**, *113*, 191–203. [CrossRef]

85. Dandy, G.C.; Simpson, A.R.; Murphy, L.J. An improved genetic algorithm for pipe network optimization. *Water Resour. Res.* **1996**, *32*, 449–458. [CrossRef]

86. Montesinos, P.; Garcia-Guzman, A.; Ayuso, J.L. Water distribution network optimization using a modified genetic algorithm. *Water Resour. Res.* **1999**, *35*, 3467–3473. [CrossRef]

87. Broad, D.R.; Dandy, G.C.; Maier, H.R. Water distribution system optimization using metamodels. *J. Water Resour. Plan. Manag. ASCE* **2005**, *131*, 172–180. [CrossRef]

88. Wu, Z.Y.; Simpson, A.R. A self-adaptive boundary search genetic algorithm and its application to water distribution systems. *J. Hydraul. Res.* **2002**, *40*, 191–203. [CrossRef]

89. Babayan, A.V.; Savic, D.A.; Walters, G.A. Multiobjective optimisation of water distribution system design under uncertain demand and pipe roughness. In *Topics on System Analysis and Integrated Water Resources Management*; Castelletti, A., Soncini-Sessa, R., Eds.; Elsevier: Amsterdam, The Netherlands, 2007; pp. 161–172.

90. Muhammed, K.; Farmani, R.; Behzadian, K.; Diao, K.; Butler, D. Optimal rehabilitation of water distribution systems using a cluster-based technique. *J. Water Resour. Plan. Manag. ASCE* **2017**, *143*. [CrossRef]

91. Morrison, R.; Sangster, T.; Downey, D.; Matthews, J.; Condit, W.; Sinha, S.; Maniar, S.; Sterling, R. *State of Technology for Rehabilitation of Water Distribution Systems*; U.S. Environmental Protection Agency (EPA): Edison, NJ, USA, 2013.

92. Kanta, L.; Zechman, E.; Brumbelow, K. Multiobjective evolutionary computation approach for redesigning water distribution systems to provide fire flows. *J. Water Resour. Plan. Manag. ASCE* **2012**, *138*, 144–152. [CrossRef]

93. Vamvakeridou-Lyroudia, L.S.; Walters, G.A.; Savic, D.A. Fuzzy multiobjective optimization of water distribution networks. *J. Water Resour. Plan. Manag. ASCE* **2005**, *131*, 467–476. [CrossRef]

94. Barkdoll, B.D.; Murray, K.; Sherrin, A.; O'Neill, J.; Ghimire, S.R. Effective-power-ranking algorithm for energy and greenhouse gas reduction in water distribution systems through pipe enhancement. *J. Water Resour. Plan. Manag. ASCE* **2015**. [CrossRef]

95. Jin, X.; Zhang, J.; Gao, J.-L.; Wu, W.-Y. Multi-objective optimization of water supply network rehabilitation with non-dominated sorting genetic algorithm-II. *J. Zhejiang Univ. Sci. A* **2008**, *9*, 391–400. [CrossRef]

96. Murphy, L.J.; Dandy, G.C.; Simpson, A.R. Optimum design and operation of pumped water distribution systems. In Proceedings of the 5th International Conference on Hydraulics in Civil Engineering, Brisbane, Australia, 15–17 February 1994.

97. Fu, G.; Kapelan, Z.; Kasprzyk, J.; Reed, P. Optimal design of water distribution systems using many-objective visual analytics. *J. Water Resour. Plan. Manag. ASCE* **2013**, *139*, 624–633. [CrossRef]

98. Prasad, T.D. Design of pumped water distribution networks with storage. *J. Water Resour. Plan. Manag. ASCE* **2010**, *136*, 129–132. [CrossRef]

99. Walters, G.A.; Halhal, D.; Savic, D.; Ouazar, D. Improved design of "anytown" distribution network using structured messy genetic algorithms. *Urban Water* **1999**, *1*, 23–38. [CrossRef]

100. Rogers, C.K.; Randall-Smith, M.; Keedwell, E.; Diduch, R. Application of Optimization Technology to Water Distribution System Master Planning. In Proceedings of the World Environmental and Water Resources Congress 2009: Great Rivers, Kansas City, MO, USA, 17–21 May 2009; Volume 342, pp. 375–384.

101. Wolpert, D.H.; Macready, W.G. No free lunch theorems for optimization. *IEEE Trans. Evol. Comput.* **1997**, *1*, 67–82. [CrossRef]

102. Cunha, M.C.; Sousa, J. Water distribution network design optimization: Simulated annealing approach. *J. Water Resour. Plan. Manag. ASCE* **1999**, *125*, 215–221. [CrossRef]

103. Eusuff, M.M.; Lansey, K.E. Optimization of water distribution network design using the shuffled frog leaping algorithm. *J. Water Resour. Plan. Manag. ASCE* **2003**, *129*, 210–225. [CrossRef]

104. Maier, H.R.; Simpson, A.R.; Zecchin, A.C.; Foong, W.K.; Phang, K.Y.; Seah, H.Y.; Tan, C.L. Ant colony optimization for design of water distribution systems. *J. Water Resour. Plan. Manag. ASCE* **2003**, *129*, 200–209. [CrossRef]

105. Geem, Z.W. Optimal cost design of water distribution networks using harmony search. *Eng. Optim.* **2006**, *38*, 259–277. [CrossRef]

106. Suribabu, C.R.; Neelakantan, T.R. Design of water distribution networks using particle swarm optimization. *Urban Water J.* **2006**, *3*, 111–120. [CrossRef]

107. Banos, R.; Gil, C.; Reca, J.; Montoya, F.G. A memetic algorithm applied to the design of water distribution networks. *Appl. Soft Comput.* **2010**, *10*, 261–266. [CrossRef]

108. Bi, W.; Dandy, G.C.; Maier, H.R. Improved genetic algorithm optimization of water distribution system design by incorporating domain knowledge. *Environ. Model. Softw.* **2015**, *69*, 370–381. [CrossRef]

109. Avila-Melgar, E.Y.; Cruz-Chávez, M.A.; Martinez-Bahena, B. General methodology for using epanet as an optimization element in evolutionary algorithms in a grid computing environment for water distribution network design. *Water Sci. Technol.* **2017**, *17*, 39–51. [CrossRef]

110. Lansey, K.E.; Duan, N.; Mays, L.W.; Tung, Y.K. Water distribution system design under uncertainties. *J. Water Resour. Plan. Manag. ASCE* **1989**, *115*, 630–645. [CrossRef]

111. Zheng, F.; Simpson, A.R.; Zecchin, A.C. A combined nlp-differential evolution algorithm approach for the optimization of looped water distribution systems. *Water Resour. Res.* **2011**, *47*. [CrossRef]

112. Tospornsampan, J.; Kita, I.; Ishii, M.; Kitamura, Y. Split-pipe design of water distribution network using simulated annealing. *Int. J. Comput. Inf. Syst. Sci. Eng.* **2007**, *1*, 153–163.

113. Cisty, M. Hybrid genetic algorithm and linear programming method for least-cost design of water distribution systems. *Water Resour. Manag.* **2010**, *24*, 1–24. [CrossRef]

114. Kessler, A.; Shamir, U. Analysis of the linear programming gradient method for optimal design of water supply networks. *Water Resour. Res.* **1989**, *25*, 1469–1480. [CrossRef]

115. Eiger, G.; Uri, S.; Ben-Tal, A. Optimal design of water distribution networks. *Water Resour. Res.* **1994**, *30*, 2637–2646. [CrossRef]

116. Loganathan, G.V.; Greene, J.J.; Ahn, T.J. Design heuristic for globally minimum cost water-distribution systems. *J. Water Resour. Plan. Manag. ASCE* **1995**, *121*, 182–192. [CrossRef]

117. Krapivka, A.; Ostfeld, A. Coupled genetic algorithm-linear programming scheme for last-cost pipe sizing of water-distribution systems. *J. Water Resour. Plan. Manag. ASCE* **2009**, *135*, 298–302. [CrossRef]

118. Creaco, E.; Franchini, M.; Walski, T. Accounting for phasing of construction within the design of water distribution networks. *J. Water Resour. Plan. Manag. ASCE* **2014**, *140*, 598–606. [CrossRef]

119. Dziedzic, R.; Karney, B.W. Cost gradient–based assessment and design improvement technique for water distribution networks with varying loads. *J. Water Resour. Plan. Manag. ASCE* **2015**. [CrossRef]

120. Dandy, G.; Hewitson, C. Optimizing hydraulics and water quality in water distribution networks using genetic algorithms. In *Building Partnerships-2000, Joint Conference on Water Resources Engineering and Water Resources Planning and Management*; Rollin, H.H., Michael, G., Eds.; ASCE: Reston, VA, USA, 2000.

121. Kang, D.; Lansey, K. Scenario-based robust optimization of regional water and wastewater infrastructure. *J. Water Resour. Plan. Manag. ASCE* **2013**, *139*, 325–338. [CrossRef]

122. Marques, J.; Cunha, M.; Savic, D. Using real options in the optimal design of water distribution networks. *J. Water Resour. Plan. Manag. ASCE* **2015**, *141*. [CrossRef]

123. Schwartz, R.; Housh, M.; Ostfeld, A. Least-cost robust design optimization of water distribution systems under multiple loading. *J. Water Resour. Plan. Manag. ASCE* **2016**, *142*. [CrossRef]

124. Roshani, E.; Filion, Y. Event-based approach to optimize the timing of water main rehabilitation with asset management strategies. *J. Water Resour. Plan. Manag. ASCE* **2014**, *140*. [CrossRef]

125. Batchabani, E.; Fuamba, M. Optimal tank design in water distribution networks: Review of literature and perspectives. *J. Water Resour. Plan. Manag. ASCE* **2014**, *140*, 136–145. [CrossRef]

126. Giustolisi, O.; Berardi, L.; Laucelli, D. Optimal water distribution network design accounting for valve shutdowns. *J. Water Resour. Plan. Manag. ASCE* **2014**, *140*, 277–287. [CrossRef]

127. Dandy, G.; Duncker, A.; Wilson, J.; Pedeux, X. An Approach for Integrated Optimization of Wastewater, Recycled and Potable Water Networks. In Proceedings of the World Environmental and Water Resources Congress 2009: Great Rivers, Kansas City, MO, USA, 17–21 May 2009; Volume 342, pp. 364–374.

128. Farley, M. *District Metering. Part 1-System Design and Installation*; WRc Engineering: Swindon, UK, 1985.

129. Schwarz, J.; Meidad, N.; Shamir, U. Water quality management in regional systems. In *Scientific Basis for Water Resources Management*; IAHS Publisher: Koblenz, Germany, 1985.

130. Dandy, G.C.; Engelhardt, M. Optimal scheduling of water pipe replacement using genetic algorithms. *J. Water Resour. Plan. Manag. ASCE* **2001**, *127*, 214–223. [CrossRef]

131. Halhal, D.; Walters, G.A.; Savic, D.A.; Ouazar, D. Scheduling of water distribution system rehabilitation using structured messy genetic algorithms. *Evol. Comput.* **1999**, *7*, 311–329. [CrossRef] [PubMed]

132. Roshani, E.; Filion, Y. Water distribution system rehabilitation under climate change mitigation scenarios in Canada. *J. Water Resour. Plan. Manag. ASCE* **2015**, *141*. [CrossRef]

133. Pellegrino, R.; Costantino, N.; Giustolisi, O. Flexible investment planning for water distribution networks. *J. Hydroinform.* **2017**. [CrossRef]

134. Zhang, W.; Chung, G.; Pierre-Louis, P.G.; Bayraksan, G.Z.; Lansey, K. Reclaimed water distribution network design under temporal and spatial growth and demand uncertainties. *Environ. Model. Softw.* **2013**, *49*, 103–117. [CrossRef]

135. Basupi, I.; Kapelan, Z. Flexible water distribution system design under future demand uncertainty. *J. Water Resour. Plan. Manag. ASCE* **2015**, *141*. [CrossRef]

136. Creaco, E.; Franchini, M.; Walski, T. Taking account of uncertainty in demand growth when phasing the construction of a water distribution network. *J. Water Resour. Plan. Manag. ASCE* **2015**, *141*. [CrossRef]

137. Marques, J.; Cunha, M.; Savić, D.A. Multi-objective optimization of water distribution systems based on a real options approach. *Environ. Model. Softw.* **2015**, *63*, 1–13. [CrossRef]

138. Lansey, K. Sustainable, robust, resilient, water distribution systems. In Proceedings of the 14th Water Distribution Systems Analysis Conference (WDSA 2012), Adelaide, Australia, 24–27 September 2012.

139. Zhuang, B.; Lansey, K.; Kang, D. Reliability/availability analysis of water distribution systems considering adaptive pump operation. In *World Environmental and Water Resources Congress 2011: Bearing Knowledge for Sustainability*; ASCE: Reston, VA, USA, 2011.

140. Babayan, A.V.; Savic, D.A.; Walters, G.A.; Kapelan, Z.S. Robust least-cost design of water distribution networks using redundancy and integration-based methodologies. *J. Water Resour. Plan. Manag.* **2007**, *133*, 67–77. [CrossRef]

141. Giustolisi, O.; Laucelli, D.; Colombo, A.F. Deterministic versus stochastic design of water distribution networks. *J. Water Resour. Plan. Manag. ASCE* **2009**, *135*, 117–127. [CrossRef]

142. Filion, Y.; Jung, B. Least-cost design of water distribution networks including fire damages. *J. Water Resour. Plan. Manag.* **2010**, *136*, 658–668. [CrossRef]

143. Andrade, M.A.; Choi, C.Y.; Lansey, K.; Jung, D. Enhanced artificial neural networks estimating water quality constraints for the optimal water distribution systems design. *J. Water Resour. Plan. Manag.* **2016**, *142*. [CrossRef]

144. McClymont, K.; Keedwell, E.; Savic, D.; Randall-Smith, M. A general multi-objective hyper-heuristic for water distribution network design with discolouration risk. *J. Hydroinform.* **2013**, *15*, 700–716. [CrossRef]

145. Gupta, R.; Dhapade, S.; Bhave, P.R. Water quality reliability analysis of water distribution networks. In Proceedings of the International Conference on Water Engineering for Sustainable Environment Organized by IAHR, Vancouver, BC, Canada, 9–14 August 2009; pp. 5607–5613.

146. Shang, F.; Uber, J.G.; Rossman, L.A. *Epanet Multi-Species Extension User's Manual*, Epa/600/s-07/021; U.S. Environmental Protection Agency: Cincinnati, OH, USA, 2011. Available online: https://cfpub.epa.gov/si/si_public_record_report.cfm?address=nhsrc/&dirEntryId=218488 (accessed on 18 August 2016).

147. Bragalli, C.; D'Ambrosio, C.; Lee, J.; Lodi, A.; Toth, P. On the optimal design of water distribution networks: A practical minlp approach. *Optim. Eng.* **2012**, *13*, 219–246. [CrossRef]

148. Deb, K. An efficient constraint handling method for genetic algorithms. *Comput. Methods Appl. Mech. Eng.* **2000**, *186*, 311–338. [CrossRef]

149. Zheng, F.; Zecchin, A.; Simpson, A.; Lambert, M. Non-crossover dither creeping mutation genetic algorithm for pipe network optimization. *J. Water Resour. Plan. Manag. ASCE* **2013**. [CrossRef]

150. Sheikholeslami, R.; Talatahari, S. Developed swarm optimizer: A new method for sizing optimization of water distribution systems. *J. Comput. Civ. Eng.* **2016**, *30*. [CrossRef]

151. Murphy, L.J.; Simpson, A.R. *Pipe Optimization Using Genetic Algorithms*; Department of Civil Engineering, University of Adelaide: Adelaide, Australia, 1992.

152. Van Dijk, M.; Van Vuuren, S.J.; Van Zyl, J.E. Optimising water distribution systems using a weighted penalty in a genetic algorithm. *Water SA* **2008**, *34*, 537–548.

153. Dandy, G.; Wilkins, A.; Rohrlach, H. A methodology for comparing evolutionary algorithms for optimising water distribution systems. In *Water Distribution Systems Analysis 2010, 12–15 September 2010*, 41203 ed.; ASCE: Tucson, Arizona, 2010.

154. Marchi, A.; Dandy, G.; Wilkins, A.; Rohrlach, H. Methodology for comparing evolutionary algorithms for optimization of water distribution systems. *J. Water Resour. Plan. Manag. ASCE* **2014**, *140*, 22–31. [CrossRef]

155. Johns, M.B.; Keedwell, E.; Savic, D. Adaptive locally constrained genetic algorithm for least-cost water distribution network design. *J. Hydroinform.* **2014**, *16*, 288–301. [CrossRef]

156. Zecchin, A.C.; Maier, H.R.; Simpson, A.R.; Leonard, M.; Nixon, J.B. Ant colony optimization applied to water distribution system design: Comparative study of five algorithms. *J. Water Resour. Plan. Manag. ASCE* **2007**, *133*, 87–92. [CrossRef]

157. Liong, S.-Y.; Atiquzzaman, M. Optimal design of water distribution network using shuffled complex evolution. *J. Inst. Eng. Singap.* **2004**, *44*, 93–107.

158. Mora, D.; Iglesias, P.L.; Martinez, F.J.; Fuertes, V.S. Statistical analysis of water distribution networks design using harmony search. In *World Environmental and Water Resources Congress 2009: Great Rivers*; American Society of Civil Engineers: Reston, VA, USA, 2009.

159. Geem, Z.W.; Kim, J.H.; Jeong, S.H. Cost efficient and practical design of water supply network using harmony search. *Afr. J. Agric. Res.* **2011**, *6*, 3110–3116.

160. Geem, Z.W. Particle-swarm harmony search for water network design. *Eng. Optim.* **2009**, *41*, 297–311. [CrossRef]

161. Geem, Z.W.; Cho, Y.H. Optimal design of water distribution networks using parameter-setting-free harmony search for two major parameters. *J. Water Resour. Plan. Manag. ASCE* **2011**, *137*, 377–380. [CrossRef]

162. Sheikholeslami, R.; Zecchin, A.C.; Zheng, F.; Talatahari, S. A hybrid cuckoo-harmony search algorithm for optimal design of water distribution systems. *J. Hydroinform.* **2016**, *18*, 544–563. [CrossRef]

163. Qiao, J.F.; Wang, Y.F.; Chai, W.; Sun, L.B. Optimal water distribution network design with improved particle swarm optimisation. *Int. J. Comput. Sci. Eng.* **2011**, *6*, 34–42. [CrossRef]

164. Aghdam, K.M.; Mirzaee, I.; Pourmahmood, N.; Aghababa, M.P. Design of water distribution networks using accelerated momentum particle swarm optimisation technique. *J. Exp. Theor. Artif. Intell.* **2014**, *26*, 459–475. [CrossRef]

165. Ezzeldin, R.; Djebedjian, B.; Saafan, T. Integer discrete particle swarm optimization of water distribution networks. *J. Pipeline Syst. Eng. Pract. ASCE* **2014**, *5*. [CrossRef]

166. Lin, M.-D.; Liu, Y.-H.; Liu, G.-F.; Chu, C.-W. Scatter search huristic for least-cost design of water distribution networks. *Eng. Optim.* **2007**, *39*, 857–876. [CrossRef]

167. Chu, C.-W.; Lin, M.-D.; Liu, G.-F.; Sung, Y.-H. Application of immune algorithms on solving minimum-cost problem of water distribution network. *Math. Comput. Model.* **2008**, *48*, 1888–1900. [CrossRef]

168. Mohan, S.; Babu, K.S.J. Water distribution network design using heuristics-based algorithm. *J. Comput. Civ. Eng.* **2009**, *23*, 249–257. [CrossRef]

169. Bolognesi, A.; Bragalli, C.; Marchi, A.; Artina, S. Genetic heritage evolution by stochastic transmission in the optimal design of water distribution networks. *Adv. Eng. Softw.* **2010**, *41*, 792–801. [CrossRef]

170. Mohan, S.; Babu, K.S.J. Optimal water distribution network design with honey-bee mating optimization. *J. Comput. Civ. Eng.* **2010**, *24*, 117–126. [CrossRef]

171. Suribabu, C. Differential evolution algorithm for optimal design of water distribution networks. *J. Hydroinform.* **2010**, *12*, 66–82. [CrossRef]

172. Sedki, A.; Ouazar, D. Hybrid particle swarm optimization and differential evolution for optimal design of water distribution systems. *Adv. Eng. Inf.* **2012**, *26*, 582–591. [CrossRef]

173. Zheng, F.; Zecchin, A.; Simpson, A. Self-adaptive differential evolution algorithm applied to water distribution system optimization. *J. Comput. Civ. Eng. ASCE* **2013**, *27*, 148–158. [CrossRef]

174. Sadollah, A.; Yoo, D.G.; Kim, J.H. Improved mine blast algorithm for optimal cost design of water distribution systems. *Eng. Optim.* **2015**, *47*, 1602–1618. [CrossRef]

175. Zhou, X.; Gao, D.Y.; Simpson, A.R. Optimal design of water distribution networks by a discrete state transition algorithm. *Eng. Optim.* **2015**, *48*, 603–628. [CrossRef]

176. Zheng, F.; Zecchin, A.; Newman, J.; Maier, H.; Dandy, G. An adaptive convergence-trajectory controlled ant colony optimization algorithm with application to water distribution system design problems. *IEEE Trans. Evol. Comput.* **2017**, *21*, 773–791. [CrossRef]

177. Maier, H.R.; Kapelan, Z.; Kasprzyk, J.; Matott, L. Thematic issue on evolutionary algorithms in water resources. *Environ. Model. Softw.* **2015**, *69*, 222–225. [CrossRef]

178. Haghighi, A.; Samani, H.M.; Samani, Z.M. Ga-ilp method for optimization of water distribution networks. *Water Resour. Manag.* **2011**, *25*, 1791–1808. [CrossRef]

179. Zheng, F.; Simpson, A.; Zecchin, A. Coupled binary linear programming-differential evolution algorithm approach for water distribution system optimization. *J. Water Resour. Plan. Manag. ASCE* **2014**, *140*, 585–597. [CrossRef]

180. Tolson, B.A.; Asadzadeh, M.; Maier, H.R.; Zecchin, A. Hybrid discrete dynamically dimensioned search (hd-dds) algorithm for water distribution system design optimization. *Water Resour. Res.* **2009**, *45*. [CrossRef]

181. Jabbary, A.; Podeh, H.T.; Younesi, H.; Haghiabi, A.H. Development of central force optimization for pipe-sizing of water distribution networks. *Water Sci. Technol.* **2016**, *16*, 1398–1409. [CrossRef]

182. Zheng, F.; Simpson, A.R.; Zecchin, A.C. A method for assessing the performance of genetic algorithm optimization for water distribution design. In *Water Distribution Systems Analysis 2010, 12–15 September 2010*, 41203 ed.; ASCE: Tucson, Arizona, 2010.

183. Zitzler, E.; Deb, K.; Thiele, L. Comparison of multiobjective evolutionary algorithms: Empirical results. *Evol. Comput.* **2000**, *8*, 173–195. [CrossRef] [PubMed]

184. Knowles, J.; Corne, D. On metrics for comparing non-dominated sets. In Proceedings of the 2002 Congress on Evolutionary Computation, Honolulu, HI, USA, 12–17 May 2002.

185. Kollat, J.B.; Reed, P.M. The value of online adaptive search: A performance comparison of nsgaii, ε-nsgaii and εmoea. In *The Third International Conference on Evolutionary Multi-Criterion Optimization (EMO 2005)*; Lecture Notes in Computer Science; Coello, C.C., Aguirre, A.H., Zitzler, E., Eds.; Springer Verlag: Guanajuato, NM, USA, 2005; pp. 386–398.

186. Zitzler, E.; Thiele, L.; Laumanns, M.; Fonseca, C.M.; da Fonseca, V.G. Performance assessment of multiobjective optimizers: An analysis and review. *Evol. Comput.* **2003**, *7*, 117–132. [CrossRef]

187. Wang, Q.; Creaco, E.; Franchini, M.; Savić, D.; Kapelan, Z. Comparing low and high-level hybrid algorithms on the two-objective optimal design of water distribution systems. *Water Resour. Manag.* **2015**, *29*, 1–16. [CrossRef]

188. Gupta, I.; Gupta, A.; Khanna, P. Genetic algorithm for optimization of water distribution systems. *Environ. Model. Softw.* **1999**, *14*, 437–446. [CrossRef]

189. Vairavamoorthy, K.; Ali, M. Pipe index vector: A method to improve genetic-algorithm-based pipe optimization. *J. Hydraul. Eng.* **2005**, *131*, 1117–1125. [CrossRef]

190. Giustolisi, O.; Laucelli, D.; Berardi, L.; Savic, D.A. Computationally efficient modeling method for large water network analysis. *J. Hydraul. Eng.* **2012**, *138*, 313–326. [CrossRef]

191. Bhave, P.R. Noncomputer optimization of single-source networks. *J. Environ. Eng. Div.* **1978**, *104*, 799–814.

192. Knowles, J. Parego: A hybrid algorithm with on-line landscape approximation for expensive multiobjective optimization problems. *IEEE Trans. Evol. Comput.* **2006**, *10*, 50–66. [CrossRef]

193. Andrade, M.; Kang, D.; Choi, C.; Lansey, K. Heuristic postoptimization approaches for design of water distribution systems. *J. Water Resour. Plan. Manag. ASCE* **2013**, *139*, 387–395. [CrossRef]

194. McClymont, K.; Keedwell, E.; Savic, D.; Randall-Smith, M. Automated construction of fast heuristics for the water distribution network design problem. In Proceedings of the 10th International Conference on Hydroinformatics (HIC 2012), Hamburg, Germany, 14–18 July 2012.

195. Wang, Q.; Guidolin, M.; Savic, D.; Kapelan, Z. Two-objective design of benchmark problems of a water distribution system via moeas: Towards the best-known approximation of the true pareto front. *J. Water Resour. Plan. Manag. ASCE* **2015**, *141*. [CrossRef]

196. Zheng, F. Comparing the real-time searching behavior of four differential-evolution variants applied to water-distribution-network design optimization. *J. Water Resour. Plan. Manag. ASCE* **2015**, *141*. [CrossRef]

197. Murphy, L.; Simpson, A.; Dandy, G.; Frey, J.; Farill, T. Genetic algorithm optimization of the fort collins–loveland water distribution system. In Proceedings of the Conference in Computers in the Water Industry, Chicago, IL, USA, 15–18 April 1996.

198. Savic, D.A.; Walters, G.A.; Randall-Smith, M.; Atkinson, R.M. Cost savings on large water distribution systems: Design through genetic algorithm optimization. In *Building Partnerships Joint Conference on Water Resources Engineering and Water Resources Planning and Management*; Rollin, H.H., Michael, G., Eds.; ASCE: Reston, VA, USA, 2000.

199. Wu, W.; Maier, H.R.; Dandy, G.C.; Leonard, R.; Bellette, K.; Cuddy, S.; Maheepala, S. Including stakeholder input in formulating and solving real-world optimisation problems: Generic framework and case study. *Environ. Model. Softw.* **2016**, *79*, 197–213. [CrossRef]

200. Walski, T.M. Real-world considerations in water distribution system design. *J. Water Resour. Plan. Manag. ASCE* **2015**, *141*. [CrossRef]

201. Zecchin, A.C.; Simpson, A.R.; Maier, H.R.; Nixon, J.B. Parametric study for an ant algorithm applied to water distribution system optimization. *IEEE Trans. Evol. Comput.* **2005**, *9*, 175–191. [CrossRef]

202. Schwarz, J. Use of mathematical programming in the management and development of israel's water resources. In *Ground Water in Water Resources Planning*; No. 142, 917–929; IAHS Publisher: Koblenz, Germany, 1983; Available online: http://hydrologie.org/redbooks/a142/142078.pdf (accessed on 20 July 2017).

203. TAHAL. *The Tekuma Model: User's Manual*; Report No. 01/81/50; TAHAL-Water Planning for Israel, Ltd.: Tel Aviv, Israel, 1981.

204. Goulter, I.C.; Lussier, B.M.; Morgan, D.R. Implications of head loss path choice in the optimization of water distribution networks. *Water Resour. Res.* **1986**, *22*, 819–822. [CrossRef]

205. Fujiwara, O.; Jenchaimahakoon, B.; Edirishinghe, N. A modified linear programming gradient method for optimal design of looped water distribution networks. *Water Resour. Res.* **1987**, *23*, 977–982. [CrossRef]

206. Lasdon, L.S.; Waren, A.D. *Grg2 User's Guide*; Department of General Business Administration, University of Texas: Austin, TX, USA, 1984.

207. IBM-ILOG-CPLEX. *V12.1: User's Manual for Cplex*; International Business Machines Corporation: Armonk, NY, USA, 2009.

208. Savic, D.A.; Bicik, J.; Morley, M.S. A dss generator for multiobjective optimisation of spreadsheet-based models. *Environ. Model. Softw.* **2011**, *26*, 551–561. [CrossRef]

209. Fujiwara, O.; Khang, D.B. Correction to "a two-phase decomposition method for optimal design of looped water distribution networks" by okitsugu fujiwara and do ba khang. *Water Resour. Res.* **1991**, *27*, 985–986. [CrossRef]

210. Sonak, V.V.; Bhave, P.R. Global optimum tree solution for single-source looped water distribution networks subjected to a single loading pattern. *Water Resour. Res.* **1993**, *29*, 2437–2443. [CrossRef]

211. Varma, K.V.K.; Narasimhan, S.; Bhallamudi, S.M. Optimal design of water distribution systems using an nlp method. *J. Environ. Eng.* **1997**, *123*, 381–388. [CrossRef]

212. Bassin, J.; Gupta, I.; Gupta, A. Graph theoretic approach to the analysis of water distribution system. *J. Indian Water Works Assoc.* **1992**, *24*, 269.

213. Walski, T.M.; Gessler, J. Improved design of 'anytown' distribution network using structured messy genetic algorithms. *Urban Water* **1999**, *1*, 265–268.
214. Walski, T.M.; Gessler, J. Erratum to discussion of the paper "improved design of 'anytown' distribution network using structured messy genetic algorithms". *Urban Water* **2000**, *2*, 259. [CrossRef]
215. Todini, E.; Pilati, S. A gradient method for the analysis of pipe networks. In Proceedings of the International Conference on Computer Applications for Water Supply and Distribution, Leicester, UK, 8–10 September 1987.
216. Morgan, D.R.; Goulter, I. Optimal urban water distribution design. *Water Resour. Res.* **1985**, *21*, 642–652. [CrossRef]
217. Gessler, J.; Sjostrom, J.; Walski, T. *Wadiso Program User's Manual*; United States Army Engineers Waterways Experiment Station: Vicksburg, MS, USA, 1987.
218. Duan, Q.; Sorooshian, S.; Gupta, V. Effective and efficient global optimization for conceptual rainfall-runoff models. *Water Resour. Res.* **1992**, *28*, 1015–1031. [CrossRef]
219. Nelder, J.A.; Mead, R. A simplex method for function minimization. *Comput. J.* **1965**, *7*, 308–313. [CrossRef]
220. Solomatine, D.P. Two strategies of adaptive cluster covering with descent and their comparison to other algorithms. *J. Glob. Optim.* **1999**, *14*, 55–78. [CrossRef]
221. Von Neuman, J. *Theory of Self-Reproducing Automata*; Burks, A.W., Ed.; University of Illinois Press: Urbana, IL, USA; London, UK, 1966.
222. Savic, D.A.; Walters, G.A.; Randall-Smith, M.; Atkinson, R.M. Large water distribution systems design through genetic algorithm optimisation. In Proceedings of the ASCE 2000 Joint Conference on Water Resources Engineering and Water Resources Planning and Management, Minneapolis, MN, USA, 30 July–2 August 2000.
223. Ostfeld, A.; Salomons, E. Optimal operation of multiquality water distribution systems: Unsteady conditions. *Eng. Optim.* **2004**, *36*, 337–359. [CrossRef]
224. Vairavamoorthy, K. *Water Distribution Networks: Design and Control for Intermittent Supply*; University of London: London, UK, 1994.
225. Vamvakeridou-Lyroudia, L.S. Optimal extension and partial renewal of an urban water supply network, using fuzzy reasoning and genetic algorithms. In Proceedings of the 30th IAHR World Congress, Thessaloniki, Greece, 24–29 August 2003.
226. Kim, J.H.; Kim, T.G.; Kim, J.H.; Yoon, Y.N. A study on the pipe network system design using non-linear programming. *J. Korean Water Resour. Assoc.* **1994**, *27*, 59–67.
227. Lee, S.-C.; Lee, S.-I. Genetic algorithms for optimal augmentation of water distribution networks. *J. Korean Water Resour. Assoc.* **2001**, *34*, 567–575.
228. Zitzler, E.; Thiele, L. Multiobjective optimization using evolutionary algorithms—A comparative case study. In *Parallel Problem Solving from Nature V (PPSN-V)*; Springer: Amsterdam, The Netherlands, 1998.
229. Martínez, J.B. Discussion of "optimization of water distribution networks using integer linear programming" by hossein mv samani and alireza mottaghi. *J. Hydraul. Eng.* **2008**, *134*, 1023–1024. [CrossRef]
230. Rubinstein, R.Y.; Kroese, D.P. *The Cross-Entropy Method: A Unified Approach to Combinatorial Optimization, Monte-Carlo Simulation and Machine Learning*; Springer: New York, NY, USA, 2004.
231. Salomons, E. *Optimal Design of Water Distribution Systems Facilities and Operation*; Technion: Haifa, Israel, 2001.
232. Dorigo, M.; Maniezzo, V.; Colorni, A. Ant system: Optimization by a colony of cooperating agents. *IEEE Trans. Syst. Man Cybern. Part B* **1996**, *26*, 29–41. [CrossRef] [PubMed]
233. Dorigo, M.; Gambardella, L.M. Ant colony system: A cooperative learning approach to the traveling salesman problem. *IEEE Trans. Evol. Comput.* **1997**, *1*, 53–66. [CrossRef]
234. Bullnheimer, B.; Hartl, R.F.; Strauss, C. A new rank based version of the ant system: A computational study. *Central Eur. J. Oper. Res. Econ.* **1999**, *7*, 25–38.
235. Stützle, T.; Hoos, H.H. Max–min ant system. *Future Gener. Comput. Syst.* **2000**, *16*, 889–914. [CrossRef]
236. United States Environmental Protection Agency (USEPA). *Epanet 2.0*; USEPA: Washington, DC, USA, 2016. Available online: http://www.epa.gov/water-research/epanet (accessed on 10 February 2016).
237. Goldberg, D.E. *Genetic Algorithms in Search, Optimization and Machine Learning*; Addison-Wesley Publishing Company: Reading, MA, USA, 1989.
238. Michalewicz, Z. *Genetic Algorithms + Data Structures = Evolution Programs*; Springer: Berlin/Heidelberg, Germany; New York, NY, USA, 2013.

239. Deb, K. *Multi-Objective Optimisation Using Evolutionary Algorithms*; John Wiley & Sons: Chichester, West Sussex, UK, 2001.

240. Van Veldhuizen, D.A. Multiobjective Evolutionary Algorithms: Classifications, Analyses, and New Innovations. Ph.D. Thesis, Department of Electrical and Computer Engineering, Graduate School of Engineering, Air Force Institute of Technology, Wright-Patterson Air Force Base, Dayton, OH, USA, 1999.

241. Ang, K.H.; Chong, G.; Li, Y. Visualization technique for analyzing non-dominated set comparison. In Proceedings of the 4th Asia-Pacific Conference on Simulated Evolution and Learning (SEAL 2002), Singapore, 18–22 November 2002.

242. Jourdan, L.; Corne, D.; Savic, D.; Walters, G. Lemmo: Hybridising rule induction and nsga II for multi-objective water systems design. In Proceedings of the 8th International Conference on Computing and Control for the Water Industry, Exeter, UK, 5–7 September 2005.

243. Corne, D.W.; Jerram, N.R.; Knowles, J.D.; Oates, M.J. Pesa-II: Region-based selection in evolutionary multiobjective optimization. In *Proceedings of the 3rd Annual Conference on Genetic and Evolutionary Computation*; Morgan Kaufmann Publishers Inc.: Burlington, VT, USA, 2001.

244. Giustolisi, O.; Doglioni, A. Water distribution system failure analysis. In *Proceedings of the 8th International Conference on Computing and Control for the Water Industry*; University of Exeter, Centre for Water Systems: Exeter, UK, 2005.

245. Giustolisi, O.; Doglioni, A.; Savic, D.A.; Laucelli, D. *A Proposal for an Effective Multi-Objective Non-Dominated Genetic Algorithm: The Optimised Multi-Objective Genetic Algorithm-Optimoga*; School of Engineering, Computer Science and Mathematics, Centre for Water Systems, University of Exeter: Exeter, UK, 2004.

246. Tolson, B.A.; Shoemaker, C.A. Dynamically dimensioned search algorithm for computationally efficient watershed model calibration. *Water Resour. Res.* **2007**, *43*. [CrossRef]

247. Reca, J.; Martínez, J.; Gil, C.; Baños, R. Application of several meta-heuristic techniques to the optimization of real looped water distribution networks. *Water Resour. Manag.* **2008**, *22*, 1367–1379. [CrossRef]

248. Jung, B.S.; Boulos, P.F.; Wood, D.J. Effect of pressure-sensitive demand on surge analysis. *Am. Water Works Assoc.* **2009**, *101*, 100–111. [CrossRef]

249. Abbass, H.A. A monogenous mbo approach to satisfiability. In *Proceeding of the International Conference on Computational Intelligence for Modelling, Control and Automation CIMCA 2001, Las Vegas, Nev.*; Canberras University Publication: Canberra, Australia, 2001.

250. Storn, R.; Price, K. *Differential Evolution—A Simple and Efficient Adaptive Scheme for Global Optimization Over Continuous Spaces*; Technical report; International Computer Science Institute: Berkeley, CA, USA, 1995.

251. Jang, H.S. Rational design method of water network using computer. *J. Korean Prof. Eng. Assoc.* **1968**, *1*, 3–8.

252. Gonçalves, G.M.; Pato, M.V. A three-phase procedure for designing an irrigation system's water distribution network. *Ann. Oper. Res.* **2000**, *94*, 163–179. [CrossRef]

253. Bansal, J.C.; Deep, K. Optimal design of water distribution networks via particle swarm optimization. In Proceedings of the International Advance Computing Conference IACC 2009, Patiala, India, 6–7 March 2009.

254. Bragalli, C.; D'Ambrosio, C.; Lee, J.; Lodi, A.; Toth, P. *Water Network Design by Minlp*; Report No. Rc24495 (wos02–056); IBM Research: Yorktown Heights, NY, USA, 2008.

255. Bonami, P.; Lee, J. *Bonmin Users' Manual*; COIN-OR Foundation: Baltimore, MD, USA, 2013; Available online: https://www.coin-or.org/Bonmin/ (accessed on 9 September 2017).

256. Tawarmalani, M.; Sahinidis, N.V. Global optimization of mixed-integer nonlinear programs: A theoretical and computational study. *Math. Program.* **2004**, *99*, 563–591. [CrossRef]

257. Sherali, H.D.; Subramanian, S.; Loganathan, G.V. Effective relaxations and partitioning schemes for solving water distribution network design problems to global optimality. *J. Glob. Optim.* **2001**, *19*, 1–26. [CrossRef]

258. Deb, K.; Pratap, A.; Agarwal, S.; Meyarivan, T. A fast and elitist multiobjective genetic algorithm: Nsga-II. *IEEE Trans. Evol. Comput.* **2002**, *6*, 182–197. [CrossRef]

259. Brumbelow, K.; Bristow, E.C.; Torres, J. Micropolis: A virtual city for water distribution systems research applications. In *AWRA 2006 Spring Specialty Conference: GIS and Water Resources IV*; American Water Resources Association: Denver, CO, USA, 2005.

260. Brumbelow, K.; Torres, J.; Guikema, S.; Bristow, E.; Kanta, L. Virtual cities for water distribution and infrastructure system research. In *ASCE World Environmental and Water Resources Congress 2007: Restoring Our Natural Habitat*; ASCE: Reston, VA, USA, 2007.

261. Burden, R.L.; Faires, J.D. *Numerical Analysis*; Brooks/Cole, Cengage Learning: Boston, MA, USA, 2005.

262. Boxall, J.; Skipworth, P.; Saul, A. A novel approach to modelling sediment movement in distribution mains based on particle characteristics. *Water Softw. Syst.* **2001**, *1*, 263–273.

263. Boxall, J.; Saul, A. Modeling discoloration in potable water distribution systems. *J. Environ. Eng.* **2005**, *131*, 716–725. [CrossRef]

264. Broad, D.R.; Maier, H.R.; Dandy, G.C. Optimal operation of complex water distribution systems using metamodels. *J. Water Resour. Plan. Manag. ASCE* **2010**, *136*, 433–443. [CrossRef]

265. Bader, J.; Deb, K.; Zitzler, E. Faster hypervolume-based search using monte carlo sampling. In *Multiple Criteria Decision Making for Sustainable Energy and Transportation Systems, Lecture Notes in Economics and Mathematical Systems*; Ehrgott, M., Naujoks, B., Stewart, T., Wallenius, J., Eds.; Springer: Berlin/Heidelberg, Germany, 2010; pp. 313–326.

266. Todini, E. Looped water distribution networks design using a resilience index based heuristic approach. *Urban Water* **2000**, *2*, 115–122. [CrossRef]

267. Farina, G.; Creaco, E.; Franchini, M. Using epanet for modelling water distribution systems with users along the pipes. *Civ. Eng. Environ. Syst.* **2014**, *31*, 36–50. [CrossRef]

268. Creaco, E.; Franchini, M. A new algorithm for real-time pressure control in water distribution networks. *Water Sci. Technol.* **2013**, *13*, 875–882. [CrossRef]

269. Taher, S.A.; Labadie, J.W. Optimal design of water-distribution networks with gis. *J. Water Resour. Plan. Manag. ASCE* **1996**, *122*, 301–311. [CrossRef]

270. Marques, J.; Cunha, M.; Savić, D.A. Using real options for an eco-friendly design of water distribution systems. *J. Hydroinform.* **2015**, *17*, 20–35. [CrossRef]

271. Walski, T.M.; Gessler, J.; Sjostrom, J.W. Water distribution systems: Simulation and sizing. In *Environmental Progress*; Wentzel, M., Ed.; Lewis Publishers: Chelsea, MI, USA, 1990.

272. Sung, Y.-H.; Lin, M.-D.; Lin, Y.-H.; Liu, Y.-L. Tabu search solution of water distribution network optimization. *J. Environ. Eng. Manag.* **2007**, *17*, 177–187.

273. Hadka, D.; Reed, P. Borg: An auto-adaptive many-objective evolutionary computing framework. *Evol. Comput.* **2013**, *21*, 231–259. [CrossRef] [PubMed]

274. Salomons, E.; Ostfeld, A.; Kapelan, Z.; Zecchin, A.; Marchi, A.; Simpson, A.R. The battle of the water networks ii. In *14th Water Distribution Systems Analysis Conference (WDSA 2012)*; Engineers Australia: Adelaide, Australia, 2012.

275. Prasad, T.D.; Park, N.-S. Multiobjective genetic algorithms for design of water distribution networks. *J. Water Resour. Plan. Manag. ASCE* **2004**, *130*, 73–82. [CrossRef]

276. Vrugt, J.A.; Robinson, B.A. Improved evolutionary optimization from genetically adaptive multimethod search. *Proc. Natl. Acad. Sci. USA* **2007**, *104*, 708–711. [CrossRef] [PubMed]

277. Deb, K.; Mohan, M.; Mishra, S. Evaluating the ε-domination based multi-objective evolutionary algorithm for a quick computation of pareto-optimal solutions. *Evol. Comput.* **2005**, *13*, 501–525. [CrossRef] [PubMed]

278. Kollat, J.B.; Reed, P.M. Comparing state-of-the-art evolutionary multi-objective algorithms for long-term groundwater monitoring design. *Adv. Water Resour.* **2006**, *29*, 792–807. [CrossRef]

279. Das, S.; Konar, A.; Chakraborty, U.K. Two improved differential evolution schemes for faster global search. In Proceedings of the 7th Annual Conference on Genetic and Evolutionary Computation (GECCO 2005), Washington, DC, USA, 25–29 June 2005.

280. Zheng, F.; Zecchin, A. An efficient decomposition and dual-stage multi-objective optimization method for water distribution systems with multiple supply sources. *Environ. Model. Softw.* **2014**, *55*, 143–155. [CrossRef]

281. Zheng, F.; Simpson, A.R.; Zecchin, A.C.; Deuerlein, J.W. A graph decomposition-based approach for water distribution network optimization. *Water Resour. Res.* **2013**, *49*, 2093–2109. [CrossRef]

282. Yang, C.H.; Tang, X.L.; Zhou, X.J.; Gui, W.H. A discrete state transition algorithm for traveling salesman problem. *Control Theor. Appl.* **2013**, *30*, 1040–1046.

283. Zhou, X.; Yang, C.; Gui, W. State transition algorithm. *J. Ind. Manag. Optim.* **2012**, *8*, 1039–1056. [CrossRef]

284. Samani, H.M.; Zanganeh, A. Optimisation of water networks using linear programming. *Water Manag.* **2010**, *163*, 475–485. [CrossRef]

285. Rubinstein, R. The cross-entropy method for combinatorial and continuous optimization. *Methodol. Comput. Appl. Probab.* **1999**, *1*, 127–190. [CrossRef]

286. Boulos, P.F.; Lansey, K.E.; Karney, B.W. *Comprehensive Water Distribution Systems Analysis Handbook for Engineers and Planners*, 2nd ed.; MWH Soft: Pasadena, CA, USA, 2006.

287. Gephi.org. Gephi, an Open Source Graph Visualization and Manipulation Software. 2017. Available online: https://gephi.org/ (accessed on 19 October 2017).

288. Coelho, S.T. *Performance Assessment in Water Supply and Distribution*; Heriot-Watt University: Edinburg, UK, 1996.

water   **MDPI**

Article

Non-Dominated Sorting Harmony Search Differential Evolution (NS-HS-DE): A Hybrid Algorithm for Multi-Objective Design of Water Distribution Networks

Jafar Yazdi [1], Young Hwan Choi [2] and Joong Hoon Kim [3,*]

[1] Department of Civil, Water and Environmental Engineering, Shahid Beheshti University, Tehran 1658953571, Iran; j_yazdi@sbu.ac.ir
[2] Department of Civil, Environmental and Architectural Engineering, Korea University, Seoul 136-713, Korea; younghwan87@korea.ac.kr
[3] School of Civil, Environmental and Architectural Engineering, Korea University, Seoul 136-713, Korea
[*] Correspondence: jaykim@korea.ac.kr

Received: 29 May 2017; Accepted: 3 August 2017; Published: 7 August 2017

Abstract: We developed a hybrid algorithm for multi-objective design of water distribution networks (WDNs) in the present study. The proposed algorithm combines the global search schemes of differential evolution (DE) with the local search capabilities of harmony search (HS) to enhance the search proficiency of evolutionary algorithms. This method was compared with other multi-objective evolutionary algorithms (MOEAs) including NSGA2, SPEA2, MOEA/D and extended versions of DE and HS combined with non-dominance criteria using several metrics. We tested the compared algorithms on four benchmark WDN design problems with two objective functions, (i) the minimization of cost and (ii) the maximization of resiliency as reliability measure. The results showed that the proposed hybrid method provided better optimal solutions and outperformed the other algorithms. It also exhibited significant improvement over previous MOEAs. The hybrid algorithm generated new optimal solutions for a case study that dominated the best-known Pareto-optimal solutions in the literature.

Keywords: MOEA; DE; HS; water distribution system; NSGA2; SPEA2; NSHSDE

1. Introduction

The field of water and environmental engineering has attracted interest in the last decade for the development and utilization of various evolutionary algorithms (EAs). These algorithms have emphasized their potential as a solution to various water-engineering problems. The popularity of EAs is likely due to their ability to solve nonlinear, nonconvex, continuous and discrete problems that classic optimization techniques have failed to solve [1]. We recognize the capability of EAs as the applications of environmental and water-resource engineering problems have become complicated in the sense that they are characterized by large decision spaces. Design of water distribution networks (WDNs) with least investment cost satisfying the pressure requirement and reliability is an example of large scale problems in the field of water resources engineering which needs to be dealt with in optimization models such as EAs. In this problem, because of the large number of network pipes, there are a huge number of designs or alternatives in front of engineers, which cannot be evaluated with only simulation models. The network design problem becomes more complicated when there are also other facilities such as pumps, gates, tanks, etc. Due to the large size of the search space and nature of the problem at hand, categorized as a Non-deterministic Polynomial-time hard (NP-hard) problem, EAs are the search algorithms that are best suited to finding optimal design.

Besides, in this problem, cost is not the only criterion which should be considered in the network design. Considering other criteria such as network reliability and resiliency in both quantitative and qualitative aspects requires the development of multi criteria decision making approaches in WDN design problems. Development of multi-objective evolutionary algorithms (MOEAs) that can efficiently search solution spaces and provide an appropriate approximation of the actual trade-off among considered criteria is thus important. For this reason, Keedwell and Khu [2] emphasized the capability of MOEAs for the design of WDNs. In the work of Montalvo et al. [3], Pareto optimal solutions which represent a compromise among different criteria were obtained. In this direction, the minimized cost and node pressure deficits were used as the main design factors in the multi-objective swarm optimization (MOSO) algorithm [3]. In designing WDNs, we mostly focused on minimizing cost and maximizing reliability as the essential design criteria.

Todini (2000) introduced a "resilience" index which expresses the hydraulic reliability based on the surplus head of each node. WDNs in general are designed by loop systems to enhance hydraulic reliability of the networks [4]. The first hydraulic resilience index was improved by Prasad and Park (2004) to consider system-redundancy concept in the loop systems [5]. They utilized the nondominated sorting genetic algorithm 2 (NSGA2) in multi-objective design of WDNs with minimized total cost and maximized system resilience. As the cost and system resilience were used as the objective functions in WDN design, some researchers compared the performance of optimization algorithms. Farmani et al. [6,7] compared NSGA2 and Strength Pareto evolutionary algorithm 2 (SPEA2) for the design of WDNs and concluded that SPEA2 has better performance than NSGA2.

Research work on multi-objective optimization design of WDNs has shifted from simple evolutionary algorithms to advanced methods with strong local and global search capabilities. For instance, some studies developed cross entropy MOEA methods and compared their performances with NSGA2 [8]. Zheng et al. [9] developed the graph decomposition approach to solve the design problem of WDN in the multi-objective optimization framework. This approach was applied for the optimization of each sub-network and it achieved the Pareto optimal solutions for each one. Pareto optimal solutions obtained from separate optimization were integrated to obtain the whole Pareto optimal solutions of the WDN design problem. Matos et al. [10] introduced an evolutionary strategy to obtain efficient and useful decision-making in engineering problems. The characteristics of this method are crossover and mutation operators that are specific to WDN optimization tasks. Bi et al. [11] developed high-quality initial populations with MOEAs for optimal design of WDNs with least cost and maximum system resilience. They proposed the Multi-Objective Pre-screened Heuristic Sampling Method (MOPHSM) that assigns pipe diameter, depending on the distance between the source and demand nodes.

In addition to improving their own algorithm performance to increase the efficiency of the search ability, many studies have developed methodologies with hybridization of different algorithms [12–18]. However, these methods have been rarely used in multi-objective optimization of WDN design. Keedwell and Khu [2] proposed CAMOGA that is a hybrid multi-objective optimization algorithm combined with cellular automaton and the genetic algorithm for WDN design in order to obtain better designs compared to NSGA2. Vrugt and Robinson [19] also developed a hybrid algorithm, called A Multi-Algorithm Genetically Adaptive Multi-objective (AMALGAM). The AMALGAM is a hybrid of four algorithms (Adaptive Metropolis Search (AMS), NSGA2, PSO and DE). Verification of the performance of AMALGAM comprised of the exploration of solutions for some mathematical benchmark multi-objective problems. The results showed the proposed algorithm has better performance for solving complex problems with respect to a few other MOEAs. Raad et al. [20] compared various hybrid optimization algorithms (AMALGAM, NSGA2, NSGA2-JG) for optimal design of WDNs. AMALGAM showed the best results. Zheng et al. [9] developed SAMODE which is a hybrid of Non-Linear Programming (NLP) and Multi-Objective Differential Evolution. SAMODE outperformed NSGA2 on three benchmark WDN case studies. Wang et al. [21] investigated and reported the best-known Pareto optimal solutions for the well-known WDN problems. The study compared five MOEAs (i.e., NSGA2, AMALGAM, Borg, ε-MOEA and ε-NSGA2) and the results showed that NSGA2 was outstanding as it generally produced the best solutions to all the problems.

The aim of this research is improving the search capabilities of the harmony search (HS) and differential evolution (DE) to attain better performances of optimization models for design of WDNs. Therefore, this research paper contributes towards the development of a new hybrid algorithm with HS and DE algorithms in a multi-objective framework to solve the WDN design problem. The hybrid approach uses DE for a global search, followed by a hybridization of DE and HS for enhancing local search capabilities, to find the approximate true Pareto fronts. The efficiency and the reliability of the hybrid algorithm were compared with those of NSGA2, SPEA2, MOEA/D and two other MOEAs, based on HS and DE. The differences were instances of the integration of non-dominance criteria and evolutionary search operators of HS and DE developed in this study. There are some reasons for using these MOEAs in our study which are enumerated below:

(1) DE and HS show high efficiency in solving numerous single-objective test problems, especially the least-cost design problems involving WDNs. Moreover, since the proposed algorithm is constituted of DE and HS operators, it is advantageous to include source algorithms in the comparison.

(2) New versions of DE and HS algorithms for multi-criteria problems [22,23] have shown better performances than other well-known algorithms, such as SPEA2 [21,24] and MOPSO [25].

(3) NSGA2 and SPEA2 are widely used in representative MOEAs in water resource and environmental engineering. More importantly, NSGA2 was reported to be the superior method relative to several contenders.

(4) MOEA/D is a new approach for solving multi-objective optimization problems (Zhang and Li, [26]), which has a specific characteristic based on the decomposition scheme to separate the original problem into several sub-problems that can be solved collaboratively and simultaneously.

The proposed hybrid algorithm is applied to four well-known benchmark WDN design problems, and four performance metrics (i.e., generational distance (*GD*), diversity (D), hypervolume (*HV*) and coverage set (CS)) are measured to evaluate the performance of the proposed method and implemented MOEAs. The best algorithm is identified based on the results obtained.

2. Methods

The non-dominance criterion is used in MOEAs to sort individuals considering objective function values. Usually MOEAs (i.e., PAES, microGA, NPGA, MOPSO, SPEA2 and NSGA2) use the non-dominated sorting approach to generate Pareto fronts. In this study, the concept of non-dominance is combined with HS and DE to explore the ability of these algorithms in solving the optimal design of WDN in a multi-objective framework. The evolutionary algorithm was developed and implemented as illustrated in the subsections that follow.

2.1. Proposed Algorithm, Non-Dominated Sorting Harmony Search Differential Evolution (NSHSDE)

DE generally has strong global search capability. It is expected that if this method is combined with a local search optimizer, the performance of the evolutionary search improves. According to a study by Yazdi et al. [27], the harmony search performs very well in finding optimal solutions located in certain parts of the search space, but cannot generate adequately diverse Pareto-optimal solutions, thus degrading the overall performance of the search algorithm. To exploit the local search advantages of HS and the exploratory strength of DE, we propose a hybrid HS-DE method here. The hybrid algorithm borrows the mutation operator from the DE algorithm to generate a new solution and uses the pitch adjustment approach of HS to modify the new solution, called "new harmony" in HS terminology. This solution is then used to update the harmony memory, which is a repository of good solutions. The hybrid method also uses the same non-dominance and crowding distance criteria as

NSGA2 to generate the Pareto-optimal curve. The main steps of the hybrid algorithm, called NSHSDE are summarized as follows:

Step 1: The initial population of the harmony memory (HM) with a predefined size, called harmony memory size (HMS), is randomly generated and stored in the harmony memory (HM).

Step 2: Some new harmonies (as many as allowed by the repository size, RS) are generated and stored in the repository. It is recommended that RS be equal to HMS.

To generate a new harmony consisting of n decision variables: $R_i = (r_{i,1}, r_{i,2}, \ldots, r_{i,n})$, the following process is carried out:

(I). A mutation vector $V_i = \{v_{i,1}, v_{i,1}, \ldots, v_{i,n}\}$ is computed according to the equation:

$$V_i = H_{C1} + F \times (H_{C2} - H_{C3}) \tag{1}$$

where H_{C1}, H_{C2} and H_{C3} are three members randomly selected from HM and F is a scaling factor in (0,1], preferably within the range 0.3–0.7.

(II). A pitch adjustment is used to enhance the diversity of the perturbed harmony vector and is considered as new harmony:

$$r_{i,j} = \begin{cases} v_{i,j} + \Delta; & \text{if } (rand_j \leq PAR) \\ v_{i,j}; & \text{if } (rand_j > PAR) \end{cases} \tag{2}$$

where *PAR* is the pitch adjusting rate, j is the variable index in the decision variable vector, $0 \leq PAR \leq 1$, and Δ is a small change computed by $\Delta = Fw \times N(0,1)$. Fw is the fret width (or band width) parameter with a value considered as a small percentage of the range of decision variable j, and $N(0,1)$ is a normal random number with mean 0 and variance 1. $r_{i,j}$ is the decision variable j of the new harmony, R_i.

Step 3: A temporary harmony memory is created by mixing the harmonies in HM and the repository.

Step 4: The temporary harmony memory is sorted according to non-domination. In this process, harmonies are allocated to several fronts $F_1, F_2, \ldots, F_l$ according to their non-domination sort order, where l is the index of the last front. Harmonies in front 1 dominate the harmonies in higher fronts; similarly, harmonies in front 2 dominate the harmonies in fronts 3, 4, ..., l, but are dominated by those in front 1, and so on.

Step 5: The harmonies in each front $F_1, F_2, \ldots, F_l$ are sorted according to the crowding distance [27].

In order to attain a better diversity in HM, among the harmonies in each front, those with fewer neighbors should have higher chance of selection in generating new harmonies. To achieve this, a greater crowding distance is allocated to the solutions with a smaller number of neighbors. Crowding distance is defined as [27]:

$$dist_i^f = \sum_{m=1}^{M} \frac{of_m^{i+1,f} - of_m^{i-1,f}}{of_m^{\max,f} - of_m^{\min,f}} \quad \forall f = 1, \ldots, F_l \tag{3}$$

where $dist_i^f$ is the crowding distance of harmony i in front of f; $of_m^{\max,f}$ and $of_m^{\min,f}$ are the maximum and minimum value of mth objective function in front f, respectively; $of_m^{i+1,f}$ and $of_m^{i-1,f}$ are respectively the value of mth objective function of the upper and lower harmonies (sorted ascending) relative to harmony i in front f and F_l is the number of non-dominated fronts in the current iteration.

Step 6: The HMS is updated by initially selecting the non-dominated solutions, starting with the first ranked non-dominated front (F_1) and proceeding with the subsequently ranked non-dominated fronts ($F_2, F_3, \ldots, F_l$) until the size exceeds the full capacity. It is necessary to reject some of the lower-ranked non-dominated solutions to reduce the total number of the non-dominated solutions to render it equal to the HMS. This is done by using the crowding distance comparison operator; using this procedure, the elements in the HM are updated.

Step 7: Steps 2 through 6 are repeated until the termination criterion (e.g., a maximum number of iterations) is satisfied.

It is worth mentioning that in this study, the value of Fw was adaptively changed during the search, as recommended by Mahdavi, Fesanghary and Damangir [28] to improve the efficiency of the local search as:

$$Fw(G) = Fw_{\max} \times e^{(c \times G)} \tag{4}$$

where G is the generation count, and constant c for each generation is given by

$$c = \frac{\ln\left(\frac{Fw_{\min}}{Fw_{\max}}\right)}{\mathrm{Max}It} \tag{5}$$

$Fw_{\min}$ and $Fw_{\max}$ are the minimum and maximum values of "fret width" respectively, and $\mathrm{Max}It$ is the number of iterations (generations).

2.2. Algorithms Implemented for Comparison

2.2.1. Non-Dominated Sorting Genetic Algorithm 2 (NSGA2)

NSGA2 is a fast and elitist multi-objective genetic algorithm that has been used successfully in many engineering optimization problems and has attained significant popularity in many disciplines. It is arguably the most popular MOEA developed so far. It uses genetic algorithm operators and two population-sorting criteria, non-dominance and crowding distance [29].

2.2.2. Non-Dominated Sorting Harmony Search (NSHS) Algorithm

Harmony search [13] is another popular EA that has garnered considerable attention in various engineering disciplines in the last decade. Several applications of this algorithm have been reported in the water engineering literature and a few multi-objective schemes have recently been developed [27,30] combining HS operators with non-domination sorting (NS) and crowding distance criteria to generate Pareto-optimal solutions in multi-objective problems. This algorithm called NSHS was compared with the NSGA2 and MOPSO algorithms based on several benchmark problems and a sewer pipe network application and the results showed that the NSHS algorithm outperformed the other two.

In this study, the effectiveness of the NSHS algorithm for WDN design problems is tested and compared against other MOEAs.

2.2.3. Non-Dominated Sorting Differential Evolution (NSDE) Algorithm

Differential Evolution, which is an improved version of the GA [31] solves global optimization problems. This algorithm has the same operators as the GA, but its performance relies more on an effective mutation operator than crossover [17]. The application of DE in single-objective designs of WDN has been reported in several studies [17,32,33]. More recently, Zheng et al. [9] developed a DE for multi-objective WDN problems by hybridizing it with NLP to estimate the Pareto front in three WDN case studies.

In this study, DE is extended to solve multi-objective problems by employing non-dominance and crowding distance criteria, as used in NSGA2, within the DE framework. The performance of the NSDE algorithm is then compared with that of other MOEAs using various design cases.

2.2.4. Improving the Strength Pareto Evolutionary Algorithm (SPEA2)

Zitzler et al. [24] developed SPEA2. This algorithm is an improved strength Pareto evolutionary algorithm (SPEA) which is maintained as an external population algorithm and updated in each generation by new non-dominated solutions. In comparison with the initial version, SPEA2 adopts a fine-grained fitness assignment strategy that incorporates density information by taking into account the dominance of a number of individuals. It also adopts a nearest neighbor density estimation technique. Finally, the clustering method is replaced by an archive truncation method that guarantees the preservation of boundary solutions. This algorithm is known as a state-of-the-art MOEA used to assess the performance of algorithms.

2.2.5. Multi-Objective Evolutionary Algorithm Based on Decomposition (MOEA/D)

In addition to the aforementioned MOEAs, we implemented MOEA/D algorithm [26] to solve WDN design problems. This algorithm was rarely used for WDN design [34]. In comparison to the MOEAs, MOEA/D is a different solving approach in which a multiobjective optimization problem is decomposed into a number of scalar optimization sub-problems that are simultaneously optimized. At each instance of generation, the population is composed of the best solution found thus far for each sub-problem. The neighborhood relations among these sub-problems are defined based on distances between their aggregation coefficient vectors. Each sub-problem (i.e., scalar aggregation function) is optimized in MOEA/D by using information only from its neighboring sub-problems. It was also reported that MOEA/D recorded lower computational complexity at each instance of generation than NSGA2 and yielded better performance on some benchmark optimization tests and real-world optimization problems [26,34]. We use the MOEA/D framework with the Tchebycheff approach [26] to decompose multi-objective optimization problems. More information about decomposition approaches and the MOEA/D method can be found in the work of Zhang and Li [26].

2.3. Performance Measures

The performance of the implemented algorithms is evaluated by four performance metrics or indices (i.e., generational distance (*GD*), diversity (D), hypervolume (*HV*) and coverage set (CS)) which are briefly explained below.

2.3.1. Generational Distance (*GD*)

This performance metric was introduced by Van Veldhuizen and Lamont [35] and is used to measure the distance between elements of the set of non-dominated vectors at any given time and those of the global Pareto-optimal set. It is defined as

$$GD = \frac{\left(\sum_{i=1}^{n} d_i{}^p \right)^{\frac{1}{p}}}{n} \tag{6}$$

where n is the number of vectors in the set of non-dominated solutions and d_i is the distance (measured in objective space) between each of these and the nearest member of the global Pareto-optimal set. The parameter P stands for the Pth norm of the distance which is assumed equal 2–i.e. Euclidean distance, in this research. An ideal value of $GD = 0$ indicates that all elements generated are in the global Pareto-optimal set. Thus, any other value indicates how "far" the generated elements are from the global Pareto front. The lower GD, the algorithm's performance better in terms of convergence.

2.3.2. Diversity (D)

This metric assesses the extent of the Pareto front curve and shows the range of diversity of the solutions obtained. It is defined as:

$$D = \prod_{j=1}^{n} \delta_j; \ \delta_j = \max(f_j) - \min(f_j) \tag{7}$$

where j is the index of jth objective function, f_j is the vector of jth objective function values of the Pareto front solutions and n is the number of objective functions. The higher the value of D metric, the better the diversity of MOEA. There is no specific lower and upper bound for this metric and its value is problem-dependent.

2.3.3. Hypervolume (*HV*)

This metric was proposed by Van Veldhuizen [35] for calculating the volume covered between the Pareto fronts and a reference point. This evaluates the performance of convergence and diversity. The reference point represents the worst objective function values. This index is calculated as:

$$HV = volume(\cup_{i=1}^{n} v_i) \tag{8}$$

where v_i is the hyper-volume constructed by solution i and the reference point in the objective function space. It should be emphasized that in this study the normalized version of *HV* called the ratio of *HV*, of the approximation set to the true Pareto-optimal front (HV_R), was used to reduce the bias arising out of the magnitudes of different objective functions and to evaluate the quality of solutions found by each MOEA. HV_R varies between zero and one with the ideal value of one.

2.3.4. Coverage Set (CS)

This index was proposed by Zitzler et al. [36] to compare the Pareto fronts through two different simulations. This metric demonstrates both the convergence and diversity abilities of the algorithm. Considering two Pareto optimal solution sets X', X'' and each set of non-domination points in $X' \cup X''$, denoted by X, CS index can be calculated between [0,1] as:

$$CS(X', X'') = \frac{|\{\alpha'' \in X''; \exists \alpha \prime \in X' : \alpha' \geq \alpha''\}|}{|X|}$$

If all points in X' dominate, or are equal to, all points in X'', by the definition given above, CS = 1. A value of 0 for the index implies the opposition. Ordinarily, both $CS(X', X'')$ and $CS(X'', X')$ need to be considered, as the set intersections are not necessarily empty. The CS metric is used to compare two Pareto sets in relative terms, without the need for the global Pareto solutions.

2.4. Multi-Objective Design of WDNs

2.4.1. Mathematical Formulation

"Network cost" and "reliability" are two major objectives that have been frequently used in the multi-objective design of water distribution systems. The same criteria are taken as the objective functions in this study, with "network resilience" indicator as the reliability measure proposed by Prasad and Park [5]. The first objective function involves economic considerations and the latter provides a measure to assess both surplus head and reliable loops in networks of varying size. The optimization problem can then be formulated as:

$$\text{Min } C = \sum_{i=1}^{np} (U_i \times L_i) \tag{10}$$

$$\text{Max } I_n = \frac{\sum_{j=1}^{nn} C_j Q_j \left(H_j - H_j^{req} \right)}{\left(\sum_{k=1}^{nr} Q_k H_k + \sum_{i=1}^{npu} \frac{P_i}{\gamma} \right) - \sum_{j=1}^{nn} Q_j H_j^{req}} \tag{11}$$

$$C_j = \frac{\sum_{i=1}^{npj} D_i}{npj \times \max\{D_i\}} \tag{12}$$

where C = total cost (monetary units), np = number of pipes, U_i = unit cost of pipe i of diameter D_i, L_i = length of pipe i, I_n = network resilience, nn = number of demand nodes, C_j, Q_j, H_j, and H_j^{req} are uniformity, demand, actual head and minimum required head respectively, of node j; nr = number of reservoirs, Q_k and H_k are discharge and actual head respectively, of reservoir k, npu = number of pumps, P_i = power of pump i, γ = specific weight of water, npj = number of pipes connected to node j, and D_i = diameter of pipe i connected to demand node j. The constraints of the optimization problem are as follows:

(1) Mass conservation constraint

For each junction node, the mass conservation law should be satisfied which can be written as:

$$\sum Q_{in} - \sum Q_{out} = Q_e \tag{13}$$

where Q_{in} and Q_{out} are in-flow and out-flow of the node, respectively, and Q_e is the external flow rate or demand at the node.

(2) Energy conservation constraint

For each loop in the network, the conservation of energy can be written as follows:

$$\sum_{k \in Loopl} \Delta h_k = 0 \qquad \forall l \in Nl \tag{14}$$

where Δh_k is the head loss in pipe k and Nl is the total number of loops in the system. The head loss in each pipe is the difference in head between connected nodes, and is a function of discharge, pipe diameter, and roughness coefficient of the pipe. Head loss is usually calculated using empirical equations, such as the Darcy-Weisbach or the Hazen-Williams equation.

(3) Minimum and maximum pressure constraints

The minimum and maximum pressure constraints on each node in the network are given by the following:

$$H_j^l \leq H_j \leq H_j^u \qquad \forall j = 1, 2, \ldots, nn \tag{15}$$

where H_j is the pressure head at node j, H_j^l is the minimum required pressure head at node j, H_j^u is the maximum allowed pressure head at node j, and nn is the number of demand nodes in the network.

(4) Pipe size availability constraint

The diameter of each pipe must belong to a commercial size set, A:

$$D_i \in \{A\} \qquad \forall i = 1, 2, \ldots, np \tag{16}$$

where D_i is the diameter of pipe i, $\{A\}$ denotes the set of commercially available pipe diameters, and np is the number of pipes.

(5) Minimum and maximum velocity constraints

The minimum and maximum velocity constraints on each pipe in the network are given by the following:

$$v_i^l \leq v_i \leq v_i^u \forall i = 1, 2, \ldots, np \tag{17}$$

where v_i Is velocity in pipe i, v_i^l and v_i^u are the minimum and maximum allowed velocity in pipe i, respectively.

To solve the optimization problem, the constrained model is converted into an unconstrained one by adding the number of constraint violations to the objective function as penalty. The constraints for the conservation of mass and energy are automatically satisfied using EPANET2.0 hydraulic solver [37].

The minimum and maximum pressure and velocity constraints, however, need to be included in the penalty functions:

$$
Cp_1 = \begin{cases} 0; & \text{if } H_j^l \leq H_j \leq H_j^u \forall j = 1,\ldots, nn \\ P \times \left[\sum_j \left(H_j^l - H_j \right) + \sum_j \left(H_j - H_j^u \right) \right]; & \text{otherwise} \end{cases} \tag{18}
$$

$$
Cp_2 = \begin{cases} 0; & \text{if } v_i^l \leq v_i \leq v_i^u \forall i = 1,\ldots, np \\ P \times \left[\sum_i \left(v_i^l - v_i \right) + \sum_i \left(v_i - v_i^u \right) \right]; & \text{otherwise} \end{cases} \tag{19}
$$

$$
Cp = Cp_1 + Cp_2 \tag{20}
$$

where P is a penalty multiplier, which was set to 10^6 in this study. Cp_1 is the summation of the penalties of all nodes with pressure violation, Cp_2 is the summation of the penalties of all pipes with velocity violation and Cp is the total penalty. Therefore, the total cost of the network is the sum of network cost C and penalty cost Cp in nodes and pipes with pressure and velocity violation, respectively.

2.4.2. Experimental Tests on WDNs

The validation of the hybrid algorithm to solve WDN design problems requires application of the algorithm to benchmark problems of different sizes which were selected from the literature including the two-loop network (small-sized problem), the Hanoi network (medium-sized), the Fossolo network (intermediate-sized) and the Balerma network (large-sized problem). These four benchmark networks are briefly described below.

1. Two-loop network (TLN)

The two-loop network (TLN) consists of eight pipes, six demand nodes, and a reservoir. The reservoir had a constant head fixed at 210 m [21]. As it is a hypothetical network, all pipes had the same length (1000 m) and a Hazen-Williams coefficient of 130. Pressure was set to at least 30.0 m at all demand nodes. Table 1 shows the commercially available diameters and the corresponding unit costs (1 in. = 0.0254 m) and Figure 1 depicts the layout of the TLN.

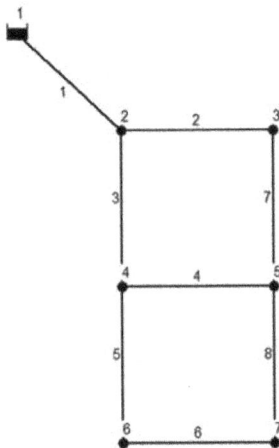

Figure 1. Layout of two-loop network, TLN [21].

Table 1. Diameter options and associated unit costs of two-loop network (TLN).

Diameter (in.)	Unit Cost ($/m)	Diameter (in.)	Unit Cost ($/m)
1	2	12	50
2	5	14	60
3	8	16	90
4	11	18	130
6	16	20	170
8	23	22	300
10	32	24	550

2 Hanoi network (HAN)

The Hanoi network (HAN) consisted of 34 pipes organized into three loops, 31 demand nodes, and a reservoir with a fixed head of 100 m [21]. The Hazen-Williams roughness coefficient for all pipes was 130. The minimum head above the ground elevation of each node was 30 m. There were six commercially available pipe sizes, ranging from 12 in. to 40 in. (1 in. = 0.0254 m). Table 2 shows the diameter options and associated unit costs and Figure 2 depicts the layout of the network.

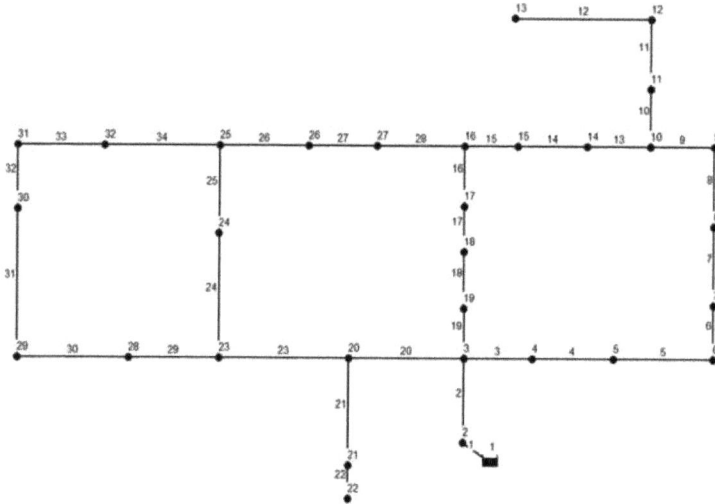

Figure 2. Layout of Hanoi network (HAN) [20].

Table 2. Diameter options and associated unit costs of Hanoi network (HAN).

Diameter (in.)	Unit Cost ($/m)	Diameter (in.)	Unit Cost ($/m)	Diameter (in.)	Unit Cost ($/m)
12	45.73	20	98.39	30	180.75
16	70.40	24	129.33	40	278.28

3 Fossolo network (FOS)

The Fossolo network (FOS) consisted of 58 pipes, 36 demand nodes, and a reservoir with a fixed head of 121.00 m [21]. The material for all pipes was polyethylene. Due to the characteristics of polyethylene, a relatively high Hazen-Williams coefficient of 150 was applied to all pipes. The minimum pressure head of the demand nodes was maintained at 40 m, whereas the maximum pressure head at each node was as specified in Table 3. Moreover, the flow velocity in each pipe was enforced at less than or equal to 1 m/s. Table 4 shows the commercially available diameters and the corresponding unit costs. Figure 3 depicts the layout of the FOS.

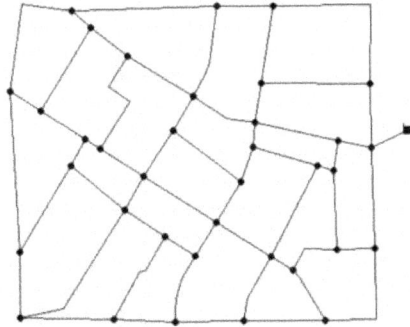

Figure 3. Layout of Fossolo network (FOS) [21].

Table 3. Maximum pressure head requirement of each node of Fossolo network (FOS).

NI	Pmax (m)	NI	Pmax (m)	NI	Pmax (m)	NI	Pmax (m)	NI	Pmax (m)	NI	Pmax (m)
1	55.85	7	53.1	13	59.1	19	58.1	25	56.6	31	56.6
2	56.6	8	54.5	14	58.4	20	58.17	26	57.6	32	56.8
3	57.65	9	55.0	15	57.5	21	58.2	27	57.1	33	56.4
4	58.5	10	56.83	16	56.7	22	57.1	28	55.35	34	56.3
5	59.76	11	57.3	17	55.5	23	56.8	29	56.5	35	55.57
6	55.60	12	58.36	18	56.9	24	53.5	30	56.9	36	55.1

Note: pp. NI = node index, which is a consecutive number starting from 1 and continuing to the total number of nodes in the network; Pmax = maximum pressure head requirement.

Table 4. Diameter options and associated unit costs of Fossolo network (FOS).

Diameter (mm)	Unit Cost (€/m)	Diameter (mm)	Unit Cost (€/m)	Diameter (mm)	Unit Cost (€/m)	Diameter (mm)	Unit Cost (€/m)
16	0.38	61.4	4.44	147.20	24.78	290.6	99.58
20.4	0.56	73.6	6.45	163.6	30.55	327.4	126.48
26	0.88	90	9.59	184.00	38.71	368.2	160.29
32.6	1.35	102.2	11.98	204.6	47.63	409.2	197.71
40.8	2.02	114.6	14.93	229.2	59.7		
51.4	3.21	130.80	19.61	257.8	75.61		

4   Balerma Irrigation Network (BIN)

The Balerma irrigation network (BIN) consisted of 454 polyvinyl chloride (PVC) pipes, 443 demand nodes, and four reservoirs with fixed heads at 112 m, 117 m, 122 m, and 127 m [21]. The Darcy–Weisbach roughness coefficient of pipes was set at 0.0025 mm for all pipes. The minimum pressure head above ground elevation was 20 m for all the demand nodes. Each pipe was allowed to select a diameter from 10 discrete values. Table 5 shows the commercially available diameters and the corresponding unit costs. Figure 4 depicts the layout of the BIN.

Figure 4. Layout of Balerma network (BIN) [21].

Table 5. Diameter options and associated unit costs of Balerma network (BIN).

Diameter (mm)	Unit Cost (€/m)	Diameter (mm)	Unit Cost (€/m)
113	7.22	226.2	28.6
126.6	9.10	285	45.39
144.6	11.92	361.8	76.32
162.8	14.84	452.2	124.64
180.8	18.38	581.8	215.85

3. Results and Discussions

This section presents the results of the computational experiments performed using the selected MOEAs on the four benchmark water distribution networks introduced in the previous section.

The MOEAs were linked to the EPANET 2.0 hydraulic solver to estimate WDN resilience and assess necessary hydraulic constraints while evaluating different network designs. Shown in Table 6 are the sizes of the search spaces, the number of decision variables, and population size as well as the computational budget in terms of the number of function evaluations (NFE) and pipe diameter options for each of the four benchmark networks. The parameters of the NSGA2 were set according to widely recommended settings in the literature; for the other algorithms, parameter values were chosen based both on values used in the literature and the results of several trial runs. The final parameters used are presented in Table 7. Each MOEA was independently run 30 times to solve each problem. The results of a typical run (not average) of each algorithm are shown in Figures 5 and 6. As can be seen, the hybrid algorithms namely, NSHSDE and NSDE, generally provided better Pareto fronts than the four other algorithms for all studied networks. The best-known Pareto fronts of the problems reported by Wang et al. [21] are also shown in the two figures; they were obtained by aggregating the Pareto-optimal solutions of different MOEAs after several executions with different population sizes and very high orders of magnitudes of NFEs.

In order to compare the algorithms more precisely, the performance metrics described in the preceding section were used. Table 8 shows the average values of the metrics obtained by 30 iterations of each of the algorithms for each benchmark network problem. In this table, the second and third metrics were normalized using those of the best-known Pareto fronts in the literature. The first metric, *GD* shows the quality of solutions in terms of non-domination strength. It can be seen that the MOEAs exhibited different behaviors according to this metric. The NSHS outperformed all others for all networks except for the TLN problem, based on the *GD* metric. The hybrid method also yielded good values in terms of the *GD* metric for two small-sized problems. However, for larger networks, its *GD* values were poor, but as explained below, this does not mean the hybrid method is inferior.

The NSDE was inferior to the proposed hybrid method in all four experimental networks in terms of *GD* values, and MOEA/D represented the worst results for the TLN, HAN, and FOS networks while its *GD* values for the largest network, i.e., the BIN problem, recorded very close to the best value. Two other algorithms, i.e., NSGA2 and SPEA2, represented the median results for the *GD* metric. Although the *GD* is an indicator of convergence, it is not a good representative of convergence strength. For example, as shown in Figure 6b, the hybrid method represented high-quality solutions in the same range as found by NSHS. It also yielded other non-dominated solutions not found by NSHS, which were more distant from the best-known Pareto front solutions. As a result, they generated a larger *GD* value for the hybrid method than the NSHS, whereas the quality of its solutions was better than those of NSHS in terms of convergence. Therefore, we see that the *GD* metric is not an appropriate criterion for measurement of the convergence strength of MOEAs.

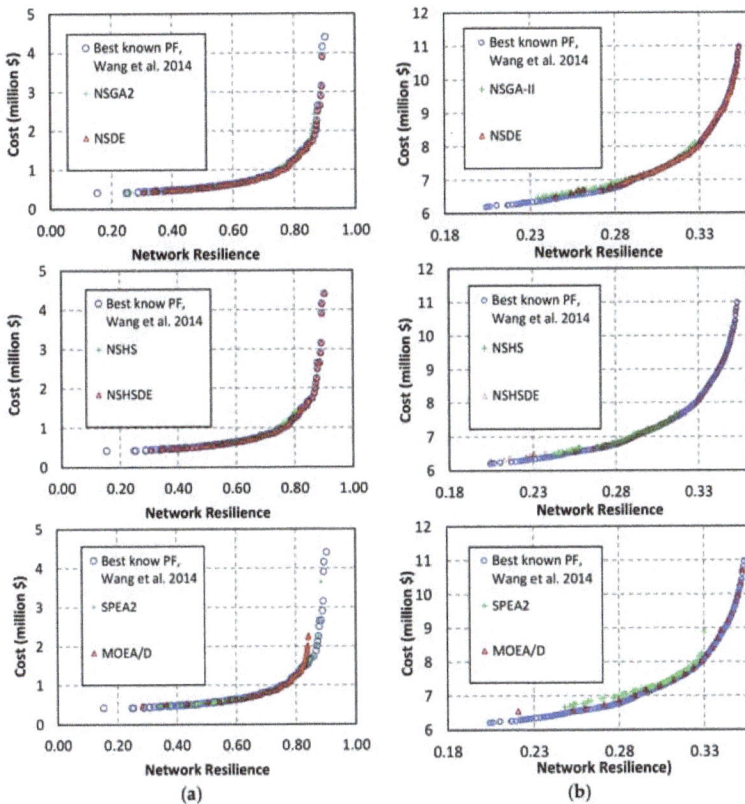

Figure 5. Pareto front (PF) of benchmark problems, (**a**) Two-loop network (TLN) problem; (**b**) Hanoi network (HAN) problem.

Table 6. Computational budgets and sizes of search space for benchmark design problems.

Problem	NFE [a]	Population Size	DV [b]	PD [c]	Search Space Size
Two-loop Network	20,000	40	8	14	1.48×10^9
Hanoi Network	50,000	60	34	6	2.87×10^{26}
Fossolo Network	200,000	100	58	22	7.25×10^{77}
Balerma Irrigation Network	1,000,000	400	454	10	1.0×10^{454}

Note: [a] NFE = number of function evaluations, [b] DV = number of decision variables, [c] PD = number of pipe diameter options.

According to the second metric—diversity (D)—the hybrid method and NSDE generated wider Pareto fronts and produced more diverse solutions than the other MOEAs, whereas NSHS had the smallest D values, showing that its performance was not global and unable to preserve diversity in the optimal solutions found. This can also be observed in the graphical results presented in Figures 5 and 6, where the solutions of the NSDE and NSHSDE algorithms are well spread on both sides of the objective function spaces, while those of the NSHS algorithm (and the other MOEAs) are concentrated in a particular part of the best-known Pareto fronts. The superior diversity of solutions observed in the NSDE and NSHSDE algorithms can be attributed to the efficient mutation operator in the DE for generating new solutions.

It is noteworthy that D metric is also not a perfect comparison measure since it does not consider the non-dominance. The spread Pareto front of an MOEA with long tails may dominated by the small-spread Pareto front of another MOEA and thus, the former would be the inferior, although has a greater D metric. This is the case for NSDE and NSHSDE algorithms and we should see what two other metrics say.

Table 7. Parameters used in multi-objective evolutionary algorithms (MOEAs).

Algorithm	Parameter	Value
NSGA2	Mutation rate	1/(no. variables)
	Crossover prob.	0.9
	Tournament size	2
	Mutation step size	$0.1 \times$ Variable range
SPEA2	Mutation rate	1/(no. variables)
	Crossover prob.	0.9
	Tournament size	2
	Mutation step size	$0.1 \times$ Variable range
NSHS	HMCR	0.98
	PAR	0.4
	Fw	$(0.05–0.005) \times$ Variable range
NSDE	F (scaling factor)	0.5
	Crossover prob.	0.7
NSHSDE	F (scaling factor)	0.5
	PAR	0.4
	Fw	$(0.05–0.005) \times$ Variable range
MOEA/D	Mutation prob.	0.3
	Mutation rate.	0.1
	T (number of neighbors)	$0.2 \times$ Pop size
	Z * (Goal point)	The ideal values for objective functions, zero for the cost and one for the resiliency index

Overall, a comparison between the *GD* and D values of the hybrid method with those of NSHS and NSDE showed that the proposed hybrid algorithm successfully exploited the mutation operator of differential evolution and the local search capability of harmony search to generate better quality solutions (in terms of convergence) than NSDE and better diversity than NSHS across all experimental networks.

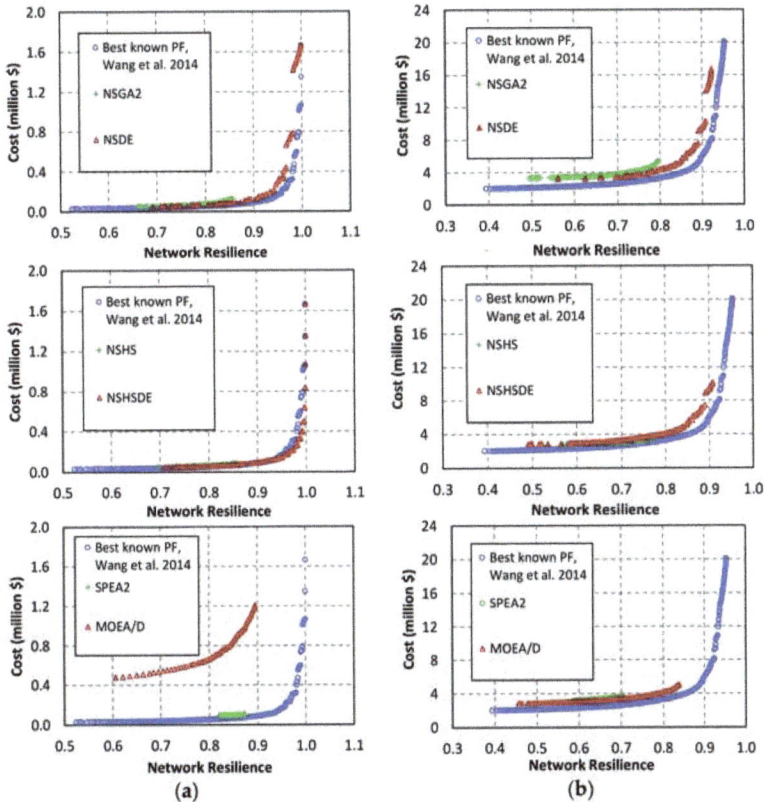

Figure 6. Pareto front (PF) of benchmark problems, (**a**) Fossolo network (FOS) problem; (**b**) Balerma network (BIN) problem.

As shown in the last column of Table 8, NSDE and NSHSDE outperformed the other methods based on the third metric, wherein their HV values were of higher order compared to those achieved from the other MOEAs for all benchmark networks. Although HV measures both convergence and diversity in MOEAs, these values are considered to be more influenced by diversity (see Table 8); thus, this metric alone could not provide consistent and unbiased evaluation. Another important observation was that, in the case of the third network problem (FOS), NSDE found some undiscovered solutions and NSHSDE found some new optimal solutions that dominated a portion of the best-known Pareto front (PF) in the literature. This was accomplished in both algorithms by incurring lower computational burden (smaller number of function evaluations) than that for the best-known PF. Comparing the coverage set (CS) metric of the best known PF and the two algorithms, NSHSDE and NSDE respectively found on an average 12% and 36% new optimal solutions in each round of execution that were not found in the best PF reported in the literature. NSDE presented more new optimal solutions; however, the analysis of the results showed that most of them are located in small clustered locations at right tail of the Pareto front and other solutions in its Pareto front are dominated by both the best known PF and NSHSDE members. In other words, new optimal solutions in NSDE are in fact undiscovered solutions which come from the strength of diversity in DE, not the strength of convergence. That is to say, they are non-dominated solutions with respect to the best known PF, but they do not dominate solutions of the best known PF. On the other hand, new solutions found by NSHSDE (12% in average) dominate the solutions of the best known PF. This means they are obtained by the strength of convergence in NSHSDE.

These results confirmed the superiority of the proposed hybrid algorithm and the improvement in its search capability. It also showed that the reported Pareto front for this network was not global. In order to complete our analysis, a pairwise comparison among the MOEAs was carried out based on the CS metric for all four experimental networks. The results are demonstrated in Table 9. CS shows what percentage of the non-dominated solutions are found by each of MOEAs, in a pairwise comparison. The pairwise comparison (diagonal values in the table) shows that NSHSDE outperformed all other MOEAs for all four WDN design problems. The performance of the other algorithms varied for different benchmark network problems; but in general, the NSDE algorithm was second best, and SPEA2 exhibited weak performance compared to the other methods. Since the CS metric is based on the non-dominance strength, unlike the three preceding metrics, CS is an unbiased and perfect convergence measure. By observing the CS values of MOEA/D, it is evident that this algorithm underperformed in the first three networks, but delivered the second best performance in the last network (BIN problem). It is likely that the MOEA/D yields good performance for large-sized problems. Moreover, based on the CS values, NSHS recorded relatively better performance than NSGA2.

Table 8. Comparison of multi-objective algorithms using the average values of generational distance (*GD*), relative diversity (D), and hypervolume (*HV*) metrics.

Problem	Algorithm	*GD*	D	*HV*$_R$
TLN	NSHSDE	193.96	0.79	1.00
	NSDE	1317.76	0.50	0.62
	NSGA2	3690.13	0.43	0.49
	NSHS	3362.55	0.19	0.26
	SPEA2	3136.40	0.51	0.72
	MOEA/D	4007.28	0.41	0.49
HAN	NSHSDE	1992.61	0.80	0.98
	NSDE	2420.65	0.66	0.96
	NSGA2	1987.79	0.22	0.30
	NSHS	1670.78	0.17	0.24
	SPEA2	1866.20	0.12	0.18
	MOEA/D	5130.38	0.81	0.86
FOS	NSHSDE	2253.75	0.5	0.96
	NSDE	10529.41	0.67	0.98
	NSGA2	213.13	0.02	0.03
	NSHS	103.00	0.00	0.01
	SPEA2	310.30	0.00	0.00
	MOEA/D	36463.37	0.33	0.37
BIN	NSHSDE	7032.75	0.32	0.36
	NSDE	28295.29	0.48	0.69
	NSGA2	2144.96	0.06	0.06
	NSHS	1697.33	0.03	0.03
	SPEA2	1762.07	0.01	0.01
	MOEA/D	1729.88	0.09	0.08

A close-up view of Pareto fronts generated by these algorithms on the FOS and BIN network problems is given in Figure 7. As shown, the solutions found by NSHS dominated the main part of the NSGA2 solutions. Compared with NSGA2, the harmony search had better local search capability with regard to the quality of solutions. This was also observed in the work of Yazdi et al. [29] for the sewer pipe network application. This algorithm, however, could not find the tails of the Pareto front, as it suffered from an inability to preserve the diversity of the population during the evolutionary search steps. Hybridizing HS with DE, which has a strong global search feature, provides considerably better performance than the MOEAs considered, as shown in Figures 5 and 6 and in Tables 8 and 9.

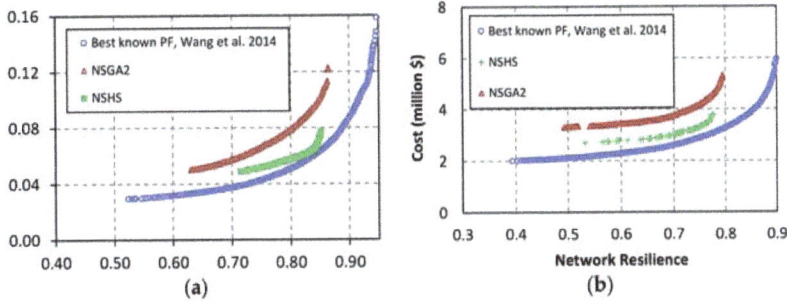

Figure 7. A close-up view of Pareto fronts (PFs) generated by Non-Dominated Sorting Harmony Search (NSHS) and nondominated sorting genetic algorithm 2 (NSGA2), (**a**) FOS problem; (**b**) BIN problem.

Table 9. Comparison of multi-objective algorithms using the coverage set (CS) metric.

Problem	Algorithm	NSHSDE	NSDE	NSGA2	NSHS	SPEA2	MOEA/D
TLN	NSHSDE	-	0.87	0.85	0.92	0.82	0.94
	NSDE	0.84	-	0.84	0.93	0.85	0.95
	NSGA2	0.74	0.76	-	0.89	0.78	0.86
	NSHS	0.31	0.32	0.35	-	0.32	0.37
	SPEA2	0.66	0.67	0.7	0.83	-	0.85
	MOEA/D	0.58	0.61	0.62	0.87	0.63	-
HAN	NSHSDE	-	0.6	0.72	0.53	0.99	0.95
	NSDE	0.54	-	0.71	0.51	0.99	0.93
	NSGA2	0.28	0.29	-	0.34	0.92	0.78
	NSHS	0.47	0.49	0.66	-	0.92	0.85
	SPEA2	0.01	0.01	0.08	0.08	-	0.48
	MOEA/D	0.06	0.07	0.22	0.15	0.52	-
FOS	NSHSDE	-	0.82	0.93	0.75	1	1
	NSDE	0.21	-	0.98	0.55	0.83	1
	NSGA2	0.07	0.02	-	0.17	0.44	0.71
	NSHS	0.25	0.45	0.83	-	0.17	0.67
	SPEA2	0	0.17	0.566	0.83	-	0.92
	MOEA/D	0	0	0.29	0.33	0.08	-
BIN	NSHSDE	-	0.59	1	0.64	1	0.53
	NSDE	0.41	-	1	0.62	0.55	0.38
	NSGA2	0	0	-	0.48	0.4	0
	NSHS	0.36	0.38	0.52	-	1	0.46
	SPEA2	0	0.45	0.59	0	-	0
	MOEA/D	0.48	0.61	1	0.54	1	-

4. Conclusions

In this research article, a hybrid multi-objective algorithm was proposed by integrating the concepts of HS and DE in a unified framework. The proposed algorithm was compared with well-known methods—NSGA2, SPEA2, MOEA/D—and two other extended versions of harmony search and differential evolution, on multi-objective problems. A quantitative comparison was made (by plotting the Pareto fronts) in terms of the generational distance (*GD*), diversity (*D*), hyper-cube volume (*HV*), and coverage set (CS) metrics. Results demonstrated that *GD* due to ignoring the cardinality (the number of solutions) of the Pareto front and D metric due to ignoring non-dominance criterion are not suitable comparative measures. While two other metrics measure both convergence and diversity, HV may have biased results because of the scaling effects of different objective functions and being more influenced by diversity. CS, which is directly calculated based on the non-dominance strength, was found to be the best comparative measure. According to HV metric, NSHSDE had the top rank in the first three considered WDN design problems and the second rank in the fourth case. Therefore, it delivered a better overall performance based on HV values. Additionally, evaluation of the NSHSDE Pareto fronts for four WDN design problems showed that the hybrid method is the superior MOEA in terms of CS metric. These results confirm the credibility of the proposed hybrid algorithm.

Of the algorithms considered, the hybrid algorithm provided on average 12% of new optimal solutions for the FOS network problem, dominating the best-known Pareto front in the literature that had been obtained by aggregating the results of several widely used algorithms. This gain, achieved by using a considerably smaller number of function evaluations, is due to the successful exploitation of the global search capability of differential evolution and the local search capability of harmony search operators. Although some limited parameter tuning was done in this work, we believe that if more computational effort is put on parameter settings, even better performance can be achieved by the hybrid method. The results also show that NSDE yielded the second best performance and SPEA2 was generally inferior to all other methods compared. Similarly, MOEA/D showed weak performance on small- and medium-sized problems, but exhibited the second best results on the large-sized Balerma network problem. The median-scoring algorithms were NSHS and NSGA2, where the former provided better local convergence but the latter provided better diversity.

In summary, the overall results of this study show that the proposed hybrid algorithm can be successfully used to solve multi-objective design problems of WDN with better efficiency. More complex networks are being studied to confirm the credibility of the proposed approach and will be reported in the future.

Acknowledgments: This subject is supported by Korea Ministry of Environment as "Projects for Developing Eco-Innovation Technologies (2016002120004)".

Author Contributions: Jafar Yazdi wrote the draft of the manuscript and provided this model. Young Hwan Choi carried out the survey of previous studies. Jafar Yazdi and Joong Hoon Kim conceived the original idea of the proposed method.

Conflicts of Interest: The authors declare no conflict of interest.

References

1. Nicklow, J.; Reed, P.; Savic, D.; Dessalegne, T.; Harrell, L.; Chan-Hilton, A.; Zechman, E. State-of-the-art for genetic algorithms and beyond in water resources planning and management. *J. Water Resour. Plan. Manag.* **2010**, *136*, 412–432. [CrossRef]
2. Keedwell, E.; Khu, S.T. Hybrid Genetic Algorithms For Multiobjective Optimisation of Water Distribution Networks. In *Genetic and Evolutionary Computation Gecco*; Springer: Berlin/Heidelberg, Germany, 2004.
3. Montalvo, I.; Izquierdo, J.; Schwarze, S.; Pérez-García, R. Multi-Objective Particle Swarm Optimization Applied to Water Distribution Systems Design: An Approach with Human Interaction. *Math. Comput. Model.* **2010**, *52*, 1219–1227. [CrossRef]
4. Todini, E. Looped water distribution networks design using a resilience index based heuristic approach. *Urban Water* **2000**, *2*, 115–122. [CrossRef]
5. Prasad, T.D.; Park, N.-S. Multiobjective Genetic Algorithms for Design of Water Distribution Networks. *J. Water Resour. Plan. Manag.* **2004**, *130*, 73–82. [CrossRef]
6. Farmani, R.; Savic, D.A.; Walters, G.A. Evolutionary Multi-Objective Optimization in Water Distribution Network Design. *Eng. Optim.* **2005**, *37*, 167–183. [CrossRef]
7. Farmani, R.; Walters, G.A.; Savic, D.A. Trade-Off between Total Cost and Reliability for Anytown Water Distribution Network. *J. Water Resour. Plan. Manag.* **2005**, *131*, 161–171. [CrossRef]

8.   Perelman, L.; Ostfeld, A.; Salomons, E. Cross-Entropy Multiobjective Optimization for Water Distribution Systems Design. *Water Resour. Res.* **2008**, *44*, W09413. [CrossRef]

9.   Zheng, F.; Simpson, A.R.; Zecchin, A.C. An Efficient Hybrid Approach for Multiobjective Optimization of Water Distribution Systems. *J. Water Resour. Res.* **2014**, *50*, 3650–3671. [CrossRef]

10. Matos, J.P.; Monteiro, A.J.; Matias, N.M.; Schleiss, A.J. Redesigning Water Distribution Networks Using A Guided Evolutionary Approach. *J. Water Resour. Plan. Manag.* **2015**, *142*, C4015004. [CrossRef]

11. Bi, W.; Dandy, G.C.; Maier, H.R. Use of Domain Knowledge to Increase the Convergence Rate of Evolutionary Algorithms for Optimizing the Cost and Resilience of Water Distribution Systems. *J. Water Resour. Plan. Manag.* **2016**, *142*, 04016027. [CrossRef]

12. Geem, Z.W. Particle Swarm Harmony Search for Water Network Design. *Eng. Optim.* **2009**, *41*, 297–311. [CrossRef]

13. Geem, Z.W.; Kim, J.H.; Loganathan, G.V. A New Heuristic Optimization Algorithm: Harmony Search. *Simulation* **2001**, *76*, 60–68. [CrossRef]

14. Geem, Z.W.; Cho, Y.H. Optimal Design of Water Distribution Networks Using Parameter-Setting-Free Harmony Search for Two Major Parameters. *J. Water Resour. Plan. Manag.* **2011**, *137*, 377–380. [CrossRef]

15. Babu, K.J.; Vijayalakshmi, D.P. Self-Adaptive PSO-GA Hybrid Model For Combinatorial Water Distribution Network Design. *J. Pipeline Syst. Eng.* **2013**, *4*, 57–67. [CrossRef]

16. Cisty, M. Hybrid Genetic Algorithm and Linear Programming Method for Least-Cost Design of Water Distribution Systems. *Water Resour. Manag.* **2010**, *24*, 24. [CrossRef]

17. Sedki, A.; Ouazar, D. Hybrid Particle Swarm Optimization and Differential Evolution for Optimal Design of Water Distribution Systems. *Adv. Eng. Inform.* **2012**, *26*, 582–591. [CrossRef]

18. Tolson, B.A.; Asadzadeh, M.; Maier, H.R.; Zecchin, A. Hybrid Discrete Dynamically Dimensioned Search (HD-DDS) Algorithm for Water Distribution System Design Optimization. *Water Resour. Res.* **2009**, *45*, W12416. [CrossRef]

19. Vrugt, J.A.; Robinson, B.A. Improved Evolutionary Optimization from Genetically Adaptive Multimethod Search. *Proc. Natl. Acad. Sci. USA* **2007**, *104*, 708–711. [CrossRef] [PubMed]

20. Raad, D.; Sinske, A.; van Vuuren, J. Robust Multi-Objective Optimization for Water Distribution System Design using a Meta-Metaheuristic. *Int. Trans. Oper. Res.* **2009**, *16*, 595–626. [CrossRef]

21. Wang, Q.; Guidolin, M.; Savic, D.; Kapelan, Z. Two-objective Design of Benchmark Problems of a Water Distribution System via MOEAs: Towards the Best-known Approximation of the True Pareto Front. *J. Water Resour. Plan. Manag.* **2014**, *141*, 04014060. [CrossRef]

22. Moosavian, N.; Lence, B.J. Nondominated Sorting Differential Evolution Algorithms for Multiobjective Optimization of Water Distribution Systems. *J. Water Resour. Plan. Manag.* **2016**, *143*, 04016082. [CrossRef]

23. Yazdi, J.; Yoo, D.G.; Kim, J.H. Comparative study of multi-objective evolutionary algorithms for hydraulic rehabilitation of urban drainage networks. *Urban Water J.* **2017**, *14*, 483–492. [CrossRef]

24. Zitzler, E.; Laumanns, M.; Thiele, L. SPEA2: Improving the Strength of Pareto Evolutionary Algorithm—Evolutionary Methods for Design, Optimisation and Control. In Proceedings of the 2001 International Center for Numerical Methods in Engineering (EUROGEN2001), Athens, Greece, 19–21 September 2001.

25. Coello, C.A.; Toscano Pulido, G.; Salazar Lechuga, M. Handling Multiple Objectives with Particle Swarm Optimization. *IEEE Trans. Evol. Comput.* **2004**, *8*, 256–278. [CrossRef]

26. Zhang, Q.; Li, H. Multiobjective Evolutionary Algorithm based on Decomposition. *IEEE Trans. Evol. Comput.* **2007**, *11*, 712–731. [CrossRef]

27. Yazdi, J.; Sadollah, A.; Lee, E.H.; Yoo, D.G.; Kim, J.H. Application of Multi-objective Evolutionary Algorithms for Rehabilitation of Storm Sewer Pipe Networks. *J. Flood Risk Manag.* **2015**. [CrossRef]

28. Mahdavi, M.; Fesanghary, M.; Damangir, E. An Improved Harmony Search Algorithm for Solving Optimization Problems. *Appl. Math. Comput.* **2007**, *188*, 1567–1579.

29. Deb, K.; Agrawal, S.; Pratap, A.; Meyarivan, T. A Fast Elitist Non-dominated Sorting Genetic Algorithm for Multi-Objective Optimization: NSGA II. In Proceedings of the International Conference on Parallel Problem Solving From Nature, Paris, France, 18–20 September 2000; Springer: Berlin/Heidelberg, Germany; pp. 849–858.

30. Kougias, I.P.; Theodossiou, N.P. Multiobjective Pump Scheduling Optimization using Harmony Search Algorithm (HSA) and Polyphonic HSA. *Water Resour. Manag.* **2013**, *275*, 1249–1261. [CrossRef]

31. Storn, R.; Price, K. *Differential Evolution: A Simple and Efficient Adaptive Scheme for Global Optimization over Continuous Space*; International Computer Science Institute: Berkeley, CA, USA, 1995.
32. Suribabu, C.R. Differential Evolution Algorithm for Optimal Design of Water Distribution Networks. *J. Hydroinform.* **2010**, *12*, 66–82. [CrossRef]
33. Zheng, F.; Zecchin, A.C.; Simpson, A.F. Self-adaptive Differential Evolution Algorithm Applied to Water Distribution System Optimization. *J. Comput. Civ. Eng.* **2013**, *27*, 148–158. [CrossRef]
34. Yazdi, J. Decomposition based multi objective evolutionary algorithms for Design of Large-Scale Water Distribution Networks. *Water Resour. Manag.* **2016**, *30*, 2749–2766. [CrossRef]
35. Van Veldhuizen, D.A. *Multiobjective Evolutionary Algorithms: Classifications, Analyses, and New Innovations*; Air Force Institute of Technology: Wright-Patterson AFB, OH, USA, 1999.
36. Zitzler, E.; Deb, K.; Thiele, L. Comparison of Multi-Objective Evolutionary Algorithms: Empirical Results. *Evol. Comput.* **2000**, *8*, 173–195. [CrossRef] [PubMed]
37. Rossman, L.A. *EPANET 2 Users Manual*; U.S. Environment Protection Agency: Cincinnati, OH, USA, 2000.

water

MDPI

Article

A Hybrid Water Distribution Networks Design Optimization Method Based on a Search Space Reduction Approach and a Genetic Algorithm

Juan Reca [1,*] (iD), **Juan Martínez** [1] and **Rafael López** [2]

[1] Department of Engineering, Universidad de Almería, Ctra. Sacramento S.N., La Cañada de S. Urbano, 04120 Almería, Spain; jumartin@ual.es
[2] Department of Applied Physics, Universidad de Córdoba, Campus Universitario de Rabanales, 14071 Córdoba, Spain; fa1lolur@uco.es
* Correspondence: jreca@ual.es; Tel.: +34-950-015-428

Received: 30 September 2017; Accepted: 29 October 2017; Published: 2 November 2017

Abstract: This work presents a new approach to increase the efficiency of the heuristics methods applied to the optimal design of water distribution systems. The approach is based on reducing the search space by bounding the diameters that can be used for every network pipe. To reduce the search space, two opposite extreme flow distribution scenarios are analyzed and velocity restrictions to the pipe flow are then applied. The first scenario produces the most uniform flow distribution in the network. The opposite scenario is represented by the network with the maximum flow accumulation. Both extreme flow distributions are calculated by solving a quadratic programming problem, which is a very robust and efficient procedure. This approach has been coupled to a Genetic Algorithm (GA). The GA has an integer coding scheme and variable number of alleles depending on the number of diameters comprised within the velocity restrictions. The methodology has been applied to several benchmark networks and its performance has been compared to a classic GA formulation with a non-bounded search space. It considerably reduced the search space and provided a much faster and more accurate convergence than the GA formulation. This approach can also be coupled to other metaheuristics.

Keywords: water distribution networks; optimization; heuristics; search space reduction; Genetic Algorithm; hybrid method

1. Introduction

The optimal design of looped water distribution networks (WDN) can be regarded as a type of complex combinatorial problem known as NP-hard (Non-deterministic Polynomial-time hard), as it is a nonlinear, constrained, non-smooth, non-convex, and, hence, multi-modal problem [1,2]. Although mathematical programming methods such as linear and nonlinear programming techniques [3,4] have been applied to solve this problem, metaheuristics methods have been preferred due to their ability to cope with global optimization problems. Genetic Algorithms (GA) and other Evolutionary Algorithms [2], Simulated Annealing (SA) [5], Shuffle Frog Leaping Algorithm [6], Iterated Local Search [7] and Particle Swarm Optimization [8] are among the most extended metaheuristic approaches applied to water distribution networks design. Genetic algorithms have been extensively applied to solve the problem of designing the optimal water distribution network ([2,9]). GAs are based on the rules of evolution and natural selection. Multi-objective heuristic approaches have also been formulated not only to minimize the network cost but to take into consideration other conflicting objectives as well, such as the reliability of the system [10–14].

Although heuristics approaches can handle global optimization problems, they do not guarantee to find the optimal solution [15]. In addition to the lack of accuracy of the solution provided, another shortcoming of these procedures is the time they take to converge. In recent years, a great deal of research work has been carried out to improve their performance; however, in spite of these efforts, heuristics methods are still relatively inefficient and time-consuming when dealing with very large water distribution networks. This inefficiency is due to the wide search space that these algorithms must explore. Since the search space is very large, general purpose heuristic algorithms waste a considerably long time evaluating unfeasible solutions. Consequently, the probability of finding the optimal solution decreases and the convergence speed increases as the size of the search space increases. Strategies for reducing the search space are thus greatly needed.

The aim of this paper is to present a new approach to increase the efficiency of the heuristics methods applied to the optimal design of water distribution networks. The proposed approach is based on bounding the search space by analyzing two opposite extreme flow distribution scenarios and then applying velocity restrictions to the flow in the network's links. The proposed methodology has been applied to minimize the cost of a well-known benchmark network. The performance of the approach presented in this paper has been compared to a classic GA formulation with a non-bounded search space.

2. Materials and Methods

2.1. Bounding Strategy

Flow distribution can be calculated in branched networks by applying flow conservation equations in the nodes of the network. From a practical standpoint, a common procedure for this type of network is to impose velocity restrictions on the flow in the pipes. Velocity limits of piping systems can vary depending on the material and diameter and other considerations. High velocities may cause pipe erosion, loud noise, and excessive head losses. Low velocities, on the contrary, may produce sedimentation and oversizing of the system. When velocity restrictions are applied, the range of possible diameters that can be selected for a specific pipe is considerably reduced, thus simplifying the complexity of the network design. However, unlike the case of branched networks, the flow distribution in looped networks is not known a priori, and, as a consequence, this procedure cannot be used.

The methodology proposed in this work is based on reducing the search space by bounding the range of possible diameters that can be selected for a specific network link. The procedure consists of generating two opposite extreme flow distribution scenarios that satisfy the nodal flow conservation equations and nodal demands. The first scenario produces the most uniform flow distribution in the network while satisfying nodal demands and flow conservation constraints. The resulting design would provide a network with high entropy and resilience. The opposite scenario features the highest flow accumulation within certain main pipes. This scenario provides a flow distribution fairly similar to the one obtained for a spanning tree of the network.

The methodology proposed in this work to calculate both extreme flow distributions is to solve a quadratic programming problem (QPP) for each of them. A quadratic programming problem involves minimizing or maximizing a quadratic function subject to linear constraints. Quadratic programming is a particular type of nonlinear programming. Although general nonlinear algorithms can be applied to solve this type of problem, there are others that are more robust, specific and efficient [16].

The objective function of the proposed QPP for the most uniform flow distribution is to minimize the sum of the square link flows of the network. These link flows have to satisfy the flow conservation equations in the nodes of the network. This set of restrictions is linear. The problem is formulated in Equation (1):

$$Min. \sum_{i=1}^{n} Q_i^2$$
$$Subject\ to:$$
$$A \cdot Q = q$$
$$Q \geq 0$$

(1)

where: n is the number of links in the network, Q_i is the flow of the link i, A is an $(m \times n)$ array, and m is the number of nodes, and q is a vector of nodal demands.

The entry a_{ij} of array A is 1 if the flow of link j goes into node i, -1 if it leaves the node, and 0 if link j is not connected to node i.

One drawback of this procedure is that the direction of the flows has to be previously defined in order to perform the calculation. For complex networks, the number of possible flow direction combinations can be very high and finding the right one is a cumbersome procedure. To overcome this limitation, we have duplicated the number of links by adding a fictitious pipe for each link of the network in such a way that two pipes with opposed flows are considered for each link of the network. Using this procedure, the number of variables is $2n$ and array A is a $(m \times 2n)$ array. The solution of the minimization QPP problem provides the right flow directions and values that minimize the sum of network flows. The so called Maximum Dispersion (MD) flow distribution is obtained in this way.

The second scenario, with a maximum flow accumulation, also termed a Maximum Concentration scenario (MC), is obtained by maximizing the objective function and solving the equivalent QPP maximization problem.

The solution of these two problems provides two vectors flows (Q_{MC} and Q_{MD}), which bounds the range of possible flows within each network link. By imposing velocity restrictions, a pair of vectors defining the range of possible diameters between the minimum ($D_{m,i}$) and the maximum ($D_{M,i}$) for each link i can be calculated in the following way:

$$D_{m,i} = \sqrt{\frac{4 \cdot Min\ (Q_{MD,i}, Q_{MC,i})}{\pi \cdot U_M}}$$
$$D_{M,i} = \sqrt{\frac{4 \cdot Max\ (Q_{MD,i}, Q_{MC,i})}{\pi \cdot U_m}}$$

(2)

2.2. Bounded Genetic Algorithm Formulation

The bounding approach developed herein can be coupled to different types of metaheuristics methods. In this work, a GA has been used to test the performance of the proposed methodology. GAs are stochastic search procedures based on the evolutionary mechanisms of natural selection and genetics [17]. GAs mimic the highly effective optimization model that has naturally evolved for dealing with large, highly complex systems.

The GA is based on the GENOME model developed by Reca and Martínez [2]. However, some modifications in the code described below have been made to implement the proposed strategy. A new software program called B-GENOME (B-GA) has been developed to implement this new approach. The program has been developed using the VBA (Visual Basic for Applications) programming language in the Excel© (Microsoft, Redmond, Washington, DC, USA) spreadsheet environment.

GENOME used an integer-coding scheme. Each solution (individual) was coded by a vector of n discrete variables (diameter sizes assigned to each link of the network). The variable was coded by an integer value ranging from one (first possible diameter for that particular link) to $n_{d,i}$ (last possible diameter). This methodology has many advantages since there are no limitations on the number of possible diameter sizes that can be assigned to a specific pipe. In the classic formulation of GENOME, the number of possible diameter sizes was equal for each link and this value was equal to the total number of diameters in the pipe database. The same coding scheme has been adopted in B-GENOME, although some modifications have been made to allow for a variable number of possible diameters for each link. The new B-GENOME algorithm used in this work has an integer coding scheme and a variable number of alleles. The number of alleles depends on the number of possible diameters comprised within the velocity restrictions of each link.

In order to test and compare the new approach to the classic GA formulation, the initial population has been obtained randomly. This initial population evolves from one generation to another by undergoing an iterative reproductive cycle. This cycle comprises three subsequent operators: selection, crossover and mutation. For the selection operator to be applied the fitness of each individual is evaluated as the sum of the cost of the pipes making up the network plus a penalty function applied to take into account nodal pressure deficits (see Equation (3)):

$$F(\boldsymbol{D}) = \sum_{i=1}^{n} c_i \cdot L_i + p \cdot \sum_{j=1}^{N} \left(max\left(hr_j - h_j\right), 0\right) \tag{3}$$

where: c_i is the pipe cost ($\text{\texteuro} \, m^{-1}$), which is a function of the diameter D_i, L_i is the length of the link i, p is a penalty multiplier, N is the number of nodes in the network, h_{rj} is the required pressure head in the node j and h_j is the actual pressure head computed by the hydraulic solver EPANET for the node j.

The value of the penalty multiplier may affect the accuracy of the solution, so it should be properly calibrated. To cope with this problem, some researchers recommend different constraint-handling techniques, such as the use of variable values or self-adaptive penalty functions [18]. However, in this work, for the sake of simplicity, and in order to compare both approaches under the same conditions, a constant penalty multiplier has been applied. A high value has been assigned to this penalty multiplier (10^9 €/m) to avoid finding solutions that violate the pressure restrictions. In order to compute the pressure deficits, the nodal pressures for each individual in the population have been computed by using a network solver. The hydraulic solver EPANET has been used for this purpose [19]. The EPANET engine is used when needed by calling on the EPANET toolkit from the VBA software code developed in this work. B-GENOME implements all the different options to perform the three basic operations that were available in GENOME [2].

2.3. Structure of B-GENOME

In addition to the GA module, B-GENOME implements a module to solve the QPPs. This module makes use of the EXCEL optimization add-on SOLVER (frontline systems). Both data input and results output modules complete the structure of the B-GENOME software model. The input module reads both the network information and the available pipe diameters database. The network information is imported from an EPANET input file (*.inp file). The pipe database is stored in a table within an Excel sheet. The output module writes the final solution found by the model (optimal vector of pipe diameters and cost of the network) and the best fitness function value for every generation in a spreadsheet.

The flowchart of the B-GENOME model is depicted in Figure 1.

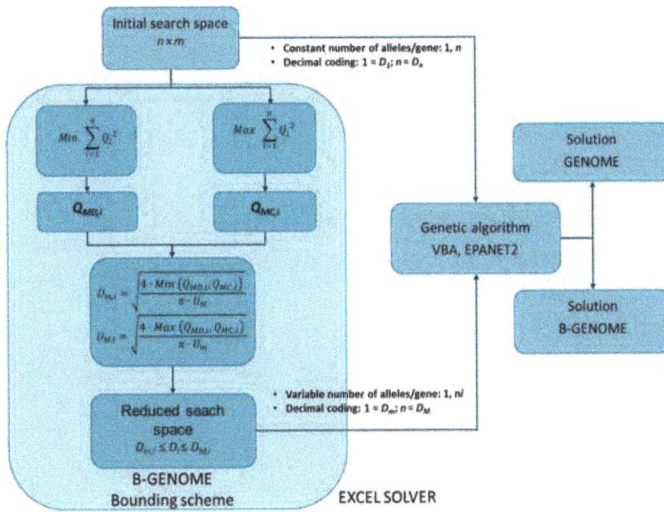

Figure 1. Flowchart of the proposed model.

2.4. Testing of the B-GENOME Model

The proposed methodology has been applied to 2 well-known benchmark networks in order to compare the performance of the classic GA formulation with the new bounded GA. The selected networks are the Alperovits and Shamir [3] network (A&S, also known as a two-loop network) and the Hanoi water distribution network. Both have been extensively used to test different water distribution design optimization algorithms, but they feature different size characteristics. While the first is a small network with seven nodes and eight pipes arranged in two loops, the latter can be considered as medium-sized, with 32 nodes and 34 pipes and 3 loops.

The A&S network layout is shown in Figure 2. The system is fed by gravity from a reservoir of 210 m fixed head. The pipes are all 1000 m long. The minimum pressure limitation is 30 m above ground level for each node. There are 14 commercial diameters to be selected. The nodal head and demands, the cost per meter for each pipe size and other data are reported by Alperovits and Shamir and other works [3].

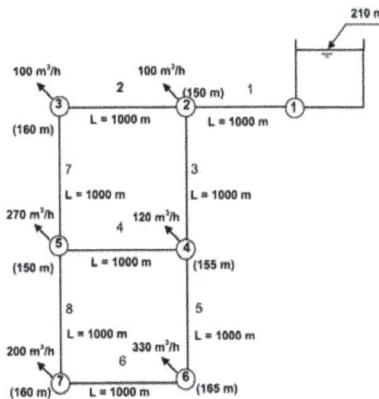

Figure 2. Layout of the Alperovits and Shamir network [3].

The Hanoi water distribution network (see Figure 3) features 32 nodes and 34 pipes organized in 3 loops. No pumping facilities are considered since only a single fixed head source at elevation of 100 m is available. The minimum head requirement at all nodes is fixed at 30 m. In this case, there is a set of 6 commercially available diameters. The cost function is nonlinear. The pipe head losses were calculated using the Hazen–Williams equation with a Hazen–Williams roughness coefficient, $C = 130$.

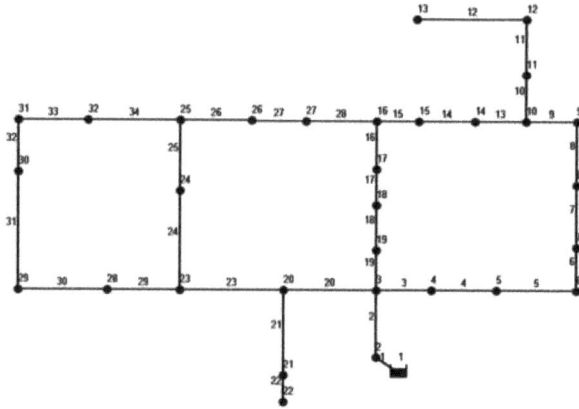

Figure 3. Layout of the Hanoi network.

To evaluate the effect of the search space reduction on the convergence speed and accuracy of the solution, both algorithms (GENOME and B-GENOME) have been applied to solve the WDN design problem of both the two-loop and Hanoi networks. To be consistent and enable comparison, the same input data and parameters and analysis options have been chosen.

The pipe head losses were calculated using the Hazen–Williams equation with a Hazen–Williams roughness coefficient, $C = 130$. The values of the other parameters of the Hazen–Williams equation were set to the defaults of the EPANET 2.0 network analysis software (USEPA, Cincinnati, OH, USA). The population size was limited to 100 individuals in the case of A&S and 200 for Hanoi. The number of generations was 200 and 300, respectively. The remaining input parameters and options for the GA algorithm are summarized in Table 1.

Table 1. Input parameters for the Genetic Algorithm.

Parameter	A&S	Hanoi
Population (np)	100	200
Generations (ng)	200	300
Crossover prob. (p_{cross})	0.9	0.9
Mutation prob. (P_{mut})	0.05	0.05
Prob. of gene crossing (r_{cross})	0.5	0.5
Reproduction plan	steady-state-delete-worst plan	steady-state-delete-worst plan
Crossover operator	uniform crossover	uniform crossover

The steady-state-delete-worst plan inserts individuals as they are bred whenever its fitness exceeds that of the least fit member of the parent population. The least fit member of the parent population is removed and replaced by the offspring. The crossover operator implies that a pair of parent chromosomes exchanges information in order to produce a pair of offspring chromosomes that inherit their characteristics. The probability of crossing two chromosomes is defined by the input parameter p_{cross}. In the uniform crossover, the parents' chromosomes exchange their genetic information gene to gene. The probability of exchanging genes is defined by the gene crossing rate (r_{cross}). Ten simulations were performed both for the new bounded algorithm and the classic GA algorithm.

3. Results

The first step in the calculation procedure is to solve the QPPs stated in Equation (1). The results of these problems provide both vectors of flows for each link of the network. Both flow values represent the limits to the flow in each link of the network and thus reduce the search space. The results provided are given for both the A&S (Table 2) and the Hanoi networks (Table 3). Table 2 (A&S) and Table 3 (Hanoi) show the flow range, the minimum and maximum diameters, and the number of possible diameters compatible with the velocity restrictions provided by the QPP problems.

Table 2. Flow range, maximum and minimum diameters and number of possible diameters for each link obtained from the Quadratic Programming Problems QPPs for the Alperovits and Shamir network.

Link	Q_{MD} (L/h)	Q_{MC} (L/h)	D_m (mm)	D_M (mm)	N°D
1	311.1	311.1	356	610	6
2	117.0	27.8	102	559	10
3	166.3	255.6	254	559	8
4	40.0	75.0	102	457	8
5	93.0	147.2	152	610	10
6	1.3	55.6	25.4	406	10
7	89.3	0.0	25.4	508	12
8	54.3	0.0	25.4	406	10

Table 3. Flow range, maximum and minimum diameters and number of possible diameters for each link obtained from the QPPs for the Hanoi network.

Link	Q_{MD} (L/h)	Q_{MC} (L/h)	D_m (mm)	D_M (mm)	N°D
1	19,940	19,940	1016	1016	1
2	19,050	19,050	1016	1016	1
3	5326	6810	1016	1016	1
4	5196	6680	1016	1016	1
5	4471	5955	1016	1016	1
6	3466	4950	762	1016	2
7	2116	3600	609.6	1016	3
8	1566	3050	508	1016	4
9	1041	2525	406.4	1016	4
10	2000	2000	609.6	1016	3
11	1500	1500	508	1016	4
12	940	940	406.4	1016	5
13	1484	0	304.8	1016	6
14	2099	615	304.8	1016	6
15	2379	895	304.8	1016	5
16	2968	1205	508	1016	4
17	3833	2070	609.6	1016	3
18	5178	3415	762	1016	2
19	5238	3475	762	1016	2
20	7637	7915	1016	1016	1
21	1415	1415	508	1016	4
22	485	485	304.8	1016	6
23	4947	5225	1016	1016	1
24	2890	3065	609.6	1016	3
25	2070	2245	609.6	1016	3
26	992	1270	406.4	1016	5
27	92	370	304.8	762	5
28	278	0	304.8	762	5
29	1011	1115	406.4	1016	5

Table 3. *Cont.*

Link	Q_{MD} (L/h)	Q_{MC} (L/h)	D_m (mm)	D_M (mm)	N°D
30	721	825	406.4	1016	5
31	361	465	304.8	1016	6
32	1	105	304.8	1016	6
33	104	0	304.8	1016	6
34	909	805	304.8	1016	6

With the aim of evaluating the accuracy of the solution and the convergence speed of the new bounded algorithm and the classic GA algorithm, ten runs were performed for each algorithm with the same input parameters, data, and analysis options. Results of these simulations are shown in Figure 4 for the A&S network and in Figure 5 for the Hanoi network.

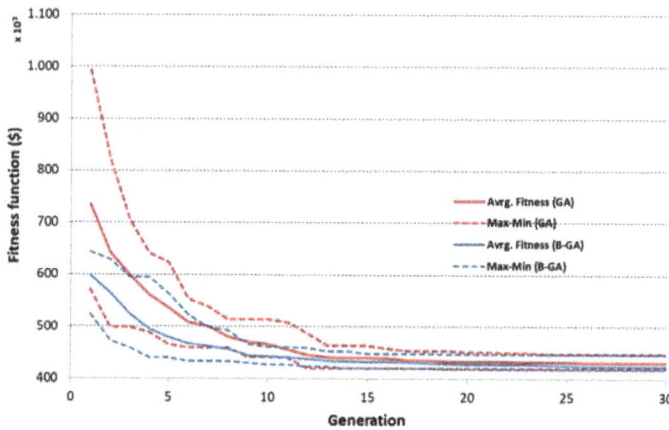

Figure 4. Evolution of the best fitness value for B-GENOME and GENOME algorithms (Alperovits and Shamir Network).

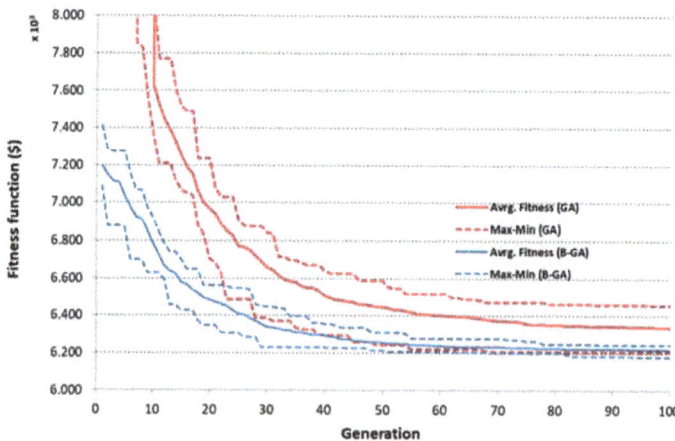

Figure 5. Evolution of the best fitness value for B-GENOME and GENOME algorithms (Hanoi Network).

Not only did the B-GA algorithm outperform the GA in convergence speed, but it also performed better when it came to the accuracy of the solution. The solutions provided by both algorithms are given in Table 4 (A$S) and Table 5 (Hanoi).

Table 4. Solutions found by both algorithms for the Alperovits and Shamir network.

Algorithm	Min Cost ($)	Max Cost ($)	Avrg. Cost ($)	Std. Dev. ($)	C. Var (%)
B-GA	419,000	447,000	424,000	9099	2.15
GA	420,000	448,000	430,900	11,344	2.63

Table 5. Solutions found by both algorithms for the Hanoi network.

Algorithm	Min Cost ($)	Max Cost ($)	Avrg. Cost ($)	Std. Dev. ($)	C. Var (%)
B-GA	6,182,006	6,242,051	6,219,390	19,831	0.32
GA	6,208,937	6,373,131	6,296,366	57,791	0.92

4. Discussion

A significant reduction of the search space was achieved with the proposed methodology. Regarding the Alperovits and Shamir network, as its pipe database is composed of 14 different diameter values and the network has eight links, the total search space in the unbounded problem is equal to $14^8 = 1.48 \times 10^9$ possible network designs. The search space for the bounded problem is reduced to 4.61×10^7, which means that the search space becomes approximately 3% of the total search space of the problem (see Table 2). In the case of the Hanoi network, the reduction is even higher. There are six possible diameters in the database and the number of links is equal to 34. The resulting number of alternative designs is 2.87×10^{26}, whereas the size of the search space in the bounded problem is 4.35×10^{16} (see Table 3). The reduction of the search space is expected to be higher for a larger number of links and the number of pipe diameters in a given problem. The velocity limits also play an important role as the search space reduction increases as the velocity limits range becomes narrower.

Another advantage of the search space reduction approach presented herein is that it is able to detect branched links in the network by comparing the flow value for these links in both flow distributions and checking if it is the same. For instance, this is the case of link 1 in the A&S network (Table 2) and links 1, 2, 10, 11, 12, 21 and 22 in the Hanoi network (Table 3). Since the flow is established in these branched links, the range of possible diameters compatible with the flow velocity restrictions is considerably reduced and so the complexity of the problem. In addition, a special treatment applying other optimization methods best suited for branched networks can be performed in these branched sub-networks.

Regarding the speed of convergence, both algorithms performed well, although the new proposed algorithm B-GA outperformed the GA for both networks. It is worth highlighting that convergence was reached rather quickly in both cases. There were no substantial differences in the convergence speed for the A&S network (both algorithms converged approximately after the 20th generation). Convergence was found later for the Hanoi network due to the larger size of the problem, and, in this case, B-GA clearly converged faster than GA (130th for the B-GA and 293th in the case of GA). In the A&S network, only 2000 function evaluations were needed to converge (100 individuals and 20 generations). In the case of the Hanoi network, 26,000 evaluations were needed in the B-GA algorithm and 58,600 in the GA algorithm. This entails a very small fraction of the total search space (see Figures 4 and 5).

The proposed B-GA algorithm not only considerably reduced the search space, but also provided a much faster and more accurate convergence than the classic GA formulation In the case of the A&S network, the best solution found by the B-GA algorithm was 419,000, which is the global optimum as reported in previous works. This minimum cost was obtained in three out of the 10 runs performed. A quasi-optimal solution (420,000) was found in another five out of 10 iterations. The average cost in the 10 simulations was close to the optimum (424,000). The GA performed slightly worse; this algorithm did not reach the minimum cost in any simulation, although the solution found came very close (420,000). The average cost was also higher (430,900) than the one obtained by B-GA (see Tables 4 and 5).

Both algorithms found solutions relatively close to the global optimum. For the Hanoi network, the best solution found by the B-GA algorithm was 6,182,006. The average cost in the 10 simulations was close to the optimum (6,219,390). Again, the GA clearly performed worse. The minimum solution found was 6,208,937 (0.44% higher). The average cost was also higher (6,296,366) than the one obtained by B-GA (1.24%). The solution found by the proposed B-GA is comparable to the optimal solution found in the literature. The lowest cost solution reported is 6,056,000 [5,20]. However, these results were not obtained using the EPANET 2.0 network solver and the coefficients of the Hazen–Williams head loss equation were slightly different. The best solution found when using the EPANET 2.0 network solver was that reported by Lansey and Eusuff [6] (6,073,000) using the Shuffled Frog Leaping Algorithm (SFLA). This solution is slightly better than the one found in this work with B-GA. Nevertheless, it should be noted that, in our study, the number of evaluations was low because the aim was not to achieve the minimum cost but to test and compare the algorithm with a classic GA algorithm under the same conditions.

As a consequence, the proposed B-GA algorithm considerably reduced the search space and provided a much faster and more accurate convergence than the classic GA formulation. It is expected that, for more complex networks (networks with a higher number of links or higher number of pipe diameters), the advantages provided by the new B-GA approach could be even greater.

Another major advantage of the proposed search space reduction is that it can be coupled to other metaheuristics. The performance of this strategy when applied to other types of metaheuristics is an issue still to be investigated.

5. Conclusions

The following conclusions can be drawn from this research work:

- A new approach based on bounding and reducing the total search space in a water distribution network design problem has been developed. This new approach reduces the search space by analyzing two opposite extreme flow distribution scenarios and then applying velocity restrictions to the pipes.
- This new approach has been coupled to a GA in order to improve its performance.
- The proposed B-GA algorithm considerably reduced the search space and provided a much faster and more accurate convergence than the classic GA formulation for a small and a medium benchmark network. It is expected that, for more complex networks, the advantages provided by the new B-GA approach could be even greater.
- This new approach could also be implemented in other types of heuristic methods. The improvements on the performance of these heuristics provided by the new approach are still to be investigated.

Acknowledgments: The authors acknowledge the support of the Spanish Ministry of Economy, Industry and Competitiveness under the contract CGL2010-21865.

Author Contributions: J. Reca and R. López conceived and developed the model, J. Reca and J. Martínez performed the simulations and analyzed the results, J. Reca wrote the paper, R. López and J. Martínez revised and corrected the final document.

Conflicts of Interest: The authors declare no conflict of interest.

References

1. Yates, D.; Templeman, A.; Boffey, T. The computational complexity of the problem of determining least capital cost designs for water supply networks. *Eng. Optim.* **1984**, *7*, 143–155. [CrossRef]
2. Reca, J.; Martínez, J. Genetic algorithms for the design of looped irrigation water distribution networks. *Water Resour. Res.* **2006**, *42*, W05416. [CrossRef]
3. Alperovits, E.; Shamir, U. Design of optimal water distribution systems. *Water Resour. Res.* **1977**, *13*, 885–900. [CrossRef]
4. Varma, K.; Narasimhan, S.; Bhallamudi, S.M. Optimal design of water distribution systems using NLP method. *J. Environ. Eng.* **1997**, *123*, 381–388. [CrossRef]
5. Cunha, M.D.; Sousa, J. Water distribution network design optimization: Simulated annealing approach. *J. Water Resour. Plan. Manag.* **1999**, *125*, 215–221. [CrossRef]
6. Lansey, K.E.; Eusuff, M.M. Optimization of water distribution network design using the Shuffled Frog Leaping Algorithm". *J. Water Resour. Plan. Manag.* **2003**, *129*, 10–25. [CrossRef]
7. De Corte, A.; Sörensen, K. An Iterated Local Search Algorithm for multi-period water distribution network design optimization. *Water* **2016**, *8*, 359. [CrossRef]
8. Montalvo, I.; Izquierdo, J.; Pérez, R.; Tung, M.M. Particle Swarm Optimization applied to the design of water supply systems. *Comput. Math. Appl.* **2008**, *56*, 769–776. [CrossRef]
9. Montesinos, P.; Garcia-Guzman, A.; Ayuso, J.L. Water distribution network optimisation using modified genetic algorithm. *Water Resour. Res.* **1999**, *35*, 3467–3473. [CrossRef]
10. Farmani, R.; Walters, G.A.; Savic, D.A. Trade-off between total cost and reliability for anytown water distribution network. *J. Water Resour. Plan. Manag.* **2005**, *131*, 161–171. [CrossRef]
11. Reca, J.; Martinez, J.; Baños, R.; Gil, C. Optimal design of gravity-fed looped water distribution networks considering the resilience index. *J. Water Resour. Plan. Manag.* **2008**, *134*, 234–238. [CrossRef]
12. Reca, J.; Martinez, J.; Gil, C.; Baños, R. Application of several meta-heuristic techniques to the optimization of real looped water distribution networks. *Water Resour. Manag.* **2008**, *22*, 1367–1379. [CrossRef]
13. Baños, R.; Gil, C.; Reca, J.; Martinez, J. Implementation of scatter search for multi-objective optimization: A comparative study. *Comput. Optim. Appl. J.* **2009**, *42*, 421–441. [CrossRef]
14. Baños, R.; Reca, J.; Martinez, J.; Gil, C.; Márquez, A.L. Resilience indexes for water distribution network design: A performance analysis under demand uncertainty. *Water Resour. Manag.* **2011**, *25*, 2351–2366. [CrossRef]
15. Glover, F.; Laguna, M.; Dowsland, K.A. *Modern Heuristic Techniques for Combinatorial Problems*; Blackwell: London, UK, 1993.
16. Vavasis, S.A. *Nonlinear Optimization: Complexity Issues*; Oxford University Press: Oxford, UK, 1991; p. 165.
17. Holland, J.H. *Adaptation in Natural and Artificial Systems*; MIT Press: Cambridge, MA, USA, 1975.
18. Wu, Z.Y.; Walski, T. Self-Adaptive Penalty Approach compared with other Constraint-Handling techniques for pipeline optimization. *J. Water Resour. Plan. Manag.* **2005**, *131*, 181–192. [CrossRef]
19. Rossman, L.A. *EPANET 2 Users Manual*; Rep. EPA/600/R-00/057; United States Environmental Protection Agency: Cincinnati, OH, USA, 2000.
20. Geem, Z.W.; Kim, J.H.; Loganathan, G.V. A new heuristic optimisation algorithm: Harmony search. *Simulation* **2001**, *76*, 60–68. [CrossRef]

water MDPI

Article

Energy Dissipation in Circular Drop Manholes under Different Outflow Conditions

Feidong Zheng [1,2], Yun Li [1,3,*], Jianjun Zhao [1] and Jianfeng An [1]

[1] Nanjing Hydraulic Research Institute, Nanjing 210029, China; feidongzheng@126.com (F.Z.); jjzhao@nhri.cn (J.Z.); jfan@nhri.cn (J.A.)
[2] College of Water Conservancy and Hydropower Engineering, Hohai University, Nanjing 210098, China
[3] State Key Laboratory of Hydrology-Water Resources and Hydraulic Engineering, Nanjing 210029, China
* Correspondence: yli_nhri@126.com; Tel.: +86-025-8582-8022

Received: 3 September 2017; Accepted: 26 September 2017; Published: 30 September 2017

Abstract: Circular drop manholes have been an important device for energy dissipation and reduction of flow velocities in urban drainage networks. The energy dissipation in a drop manhole depends on the manhole flow patterns, the outflow regimes in the exit pipe and the downstream operation conditions, and is closely related to the hydraulic and geometric parameters of the manhole. In the present work, the energy dissipation of a drop manhole with three drop heights was experimentally investigated under free outflow conditions and constrained outflow conditions. The results demonstrate that the local head loss coefficient is solely related to the dimensionless drop parameter for free surface outflow without a downstream backwater effect, whereas it depends on the dimensionless submerge parameter for constrained outflow. Moreover, it is concluded that the energy dissipation is largely promoted when outlet choking occurs.

Keywords: drop manhole; energy dissipation; free outflow conditions; constrained outflow conditions; outlet choking

---

1. Introduction

Drop manholes are hydraulic features that are widely implemented in urban drainage networks for steep catchments. The energy dissipation of plunge flow in drop manholes is one of the major concerns for urban system drainage designers. As pointed out by Christodoulou [1] and Granata et al. [2], adequate energy dissipation in drop manholes should be achieved in order to avoid excessive flow velocities—and, thus, erosion—in the exit pipe. However, this cannot be always attained, because of the wide range of discharges experienced in sewer systems during a flood event [3].

The energy dissipation of a drop manhole is related to many factors, which can be grouped into four categories: (a) the approach flow conditions associated with the filling ratio of the upstream pipe and the approach flow Froude number; (b) the outlet flow conditions, such as free outflow conditions, including free surface flow and pressurized flow (in the condition of outlet choking), and constrained outflow conditions, in which backwater effects downstream of the exit pipe are imposed on the outlet flow; (c) manhole configurations and dimensions, such as inlet or outlet entrance configurations, baffles in the manhole, drop height, and manhole diameter, etc.; and (d) air supply conditions. In the hydraulic studies of circular drop manholes [1,2,4–7] and rectangular drop manholes [3,8–10], the energy dissipation was investigated under free surface outflow conditions for different approach flow conditions or manhole configurations. In some other studies, the interest has been focused on the effects of ventilation absence on the sub-atmospheric pressure and pool depth, which can strongly influence the interaction between water and airflow and the dissipation energy in drop manholes [11]. However, the energy dissipation of a circular drop manhole has not yet been investigated under constrained outflow conditions or outlet choking, although the drop manhole has to operate under surcharged conditions in many instances, i.e., severe rain events.

This work aims to investigate the energy dissipation of the flow inside the circular drop manhole and its relation to the outflow conditions. Results of laboratory experiments are presented and analyzed for three manhole models of different drop heights.

2. Experimental Set-Up and Experiments

The experimental arrangement is schematically shown in Figure 1. The experimental facility consisted of a head tank, plexiglass circular manhole models and a downstream pool. The plunge flow was created by the flow from the upstream horizontal inlet pipe with an internal diameter D_{in} = 200 mm. The inflow pipe was connected to the manhole with a straight inlet entrance, and the flow to this pipe was provided from a head tank. A manhole model with an internal diameter of D_M = 0.54 m with drop heights of s = 0.93, 1.50 and 2.40 m was used in the tests performed. The lower part of the vertical dropshaft was connected to a horizontal outlet pipe with an internal diameter D_{out} = 200 mm. The shaft pool height P was 35 mm. The flow from the outflow pipe was drained into a downstream pool that was connected to the laboratory sump. The pool was considered to be a pressurized system downstream from the manhole, which allowed for analyzing a wide range of back pressures from the exit pipe, taking into account the various work conditions.

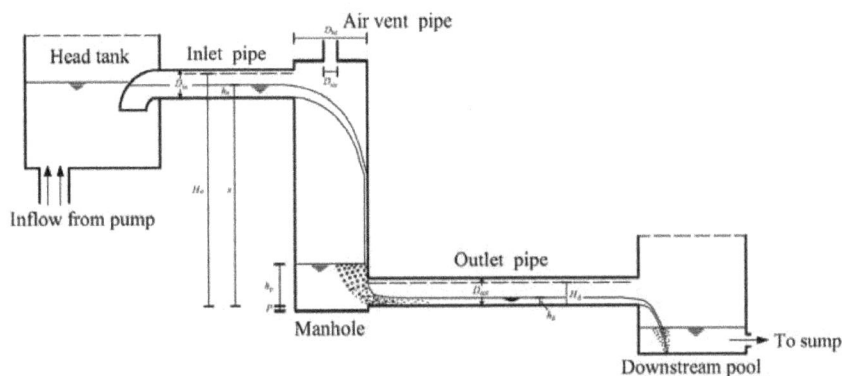

Figure 1. Sketch of experimental setup.

A pump was used to supply water from the main laboratory sump, and the discharges were measured with an ultrasonic flowmeter. Flow depths were recorded with piezometers in the upstream and downstream pipes, while the time-average pool depth h_p was measured by using a set of piezometers connected to the manhole bottom. The approach flow depth h_o was measured 1.2 m from the manhole inlet, where the flow has a horizontal surface and the pressure distribution is almost hydrostatic. The downstream flow depth h_d was recorded 2 m from the manhole outlet, where the flow is gradually varied and the air entrained by the manhole almost detrained. The air demand tests were performed by connecting the inlet pipe to the head tank with a same diameter elbow and sealing the manhole with a plexiglass cover on the manhole top, thus the air was supplied only through a 50 mm diameter pipe fitted to the cover. The airflow into the manhole was calculated by measuring the mean airflow velocity with a thermal anemometer.

Overall, seven series of experiments were run to investigate the energy dissipation in circular drop manholes under different outflow conditions (see Table 1). The first three series were run under free surface outflow conditions without backwater effects from the downstream pool. In series four, experiments were performed to investigate the outlet choking, i.e., the sudden transition from free surface to pressurized flow in the exit pipe. Series 5–7 were run under constrained outflow conditions, in which back pressures from the downstream pool were imposed on the outlet flow.

Table 1. List of experiments.

Series	s (m)	Inflow Condition	Outflow Condition
1	0.93	Free surface	Free surface
2	1.50	Free surface	Free surface
3	2.40	Free surface	Free surface
4	0.93	Pressurized (full pipe)	Free surface, Pressurized
5	0.93	Free surface	Constrained (backwater effect)
6	1.50	Free surface	Constrained (backwater effect)
7	2.40	Free surface	Constrained (backwater effect)

3. Results and Discussion

3.1. Flow Patterns

For a manhole under free outflow conditions, the energy dissipation is strongly related to the flow patterns describing the drop manhole flow and the outflow regimes in the exit pipe. Chanson [8–10] defined three basic flow regimes of the drop manhole flow for rectangular drop manholes, namely, flow Regimes R1, R2, and R3, based on the free falling nappe impact location. Regime R1 usually occurs at low flow rates with the falling jet directly impacting into the manhole pool. For Regime R2, the falling nappe impacts the manhole outlet zone. With increasing discharges, this flow regime transforms into Regime R3, with the falling jet impacting on the manhole inner sidewall. For Regime R3, a water veil spreads down the manhole wall beyond impingement, forming a water curtain at the manhole outlet. For larger discharges, a roller tends to form at the top of the impact region, as observed by Rajaratnam et al. [4]. At high flow rates, outlet choking occurs when the free surface flow in the exit pipe transits to pressurized flow [12].

The classification of drop manhole flow regimes was extended by de Marinis et al. [13] and Granata et al. [2,6], taking into account additional effects present for circular drop manholes. According to their investigations, three subregimes for Regime R2 and two subregimes for Regime R3 were proposed, as indicated in Figures 2 and 3, respectively. Regime R2a occurs if the jet impacts the zone between the manhole bottom and the manhole outlet. This flow regime transforms to Regime R2b, with the entire falling jet impacting the outlet pipe invert. Regime R2c can be observed if the jet partially impacts on the outlet pipe obvert. Compared with Regime R3a, the flow jet for Regime R3b impacts the manhole sidewall at a higher Froude number, leading to the formation of a radially spreading water jet. Based on the experimental observations of this study, different shapes of the free falling nappe before impingement were observed (see Table 2). For a manhole with large drop height at low flow rates, the side edges of the nappe intersect and form a 'central ridge', while for smaller drop height or high flow rates, the nappe usually exhibits with a shape of horseshoe. These were consistent with the earlier observations of Chanson [9].

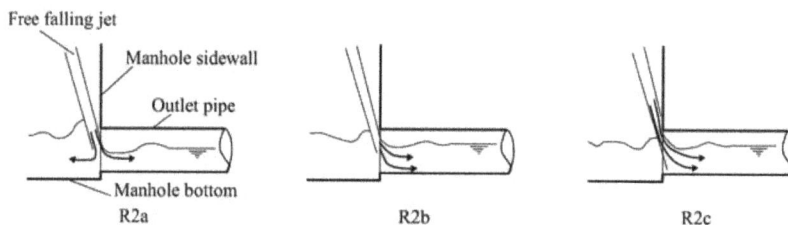

Figure 2. Regime R2 with subregimes.

Figure 3. Regime R3 for *s* = 1.5 m with Regimes (**a**) R3a; and (**b**) R3b.

Table 2. Shape of free flow jet before impingement.

s (m)	Shape of Flow Jet for Different Regimes		
	R1	R2	R3
0.93	Central ridge, Horseshoe	Horseshoe	Horseshoe
1.50	Central ridge	Horseshoe	Horseshoe
2.40	Central ridge	Central ridge	Central ridge, Horseshoe

For a manhole under constrained outflow conditions, the energy dissipation also depends on the downstream operation conditions, i.e., water depths in the downstream pool. At low flow rates, no obvious hydraulic jump forms in the exit pipe in the variation range of water depths in the downstream pool. For large discharges, with the increase of water depth in the downstream pool a hydraulic jump starts in the exit pipe and then moves toward the outlet entrance until it becomes a critical hydraulic jump at the outlet entrance, and behaves as a submerged jump near the out entrance at last.

3.2. Energy Dissipation

3.2.1. Definition

For a manhole under free outflow conditions, the energy dissipation in regime R1 was caused by the direct impact of the free falling nappe on the bottom of the shaft, inducing zones of large velocity gradients and increase of flow turbulence. A poor energy dissipation of the drop manhole may occur if the falling jet collided with the invert of downstream sewer (Regime R2). In this regime, most of the flowrate was conveyed to downstream pipe directly, causing undesirable downstream conditions. In regime R3, the energy dissipation occurred when the falling jet impinged on the inner side of the manhole, leading to the formation of a splash jet directed upwards, and a downward spreading jet. This energy dissipation was promoted by the frictional resistance of the spreading flow due to the roughness of the manhole wall, and by the mixing of the spreading jet with the water in the manhole pool. The energy dissipation can vary within large limits when the free surface flow in the exit pipe transits to pressurized flow. For a manhole under constrained outflow conditions, the energy dissipation also depends on the water depth in the downstream pool. Figure 4 shows two definition sketches of drop manholes under free surface and pressurized outflow conditions. The relative energy loss is defined as

$$\eta = \frac{H_o - H_d}{H_o} \tag{1}$$

where the approach flow energy head is $H_o = s + h_o + V_o^2/2g$, and the outflow energy head is $H_d = h_d + V_d^2/2g$ for free surface conditions, while $H_d = p_d/\rho g + V_d^2/2g$ for pressurized conditions.

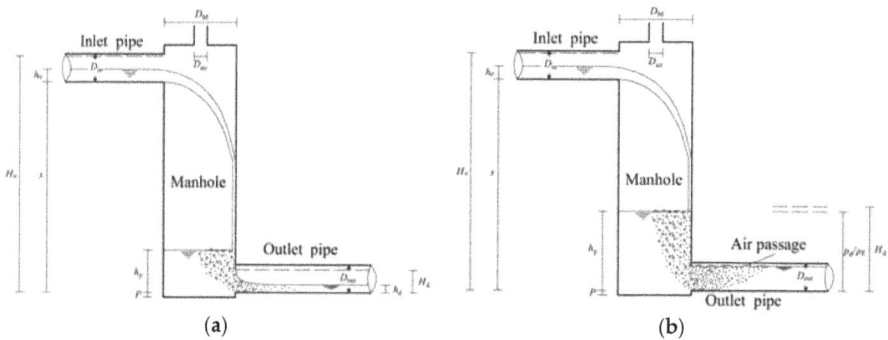

Figure 4. Definition sketch for (**a**) free surface outflow; and (**b**) pressurized outflow.

3.2.2. Free Outflow Conditions

(1) Free surface outflow

Plots of η versus the dimensionless flow rate Q^* for free surface outflow are presented in Figure 5 under different drop heights s, in which Q^* is defined by the equation

$$Q^* = \frac{Q_w}{\sqrt{gD_{in}^5}} \tag{2}$$

where Q_w = the volumetric flow rate and g = acceleration due to gravity.

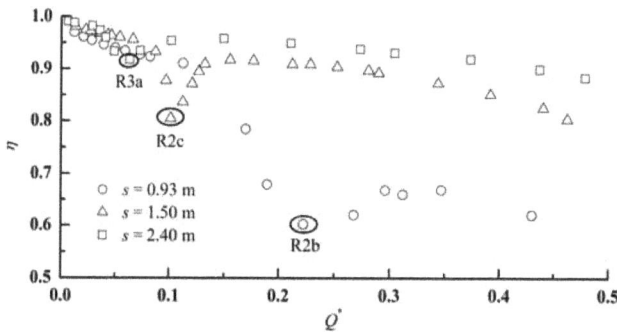

Figure 5. Variation of η with Q^* for free surface outflow.

It can be seen from this figure that the largest energy losses are observed in Regime R1 for each of the drop heights s. Values of η drop rapidly with an increase in Q^*, and pass through their minimum values at different Regimes, i.e., at Regime R2 for $s = 0.93$ m and $s = 1.50$ m, and at Regime R3 for $s = 2.40$ m. This may be attributed to the effect of the shape of the free-falling nappe before impingement; the nappe has a horseshoe shape for $s = 0.93$ m and $s = 1.50$ m, while the side edges of the nappe intersect for $s = 2.40$ m. Subsequently, values of η increase first, and then decrease for different s at high flow rates.

The local head loss in a drop manhole is generally estimated regardless of flow regimes by

$$\Delta H = K \frac{V_o^2}{2g} \tag{3}$$

where K is the local head loss coefficient. Dissipative studies of circular drop manholes have convinced us that the local head loss coefficient was based solely on the drop parameter $D = (gs)^{0.5}/V_o$ [1,6].

Granata et al. [6] found that the local head loss coefficient of a drop manhole under free surface outflow conditions, K_f, depends on the drop parameter D in the range $0 < D < 8$ as

$$K_f = 0.25 + 2D^2 \tag{4}$$

It is shown in Figure 6 that the present test date agrees well with Equation (4), extending the range of agreement up to $D = 21.6$.

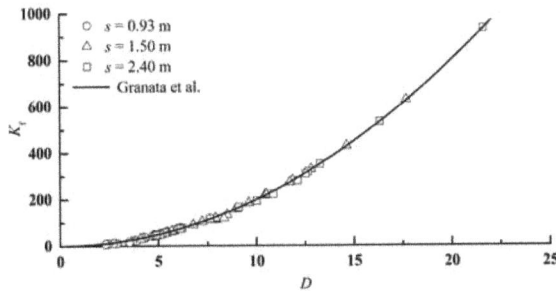

Figure 6. K_f versus D for entire present test range.

(2) Pressurized outflow

Figure 7 shows that the energy dissipation for drop manhole increases notably when the free surface flow in the exit pipe shifts to pressurized flow, indicating a large dissipation effect in choking flow. This may be explained by the strong mixing mechanism between the entrained air and the water flow in the outlet pipe, which can be illustrated by the remarkable increase in dimensionless air demand Q_a/Q_w in Figure 8. The dimensionless pool height h_p/D_{out} is plotted as a function of Q^* in Figure 9. Once the flow in the exit pipe chokes, an abrupt drop in manhole pool height and large unsteady fluctuations in pool water surface illustrated by large standard deviations occur. These results indicate that the outlet choking is advantageous in promoting energy dissipation and reducing manhole pool height, which has to be smaller than the drop height to avoid undesired backwater effects to the approaching flow.

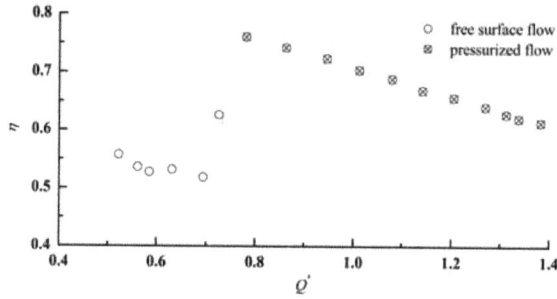

Figure 7. Variation of η with Q^* for $s = 0.93$ m.

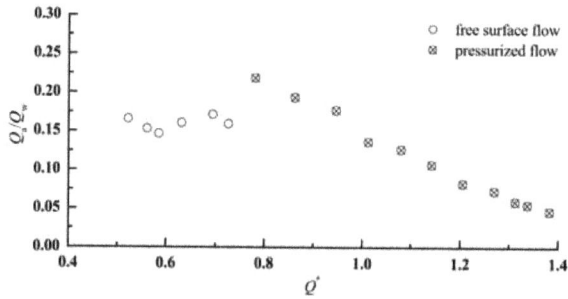

Figure 8. Variation of Q_a/Q_w with Q^* for $s = 0.93$ m.

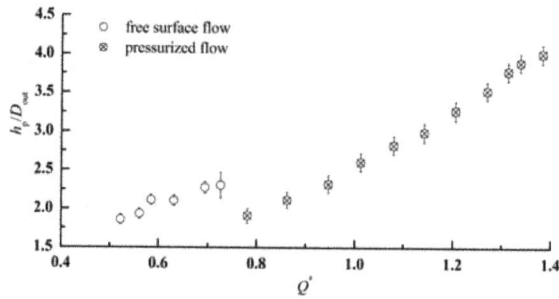

Figure 9. Variation of h_p/D_{out} with Q^* for $s = 0.93$ m.

3.2.3. Constrained Outflow Conditions

For a drop manhole under constrained outflow conditions, the energy dissipation is difficult to estimate when the hydraulic jump passes the measuring location because of the uncertainly in the calculation of the residual energy in the exit pipe, given the unstable water depth and pressure. Hence, for a given large flow rate, the minimum water depth in the downstream pool is set to the value at which a critical hydraulic jump or submerged hydraulic jump occurs. The head loss coefficient of a drop manhole under constrained outflow conditions due to backwater effects from the downstream pool, K_c, can be calculated using Equation (3).

For a given drop height s and specified approach flow condition, if H_d is proportional to the manhole pool height h_p, that is

$$H_d = \alpha h_p \tag{5}$$

where α is a dimensionless parameter, then K_c can be described by

$$K_c = \frac{2gH_o}{V_o^2} - \alpha\frac{2gD_{out}}{V_o^2}\frac{h_p}{D_{out}} \qquad (6)$$

Equation (6) is evaluated with experimental results for different manhole drop heights s in Figure 10. It is apparent that a definite linear relationship exists between K_c and the dimensionless pool height h_p/D_{out} for each drop height s and drop parameter D, indicating that α is approximately constant if the manhole configuration and approach flow condition remain unchanged.

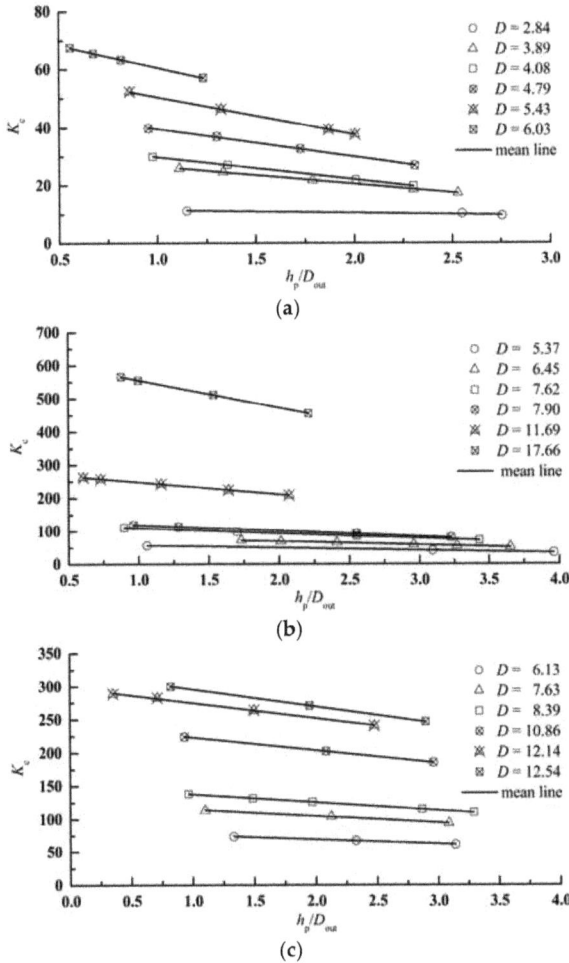

Figure 10. Variation of K_c with h_p/D_{out} for (a) $s = 0.93$ m, (b) $s = 1.50$ m and (c) $s = 2.40$ m.

Equation (6) could be written in a different form:

$$C = \frac{2gH_o}{V_o^2} - K_c = \alpha\frac{2gh_p}{V_o^2} \qquad (7)$$

where C is a dimensionless parameter. If α is a constant for different s and D, then C will be a linear function of $2gh_p/V_o^2$. In Figure 11, experimental results for constrained outflow conditions are shown plotted with C against $2gh_p/V_o^2$ for a number of discharges and three manhole drop heights. It is interesting that a definite linear relation appears to exist between these two parameters, indicating that α is approximately constant. This can be seen in Figure 12, where α is plotted against $2gh_p/V_o^2$, especially for $2gh_p/V_o^2$ greater than 25. Using the data of Figure 11, the best fitting is

$$C = \frac{2gh_p}{V_o^2} - 0.36 \tag{8}$$

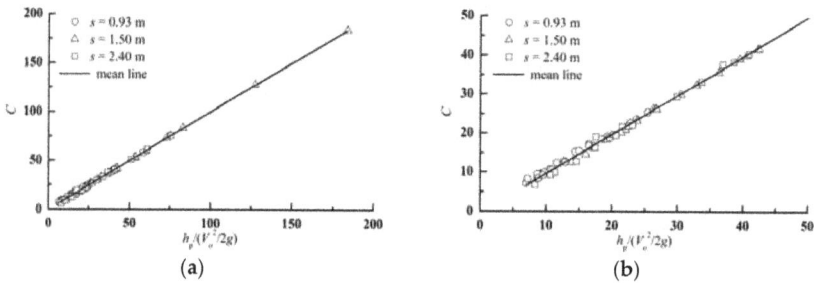

(a)

(b)

Figure 11. Rating curve between C and $h_p/(V_o^2/2g)$ for (**a**) entire present test range; (**b**) test data for $0 < h_p/(V_o^2/2g) < 50$.

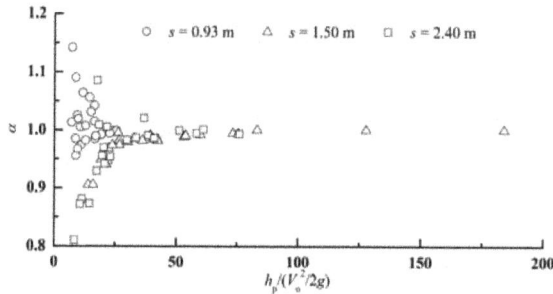

Figure 12. Variation of α with $h_p/(V_o^2/2g)$.

Hence, Equation (7) reduces to

$$K_c = \frac{2g(s + h_o - h_p)}{V_o^2} + 1.36 \tag{9}$$

Combination of Equations (4) and (9) generates

$$K_c = K_f + \left[1.11 - \frac{2g(h_p - h_o)}{V_o^2} \right] \tag{10}$$

The head loss coefficient K_c can be considered as the sum of that for free surface outflow conditions (K_f) and that due to the back pressure from the exit pipe (terms in brackets), which is related to the approach flow conditions and pool height. Similar to D, a dimensionless submerge parameter, $D' = [g(s + h_o - h_p)]^{0.5}/V_o$, can be defined for constrained outflow conditions, in which the term in parentheses is the difference in elevation between the water surfaces of the approach flow and the manhole pool. Thus, the head loss coefficient for constrained outflow conditions can be solely characterized by the submerge parameter D', for $2.0 < D' < 16.8$ resulting in

$$K_c = 2D'^2 + 1.36 \tag{11}$$

4. Conclusions

Drop manholes are effective energy dissipaters widely employed in urban drainage networks. In the present study, the hydraulic performance of circular drop manholes was investigated experimentally with respect to their energy dissipation in three models of different drop heights. It is concluded that the local head loss coefficient is solely dependent on the dimensionless drop parameter $(gs)^{0.5}/V_o$ for free surface outflow without a downstream backwater effect, while it can be solely characterized by the dimensionless submerge parameter, defined as $[g(s + h_o - h_p)]^{0.5}/V_o$, for constrained outflow due to the downstream backwater effect. Based on experimental results, empirical equations for different outflow conditions are proposed for practical applications. Furthermore, the mixing between airflow and water flow is largely intensified when outlet choking occurs, resulting in abrupt increases in air demand and energy dissipation.

Acknowledgments: This research was supported by the National Natural Science Foundation of China (Grant No. 51509162).

Author Contributions: Feidong Zheng performed the research and wrote the article; Yun Li led the work performance; Jianjun Zhao and Jianfeng An collected data through review of papers.

Conflicts of Interest: The authors declare no conflict of interest.

References

1. Christodoulou, G.C. Drop manholes in supercritical pipelines. *J. Irrig. Drain. Eng. ASCE* **1991**, *117*, 37–47. [CrossRef]
2. Granata, F.; de Marinis, G.; Gargano, R.; Hager, W.H. Energy loss in circular drop manholes. In Proceedings of the 33rd IAHR Congress (CD-Rom), Vancouver, BC, Canada, 9–14 August 2009.
3. Carvalho, R.F.; Leandro, J. Hydraulic characteristics of a drop square manhole with a downstream control gate. *J. Irrig. Drain. Eng. ASCE* **2011**, *138*, 569–576. [CrossRef]
4. Rajaratnam, N.; Mainali, A.; Hsung, C.Y. Observations on flow in vertical dropshafts in urban drainage systems. *J. Environ. Eng. ASCE* **1997**, *123*, 486–491. [CrossRef]
5. Jalil, A. Experimental and Numerical Study of Plunging Flow in Vertical Dropshafts. Ph.D. Thesis, Univerity of Alberta, Edmonton, AB, Canada, 2009.
6. Granata, F.; de Marinis, G.; Gargano, R.; Hager, W.H. Hydraulics of circular drop manholes. *J. Irrig. Drain. Eng. ASCE* **2011**, *137*, 102–111. [CrossRef]
7. Granata, F.; de Marinis, G.; Gargano, R. Flow-improving elements in circular drop manholes. *J. Hydraul. Res.* **2014**, *52*, 347–355. [CrossRef]
8. Chanson, H. Hydraulics of Roman aqueducts: Steep chutes, cascades, and dropshafts. *Am. J. Archaeol.* **2000**, *104*, 47–72. [CrossRef]
9. Chanson, H. An experimental study of Roman dropshaft hydraulics. *J. Hydraul. Res.* **2002**, *40*, 3–12. [CrossRef]
10. Chanson, H. Hydraulics of rectangular dropshafts. *J. Irrig. Drain. Eng. ASCE* **2004**, *130*, 523–529. [CrossRef]
11. Granata, F.; de Marinis, G.; Gargano, R. Air-water flows in circular drop manholes. *Urban Water J.* **2015**, *12*, 477–487. [CrossRef]
12. Camino, G.A.; Rajaratnam, N.; Zhu, D.Z. Choking conditions inside plunging flow dropshafts. *Can. J. Civ. Eng.* **2014**, *41*, 624–632. [CrossRef]
13. De Marinis, G.; Gargano, R.; Granata, F.; Hager, W.H. Circular drop manholes: Preliminary experimental results. In Proceedings of the 32rd IAHR Congress (CD-Rom), Venice, Italy, 1–6 July 2007.

water

MDPI

Article

PEPSO: Reducing Electricity Usage and Associated Pollution Emissions of Water

S. Mohsen Sadatiyan A. [1,*] and Carol J. Miller [2]

1   CDM Smith Inc., Detroit, MI 48226, USA
2   Department of Civil and Environmental Engineering, Wayne State University, Detroit, MI 48202, USA;
    ab1421@wayne.edu
*   Correspondence: sadatiyanasm@cdmsmith.com or mohsen@wayne.edu; Tel.: +1-313-899-0365

Received: 1 July 2017; Accepted: 21 August 2017; Published: 26 August 2017

Abstract: Using metaheuristic optimization methods has enabled researchers to reduce the electricity consumption cost of small water distribution systems (WDSs). However, dealing with complicated WDSs and reducing their environmental footprint remains a challenge. In this study a multi-objective version of Pollution Emission Pump Station Optimization tool (PEPSO) is introduced that can reduce the electricity cost and pollution emissions (associated with the energy consumption) of pumps of WDSs. PEPSO includes a user-friendly graphical interface and a customized version of the non-dominated sorting genetic algorithm. A measure that is called "Undesirability Index" (UI) is defined to assist the search for a promising optimization path. The UI also ensures that the final results are desirable and practical. The various features of PEPSO are tested under six scenarios for optimizing the WDS of Monroe City, MI, and Richmond, UK. The test results indicate that in a reasonable amount of time, PEPSO can optimize and provide practical results for both WDSs.

Keywords: optimization; water distribution network; pump schedule; genetic algorithm; energy; pollution emissions

---

1. Introduction

In the modern world, many systems are designed based on scientific analysis and engineering techniques, but it does not mean that these systems are developed and operated in an optimal way. In recent decades, due to improvement in the computational power of machines and development of new optimization techniques, engineers have focused more on using computer models and deterministic or meta heuristic optimization techniques to optimize the design and operation of systems. There are many optimization efforts related to water systems including piping design optimization, pump operation optimization, sensor placement improvement, model calibration, leakage detection and reduction, system reliability, etc. [1].

About 4% of electricity usage in the US is attributed to the supply, conveyance, and treatment of water and wastewater at the cost of approximately 4 billion US dollars per year. Moreover, due to increasing urban and industrial water demands and a decrease in access to high-quality water resources, it is predicted that the energy consumption of this sector will increase more than 50% by 2050 [2]. According to the US Department of Energy, approximately 75% of the operating costs of municipal water supply, treatment and distribution facilities is attributed to electricity demand [3]. As noted by several researchers, optimizing pump operation has a considerable effect on the water industries, which can offer a reduction of up to 10% in the annual expenditure of energy and other related costs [4,5]. Using hydraulic models to investigate the potential of energy usage and associated pollution emission reduction in water systems has been studied by different researchers. For instance, Perez-Sanchez and his colleagues, by using EPANET model of an irrigation system, showed that theoretically 188.23 MWh/year energy-equivalent to 137.4 ton CO_2/year-can be recovered from the system [6,7]. León-Celi, C et al. also used EPANET toolkit and two optimization algorithms to find the optimum flowrate distribution in water systems with multiple pump stations and minimize energy usage and potential leakage [8].

Time-of-use tariff and change in sources of energy in time may increase or decrease the electricity cost or pollution emissions (associated with the generation of energy) of the system, even if the total energy consumption of the system does not change. Elevated storage tanks in the system provide flexibility for operators to shift energy usage of the system. Shifting energy consumption may allow the operator to take advantage of cheaper energy and less polluting generator sources.

Uncertainties in demand of the system and complexity of the possible combination of pump status that can potentially answer operational requirements of the WDS, increase the tendency of operators to maintain water pressure in the system higher than the minimum required pressure. This increases energy usage, water leakage and consequently water and energy waste. Therefore, developing an optimization tool that can automatically react to changes in various inputs and generate a near optimum pump schedule may decrease electricity cost and the environmental footprint of the system.

2. Literature Review

About four decades ago, when researchers started to work on optimization of WDSs, most of them focused on construction cost (reducing the cost of piping) and operation cost (minimizing the cost of energy usage and the power demand of the pump station). However, after a while, other objectives such as increasing reliability and water quality or decreasing environmental footprint were included in the optimization process. In the last decade, the attention toward the environmental effect of energy usage and sustainability of WDSs increased due to increase in public and scientific awareness of climate change and the effect of pollutant emissions from power generation [9]. At first, most researchers considered the WDS optimization problem as a single objective problem. However, some researchers adopt multi-objective methods for optimization of the operation of the WDS [10].

One of the main objectives of the pump operation optimization is reducing the operation cost of pumps. The real electricity tariffs, in many cases, include a peak power demand charge ($/kW) in addition to the energy consumption charge ($/kWh). So, it is evident that a useful optimization tool should be able to use complicated electricity tariffs including both energy consumption and peak power demand costs. There are different examples in previous research of energy consumption charge and peak power demand charge being used. Wang et al. used a time-of-use electricity tariff in their optimization study [10]. Baran et al. also used a time-dependent electricity tariff that was defined based on on-peak and off-peak hours [11]. Shamir and Salomon used a more complicated electricity tariff. They used the real and complex electricity tariff of Haifa city, Israel, which includes three time periods, representing high, medium, and low energy costs. The tariff was different for the weekend and holidays and the various seasons of the year [12]. Working multiple pumps at the same time may cause an increase in required power. This may increase the total electricity cost of the system. There are some examples of researchers taking the power demand charge into account. For instance, Fracasso and Barnes included the amount of peak power demand (kW) as an objective of the optimization process [13].

In addition to the electricity cost, pollutant emissions associated with the electricity consumption is another objective that needs to be optimized to have a sustainable WDS. Wu et al. included the effect of variable emission rates and electricity tariffs in their WDS design optimization study [14]. Stokes et al. also suggested a framework for the modeling and optimization of Greenhouse Gases (GHG) emissions associated with energy usage and pump operation [15]. In most of these efforts, the emission rate of energy usage was considered as a constant value and was linearly related to the amount of consumed energy. However, most of the time, the source of electrical energy is a mix of various types of power generators. As this combination of generators may change in time, the emitted amount of pollutants per unit of consumed energy may change. Researchers at Wayne State University developed the LEEM methodology to calculate the amount of pollutant emissions associated with energy generation at different points in space and time. LEEM is an acronym for Locational Emissions Estimation Methodology and provides real-time and predicted marginal emission factors (kg/MWh) based on location and time of energy consumption [16].

Besides the two above-mentioned objectives, some constraints help to direct the algorithm to solutions that satisfy operational requirements of the WDS. For instance, frequent pump starts can increase the maintenance costs of the system [17]. Some researchers placed some limits on the maximum number of pump starts. Similarly, water pressures at junctions or water flow rate in pipes can be bounded. Constraints can be handled explicitly or can be converted to objectives and handled implicitly during the optimization process. One of the common methods of converting a constraint to an objective is using penalty formula. By this approach, violation from a constraint can be converted to a penalty value and reducing the penalty can be considered as an objective. Zecchin et al. used the pressure penalty to add a pressure constraint to the objective function of the ant colony (AC) algorithm that they used for WDS design optimization [18]. Lopez-Ibanez investigated the effect of constraint on the maximum number of pump starts. He found that a lower limit of the maximum number of pump starts that does not hinder the search for an optimum solution is related to characteristics of the network [19].

In addition to the maximum number of pump starts, other constraints such as minimum and maximum allowed water level in tanks, maximum and minimum allowed pressure at different points of the water network and maximum and minimum allowed velocity of water in different pipes can be considered during optimization. The effect of all of these constraints can be translated to penalty values. Reducing total penalties of a pump operation schedule can be formulated as an objective of optimization. So, reducing electricity usage cost, pollution emissions (associated with electricity usage) and penalties can be considered as three objectives of a pump operation optimization problem.

A multi-objective optimization problem can be solved with multi-objective methods or can be converted to a single objective problem and solved with a single objective optimization algorithm. For instance, Wu and Behandish calculated the amount of the objective function by the total weighted cost of energy and amount of three penalties [20]. Abiodun and Ismail completed a bi-objective optimization that aimed to reduce the electricity cost and maintenance problems [5]. In other studies, researchers used the multi-objective optimization method to solve a multi-objective problem directly and find the Pareto frontier of solutions. For instance, Fu and Kapelan used a multi-objective optimization method for finding the best WDS design based on pipe cost and system robustness [21]. Pollutant Emission Pump Station Optimization (PEPSO) is a platform developed by the water research team at Wayne State University for optimizing the pump schedule of the WDS [16]. The initial version of PEPSO used weighting factors to calculate a single combined objective from electricity usage, pollutant emissions and penalties [22]. However, the newer version of this tool is equipped with a multi-objective optimization algorithm to optimize each objective independent of others and find the Pareto frontiers of solutions.

Converting the multi-objective problem to a single objective problem increases the simplicity of the optimization algorithm. Also, the optimum result is a single solution that can be used directly. On the other hand, by using multi-objective methods, finding optimum solutions with respect to one objective does not have any effect on the process of finding the optimum value of other objectives. Also, there is not any need for normalizing and weighting operations. Defining a meaningful method to combine different objectives such as the cost of electricity usage and weight of pollution emissions and coming up with a single objective is a challenging process. Additionally, the multi-objective approach creates a range optimal solutions as a Pareto frontier that provides some flexibilities for users to select the preferred solution based on their requirements. Finally, a multi-objective algorithm can search the solution space of a multi-objective optimization problem with more freedom. This cannot be achieved with a single objective algorithm and when the effect of one objective on the combined objective is much more considerable than the effect of other objectives.

In the last two decades, many researchers have shifted the focus of WDS optimization from traditional and deterministic techniques, based on linear and nonlinear programming, to the implementation of methods that were generally based on heuristics derived from nature [18,23]. In recent years, Evolutionary Computation has proven to be a powerful tool to solve optimal pump-scheduling problems [11]. The great advantage of metaheuristic algorithms over deterministic methods is that they can be used for almost all types of optimization problems without considering the linearity or convexity of the problem. However, while using metaheuristic algorithms, constraints related to the hydraulic behavior of the solution must be checked separately or should be converted to an objective [24].

Genetic Algorithm (GA) is one of the most used algorithms in the optimization field and especially in water-related problems [25,26]. At first, Simpson et al. suggested using GA in the mid-90s for WDS optimization [27]. Lopez-Ibanez investigated various representations of pump schedule in his thesis and suggested that time-controlled trigger-based representation can lead to a better result and ensure maximum limit of switches per pump in comparison with level-controlled trigger representation. However, his result also showed that time-controlled trigger-based representation did not have considerable advantages on the common binary representation [19]. In this study, we used a customized version of the multi-objective Non-Dominated Sorting Genetic Algorithm (NSGA II) with the binary representation of solutions to develop that WDS or water transmission lines. These networks have a handful of pipes, junctions, pumps and occasionally one or two optimization tool.

One of the most famous free and publicly available software for modeling the WDS is EPANET2 that is published by the US EPA [7]. Lopez-Ibanez reviewed about 20 articles from 1995 to 2004 and reported that most of the researchers used complete hydraulic simulation to evaluate the effect of decision variables on the status of the hydraulic network [19]. We also used EPANET 2.00.12 as the hydraulic solver of the optimizer tool [7].

Most of the previous studies focused on a small-scale elevated tanks [5,10,28]. A small portion of real systems are similar to small test networks of these researches, but most of the time we face large networks with a couple of hundred pipes, junctions, and a considerable number of pumps, valves, tanks, etc. There are a few studies that tried to optimize a real and large-size WDS [4,29,30]. Most of the systems that were used in WDS optimization studies do not have variable speed pumps. The WDS of Monroe City, MI, previously used for comparing three pump operation optimization tools, has both fixed and variable speed pumps [31]. The WDS of Richmond, UK, is also used in several types of research. This model is publicly available for the researcher and is suggested to be used for operation optimization studies of WDSs [32]. In this study, we used both Monroe and Richmond WDS models to test the developed optimization tool.

3. Tool Development and Methodology

3.1. Transition from the First Version of PEPSO to the Second Version

The optimization tool that is introduced in this article is the second version of PEPSO [33]. The first version of PEPSO was the only tool in this field which was able to optimize the pump schedule of a WDS to reduce pollution emissions of the system based on location and time of energy consumption.

The first version of PEPSO was compared with other optimization methods, including the Markov Decision Process (MDP) and Darwin Scheduler (DS) [31]. This comparison showed that PEPSO was as good as other tools on the market and its unique emission optimization capability made it an exceptional tool. However, using a single objective optimization technique limited the capability of this tool for searching a wide area of the solution space. Besides, the first version of PEPSO did not have enough options to control the water level in tanks of the WDS effectively. It could use a time-of-use electricity tariff for the whole system, but it was not able to use separate tariffs for different electricity meters and calculate peak power demand cost. Also, this tool could not effectively control the number of pump starts during an operation cycle.

Despite all unique features of the first version of PEPSO, all of the limitations mentioned above prevent its use for optimizing the pump schedule of the WDS outside of the research environment. The second version of PEPSO that is introduced in this article was developed to alleviate all of these shortages. Also, some fundamental changes in the optimization algorithm of PEPSO increased the efficiency of the optimization process, resulting in the generation of a more practical solution in a shorter period.

3.2. Introducing the New PEPSO

PEPSO is a free and publicly available modular optimization program with a graphical interface that uses a customized NSGA II algorithm for optimization. Different qualities and characteristics, including clarity, familiarity, responsiveness, efficiency, consistency, aesthetics and forgiveness were integrated into the graphical user interface of PEPSO. Users can define a detailed electricity tariff for each pump including time-of-use energy consumption charge ($/kWh) and power demand charge ($/kW). They can also select a desired pollution level or a combination of pollution levels for optimization. PEPSO can connect to the LEEM server or use offline data to get the location and time-dependent emission factors (kg/MWh) that are required for pollution emission optimization. Various types of hard and soft constraints can be imposed on pumps, tanks, junctions and pipes of the WDS. In PEPSO, a wide range of optimization options including five different stopping criteria, exploration and exploitation rates, initial conditions, etc., can be defined. Users can also select any combination of three objectives (electricity cost, pollutant emissions, total penalty) for optimization. Finally, this tool provides a broad range of reports in the format of text (tabular data) and/or 2D and 3D graphics (charts and plots). All of these features can be accessed through the Graphical User Interface (GUI) or can be defined and edited directly on the PEPSO project file by using a simple text editor [33].

It was mentioned that PEPSO uses a binary coding scheme for storing pump schedules. It means that the pump schedule is stored as a table, each row of which shows the operational status of a pump and each column corresponds to a time block (usually a one-hour block). For fixed speed pumps (FSP), each cell of the pump schedule table can store the value of 0 or 1 which refer to the OFF or ON status of the pump. For variable speed pumps (VSP), the value of the cell can be 0 (OFF) or a number between the minimum relative rotational speed and 1.

The new PEPSO introduces the Undesirability Index (UI) to improve the crossover and mutation operations of GA. The UI is a measure which shows the relative level of the undesirability of the operational status of a pump at a time block. So, pump operation with a high UI are good candidates for modification, and changing them may help to get closer to the optimum pump schedule and get more practical solutions. The UI of a pump schedule is stored in a table similar to the pump operation schedule table. PEPSO solves the hydraulic model of the WDS based on a proposed pump schedule. Then, for each cell of the UI table, PEPSO checks if the pump operation status causes excess or deficit pressure at junctions or water level in tanks. It also checks if the pump operation status caused "negative pressure warning", "pump cannot deliver head or flow warning" and "system disconnected warning". All of this information will be used to calculate the UI of a pump at a time block. For instance, if at a time block, the program sees that there is some excess pressure at junctions and a high level of water in tanks, it shows that energy in the system is probably more than the minimum required level. Therefore, some pumps that are ON during that time block can be turned OFF. Here, PEPSO increases the UI of the ON pumps at that time block, during which the mutation process helps to identify suitable mutation candidates (pumps that can be turned OFF) and generate a better pump schedule. So, instead of randomly changing the pump operation status at each iteration, PEPSO uses the UI values and finds a promising part of the pump operation schedule that can be changed to create a better pump schedule in a more efficient way. The process of calculating the UI is shown by a flowchart in Figure 1. Up and down arrows in the algorithm show operations that change the UI value of a pump at a time step in a way that increases or decreases the probability of turning on the pump, respectively (or increase/decrease the rotational speed in the case of variable speed pump). More details about calculating the UI and its usage during crossover, mutation and the elitism process is beyond the scope of this article and can be found in Sadatiyan's thesis [34].

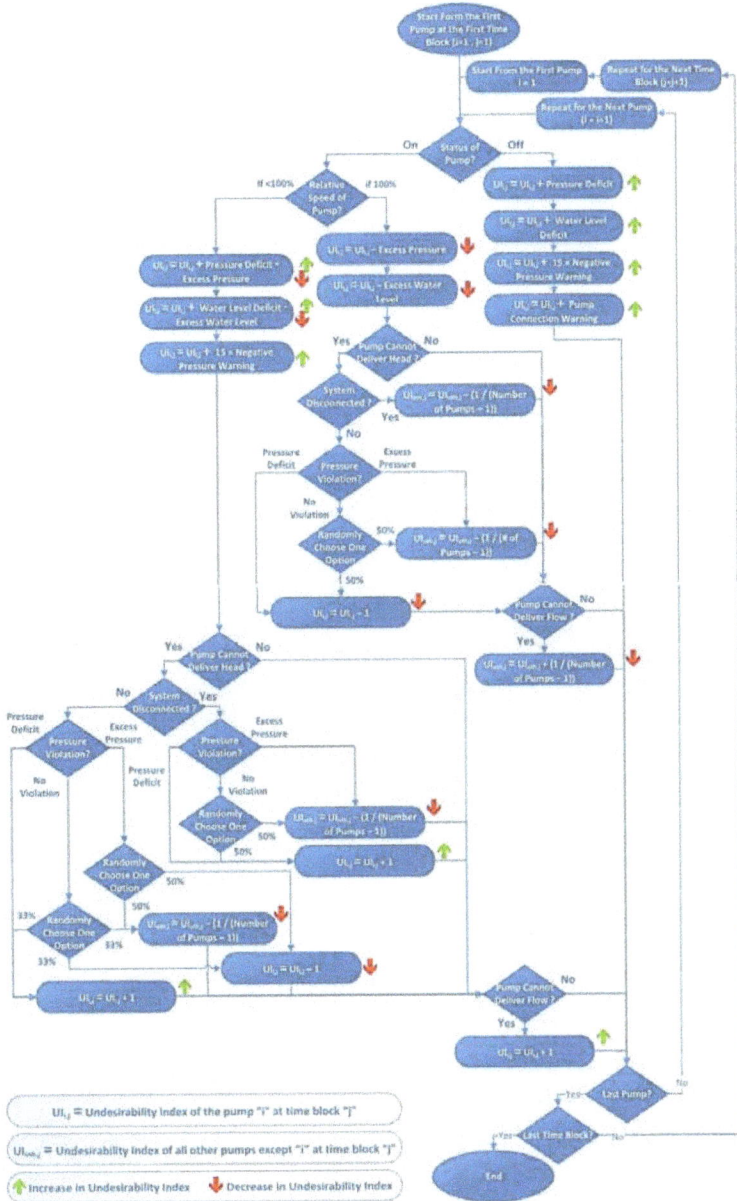

Figure 1. Undesirability Index (UI) calculation algorithm of Pollutant Emission Pump Station Optimization (PEPSO) [34].

To produce a new generation during the optimization process, at first a user-defined portion of the population is selected as parents for crossover and mutation steps. PEPSO uses the roulette wheel sampling method [35] for this purpose, and the probability of choosing a pumping schedule for crossover, mutation and elitism steps is proportional to its non-dominated rank. The customized crossover operator of PEPSO acts on a whole time block of the selected pump schedule instead of an individual pump status at a time block (a single gene). After choosing a pump schedule as the primary parent, Time Step Undesirability Indexes (TSUIs) of the parent will be calculated. The TSUI of a time step is the total absolute values of the UI of all pumps at the time step. A time block with a high TSUI indicates that the combined effect of the operational status of all pumps at the time block is not desirable. It indicates that the time block is a suitable candidate to be changed and create a better pump schedule. Therefore, the operational status of all pumps at the selected time block will be replaced with the operational status of all pumps from the same time block of another solution with a lower TSUI.

After the crossover step, a user-defined portion of the population is selected for mutation. During mutation, a user-defined portion of the pump status at different time blocks (genes) is selected to be changed. For constant speed pumps, the status of the selected gene is modified from ON to OFF and vice versa. For variable speed pumps, the UI is used to determine if it is better to increase the relative rotation speed of the pump or decrease it to make the UI value closer to zero. The probability of selecting a cell or a time block of a pump schedule in mutation or crossover steps is proportional to their TSUI and UI respectively. Before using the roulette wheel method, PEPSO adjusts the selection probability values of all pump schedules. The amount of water level deficit in tanks at the end of the operation cycle and the number of times that the proposed schedule causes negative pressure warning in the system are used to reduce the probability of selecting a pump schedule. This adjustment reduces the probability of selecting a pump schedule which is not practically acceptable for mutation, crossover and elitism steps [35].

All generated children are added to the previous generation. The combined population is ranked and sorted. By using the roulette wheel sampling method, the required number of pump schedules is selected to create the next generation.

PEPSO uses a wide variety of stopping criteria to determine when to stop the optimization process and report the final result. The optimization process can be stopped based on (1) the maximum time of optimization, (2) the maximum number of iterations, (3) the maximum number of solution evaluations, (4) the maximum number of stagnant iterations, and (5) reaching a goal for each optimization objective. The stagnant term relates to the change in the value of the objectives of the solution. If during an iteration, the value of objectives of the best solution does not change more than a defined minimum amount, the iteration will be considered as a stagnant iteration.

It was explained that PEPSO is a multi-objective optimizer, so the final result of the optimization process is a Pareto frontier of non-dominated solutions (pump schedules) [36]. However, in practice, we can use just one schedule for operating pumps. Here, PEPSO is equipped with an algorithm which helps users to select the final pump schedule among the solutions of the Pareto frontier. PEPSO, at first, solves the hydraulic model of the WDS by using all pump schedules in the Pareto frontier. If a pump schedule causes negative pressure warning, then that pump schedule will be filtered out. Similarly, if the water level in tanks or the pressure at junctions of the WDS goes beyond the hard constraints, that pump schedule will be filtered out. The remaining pump schedules are non-dominated solutions which are feasible and practical. So, by using weighting factors that are defined by users, the final solution will be selected. It should be noted that the weighting factors are not used during the optimization process and are just used as an indicator of user preference for selecting the desired solution among the solutions of the final Pareto frontier.

4. Experimental Demonstration

Multiple optimization scenarios were considered for two networks to evaluate the performance of the developed optimizer tool. These test cases, scenarios, and the result are explained in the following sections.

Design of Experiment

The EPANET hydraulic model of two WDSs is used for the optimization test. The first case is the detailed model of the Monroe WDS in Michigan and the second case is the skeletonized model of the Richmond WDS, UK [37,38]. The information summary of both models is presented in Table 1. The WDS of Monroe has more components, and its hydraulic simulation is more computationally intensive than the Richmond WDS. In this research, a Lenovo ThinkPad W520 workstation was used for conducting the tests [39]. The CPU time of a 24 h hydraulic simulation of the Monroe WDS with this computer is 14.95 milliseconds. This time for the skeletonized version of the Richmond network is 5.54 milliseconds. In addition to the complexity of the hydraulic model, the Monroe WDS has six more pumps than the Richmond WDS. This considerably increases the number of possible pump combinations and size of solution space of the Monroe WDS. The water storage capacity of the Monroe WDS is considerably lower than the storage capacity of the Richmond WDS. So, the Richmond WDS has more flexibility regarding shifting energy consumption of the system by storing water in elevated tanks.

Table 1. Summary of the detailed model of Monroe and Skeletonized model of Richmond water distribution systems (WDSs).

Item	Monroe	Richmond Skeletonized
No. of Fixed Speed Pumps	11	7
No. of Variable Speed Pumps	2	0
No. of Pump Stations	2	6
No. of Tanks	3	6
No. of Water Sources	1	2
No. of Pipes	1945	44
No. of Junctions	1531	41
Total Length of Pipes (km)	450	22.69
Pipe Size Range (mm)	50–910	76–300
Total Demand (m^3/day)	36,500	3921
Storage volume (m^3)	3974	2598
Storage to Daily Demand Ratio	11/100	66/100
Range of Power of Pumps (kW)	36–220	3–60
Max. Static Water Head (m)	60	199
Demand Pattern Duration (hr)	24	24
Demand Pattern Time Step (hr)	1	1
Min and Max. Demand Multiplier	0.67–1.19	0.39–1.53

Constraint on the water level in tanks and pressure at junctions of both the Monroe and Richmond WDS are defined in Tables 2 and 3. Some strategic junctions are selected in both networks, and the upper and lower bound of their desired water pressure ranges are shown in Table 2. In addition, minimum and maximum allowed pressure of 0 and 200 m are respectively defined as hard constraints and allowed pressure limits for all junctions. Similarly, for all tanks of both WDSs, the upper and lower bound of the desirable range of water level are presented in Table 3. These limits are soft constraints and violating them increases the total penalty associated with the pump schedule. For calculating penalties, at first, the amount of violation from the upper and lower limits is calculated and then the calculated violation is raised to the power of 1.5. Penalty calculation formulae that are implemented in the PEPSO algorithm are listed below.

$$P_P = \sum_{i=1}^{I} \sum_{j=1}^{J} \begin{cases} (p_{ij} - p_{jmax})^x, & p_{ij} > p_{jmax} \\ (p_{jmin} - p_{ij})^x, & p_{ij} < p_{jmin} \end{cases} , \tag{1}$$

$$P_T = \sum_{i=1}^{I} \sum_{k=1}^{K} \begin{cases} (h_{ik} - h_{kmax})^x, & h_{ik} > h_{kmax} \\ (h_{kmin} - h_{ik})^x, & h_{ik} < h_{kmin} \end{cases} , \tag{2}$$

$$Total\ Penalty = P_P + P_T, \tag{3}$$

where

P_P : Penalty associated with water pressure violation at junctions
P_T : Penalty associated with water level violation at tanks
i : Time step index starts from the 1st time block and goes to the Ith time block
j : Junction index starts from the 1st junction and goes to the Jth junction
k : Tank index starts from the 1st tank and goes to the Kth tank
x : A power defined to increase the penalty by increasing the amount of violation. $x = 1.5$ is used.
p_{ij} : Water pressure of junction j at time block i
p_{jmax} : Maximum allowed water pressure of junction j
p_{jmin} : Minimum allowed water pressure of junction j
h_{ik} : Water level of tank k at time block i
h_{kmax} : Maximum allowed water level of tank k
h_{kmin} : Minimum allowed water level of tank k

Table 2. Water pressure constraints for strategic junctions of Monroe and Richmond WDS models.

Test Case	Strategic Junction ID	Min. Water Pressure (psi)	Max. Water Pressure (psi)
Monroe	J-6	42	52
	J-27	32	46
	J-131	28	42
	J-514	42	56
Richmond	42	20	140
	1302	0	100
	10	0	100
	312	0	100
	325	0	100
	701	0	100
	745	20	100
	249	20	100
	753	20	100
	637	20	140

Table 3. Water level constraints for tanks of Monroe and Richmond WDS models.

Test Case	Tank ID	Min. Water Level (m)	Max. Water Level (m)
Monroe	T-2	1.56	8.12
	T-3	1.41	7.28
	T-5	1.78	8.66
Richmond	A	0.30	1.70
	B	0.50	2.86
	C	0.32	1.79
	D	0.55	3.10
	E	0.44	2.29
	F	0.33	1.86

The energy consumption charge component of an electricity tariff of the Monroe WDS is time dependent and has two off-peak and on-peak rates. The on-peak period starts from 11:00 and finishes by 19:00 and its energy charge is 0.04408 ($/kWh). The Energy charge of other Off-peak hours is 0.04108 ($/kWh). Daily peak power demand charge of this system is 0.48 ($/kW). The Richmond WDS has six electricity tariffs for six pump stations (see Table 4). These are also time-of-use tariffs and just have the energy charge component (not peak power demand charge). The on-peak hours start from 07:00 and ends by 24:00 for all tariffs.

Table 4. Electricity tariffs of the Richmond WDS.

Pump Station	On-Peak Rate ($/kWh)	Off-Peak Rate ($/kWh)
A	0.0679	0.0241
B	0.0754	0.0241
C	0.1234	0.0246
D	0.0987	0.0246
E	0.1122	0.0246
F	0.1194	0.0244

Emission factors (kg/MWh) that are needed for the calculating pollution emissions of each solution are obtained from the LEEM server. Table 5 shows emission factor of CO_2 that was employed in all optimization tests of this study [40]. Based on the hydraulic models, duration of an optimization run is 24 h with one-hour time step. The same set of values for parameters of optimization algorithm was used for all tests that are listed in Table 6. The crossover and mutation percentage define the portion of the population which should be selected for crossover and mutation steps respectively. The Crossover and mutation rate shows the portion of selected solution which should be modified during crossover and mutation steps.

Table 5. Emission factors of CO_2 obtained from the Locational Emission Estimation Methodology (LEEM) server.

Time	CO_2 Emission Factor (kg/MWh)	Time	CO_2 Emission Factor (kg/MWh)
00:00	767.771	12:00	662.793
01:00	738.324	13:00	630.703
02:00	702.904	14:00	630.531
03:00	702.904	15:00	628.591
04:00	702.904	16:00	628.882
05:00	767.771	17:00	666.549
06:00	781.469	18:00	693.607
07:00	808.212	19:00	665.274
08:00	764.333	20:00	730.766
09:00	719.768	21:00	790.628
10:00	719.768	22:00	808.212
11:00	695.334	23:00	780.477

Table 6. The optimization parameters used for all tests.

Parameter	Value
Max. No. of Solution Evaluations	16,600
Population Size	100
Percentage of Elite Solution	20%
Crossover Percentage	50%
Crossover Rate	50%
Mutation Percentage	5%
Mutation Rate	10%

Six optimization scenarios were defined to test different aspects of the optimization process. In the base scenario (S0), WDSs were optimized based on the electricity cost and total penalties (penalties of water level in tanks and pressure at junctions). The result of an optimization run may change based on the initial population and stochastic characteristics of the optimization operators. So, each scenario was repeated five times and the average results of five repeated tests of each scenario and its standard error of means (SEM) are reported. In the first scenario (S1), WDSs were optimized to reduce all three objectives (electricity cost, CO_2 emissions, and total penalty). The next scenario (S2) is defined to evaluate the effect of optimizing based on the electricity cost and CO_2 emissions, so it is just optimized based on penalties. This scenario is similar to the base scenario, but it uses the total penalty as the only optimization objective.

The third scenario (S3) was defined to test the effect of using the UI in the optimization process. So, this scenario (S3) is similar to the base scenario without calculating and using the UI. The fourth scenario (S4) is defined to investigate the effect of water level constraints on the final results. This scenario does not have any water level constraint in the tanks. Finally, the fifth scenario (S5) is defined to see the effect of time-of-use electricity. This scenario is like the base scenario but uses a fixed rate electricity charge ($/kWh) for the whole period of operation and does not include the peak power demand charge ($/kW).

5. Results and Discussion

In total, 60 optimization runs have been done on two WDS models. The required time for completing an optimization run of the Monroe WDS is 02:14:44 ± 00:03:43. This time, for the skeletonized version of the Richmond WDS model, is 00:35:38 ± 00:01:36. PEPSO reports the electricity cost of the final solution. However, before comparing the electricity cost of different solutions, we should consider that the final volume of stored water in the system might not be equal in all solutions. Although the final level of water was in the acceptable range from a system operation point of view, filling or draining an elevated tank can be regarded as storing energy into or draining energy from the system. Therefore, the net energy consumption of the system is calculated considering the change in volume of stored water. Similarly, the net electricity cost and net CO_2 emissions of each solution is calculated before comparing the results. It is assumed that a long-run deficit or surplus water volume at the end of each day will be balanced by the change of operation in different hours of upcoming days. Therefore, the average electricity charge ($/kWh) and CO_2 emission factor (kg/MWh) were used to take into account the effect of this deficit or surplus water volume and calculate the net electricity cost and net CO_2 emissions.

Net electricity cost ($) (left), net CO_2 emissions (kg) (middle) and total penalty (right) of all five scenarios of the Monroe WDS (top) and the Richmond skeletonized WDS (bottom) tests are displayed in bar charts in Figure 2. Each column shows the average result of five repeated tests and the error bar on top of it displays the SEM value. Except for columns that show high total penalty values, the SEMs of all the other results are relatively small. This shows the consistency in the outcome of PEPSO runs. Since penalty values are related to the amount of violation raised to the power of 1.5, it is expected to see that the moderate change in violation value results in a more severe change in penalty values.

Comparing results of scenarios S0, S1 and S2 showed, in both WDSs, optimizing based on three objectives (S1) is the most effective strategy for reducing objectives and obtaining a practical result. Theoretically, we expect to see the lowest amount of electricity cost in the result of the S0 scenario, but the result showed that in the majority of tests, both the electricity cost and CO_2 emissions of the S1 scenario are less than S0. On average, the electricity cost and CO_2 emissions of S1 scenarios are 12.9% and 11.7% in Richmond tests and 1.7% and 1.7% in Monroe tests less than S0 results respectively. Since, in most cases, reducing energy usage decreases both the electricity cost and pollution emissions, optimizing based on all three objectives (S1) helps PEPSO to better explore the solution space. So, despite our theoretical expectation to see the minimum electricity cost in the result of the S0 scenario, in practice, the S1 scenario is more efficient at finding low energy consumption solutions in a limited amount of time.

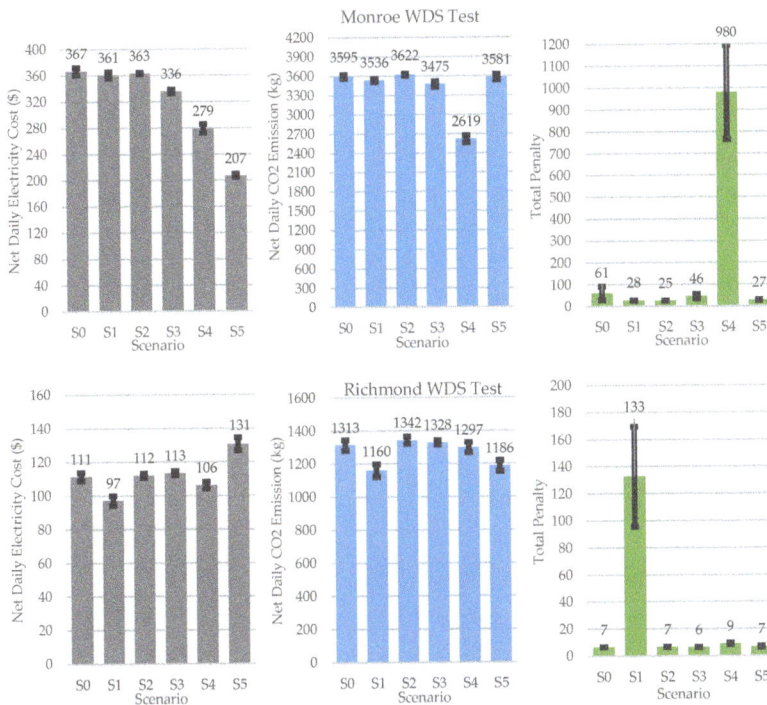

Figure 2. Electricity cost (left), CO_2 emissions (middle) and total penalty (right) results of five scenarios of Monroe WDS (top) and Richmond skeletonized WDS (bottom) tests.

As expected, optimizing just based on total penalty (S2) results in less penalty with respect to the outcome of both S0 and S1 scenarios. The total penalty of the S1 scenario of the Richmond test was considerably higher than the S0 and S2 scenarios. Investigating the detailed results, in this case, showed that there are two solution groups that can be selected as the optimum solution. In the first group, pumps are using more energy and pressure at junctions and water levels in tanks are slightly below the upper boundary of the desirable range. So, total penalties of this solution group are low. The second solution group has considerably less energy consumption and correspondingly less CO_2 emissions. However, in these solutions, pressure at a couple of junctions and water level at some tanks are below the desired level which increases the total penalty of these solutions. Although the violations in these cases are not beyond the acceptable range, from the optimizer perspective, these are dominated solutions when there is only one objective (total penalty). So, PEPSO does not choose the final solution from the second group. However, in the S1 scenario, when all three objectives are considered, a solution from the second group, which has some penalties but has a considerably lower electricity cost and CO_2 emissions, is reported as the optimum solution.

Comparing results of the S0 scenario with those of the S3 scenario showed that, when the UI is used in the optimization run of the Monroe WDS, on average, the net electricity cost is reduced by 8.5%. Although, at first sight, it seems that using the UI reduced the effectivity of the optimization algorithm, a closer look at the results revealed that the result of the S0 scenario is more practical than the S3 scenario. During the whole operation period, the stored volume of water in tanks in the S3 scenario is, on average, 5.6% lower than S0. The final volume of stored water in tanks for the S3 scenario is 10.9% lower than the final volume of stored water in the S0 scenario. Also, solutions of the S0 scenario, on average, have less than two warnings about pumps that cannot deliver head, but S3 results, on average, have about four and one warnings for pumps that cannot deliver head and flow respectively. Figure 3 displays the water level pattern in tanks of typical results of S0 (top left) and S3 (top right) scenarios. It can be seen that the solution of S3 tends to drain tanks more than that of S0.

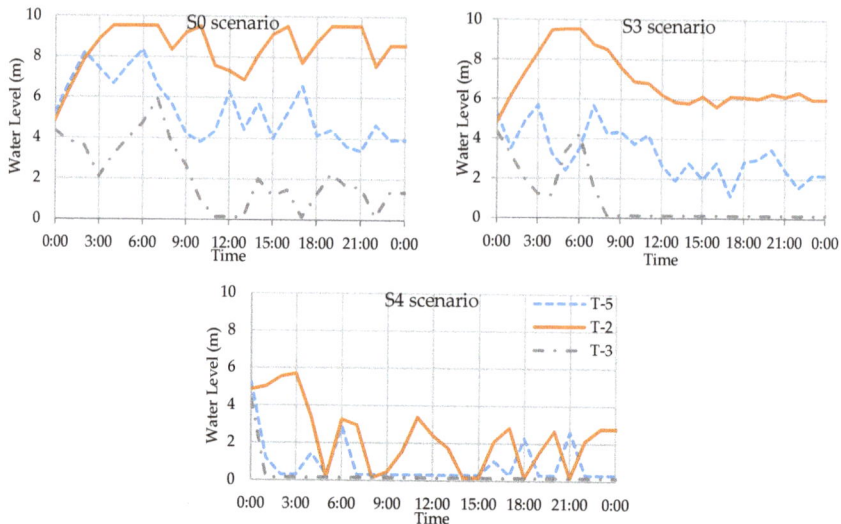

Figure 3. Typical water level pattern in tanks of S0 (top left), S3 (top right) and S4 (bottom) scenarios of the Monroe WDS.

The hydraulic model of the skeletonized version of the Richmond WDS was simpler than the Monroe WDS. So, in this case, optimizing with or without the undesirability calculation did not considerably change the results. Results of both S0 and S3 scenarios are close with respect to total penalty, electricity cost and the number of warnings. It seems that the undesirability calculation helped a little bit to find solutions with slightly lower CO_2 emissions (2.2%). However, it should be considered that calculating undesirability is an additional computation load on the optimization process. On average, calculating and using the UI in the optimization process of the Monroe WDS increased the required time for the optimization run by 8.9%. Based on these results, we can say that calculating the UI increased the required time for 16,600 solution evaluations in an optimization run. However, the final result was more practical and of higher quality. Obtaining a final solution with the same level of quality without using the UI needs more iterations and solution evaluations that increase the length of the optimization process. We expected that using the UI, by quantifying positive and negative effects of pump statuses on hydraulic responses of the water network, adds some intelligence to the process of producing the next generation and makes possible more purposeful crossover, mutation and elitism steps. Although calculating the UI increases the computational load of each iteration, we expected to see that within the same number of iterations, using the UI can provide better results. The outcome of these tests showed promising results regarding the use of the UI. However, this area still needs further research. More studies on complicated networks with vast solution space can help to show and quantify the level of effectiveness of the UI. It is possible that, in the case of a complex system with multiple pumps and vast solution space, traditional blind crossover, mutation and elitism steps (without using UI) cannot find an acceptable solution within a reasonable number of iterations. Results of the S4 scenario showed that giving PEPSO the possibility to operate pumps without tank level constraints, on average, reduces the electricity cost and CO_2 emissions of the system by 24.0% and 27.2%.

Despite the fact that removing water level constraints reduces the electricity cost and CO_2 emissions, it considerably increased the water level violation of tanks and water pressure violation at strategic junctions. In the S4 scenario, the pressure of junctions has some fluctuations that caused considerable low and high-pressure penalties. The water level penalty of tanks of the S4 scenario is four times more than for the S0 scenario. Comparing patterns of the water level in tanks (see Figure 3) and water pressure at junctions (see Figure 4) of the S4 and S0 scenarios can clearly show these differences.

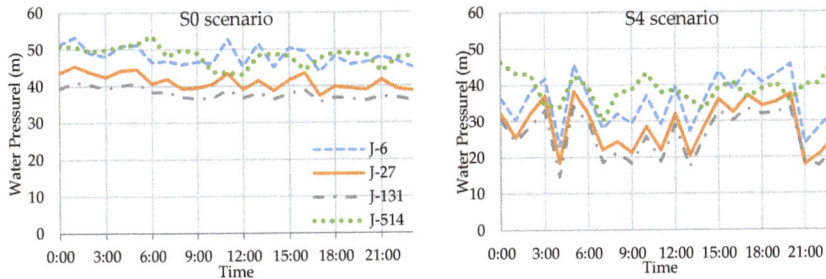

Figure 4. Typical water pressure pattern at junctions of S0 (left) and S4 (right) scenarios of the Monroe WDS.

On average, constraining the water levels of tanks during the operation cycle led to a reduction of more than 32% in the final volume of stored water. Monroe test results indicate that water level constraint effectively helps to keep the final tank level balanced and prevents tanks from draining during the whole operation period. Similarly, optimizing the pump operation of the Richmond WDS without water level constraints for tanks (S4), on average, reduces the net electricity cost and CO_2 emissions by 4.8% and 1.2% respectively. However, this increases the total penalty by 35.1%.

Results of the test on the Monroe WDS showed that having a flat rate electricity tariff, on average, can lead to a 9.7% increase in peak power demand (kW) while the total consumed energy (kWh) is almost unchanged. Although the total energy consumption in both S0 and S5 scenarios is almost unchanged, 2.1% of the total energy consumption in the S5 scenario shifted from off-peak hours to on-peak hours. These results confirm that the power demand charge and time-of-use electricity tariffs will force PEPSO to find an optimized solution with more energy consumption during off-peak times and with a reduced peak power demand.

The overall electricity cost in the Richmond system is related to the time-dependent energy consumption charge. However, in the S5 requirement of the system. PEPSO uses a customized version of the NSGA II to find the Pareto frontier and then select the best solution as the optimum pump scenario, the flat rate electricity tariff, on average, reduced the total energy use (kWh) from the off-peak hours by 3%and added half of that to the on-peak hours. By this change, the remaining 1.5% of energy is saved. Previously, due to the use of a time-of-use electricity tariff, PEPSO needed to shift energy usage to reduce the electricity cost of the system. This shift of energy usage caused a 1.5% energy loss due to head losses while filling and draining tanks. It is interesting to see that the solution of the S0 scenario drained 21.6% of the stored volume of water in tanks of the Richmond WDS. While the S5 scenario just drained 12.0% of this volume.

6. Conclusions

The new version of PEPSO, which is introduced in this article, is a multi-objective optimization tool. It can be used to find a pump schedule for the WDS to reduce the electricity cost and corresponding pollution emissions while satisfying the required operational schedule. It uses EPANET toolkit for hydraulic stimulation. The Undesirability Index is a measure that enables PEPSO to find promising ways of modifying the solution to get closer to the global optimum solution and create practical solutions. Test results on the Monroe WDS and skeletonized version of the Richmond WDS model showed that PEPSO could optimize the detailed model of the Monroe WDS effectively with 13 pumps in about 2 h with a computer system that can be found in a typical WDS design or operation offices (for more detail, see Section 5). The time required to optimize the skeletonized version of the Richmond WDS model was about half an hour.

- Optimizing based on all three objectives (S1) reduces the CO_2 emissions of the Monroe and Richmond WDSs by 1.3–3.4%. Optimizing based on all three objectives at the same time is more effective than optimizing based on only the electricity cost or total penalty.

- Optimizing based on just penalty (S2 scenario) reduced the total penalty on Monroe and Richmond WDSs by 10 and 5.8% respectively.

- Calculating the Undesirability Index helped PEPSO to find more practical optimized solutions with fewer EPANET warnings and less tank drainage. However, on average, the undesirability calculation increased the required optimization time by 8.9%. The effect of the UI on finding high-quality solutions for a complex system with vast solution space needs to be evaluated.

- In the S4 scenario, the Monroe WDS was optimized without tank level constraints. The water level penalty of tanks of the S4 scenario is more than four times the water level penalties of the base scenario (S0). Like the Monroe WDS, optimizing without tank level constraints reduced the electricity cost and CO_2 emissions of the Richmond WDS. However, it considerably increases the water level penalty of tanks (35.1%). Removing water level constraints increases both water level and water pressure penalties and led to impractical and unacceptable solutions.

- The time-of-use electricity tariff forces PEPSO to shift 1.7% of energy consumption from on-peak hours to off-peak hours. Including the power demand charge in the electricity tariff also, on average, reduces the peak power demand of the Monroe WDS by 9.7%. In the Richmond test, using a flat rate energy consumption charge enables PEPSO to consume energy at the time of high demand. This eliminated the need to store more water during off-peak hours which was causing 1.5% energy losses. In addition, by this method, PEPSO reduced tank drainage by about 10%.

- PEPSO used a multi-objective optimization algorithm to optimize three objectives independent of each other and report the final Pareto frontier that can be used in system studies and research. However, for practical use, one of the solutions among the Pareto frontier should be selected for operation. This selection is made by considering user preference based on user-defined weighting factors and also by removing impractical solution from the Pareto frontier (e.g., a solution with zero energy usage but high penalties). Defining different weighting factors can change the selected solution. Weighting factors are dependent on geographical, social, economic, etc., characteristics of the water system, defined constraints and practical preferences of operators. This area needs to be studied further to create a guideline that can help users to define weighting factors in such a way that results in the selection of the most desirable solution from the Pareto frontier.

- In this study, the net electricity cost and net CO_2 emissions are calculated to take into account the effect of deficit or surplus water volume of tanks within the acceptable range. However, using the average electricity charge ($/kWh) and CO_2 emission factor (kg/MWh) might not match real operation conditions. Therefore, we suggest running tests and simulations for a longer period (e.g., a week instead of 24 h) or using better calculation methods to take into account the effect of tank level changes at the end of simulation in a more accurate way.

Acknowledgments: This study was made possible through the support of Great Lake Protection Fund. The GLPF did not influence the study design, interpretation of data, or preparation of this article. We would like to thank the LEEM team and water research teams of Wayne State and Dayton Universities. We also want to express our gratitude to Barry LaRoy from the water department of City of Monroe and the Yorkshire Water for sharing hydraulic model of their water networks.

Author Contributions: Mohsen Sadatiyan and Carol Miller conceived and designed the research plan and experiments; Mohsen Sadatiyan developed the required software and performed the experiments; Mohsen Sadatiyan and Carol Miller analyzed the data; Mohsen Sadatiyan wrote the paper and Carol Miller reviewed and edited its content.

Conflicts of Interest: The authors declare no conflict of interest. The founding sponsors had no role in the design of the study; in the collection, analyses, or interpretation of data; in the writing of the manuscript, and in the decision to publish the results.

References

1. Nicklow, J.; Reed, P.; Savic, D.; Dessalegne, T.; Harrell, L.; Chan-Hilton, A.; Karamouz, M.; Minsker, B.; Ostfeld, A.; Singh, A.; Zechman, E. State of the Art for Genetic Algorithms and Beyond in Water Resources Planning and Management. *J. Water Res. Pl-Asce* **2010**, *136*, 412–432. [CrossRef]
2. Giacomello, C.; Kapelan, C.; Nicolini, M. Fast hybrid optimization method for effective pump scheduling. *J. Water Res. Pl-Asce* **2013**, *139*, 175–183. [CrossRef]
3. U.S. Department of Energy. *Energy Demands on Water Resources: Report to Congress on the Interdependency of Energy and Water*; U.S. Department of Energy: Washington, DC, USA, 2006.
4. Jamieson, D.G.; Shamir, U.; Martinez, F.; Franchini, M. Conceptual design of a generic, real-time, near-optimal control system for water-distribution networks. *J. Hydroinform.* **2007**, *9*, 3–14. [CrossRef]
5. Abiodun, F.T.; Ismail, F.S. Pump Scheduling Optimization Model for Water Supply System Using AWGA. In Proceedings of the 2013 IEEE Symposium on Computers & Informatics (ISCI), Langkawi, Malaysia, 7–9 April 2013; IEEE: Langkawi, Malaysia. [CrossRef]
6. Pérez-Sánchez, M.; Sánchez-Romero, F.; Ramos, H.; López-Jiménez, P. Modeling Irrigation Networks for the Quantification of Potential Energy Recovering: A Case Study. *Water* **2016**, *8*, 234. [CrossRef]
7. Rossman, L.A. *Water Supply and Water Resources Division*; United States Environmental Protection Agency: Cincinnati, OH, USA, 2000.
8. León-Celi, C.; Iglesias-Rey, P.L.; Martínez-Solano, F.J.; Mora-Melia, D. A Methodology for the Optimization of Flow Rate Injection to Looped Water Distribution Networks through Multiple Pumping Stations. *Water* **2016**, *8*, 575. [CrossRef]
9. Wu, W.Y.; Maier, H.R.; Simpson, A.R. Multi-objective optimization of water distribution systems accounting for economic cost, hydraulic reliability, and greenhouse gas emissions. *Water Resour. Res.* **2013**, *49*, 1211–1225. [CrossRef]
10. Wang, J.Y.; Chen, F.G.; Chen, J.S. A green pump scheduling algorithm for minimizing power consumption and land depletion. *Concurr. Eng-Res. A* **2013**, *21*, 121–128. [CrossRef]
11. Barán, B.; Lücken, C.; Sotelo, A. Multi-objective pump scheduling optimization using evolutionary strategies. *Adv. Eng. Softw.* **2005**, *36*, 39–47. [CrossRef]
12. Shamir, U.; Salomons, E. Optimal Real-Time Operation of Urban Water Distribution Systems Using Reduced Models. *J. Water Res. Pl-Asce* **2008**, *134*, 181–185. [CrossRef]
13. Fracasso, P.T.; Barnes, F.S.; Costa, A.H.R. Optimized Control for Water Utilities. *Procedia Eng.* **2014**, *70*, 678–687. [CrossRef]
14. Wu, W.Y.; Simpson, A.R.; Maier, H.R. Sensitivity of Optimal Tradeoffs between Cost and Greenhouse Gas Emissions for Water Distribution Systems to Electricity Tariff and Generation. *J. Water Res. Pl-Asce* **2012**, *138*, 182–186. [CrossRef]
15. Stokes, C.S.; Simpson, A.R.; Maier, H.R. An Improved Framework for the Modeling and Optimization of Greenhouse Gas Emissions Associated With Water Distribution Systems. 2012. Available online: http://www.iemss.org/sites/iemss2012//proceedings/C3_0560_Maier_et_al.pdf (accessed on 26 August 2017).
16. Rogers, M.M.; Wang, Y.; Wang, C.S.; McElmurry, S.P.; Miller, C.J. Evaluation of a rapid LMP-based approach for calculating marginal unit emissions. *Appl. Energy* **2013**, *111*, 812–820. [CrossRef]
17. Wang, J.Y.; Chang, T.P.; Chen, J.S. An enhanced genetic algorithm for bi-objective pump scheduling in water supply. *Expert Syst. Appl.* **2009**, *36*, 10249–10258. [CrossRef]
18. Zecchin, A.C.; Maier, H.R.; Simpson, A.R.; Leonard, M.; Nixon, J.B. Ant Colony Optimization Applied to Water Distribution System Design: Comparative Study of Five Algorithms. *J. Water Res. Pl-Asce* **2007**, *133*, 87–92. [CrossRef]

19. Lopez-Ibanez, M. Operational Optimisation of Water Distribution Networks. Ph.D. Thesis, Edinburgh Napier University, Scotland, UK, November 2009.

20. Zheng, Y.W.; Behandish, M. Comparing methods of parallel genetic optimization for pump scheduling using hydraulic model and GPU-based ANN meta-model. In Proceedings of the 14th Water Distribution Systems Analysis Conference, Adelaide, Austrilia, 24–27 September 2012; pp. 233–248.

21. Fu, G.T.; Kapelan, Z. Embedding Neural Networks in Multi-objective Genetic Algorithms for Water Distribution System Design. In *Water Distribution Systems Analysis 2010*; American Society of Civil Engineers: San Francisco, CA, USA, 2011; pp. 888–898.

22. Sadatiyan Abkenar, S.M.; Stanley, S.D.; Miller, C.J.; Chase, D.V.; McElmurry, S.P. Evaluation of genetic algorithms using discrete and continuous methods for pump optimization of water distribution systems. *Sustain. Comput. Inform. Syst.* **2015**, *8*, 18–23. [CrossRef]

23. Bi, W.; Dandy, G.C. Optimization of Water Distribution Systems Using Online Retrained Metamodels. *J. Water Res. Pl-Asce* **2014**, *140*, 04014032. [CrossRef]

24. Marchi, A.; Dandy, G.; Wilkins, A.; Rohrlach, H. Methodology for Comparing Evolutionary Algorithms for Optimization of Water Distribution Systems. *J. Water Res. Pl-Asce* **2014**, *140*, 22–31. [CrossRef]

25. Zheng, F.F. Advanced Hybrid Approaches Based on Graph Theory Decomposition, Modified Evolutionary Algorithms and Deterministic Optimization Techniques for the Design of Water Distribution Systems. Ph.D. Thesis, University of Adelaide, Adelaide, Australia, 2013.

26. Wang, H.H.; Liu, S.M.; Meng, F.L.; Li, M.M. Gene Expression Programming Algorithms for Optimization of Water Distribution Networks. *Procedia Eng.* **2012**, *37*, 359–364. [CrossRef]

27. Simpson, A.R.; Dandy, G.C.; Murphy, L.J. Genetic Algorithms Compared to Other Techniques for Pipe Optimization. *J. Water Res. Pl-Asce* **1994**, *120*, 423–443. [CrossRef]

28. Rao, Z.F.; Alvarruiz, F. Use of an artificial neural network to capture the domain knowledge of a conventional hydraulic simulation model. *J. Hydroinform.* **2007**, *9*, 15–24. [CrossRef]

29. Martínez, F.; Hernández, V.; Alonso, J.M.; Rao, Z.F.; Alvisi, S. Optimizing the operation of the Valencia water-distribution network. *J. Hydroinform.* **2007**, *9*, 65–78. [CrossRef]

30. Zheng, Y.W.; Morad, B. Real-time pump scheduling using genetic algorithm and artificial neural network based on graphics processing unit. In Proceedings of the 14th Water Distribution Systems Analysis Conference, Adelaide, Austrilia, 24–27 September 2012; pp. 1088–1099.

31. Alighalehbabakhani, F.; Miller, C.J.; Sadatiyan Abkenar, S.M.; Fracasso, P.T.; Jin, S.X.; McElmurry, S.P. Comparative evaluation of three distinct energy optimization tools applied to real water network (Monroe). *Sustain. Comput. Inform. Syst.* **2015**, *8*, 29–35. [CrossRef]

32. Centre for Water Systems, 2014, Benchmarks. Available online: http://emps.exeter.ac.uk/engineering/research/cws/resources/benchmarks/ (accessed on 8 June 2014).

33. Sadatiyan, S.M.; Miller, C.J. *PEPSO II User Manual*; W. S. University: Detroit, MI, USA, 2016; p. 85. Available online: http://engineering.wayne.edu/wsuwater/hydraulics/pepso.php (accessed on 15 June 2016).

34. Sadatiyan, S.M. Enhanced Pump Schedule Optimization for Large Water Distribution Networks to Maximize Environmental and Economic Benefits. Ph.D. Thesis, Wayne State University, Detroit, MI, USA, January 2016.

35. Deb, K. *Multi-Objective Optimization Using Evolutionary Algorithms*; John Wiley & Sons: New York, NY, USA, 2001.

36. Narzisi, G. *Multi-Objective Optimization, A Quick Introduction*; Courant Institute of Mathematical Sciences, New York University: New York, NY, USA, 2008; p. 35. Available online: http://cims.nyu.edu/~gn387/glp/lec1.pdf (accessed on 8 August 2017).

37. Alighalehbabakhani, F.; McElmurry, S.; Miller, C.J.; Sadatiyan Abkenar, S.M. A Case Study of Energy Cost Optimization in Monroe Water Distribution System. In Proceeding of the 2013 International Green Computing Conference (IGCC), Arlington, VA, USA, 27–29 June 2013; Volume 5, pp. 1–5. [CrossRef]

38. Van Zyl, J.E. A Methodology for Improved Operational Optimization of Water Distribution Systems. Ph.D. Thesis, School of Engineering and Computer Science, University of Exeter, England, UK, September 2001.

39. ThinkPad W520, Lenovo, Document ID: PD015361, Last Updated: 2011/8/24. Available online: https://pcsupport.lenovo.com/us/en/products/laptops-and-netbooks/thinkpad-w-series-laptops/thinkpad-w520/parts/PD015361 (accessed on 24 August 2017).
40. LEEM Server. Available online: http://leem.today/ (accessed on 20 July 2017).

water

MDPI

Article

Enabling Efficient and Sustainable Transitions of Water Distribution Systems under Network Structure Uncertainty

Jonatan Zischg *, Michael Mair ⓘ, Wolfgang Rauch and Robert Sitzenfrei ⓘ

Unit of Environmental Engineering, University of Innsbruck, 6020 Innsbruck, Austria;
michael.mair@uibk.ac.at (M.M.); wolfgang.rauch@uibk.ac.at (W.R.); robert.sitzenfrei@uibk.ac.at (R.S.)
* Correspondence: jonatan.zischg@uibk.ac.at; Tel.: +43-512-507-62160

Received: 23 July 2017; Accepted: 11 September 2017; Published: 18 September 2017

Abstract: This paper focuses on the performance of water distribution systems (WDSs) during long-term city transitions. A transition describes the pathway from an initial to a final planning stage including the structural and functional changes on the infrastructure over time. A methodology is presented where consecutive WDSs under changing conditions are automatically created, simulated and then analyzed at specific points in time during a transition process of several decades. Consequential WDS analyses include (a) uncertain network structure, (b) temporal and spatial demand variation and (c) network displacement. With the proposed approach, it is possible to identify robust WDS structures and critical points in time for which sufficient hydraulic and water quality requirements cannot be ensured to the customers. The approach is applied to a case study, where a WDS transition of epic dimensions is currently taking place due to a city relocation. The resulting necessity of its WDS transition is modelled with automatically created planning options for consecutive years of the transition process. For the investigated case study, we tested a traditional "doing-all-at-the-end" approach, where necessary pipe upgrades are performed at the last stages of the transition process. Results show that the sole design of the desired final-stage WDS is insufficient. Owing to the drastic network deconstruction and the stepwise "loss of capacity", critical pipes must be redesigned at earlier stages to maintain acceptable service levels for most of the investigated future scenarios.

Keywords: hydraulic simulation; network structure uncertainty; performance assessment; scenario analysis; water distribution benchmarking

1. Introduction

Most engineered water distribution systems (WDSs) in urban areas are facing multiple internal and external development pressures during their lifespan, and have to be continuously adapted in order to guarantee a sufficient high level of service at all times [1]. Therefore, future changes in demographic, climatic and socioeconomic developments are going to be the key drivers for changing the system's structure and operation [2]. In this context, the term "transition" is used to describe the pathway from an initial (current) to a final (planning) development stage in a WDS, including its structural and functional changes over time [3]. The intermediate development stages (i.e., specific points in time during the transition process), also including the initial and final development stage, are hereinafter defined as transition stages.

Fast ongoing system transitions (e.g., urban development) and the contradicting long lifespan of WDS components of several decades, stress existing infrastructure and require new approaches on how WDSs are designed and operated [4]. State-of-the-art strategies address flexible infrastructure design, where planners can react to future uncertainties [5]. Basupi and Kapelan [6] introduced a flexible design method under consideration of future demand uncertainties to provide cost-efficient solutions to decision makers. Creaco et al. [7] defined the gradual and optimal WDS growth over time as phasing of construction, where structural network expansions are investigated and optimized at several time steps instead of considering only a single design phase, also taking into account demand uncertainties [8]. As such, this strives for the optimal scheduling of pipe upgrade works. Beh et al. [9] investigated the augmentation of WDSs from a water resource perspective, proving the benefits of adaptive plans compared with those fixed at the initial planning stage. However, from a network perspective these studies assume a certain development of future network structure and are mostly limited to small test cases. Conversely, this study focuses on an exploratory modelling approach under deep uncertainties in network structure and a case study problem of a larger scale (several thousand pipes and junctions), without seeking for optimal and computational expensive solutions [10]. Furthermore, studies investigating the combined effects of network construction and deconstruction at the same time are lacking and therefore further investigations are necessary.

The objective of this work is to assist decision makers in testing various planning options (e.g., changing the future city layout) and design strategies (e.g., single-stage design). This paper introduces a holistic modelling framework to assess system performances of WDSs during long-term transitions. A methodology is presented to generate and assess water networks at specific points in time by considering the uncertainty in the future development. The transition process describes the disconnection and addition of pipes to the network, including the related shift in water demand. The uncertainties are addressed with the automatic creation of planning options (stochastic future network structures) and different development scenarios affecting the total water consumption. Each examined point in time during the transition process (transition stage) is evaluated with performance indicators to describe the hydraulic and water quality states of the system. They enable the detection of weak points and critical transition stages [11]. This work includes hydraulic assessments (pressure head), qualitative statements (water age) and a capacity index (pressure surplus).

The novelty of the proposed methodology is the automatic creation of planning options, which are suitable if data for modelling are not available, limited or of poor quality. Frequently, data availability is limited by legal restrictions but also existing data of a good quality might not be suitable for the desired modelling aim. To overcome this problem, an approach using data of alternative systems (e.g., street network data) with strong structural similarities to the WDSs, represents a good alternative to complement or even compensate missing data and test developed models on a less case-study-specific perspective [12–14]. Furthermore, newly planned networks can be developed by creating a variety of WDSs with little effort, considering different structural aspects and development scenarios for the purpose of identifying the most robust system, either for newly built WDS or existing WDS parts in poor condition, requiring redesign [7,15].

In this paper, the proposed methodology is applied to the Swedish town Kiruna, where a major transition of the city is taking place until the end of the century. Substantial parts of the town, including its water infrastructure, have to be moved due to expanding underground mining activities. Various planning options for the future WDS (network structure and loop degree) are tested for different scenarios and one design strategy. The results prove the proposed method is capable of identifying critical WDS stages during a transition process. The outcome can further support engineers and planners to evaluate risks and opportunities for planning and scheduling cost-efficient pipe upgrades. In this work, we tested a simple design strategy of a traditional "single-stage-design" and a "doing-all-at-the-end" approach, where necessary pipe upgrades are performed at the final stages of the transition process. For the investigated case study, we found that the sole design of the final-stage WDS is insufficient for most of the future scenarios and planning options. Owing to the drastic network deconstruction and the stepwise "loss of capacity", critical pipes must be redesigned at earlier stages for the scenarios of constant demand and demand increase (e.g., population growth). The methodology can also be applied to investigate uncertain spatial city development and testing different expansion or deconstruction scenarios for any city.

2. Modelling Framework

The future development of water consumption is highly uncertain. It depends on multiple factors, such as changes in population and consumption patterns, climate change, variations of land use, tourism and economic trends [16]. All these global pressures can have a significant impact on the performances of existing WDSs during their lifespan and can be investigated with a scenario analysis [17]. Uncertainties in the network structure occur when planning new WDSs [7]. Also in existing WDSs, the exact physical location of network parts (mostly secondary pipes) is sometimes incomplete or even unknown [14]. Furthermore, the variation of future network structure is considered when retrofitting existing systems (e.g., provision of alternative flow paths) to enhance the reliability and robustness of the WDS. To tackle these challenges, uncertainties in the future network are considered with stochastically generated WDS structures [18] (e.g., varying minimum spanning tree and loop degree in the network) and referred to as possible planning options for the future. In this context, the minimum spanning tree is an undirected graph (e.g., WDS network) with no loops, connecting all demand nodes with the water source(s) and the sum of all edge weights (e.g., pipe length) is minimal [19]. The loop degree is introduced by the cycle index (CI). It defines the length ratio of alternative paths between two nodes (shortcuts) and the corresponding path obtained from the minimum spanning tree (CI = 0 is a fully branched system, for CI > 0 loops are created) [14].

The following presents the development of a modelling tool to automatically evaluate network performances during a WDS transition process. In Figure 1 the modelling framework is outlined in four sequential tasks and described in detail in the following paragraphs.

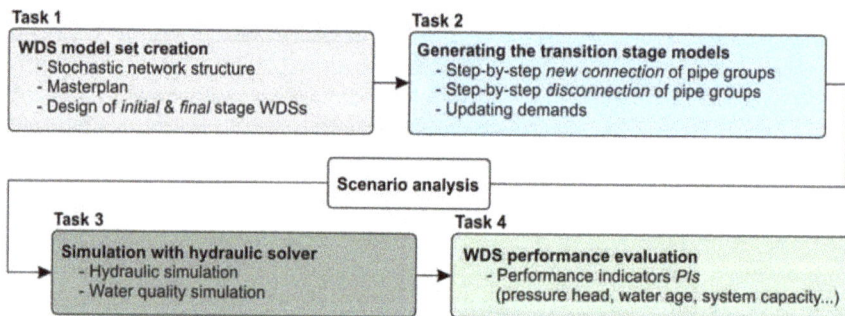

Figure 1. Description of the modelling framework.

Task 1 describes the generation of networks according to the approach presented by Mair et al. [18]. The generation algorithms are implemented in C++ and distributed as open source. (Available online: https://github.com/iut-ibk/DynaVIBe (accessed on 15 August 2017)) The software includes: 1) a spanning tree-based algorithm for network structure design, 2) algorithms for future demand projection and 3) an automated pipe-sizing algorithm to create WDSs based on GIS data [20,21]. Previous studies showed the high colocation of street and water distribution networks [14], which offers the opportunity to use that information for the WDS design. The required input data are the digital elevation map (DEM) of the town, the positions and the supply ratio (%) of the water sources, the street network, the nodal demands and – optional - the known pipe sections (see "set" pipe in Figure 2). In case the WDS and/or the nodal demands are incomplete or even unknown, the missing information represent variables in the stochastic WDS generation process. This principle applies for all unknown WDS parts, regardless of existing and/or future WDS. For this reason, the WDS is designed for the initial (existing) and the final (planning) stage. For example, in a WDS only the pipes with a diameter greater than 200 mm are known. In that case, the missing pipe connections to demand nodes can be stochastically generated on basis of the overlying street network, by assuming the WDS network $|P|$ being a subset ($|P| \subseteq |S|$) of the street network $|S|$. Resulting WDSs, using different a different level of detail in the input data, are shown in Figure 2 on the right. With this modelling approach, a robust network structure can be identified for the future, not only for newly planned WDSs, but also for the redesign of existing systems (e.g., closing loops for redundancy).

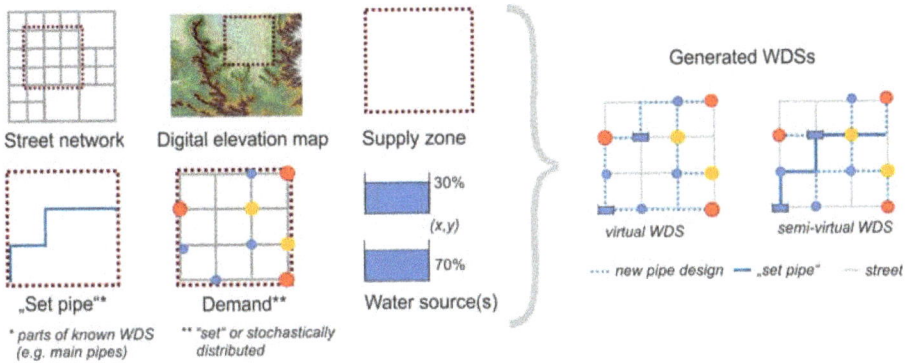

Figure 2. Concept of the stochastic generation of water distribution systems (WDSs). Following the idea from Sitzenfrei et al [22].

When considering the dynamics of long-term WDS transitioning (e.g., network expansion, demand shifting, network shrinking) with this methodology, a city's master plan must be integrated. For example, expanding future street network data derived from architect plans are used to generate future WDSs. Figure 3 illustrates the initial (existing) stage i and final (planning) stage f of the WDS during the transition process, which are generated and fulfil the boundaries of the master plan (future street network, deconstruction and construction zones). Unknown pipe diameters are independently designed at both stages, using a pipe-sizing algorithm based on the approach of Saldarriaga et al. [23]. With this algorithm, the pipe diameters are calculated based on a) the maximum hourly demand $Q_{h,max}$ as design value and b) an assumed pressure surface inclination of 5 m/km. The pipe diameters are then divided into discrete diameter classes. Let $|P^i| = \{p_1, p_2, \ldots, p_{ni}\}^i$ be the set of all pipes at initial stage i, where ni is the number of pipes at stage i. Conversely, $|P^f| = \{p_1, p_2, \ldots, p_{nf}\}^f$ is the set of all pipes at final stage f, where nf is the number of pipes at stage f. Then the intersection of the sets $|P^i| \cap |P^f|$ is built and the designed diameters compared. Pipes with a changing diameter are added to the new pipe set $|P^{up}| = \{\}$, representing the necessary pipe upgrades. According to Mair et al. [14] the generated WDS set is classified with the cycle index (CI), describing the degree of loops (alternative flow paths) in the network. With this approach sufficiently working "engineering" solutions, rather than optimal WDS are generated with adequate computational capacity. For the target application to large case studies (with several thousand elements), the reduced computational effort is a compelling argument.

Task 2 presents the creation of transition stage models to describe the detailed step-by-step progress of construction and deconstruction phases on basis of the master plan. In doing so, specific points in time between initial stage i and final stage f, i.e., intermediate stages, are investigated. Figure 4 shows an example of an intermediate stage j where some parts of the initial WDS are disconnected, while other parts are added at the same time. Not only the pipe structure is changed, also a shifting of the nodal demand from the disconnected to the new connected WDS parts occurs. The transition stage models comprise the initial, intermediate and final stages of the WDS. The number of intermediate stages is a model input parameter and depends on the temporal definition within the master plan (e.g., phased construction and deconstruction zones of new building blocks).

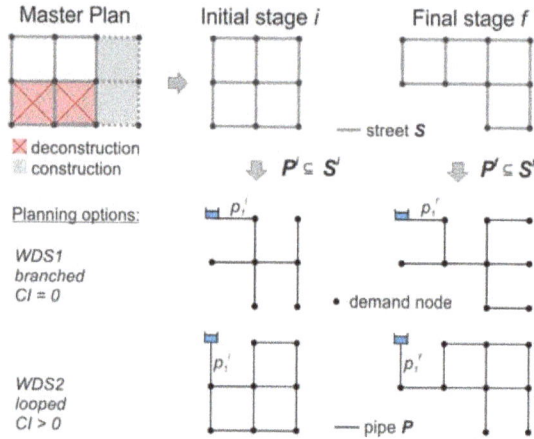

Figure 3. Long-term network dynamics in the concept of network generation by integrating stages in the design of the WDS.

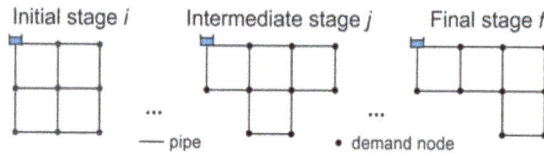

Figure 4. Creation of intermediate stage WDSs on basis of the master plan and the initial and final stage.

The disconnection of pipes, along with a demand shifting, changes the flow pattern of the WDS but does not necessarily cause performance reductions. Possible examples are: (1) the disconnection of final branches or (2) the shifting of demand nodes towards the water source, which can even increase the WDS performance (e.g., minimum pressure). The zonal disconnection of pipes from the main WDS can cause an interruption (isolation) of certain areas from the water source. Such parts of the network are identified and must be removed or reconnected to the main WDS to ensure a hydraulic supply (**Task 3**). In the current work, the physical location and attributes of WDS components that are unaffected from the deconstruction and the construction processes are maintained over time, i.e., the intermediate and final-stage WDS are dependent on their previous transition stages and originate from the initial stage (network structure and pipe diameter). The scheduling of the determined pipe upgrades ($|P^{up}|$) to achieve high performances at final stage is part of a design strategy. In this work, we test a simple "doing-all-at-the-end" approach, where necessary pipe upgrades are performed at the final stages of the transition process.

In **Task 3**, a model interface to a hydraulic solver is implemented where each transition stage model (see **Task 2**) is simulated under different scenarios. In WDS modelling, it is state-of-the-art to use extended period simulations to consider the diurnal demand patterns of several representative days [24]. In this work, the hydraulic solver EPANET 2 [25] is used, where each transition stage model (**Task 2**) is simulated under different scenarios. In a first simulation run, the system is solved for one day with high water consumption to identify supply problems. Then, the simulation is repeated for a period of low water usage (six consecutive days of low water consumption are simulated to determine maximum water age and potential stagnation problems). The water age is calculated from the residence time of the storage tank and the travel time in the network from the source node(s) to the demand nodes at low flow conditions. For this purpose, we used the water quality analysis tool of EPANET 2 [25].

Task 4 describes the performance evaluation of the WDS. Helpful tools to assess hydraulic and quality requirements of WDSs are global performance indicators (*PIs*). By definition, the *PIs* take values in the interval from 0 (worst performance) to 1 (best performance), depending on predefined threshold values and a performance criterion. Furthermore, statistical values complement the investigation. First, the nodal performances PI_k are determined for each node k before they are averaged and weighed to one global representative value *PI*. For this study, we analyzed a minimum performance indicator at peak demand and the mean pressure head at average demand under normal operation conditions (e.g., no pipe breaks). The threshold values for *PIs* differ among case studies and design guidelines of sufficient performance, and have a strong impact on the overall performance. Therefore, they must be defined by the user [20]. The minimum pressure performance, including the selected threshold value in accordance with the Austrian Standard ÖNORM B2538 [26], is defined as:

$$PI_k^{min.\ pressure} = \begin{cases} 0, & p_{k,min} \leq 0, \\ 1, & p_{k,min} \geq 30, \\ \dfrac{p_{k,min}}{30}, & p_{k,min} > 0 \land p_{k,min} < 30, \end{cases} \tag{1}$$

where $p_{k,min}$ is the nodal pressure head in meters. Furthermore, the water quality of the WDS model is described by the maximum water age, which is a driving factor related to microorganism growth [27]. The maximum nodal water age $w_{k,max}$ in hours is calculated based on flow velocities and pipe lengths and assessed after a low demand period of 144 h and contains the initial water age from the storage tank. In this work, the nodal water age performance is defined as:

$$PI_k^{max.\ water\ age} = \begin{cases} 0, & w_{k,max} \geq 120, \\ 1, & w_{k,max} \leq 24, \\ 1 - \dfrac{w_{k,max} - 24}{120 - 24}, & w_{k,max} > 24 \land w_{k,max} < 120. \end{cases} \tag{2}$$

Additionally, the respective nodal performance indicators PI_k's are averaged and weighed with the nodal demand d_k to consider the hydraulic importance of node k (e.g., number of supplied customers) as follows:

$$PI = \frac{\sum_{k=1}^{N_n} d_k * PI_k}{d_{tot}}. \tag{3}$$

To quantify properties of robustness and fault tolerance in the system, a network capacity index I_r is assessed [28,29]. The index is based on the power balance of the network and gives information about how much pressure surplus is available at each network node, compared to a minimum required head [30]. The pressure surplus can be seen as a "buffer capacity" that can be used under critical operation conditions (e.g., pipe breaks), when the internal energy dissipation increases. The capacity index is assessed at peak demand and defined as follows:

$$I_r = 1 - \frac{P_D}{P_{D,max}}, \tag{4}$$

where $P_D = \sum_{j=1}^{N_p} q_j \Delta H_j$ is the dissipated power, and $P_{D,max} = \sum_{s=1}^{N_r} q_s H_s - \sum_{k=1}^{N_n} d_k \overline{H_k}$ describes the maximum dissipated power to meet the minimum head constraints $\overline{H_k}$ at node i. For this case study, the minimum required nodal head $\overline{H_k}$ is adopted with the nodal elevation plus an additional pressure head of 15 metres at node k. H_s refers to the nodal head of supply source s and ΔH_j is the head loss along pipe j. The inflow from source s is described with q_s, d_k represents the nodal demand and q_j

is the flow in pipe j. N_r, N_n and N_p state the number of supply sources, nodes, and pipes, respectively. A more detailed description can be found in Di Nardo et al. [31].

3. Case Study Application and Numeric Results

The developed methodology is applied to the city transition of Kiruna, where significant structural network changes are currently (and during the next decades) taking place, since the city has to make way for an underground expanding ore mine. Kiruna, a small city with about 20,000 inhabitants, is the northernmost city in Sweden and became one of the major centres of the mining industry due to iron ore extraction. As can be seen from Figure 5a, in the southwestern area of the town, the world's largest underground iron ore mine is located and operated. Due to expanding excavations, however, current parts of town and its water infrastructure are threatened by subsidence and erosion. Nevertheless, mining activities will continue. Therefore, a master plan was developed by decision makers (Kiruna municipality, mining company LKAB and architects White), which provides a step-by-step resettlement of the inhabitants living in defined deformation zones (red lines in Figure 5b) to a newly built city centre about 4 km eastward (see Figure 5b). The transition process is expected to be completed by the end of this century [32].

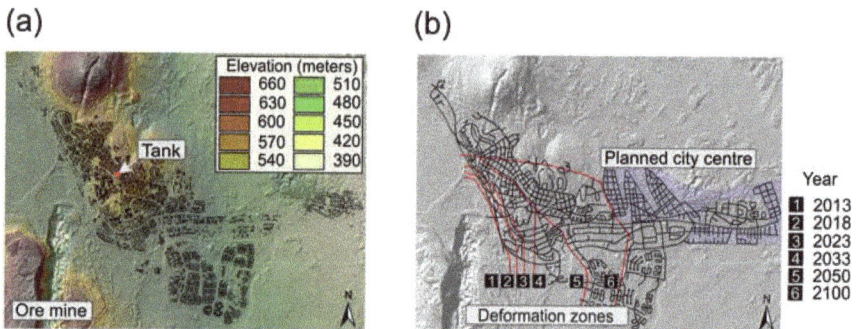

Figure 5. (**a**) Current Kiruna with digital elevation map (DEM); (**b**) Current and future street networks (planned) with the step-by-step expansion of deformation zones, according to the master plan.

In the city's master plan, the structural changes of the city are very certain but no specific planning options for the WDS transition are defined, and therefore this case study is well suited to present the developed methodology. As a result, statements about the WDS efficiency and performance trends during this long-term transition process from the initial stage in 2012 to the final stage in 2100 are possible and many future WDS structures and scenarios are tested and investigated for strengths and weaknesses.

To deal with the demand uncertainties, three simplified scenarios based on different future developments of water consumption are examined as follows (see Figure 6a):

- The basic assumption for the scenario "Baseline" is that the total demand remains constant within the transition period. It represents no change in population but a change of its location ($Q_{h,max,2100} = 128$ L/s).
- The scenario "Growth" implies a linear increase of water usage of 30 percent until the end of the century [33]. This represents a population or demand per capita growth ($Q_{h,max,2100} = 166$ L/s).
- The scenario "Stagnation" describes an economic decay, where migration of labour occurs due to an assumed reduction of mining activities [33]. The water demand is taken to gradually decrease by 30 percent until 2100 ($Q_{h,max,2100} = 90$ L/s).

In the context of network structure uncertainty, 30 possible WDSs (planning options) with different properties are automatically created by the variation of the cycle index (CI), which defines the degree of alternative flow paths [14]. Two possible planning options with looped and branched WDS structure, generated on basis of the street network, are illustrated in Figure 6b,c.

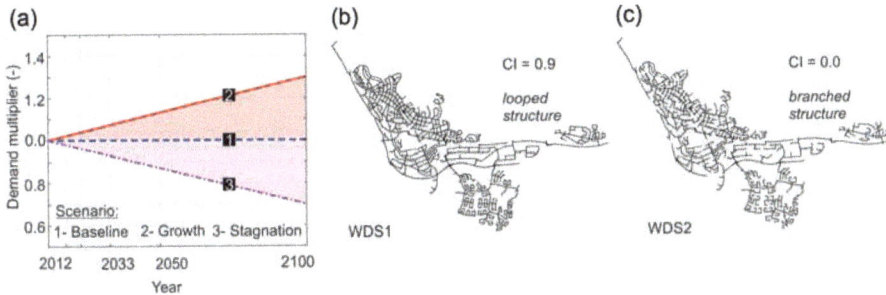

Figure 6. Future demand scenarios (**a**) and two possible planning options (WDS1 & WDS2) with different network structure (**b**) and (**c**).

There are restrictions on the hydraulic modelling data of Kiruna's actual WDS and, as such, are not allowed to be published. Nevertheless, with the proposed method each stage of the WDS can and has been approximated by a set of stochastic generated WDSs, based on freely available real street network data (see Modelling framework: **Task 1**). This gives the opportunity to apply and test the approach also to variety of initial WDS structures. However, more detailed and complementary information (e.g., actual WDS) can easily be integrated, if available, and allowed in terms of legal aspects.

Task 1 describes the creation of the WDS model set. For this application the required input data are the digital elevation map (DEM) of the town, the position of the current and future water sources, and the real street networks for the initial (year 2012) and final stages (year 2100). The future street network data is taken from architecture plans. Due to the data restrictions, the population density and nodal demand distribution are assumed to be uniform over the area of interest with 1,000 implemented demand points. Elevation differences in the investigation area allow for the construction of a functioning gravity driven WDS and hence, no pumping is required (see Figure 5a). The generated WDS includes the positions of the reservoir, tank and pressure reduction valves (PRVs) and flow control valves (FCVs). The locations of the valves within the 30 stochastic WDS structures are unchanged. The FCVs are positioned between reservoir and tank, while the locations of the PRVs are chosen based on minimum and maximum pressure head requirements (30–100 m) within each pressure zone. Regarding the size of the generated networks, they consist of approximately 100 km pipe length (~4000 pipes) at the initial stage and about 75 km (~2000 pipes) at the final stage. This shows that the future network is planned to be denser as compared to the existing one at all demand projections.

Figure 7 illustrates one of the thirty generated planning option of the WDS (different network structure) at the initial stage (year 2012) and the final stage (year 2100). When comparing the initial and final-stage WDS, not only a huge change in the pipe network can be observed, but also the water tank of the initial stage is situated in the deformation zone 2100 and thus has to be relocated. Additionally, two new main pipes from the new tank to the separated networks and a new position of the PRV are necessary (see Figure 7b). According to the Austrian Standard ÖNORM B2538 [26], additional loads for firefighting can be neglected in the design process due to the size of the supply zone (population $\geq$ 20,000). The intersection of the pipe sets $|P^i| \cap |P^f|$ is about half the size of $|P^i|$, which implies that 50% of pipes at the initial stage WDS keep their physical location in the final-stage WDS. Therefrom approximately 10% of the pipe diameters have to be upgraded due to changing flow conditions and to maintain a connected network (e.g., new water source). This builds the pipe set describing necessary upgrades ($|P^{up}|$).

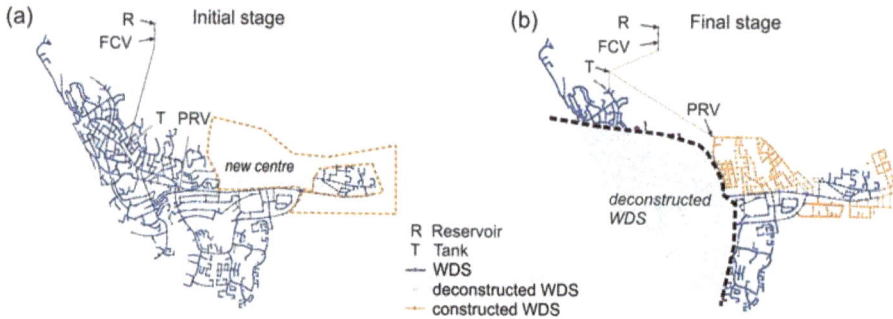

Figure 7. One of thirty created planning options representing (**a**) the initial and (**b**) the final stage.

Task 2 describes the generation of the transition stage models. For this purpose, the master plan defining the phased transition process of the town is used. The transition stages represent the years 2012, 2013, 2018, 2023, 2033, 2050 and 2100. Figure 8 presents an example of the creation of six transition stage models for one planning option. It can be seen that the town is step-by-step moved by a simultaneous deconstruction of the initial WDS and construction of a new piping system. The resettlement progress of people living in the deformation zones is modelled by transferring the demand nodes to the new city centre, assuming the same uniform spatial distribution as for the final design stage. The total demand is dependent on the investigated scenario.

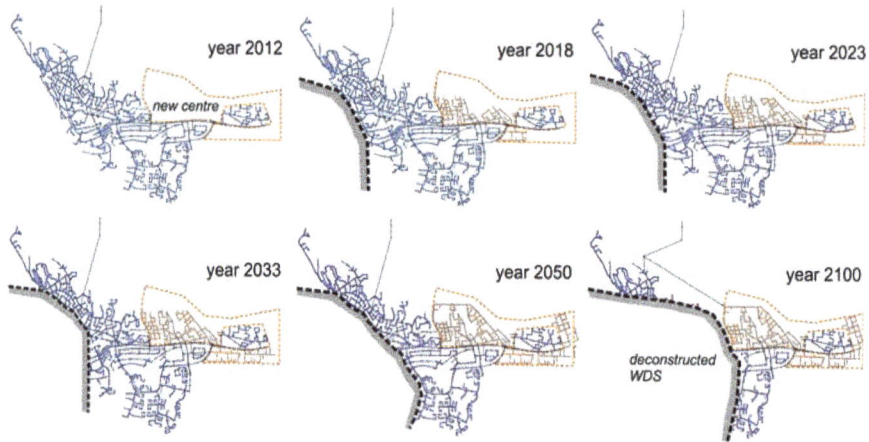

Figure 8. Transition stage models for Kiruna at six stages from year 2012 to year 2100.

Based on the determined set of necessary pipe upgrades ($|P^{up}|$) from initial and final network design (see **Task 1**), a simple "doing-all-at-the-end" approach is investigated: Pipe upgrades for achieving the efficient final-stage WDS are performed at the stages in 2050 (pipe replacement rate on average 2%) and the final stage 2100 (pipe replacement rate on average 8%). This means that until stage 2050, the pipe diameters of the WDS remain unchanged, while all of the newly constructed pipes are designed for the final stage. With this approach it can be determined, (a) whether the WDS can tolerate the occurring changes or not, and (b) at which transition stage additional redesign (e.g., pipe replacements) might be necessary for the 3 scenarios and the 30 planning options.

The transition stage models are automatically generated for each of the 30 network structures (planning options). Altogether, for the case study analysis 1260 WDSs are created and simulated, containing 30 variations of network structure, each evaluated for 7 stages, under 2 hydraulic loads (high and low daily demand) and 3 future demand scenarios (constant demand and demand increase/decrease).

Tasks 3 and **4** present the hydraulic simulations and the performance evaluations of the WDSs. The application of the developed approach is shown with the outcome of the performance analysis during the WDS transition of Kiruna as a model. In Figure 9, we firstly show example contour plots of the pressure distributions at peak demand for three different stages (2012, 2033 & 2100) and two scenarios (Baseline & Growth).

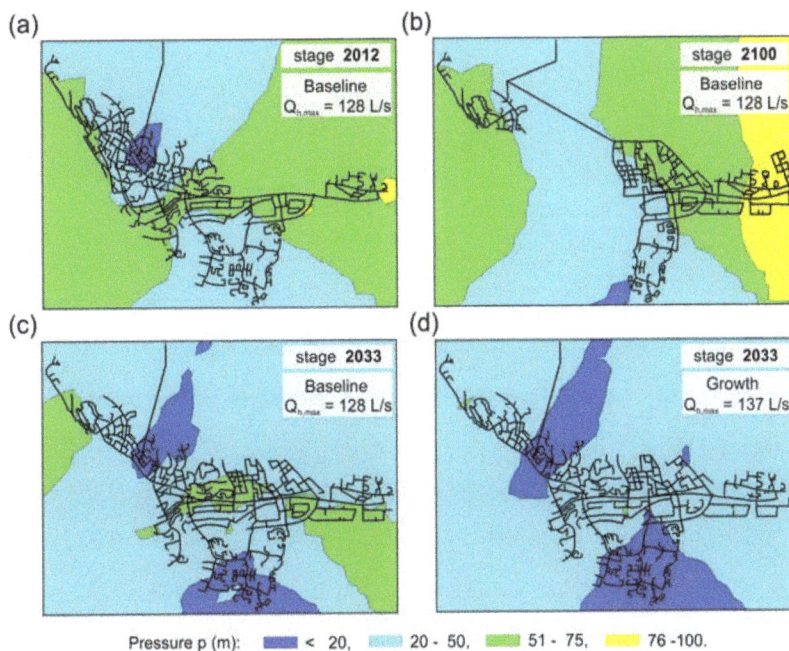

Figure 9. Pressure distributions at peak demand for the Baseline scenario for the initial stage 2012 (**a**), the final stage 2100 (**b**), the transition stage 2033 (**c**), and the Growth scenario for transition stage 2033 (**d**).

In Figure 10 the findings of three performance indicators (PIs of system capacity, minimum pressure and water age) and one statistical value (mean pressure) are presented for all WDS stages. Therein a bandwidth of performance developments for the 30 planning options (possible WDS structures) is shown for the three scenarios. While the scenario "Baseline" is plotted in dashed lines, the scenario "Growth" is represented by continuous lines. The scenario "Stagnation" is illustrated by dot-dashed lines.

The three main findings of the WDS transition for the case-study application are discussed in the following: First, the performance drops are highly correlated with the future water consumption. The scenario of a linear increase in water consumption reveals lower network performances than the assumption of decreasing water consumption, with the exception of water age where the opposite behaviour is observed. Sufficient minimum pressure performances can only be guaranteed for scenario "Stagnation" (demand reduction) and utilizing the pressure surplus ("buffer capacity") of the WDS (see drop of capacity index). However, problems with water age occur for this scenario. For scenario "Baseline" and scenario "Growth", the capacity of the (remaining) WDS parts is not high enough to cope with the structural ("loss of capacity") and demographic ("increase demand") changes after stage 2023, while guaranteeing sufficient minimum pressure heads. The performance improvement at the final stage is due to the pipe upgrades (10% percent pipe replacement, see **Task 1**).

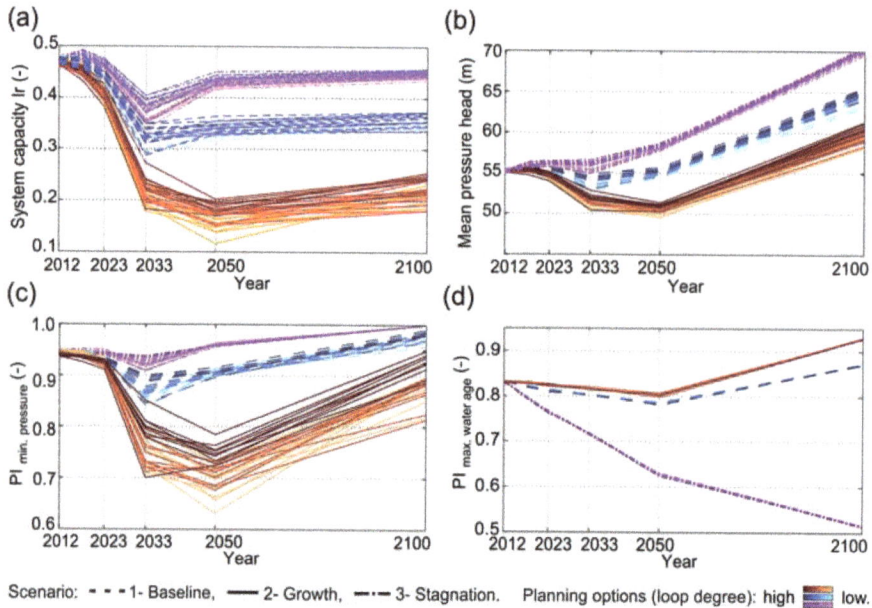

Figure 10. Performance developments for three demand scenarios and WDS structure variations; (a) System capacity, (b) Mean pressure head, (c) Minimum pressure performance and (d) Maximum water age performance.

Second, influences of the 30 planning options with different network structure (degree of loops) on the system performances are proven (light shading in Figure 10 indicates a low loop degree and dark shading a high loop degree). Especially the performances of minimum pressure and system capacity are favoured by a higher loop degree. The drops in system capacity, indicating a stronger usage of additional pressure surplus, are lower for highly looped networks due to more alternative flow paths. The WDS with the best performances is identified to be one of the highly looped planning options. However, not every highly looped planning option revealed higher performances (e.g., see crossing lines with different shading in Figure 10), but in terms of robustness and redundancy, a higher degree of loops is advisable when facing structural and demand uncertainties.

Third, critical points are identified during the WDS transition. The first four stages of the transition processes, reveal only slight changes in the WDS performances. Up to year 2023, the changing flow patterns are compensated by the WDS. However, after year 2023, partially inefficient WDSs occur. The performance drops at the stages 2033 and 2050 are related to a severe "loss of capacity" within the remaining WDS. This outcome demonstrates that the sole "single-stage-design" of the final-stage WDS (including the "doing-all-at-the-end" pipe upgrades) is insufficient. Owing to the network deconstruction, pipes that are already disconnected at final stage f ($|P^i| \setminus |P^f|$), must be redesigned at the intermediate stages 2033 and 2050 to maintain acceptable service pressures.

4. Discussion

The aim of this work is to provide a performance assessment tool assisting decision makers during long term network transitions processes of WDSs over time and to determine the sensitivity of different future uncertainties, like demographic changes, network structure uncertainties and shifting network layout (e.g., WDS expansion). The proposed approach aims to tackle the entire complexity of the stated problem with the implication, that for some specific details, assumptions and simplifications are necessary. Once successfully setup, in future studies, analyzes focusing on detailed questions (e.g., redesign and rehabilitation of the existing WDS) will be addressed.

Based on that approach different design strategies can be tested. This work provides a helpful approach to identify critical stages (time points) and locations (weak points) during the WDS transitioning. Weak points are pipes whose diameters become inappropriate due to the changing flow patterns that originate from pipe disconnections, shifting demand nodes and total future water consumptions. Indicators for inefficient pipe diameters are high flow velocities v and high unit head losses h. Figure 11a presents the pipe velocities and unit head losses (both length-weighted) for the case study application at stage 2033 and for scenario "Growth". The network deconstruction at that stage causes a "loss of capacity" and therefore increased velocities and unit head losses in specific pipes. Exemplarily, a velocity threshold of 1.5 m/s is exceeded in about 5% of the pipes at stage 2033. These identified weak points have to be addressed when redesigning the WDS. Problems related to water age are marginal for the first two scenarios and occur mainly in parts of the new city centre, where the design capacity of some pipes is not reached at all stages. However, by neglecting the initial water age from the tank (e.g., due to operational measures), the mean water network travel time is below 24 h for scenario "Baseline" and scenario "Growth" at all of the stages of the WDS transition. In Figure 11b the maximum water age distribution at stage 2033 and scenario "Baseline" after an extended period simulation is presented. It can be seen that the water age is below 48 h for most junctions (including the initial tank water age of 19 h) and only minor parts of the new city centre show values above 48 h. In the literature, average WDS retention times are between 12 and 24 h for cities of a similar size [34]. However, a future demand reduction all over the city (see scenario "Stagnation") reduces the water age performance. In this case additional operational measures, like regular pipe flushing, would be necessary.

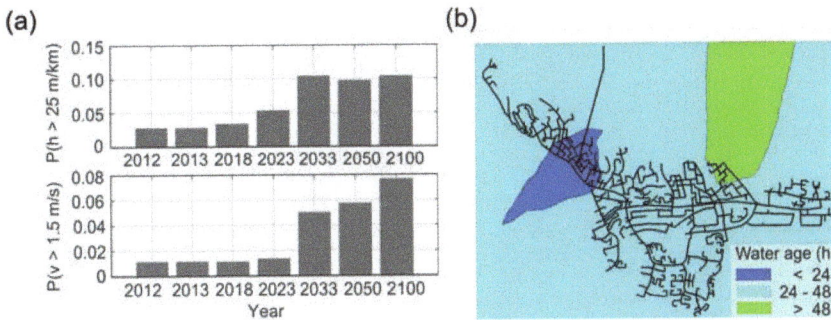

Figure 11. (a) Threshold exceedances of unit head loss (top) and velocity (bottom) at peak load for scenario "Growth", (b) WDS with maximum water age for scenario "Baseline" at transition stage 2033.

The case study application showed that up to stage 2033 sufficient hydraulic and quality requirements are guaranteed under all structural and demand uncertainties. Due to the "final-stage-design" also the performances at the final stage are sufficient (with the exception of water age under total demand decrease scenario). To improve the minimum pressure performance at the intermediate stages 2033 and 2050 (for constant demand and demand increase), other design strategies will be pursued in further investigations: (1) anticipated pipe replacements, rather than "doing-all-at-the-end" upgrades of the identified pipe upgrades $|P^{up}|$, and/or (2) network redesign of overloaded WDS components that are already deconstructed at the final-stage WDS ($|P^i|\setminus|P^f|$). Future work will also address the integration of the real WDS in the network generation procedure.

5. Conclusions

In this work, a novel modelling approach was presented to assess the performance of structural and functional long-term WDS transitions, e.g., network displacement and population growth. All these changing boundary conditions were applied to a set of automatically created planning options with different network structure. The main advantages of the method are the automated creation of hydraulic models at different points in time during the WDS transition (stages), the low computational costs and the applicability of the method to a variety of WDSs, representing different planning options. With the method different scenarios can be tested (e.g., network displacement and population growth) and critical time points determined, where predefined performance criteria cannot be guaranteed anymore.

The benefit of the methodology is the fast generation of WDS stages based on surrogate information such as the street network and the city's master plan. The methodology can handle a different degree of information content; known information of the WDS (e.g., physical location and attributes of pipes, demand distribution) can easily be considered as "fixed" parameters in the WDS generation, while all unknown information is supplemented by stochastic approximations (e.g., expanding WDS structure on basis of future street network and/or random demand distributions). As a result, different planning options are put under stress and evaluated with performance indicators.

As a case study, the city of Kiruna is used where a city expansion and a step-by-step destruction (causing the relocation of people) take place at the same time. The proposed approach was applied to identify critical points in space and time, and to assist decision makers in such a city (and infrastructure) transition. For the model application to the case study of Kiruna, several stresses were applied to the WDS at different points in time on basis of the city's master plan, including the simultaneous network disconnection and new connection, shifting demand nodes and different future demand developments. The performance drops after transition stage 2023, showed that the approach of the "doing-all-at-the-end" pipe upgrades on the basis of the sufficiently working initial and final-stage WDS ($|P^i| \cap |P^f|$) was insufficient. The performance drops revealed that the pressure surplus (quantified by the capacity index) of the remaining WDS after a major disconnection was not high enough to cope with the occurring changes ("loss of capacity"). At the stages of year 2033 and 2050, an improved design strategy with additional pipe redesign in the pipe set ($|P^i| \setminus |P^f|$) must be performed for the future scenarios of constant demand and demand increase (e.g., population growth).

Acknowledgments: This research is partly funded by the Joint Programming Initiative Urban Europe on behalf of the Austrian Ministry for Transport, Innovation and Technology (BMVIT) in the project Green/Blue Infrastructure for Sustainable, Attractive Cities (project number 839743), by the Austrian Research Promotion Agency (FFG) within the research project ORONET (project number: 858557) and by the University of Innsbruck.

Author Contributions: All authors substantially contributed in conceiving and designing the model and realizing this manuscript. Michael Mair and Jonatan Zischg prepared the input data and performed the model implementation. Robert Sitzenfrei and Jonatan Zischg worked on the analysis and presentation of the results. Wolfgang Rauch supervised the entire research. Jonatan Zischg wrote the paper. All authors have read and approved the final manuscript.

Conflicts of Interest: The authors declare no conflicts of interest. The founding sponsors had no role in the design of the study; in the collection, analyses, or interpretation of data; in the writing of the manuscript, and in the decision to publish the results.

References

1. Klassert, C.; Sigel, K.; Gawel, E.; Klauer, B. Modeling residential water consumption in amman: The role of intermittency, storage, and pricing for piped and tanker water. *Water* **2015**, *7*, 3643–3670. [CrossRef]
2. Ahiablame, L.; Engel, B.; Venort, T. Improving water supply systems for domestic uses in urban togo: The case of a suburb in lomé. *Water* **2012**, *4*, 123–134. [CrossRef]

3. Sempewo, J.; Vairavamoorthy, K.; Grimshaw, F. Transitioning of urban water distribution systems. In *World Environmental and Water Resources Congress 2010: Challenges of Change, Proceedings of the World Environmental and Water Resources Congress 2010, Providence, RI, USA, 16–20 May 2010*; American Society of Civil Engineers (ASCE): Reston, VA, USA, 2010; pp. 3662–3674.

4. Sitzenfrei, R.; Möderl, M.; Mair, M.; Rauch, W. Modeling dynamic expansion of water distribution systems for new urban developments. In *World Environmental and Water Resources Congress 2012: Crossing Boundaries, Proceedings of the World Environmental and Water Resources Congress 2012, Albuquerque, NM, USA, 20–24 May 2012*; American Society of Civil Engineers: Clearwater Beach, FL, USA, 2012; pp. 3186–3196.

5. Fletcher, S.M.; Miotti, M.; Swaminathan, J.; Klemun, M.M.; Strzepek, K.; Siddiqi, A. Water supply infrastructure planning: Decision-making framework to classify multiple uncertainties and evaluate flexible design. *J. Water Res. Pl-Asce.* **2017**, *143*, 04017061. [CrossRef]

6. Basupi, I.; Kapelan, Z. Flexible water distribution system design under future demand uncertainty. *J. Water Res. Pl-Asce.* **2013**, *141*, 04014067. [CrossRef]

7. Creaco, E.; Franchini, M.; Walski, T. Accounting for phasing of construction within the design of water distribution networks. *J. Water Res. Pl-Asce.* **2013**, *140*, 598–606. [CrossRef]

8. Creaco, E.; Franchini, M.; Walski, T. Taking account of uncertainty in demand growth when phasing the construction of a water distribution network. *J. Water Res. Pl-Asce.* **2015**, *141*, 04014049. [CrossRef]

9. Beh, E.H.; Maier, H.R.; Dandy, G.C. Adaptive, multiobjective optimal sequencing approach for urban water supply augmentation under deep uncertainty. *Water Resour. Res.* **2015**, *51*, 1529–1551. [CrossRef]

10. Urich, C.; Rauch, W. Exploring critical pathways for urban water management to identify robust strategies under deep uncertainties. *Water Res.* **2014**, *66*, 374–389. [CrossRef] [PubMed]

11. Alegre, H.; Baptista, J.M.; Cabrera, E.J.; Cubillo, F.; Duarte, P.; Hirner, W.; Merkel, W.; Parena, R. *Performance Indicators for Water Supply Services*; IWA Publishing: London, UK, 2006; p. 312.

12. De Corte, A.; Sörensen, K. Hydrogen: An artificial water distribution network generator. *Water Resour. Manag.* **2014**, *28*, 333–350. [CrossRef]

13. Möderl, M.; Sitzenfrei, R.; Fetz, T.; Fleischhacker, E.; Rauch, W. Systematic generation of virtual networks for water supply. *Water Resour. Res.* **2011**, *47*. [CrossRef]

14. Mair, M.; Zischg, J.; Rauch, W.; Sitzenfrei, R. Where to find water pipes and sewers?—On the correlation of infrastructure networks in the urban environment. *Water* **2017**, *9*, 146. [CrossRef]

15. Zischg, J.; Mair, M.; Rauch, W.; Sitzenfrei, R. Stochastic performance assessment and optimization strategies of the water supply network transition of Kiruna during city relocation. In *World Environmental and Water Resources Congress 2015: Floods, Droughts, and Ecosystems, Proceedings of the World Environmental and Water Resources Congress 2015, Austin, TX, USA, 17–21 May 2015*; American Society of Civil Engineers: Reston, VA, USA, 2015; pp. 853–862.

16. Kang, D.; Lansey, K. Multiperiod planning of water supply infrastructure based on scenario analysis. *J. Water Res. Pl-Asce.* **2014**, *140*, 40–54. [CrossRef]

17. Scott, C.; Bailey, C.; Marra, R.; Woods, G.; Ormerod, K.J.; Lansey, K. Scenario planning to address critical uncertainties for robust and resilient water–wastewater infrastructures under conditions of water scarcity and rapid development. *Water* **2012**, *4*, 848–868. [CrossRef]

18. Mair, M.; Rauch, W.; Sitzenfrei, R. Spanning tree-based algorithm for generating water distribution network sets by using street network data sets. In *World Environmental and Water Resources Congress 2014: Water without Borders, Proceeding of the World Environmental and Water Resources Congress 2014, Oregon, Portland, 1–5 June 2014*; American Society of Civil Engineers: Reston, VA, USA, 2014; pp. 465–474.

19. Kruskal, B.J.J. On the shortest spanning subtree of a graph and the traveling salesman problem. *P. Am. Math. Soc.* **1956**, *7*, 48–50. [CrossRef]

20. Sitzenfrei, R. Stochastic generation of urban water systems for case study analysis. Ph.D. Thesis, University of Innsbruck, Innsbruck, Austria, October 2010.

21. Sitzenfrei, R.; Möderl, M.; Rauch, W. Automatic generation of water distribution systems based on GIS data. *Modell. Softw.* **2013**, *47*, 138–147. [CrossRef] [PubMed]

22. Sitzenfrei, R.; Mair, M.; Diao, K.; Rauch, W. Assessing model structure uncertainties in water distribution models. In *World Environmental and Water Resources Congress 2014: Water without Borders, Proceeding of the World Environmental and Water Resources Congress 2014, Oregon, Portland, 1–5 June, 2014*; American Society of Civil Engineers: Reston, VA, USA 2014; pp. 515–524.

23. Saldarriaga, J.; Takahashi, S.; Hernández, F.; Escovar, M. Predetermining pressure surfaces in water distribution system design. In *World Environmental and Water Resources Congress 2011: Bearing Knowledge for Sustainability, Proceedings of the World Environmental and Water Resources Congress 2011, Palm Springs, CA, USA, 22–26 May 2011*; American Society of Civil Engineers: Reston, VA, USA, 2011; pp. 93–102.
24. Walski, T.M. *Advanced Water Distribution Modeling and Management*; Haestad Press: Waterbury, CT, USA, 2003.
25. Rossman, L.A. The EPANET programmer's toolkit for analysis of water distribution systems. In Proceedings of the 29th Annual Water Resources Planning and Management Conference, Tempe, AZ, USA, 6–9 June, 1999; pp. 1–10.
26. ÖNORM (AUSTRIAN STANDARDS). Available online: https://www.austrian-standards.at (accessed on 16 September 2017).
27. Machell, J.; Boxall, J. Modeling and field work to investigate the relationship between age and quality of tap water. *J. Water Res. Pl-Asce.* **2014**, *140*, 04014020. [CrossRef]
28. Todini, E. Looped water distribution networks design using a resilience index based heuristic approach. *Urban Water* **2000**, *2*, 115–122. [CrossRef]
29. Creaco, E.; Fortunato, A.; Franchini, M.; Mazzola, M.R. Comparison between entropy and resilience as indirect measures of reliability in the framework of water distribution network design. *Prced. Eng.* **2013**, *70*, 379–388. [CrossRef]
30. Yazdani, A.; Otoo, R.A.; Jeffrey, P. Resilience enhancing expansion strategies for water distribution systems: A network theory approach. *Environ. Modell. Softw.* **2011**, *26*, 1574–1582. [CrossRef]
31. Di Nardo, A.; Di Natale, M.; Greco, R.; Santonastaso, G.F. Resilience and entropy indices for water supply network sectorization in district meter areas. In Proceedings of the International Conference on Hydroinformatics, Tianjin, China, 7–10 September 2010.
32. Zischg, J.; Goncalves, M.L.; Bacchin, T.K.; Leonhardt, G.; Viklander, M.; van Timmeren, A.; Rauch, W.; Sitzenfrei, R. Info-gap robustness pathway method for transitioning of urban drainage systems under deep uncertainties. *Water Sci. Technol.* **2017**, wst2017320. [CrossRef] [PubMed]
33. Gunn, A.; Rogers, B.; Urich, C. *Identification and Assessment of Long-term Green/Blue Drainage Strategies for Kiruna, Sweden*; Monash University: Melbourne, Australia; Unpublished work, 2016.
34. American Water Works Association (AWWA) and Economic and Engineering Services, Inc (EES, I.). Effects of Water Age on Distribution System Water Quality. Available online: http://water.epa.gov/lawsregs/rulesregs/sdwa/tcr/upload/2007_05_18_disinfection_tcr_whitepaper_tcr_waterdistribution.pdf (accessed on 7 October 2015).

water

MDPI

Article

Implementation of DMAs in Intermittent Water Supply Networks Based on Equity Criteria

Amilkar E. Ilaya-Ayza [1],*, Carlos Martins [2], Enrique Campbell [3] and Joaquín Izquierdo [2] ![orcid]

[1]   Facultad Nacional de Ingeniería, Universidad Técnica de Oruro, Ciudad universitaria s/n, 49 Oruro, Bolivia

[2]   FluIng-IMM, Universitat Politècnica de València, Camino de Vera s/n, Edif. 5C, 46022 Valencia, Spain; carlos.martins.a@gmail.com (C.M.); jizquier@upv.es (J.I.)

[3]   Berliner Wasserbetriebe, Cisero Straße 40, 10107 Berlin, Germany; Enrique.Campbellgonzales@bwb.de

*   Correspondence: amilkar.ilaya@uto.edu.bo; Tel.: +591-71233652

Received: 26 September 2017; Accepted: 31 October 2017; Published: 3 November 2017

Abstract: Intermittent supply is a common way of delivering water in many developing countries. Limitations on water and economic resources, in addition to poor management and population growth, limit the possibilities of delivering water 24 h a day. Intermittent water supply networks are usually designed and managed in an empirical manner, or using tools and criteria devised for continuous supply systems, and this approach can produce supply inequity. In this paper, an approach based on the hydraulic capacity concept, which uses soft computing tools of graph theory and cluster analysis, is developed to define sectors, also called district metered areas (DMAs), to produce an equitable water supply. Moreover, this approach helps determine the supply time for each sector, which depends on each sector's hydraulic characteristics. This process also includes the opinions of water company experts, the individuals who are best acquainted with the intricacies of the network.

Keywords: intermittent water supply; cluster analysis; graph theory; DMA; equity

1. Introduction

In developing countries, water supply continuity is threatened by the reduction of available water resources due to pollution, climate change, urban population growth, and management deficiencies in water supply systems. In this context, intermittent water supply becomes an alternative, in which water is delivered for a few hours a day.

There are several studies that analyze the various deficiencies of intermittent supply, since it causes problems in the system infrastructure itself [1–5], produces health risks for users [6–14], and generates supply inequity [15]. Nevertheless, water is currently delivered to millions of people around the world under intermittent supply conditions.

Galaitsi et al. [16], based on the influence on the living conditions of users, classify intermittency in water supply as predictable, irregular, or unreliable. Predictable intermittency is the only option that has a defined supply schedule. In this paper, we deal with predictable supply.

Intermittent supply networks can either work in their entirety, or by sectors [17], also called district-metered areas (DMAs). Sectors are useful in extensive intermittent supply networks, since supply schedules can be more easily established [18]. In this situation, however, setting and sizing the sectors does not always assure equitable supply, because sectors are designed with empirical or continuous-supply based criteria.

A sector is a restrained water supply network area, whose hydraulic behavior can be permanently or temporarily isolated [19]. A sector can be set by installing isolation valves in sector-connecting pipes. In some cases, sectors can be permanently disconnected [20]. Technical management of extensive supply networks is a complex task. Thus, network reduction into connected sectors becomes a very useful strategy [21].

Although installing flowmeters at the incoming pipes of each sector is common for leak control [22], sectors without measurement can exist in intermittent supply networks, since their main goal is to deliver water at differentiated schedules [18].

For DMA implementation in networks with continuous water supply, there is a general trend to use optimization techniques to achieve an adequate service level [19,21,23–26]. Several authors also suggest graph theory for the sectorization process [25,27,28]. Although sector importance in intermittent water supply is acknowledged [3,29], there are no specific tools for designing sectors in intermittent supply networks.

Upgrading the infrastructure to provide continuous water supply is an initial option for improving intermittent supply systems [30]. This option is usually hard to achieve. Moreover, if transition conditions are not feasible, it must be recognized that supply will always be intermittent. Consequently, more proactive management tools that minimize the negative effects caused by this type of supply are required [15,31,32]. This paradigm enables improving the living conditions of people who dwell in intermittently supplied areas, and achieves predictable intermittent supply systems [16].

In both supply system improvement perspectives, network sectorization is a fundamental step. Sectors are also important in transition processes to continuous supply [17], and crucial for intermittent supply system management that aims to improve supply equity. Moreover, sectorization under an intermittent-supply based perspective may be useful for vulnerable continuous supply systems. In 2016, for instance, the continuous supply network of La Paz (Bolivia) had to become temporarily intermittent due to insufficient water in its supply sources [33].

If an intermittent supply network is not sectorized, the peak flow demand during supply hours is very high, since water demand occurs simultaneously for the entire network. Thus, high water demand results in low service level conditions and may produce deficient pressure areas, which then produces supply inequity. Network sectorization and supply schedule setting help reduce this high peak demand.

In this paper, an approach based on the theoretical maximum flow concept, which uses soft computing tools from graph theory and cluster analysis [34,35], is developed to define sectors to produce equitable water supply. For node clustering, this process also includes water company expert opinions, from the individuals who best know network details.

Unlike continuous supply systems, the DMA implementation process in intermittent supply systems also includes criteria that assure supply equity, such as the restrained maximum pressure difference. Moreover, this approach helps determine the supply time for each sector based on their hydraulic characteristics.

2. Methodology

Sector implementation is based on three main goals: achieve supply equity; consider water company expert opinion; and determine adequate supply times for the sector (since these supply times are crucial for good management of intermittent supply networks). Sectors in intermittent supply networks are usually designed using continuous supply criteria. Those criteria are not considered in this paper.

2.1. Water Supply Equity

Equity in intermittent water supply aims to achieve a fair distribution of the limited amount of water available during the few hours of supply [15].

Inequity in intermittent water supply is related to water wastage at the highest pressure nodes and scarcity at the lowest pressure nodes. Accordingly, a network with supply equity is a system that restrains these extreme situations. Therefore, pressure is important for achieving equity in supply, and differences between maximum and minimum pressures must be small.

Home storage, which is very common in intermittent supply networks, make users compete for water supply, since their goal is to collect as much water as possible in a short period of time [36]. This competition also creates water supply inequity.

The essential difference between designing continuous and intermittent supply systems lies in including, or not, equity as a design principle [15,37]. If supply equity is considered a design criterion, water scarcity impact may be substantially reduced [38].

The main intervening factors in equitable supply are: pressure at the nodes; supply flows; velocities; elevation differences; supplied area size [39]; network topology; supply source location [37]; and network capacity [30]. Moreover, Vairamoorthy et al. [31] include the following elements to improve supply equity: supply duration; connection type; and connection location.

One of the most important components of intermittent supply systems is the distribution network itself. If the network has deficiencies, it may impose inequitable supply conditions and thus cause water wastage in high-pressure areas, as well as a lack of water in others [17]. Sector implementation may correct these deficiencies and help achieve supply equity. An appropriate criterion to evaluate supply equity is by controlling the pressure difference between the highest and the lowest pressure nodes. In this paper, values between 3 and 5 m are adopted, as recommended by CPHEEO (Central Public Health and Environmental Engineering Organization) [2].

2.2. Supply Time

Water supply time, or supply period, is an intrinsic characteristic of intermittent water supply systems. Nevertheless, it is usually adopted without rigorous technical criteria and usually produces supply inequity.

Inequity in water supply not only occurs in space but also time. Users in advantageous locations in the network receive water almost immediately after supply starts. In contrast, users in less fortunate locations must wait much longer [40].

Supply time definition, which is based on the hydraulic characteristics of network and sectors, helps achieve better planning and management of intermittent water supply systems. We address this question after describing our sectorization approach.

2.3. Theoretical Maximum Flow

The theoretical maximum flow, Q_{maxt}, or network capacity defines the maximum flow that a network can supply with at least a minimum pressure, P_{min}, at every node. The lowest pressure node must have the predefined minimum pressure [30]. The theoretical maximum flow value is determined through a demand-driven-analysis (DDA) hydraulic modeling of the network, in which nodes are associated with a given average demand. For this determination, several working conditions are evaluated and the peak factor is modified until the minimum pressure at the most unfavorable network node is guaranteed.

For this purpose, a setting curve—a network-H-Q curve that guarantees the minimum pressure at the lowest pressure node—is used. In a tank supplied network, for example, (see Figure 1), the intersection between the setting curve and the source water level determines the theoretical maximum flow. If due to minimum pressure reduction, the setting curve runs lower, then network capacity is increased.

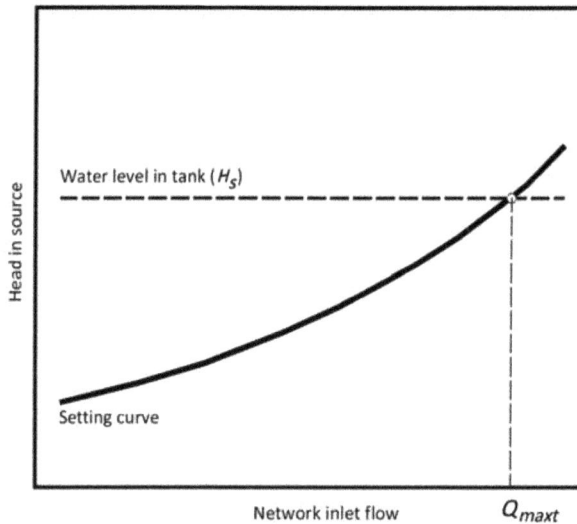

Figure 1. Theoretical maximum flow for a tank fed network [30].

2.4. Sector Development

Our sectorization process, which is fully described in this subsection, is schematized in Figure 2. This figure must be understood as a high-level pseudocode, with appropriate references to the equations in this subsection, accompanied by conceptual descriptions for the various sub-processes that integrate the entire process and which are described in detail below. These sub-processes are:

- Calculation of weights in pipes and nodes
- Calculation of criteria weights
- Critical node selection
- Shortest path between critical node and source
- Node clustering
- Hydraulic calculation and verification of water supply equity

For the stages that require hydraulic calculation we use EPANET 2.0 [41].
To better follow this subsection, Table 1 provides a list of the variables used.

Table 1. List of used variables.

Variable	Definition	Unit
R	Graph of whole network, used as hydraulic model	-
$V(R)$	Set of whole network nodes	-
$E(R)$	Set of whole network pipes	-
I_R	Incidence relation of graph	-
Q_{maxt}	Theoretical maximum flow or network capacity	L/s
n	Node	-
p	Pipe	-
P_n^{Qmaxt}	Pressure at nodes at theoretical maximum flow working condition	m
Q_p^{Qmaxt}	Flow at pipes at theoretical maximum flow working condition	L/s
h_p^{Qmaxt}	Head loss at pipes at theoretical maximum flow working condition	m
w_n	Weight in node n	m
w_p	Weight in pipe p	m^{-1}
z_1	Weight for east coordinate criterion	-
z_2	Weight for north coordinate criterion	-
z_3	Weight for elevation criterion	-
z_4	Weight for service pressure criterion	-
$n_{crit,i}$	Critical node at the developing sector i	-
i	Developing sector	-
C_i	Subset of selected nodes or developing sector	-
V_i	Set of remaining nodes	-
E_i	Set of remaining pipes	-
S_i	Subset of shortest path nodes	-
F_i	Subset of shortest path pipes	-
$d(\mu_c,x_j)$	Similarity distance	-
m	Number of criteria weight, $m = 1$ for east coordinate, $m = 2$ for north coordinate, $m = 3$ for elevation, and $m = 4$ for service pressure	-
μ_{cm}	Centroid depending on the m criteria	-
x_{nm}	Normalized value for each n node, depending on the m criteria	-
g_n	Node degree	-
w_{g_n}	Weight for node degree	-
M	Constant depending on the node degree importance	-
n_{sel}	Selected node	-
p_{sel}	Selected pipe	-
q	Node of subset C_i	-
x_{qm}	Normalized value for each q node, depending on the m criteria	-
B_i	Node subset used for hydraulic calculations	-
N_c	Total node number of a sector	-
H_i	Graph of developing sector i, used as hydraulic model	-
u	Working condition for developing sector	-
Q_{maxt}^u	Theoretical maximum flow for working condition u	L/s
k_u	Peak factor for working condition u	-
P_{min}	Minimum pressure in subset B_i	m
P_{max}	Maximum pressure in subset B_i	m
j	Node of subset B_i	-
Q_j	Average demand for each j node in subset B_i	L/s
ns	Total node number of subset B_i	-
t_s	Supply time	h
V_s	Total supplied water volume in continuous and intermittent supply	m^3
ΔP	Pressure difference	m
P_{eq}	Limit value of pressure difference that assures water supply equity	m
t_{min}	Minimum supply time, depending on the network capacity	h
Q_{int}	Average flow in intermittent water supply	L/s
H_s	Water level in tank or supply source	m

- Graph of the network: $R(V(R), E(R), l_R)$ (1)
- Calculation of theoretical maximum flow Q_{maxt}
- Calculation of weights in nodes related to pressure $\rightarrow w_n$ (2)
- Calculation of weights in pipes related to dissipated energy $\rightarrow w_p$ (3)
- Calculation of degree weights in nodes $\rightarrow w_{gn}$ (7)

Calculation of weights in pipes and nodes

- Water company experts' opinion
- Criteria: east coordinate, north coordinate, elevation, pressure at theoretical maximum flow working condition.
- Calculation of criteria weights: z_1, z_2, z_3, z_4

Calculation of criteria weights

- Select the critical node $n_{crit,i}$ in set V_i
$$n_{crit,i} = \text{argmin}\{w_n : \forall n \in V_i\} \quad (4)$$
$$n_{crit,i} \in C_i$$
$$V_{i+1} = V_i - \{n_{crit,i}\} \quad (5)$$

Critical node selection

- Shortest Path (Dijkstra algorithm) between $n_{crit,i}$ and source, based on weight w_p
- Nodes shortest path $\in S_i$
- Edges shortest path $\in F_i$

Shortest path between critical node and source

- Calculation of centroid μ_{cm} (17) of set C_i
- Calculation of distance $d(\mu_c, x_n)$ between μ_{cm} and nodes of set V_i (6)

- Select the node with the lowest distance to V_i
$$n_{sel} = \text{argmin}\{d(\mu_c, x_n) : \forall n \in V_i\} \quad (8)$$

Are there pipes linking n_{sel} with B_i?   No

No

Yes

Node clustering

$$n_{sel} \in C_i$$
$$B_i = C_i \cup S_i \quad (10)$$
$$V_{i+1} \rightarrow V_{i+1} - \{n_{sel}\} \quad (12)$$

- Select the pipe with lower weight w_p
$$p_{sel} = \text{argmin}\{w_p : \forall p \in E_i\} \quad (9)$$
$$p_{sel} \in F_i$$
$$E_{i+1} \rightarrow E_{i+1} - \{p_{sel}\} \quad (13)$$

- Calculation of Q''_{maxt} (14) and k_u based on graph of developing sector: $H_i = (B_i(H_i), F_i(H_i), l_R)$ (11).
- All other pipes must be closed
- Calculation of P_{max}, P_{min} and t_s (15) in set B_i

$V_{i+1} = \emptyset$ ←No— $\Delta P = P_{max} - P_{min} \leq P_{eq}$ (16) —Yes→

Are there more pipes linking n_{sel} with B_i?

Yes

No

Hydraulic calculation and verification of water supply equity

—Yes— Sectorized network ←—Yes— $V_{i+1} = \emptyset$

No

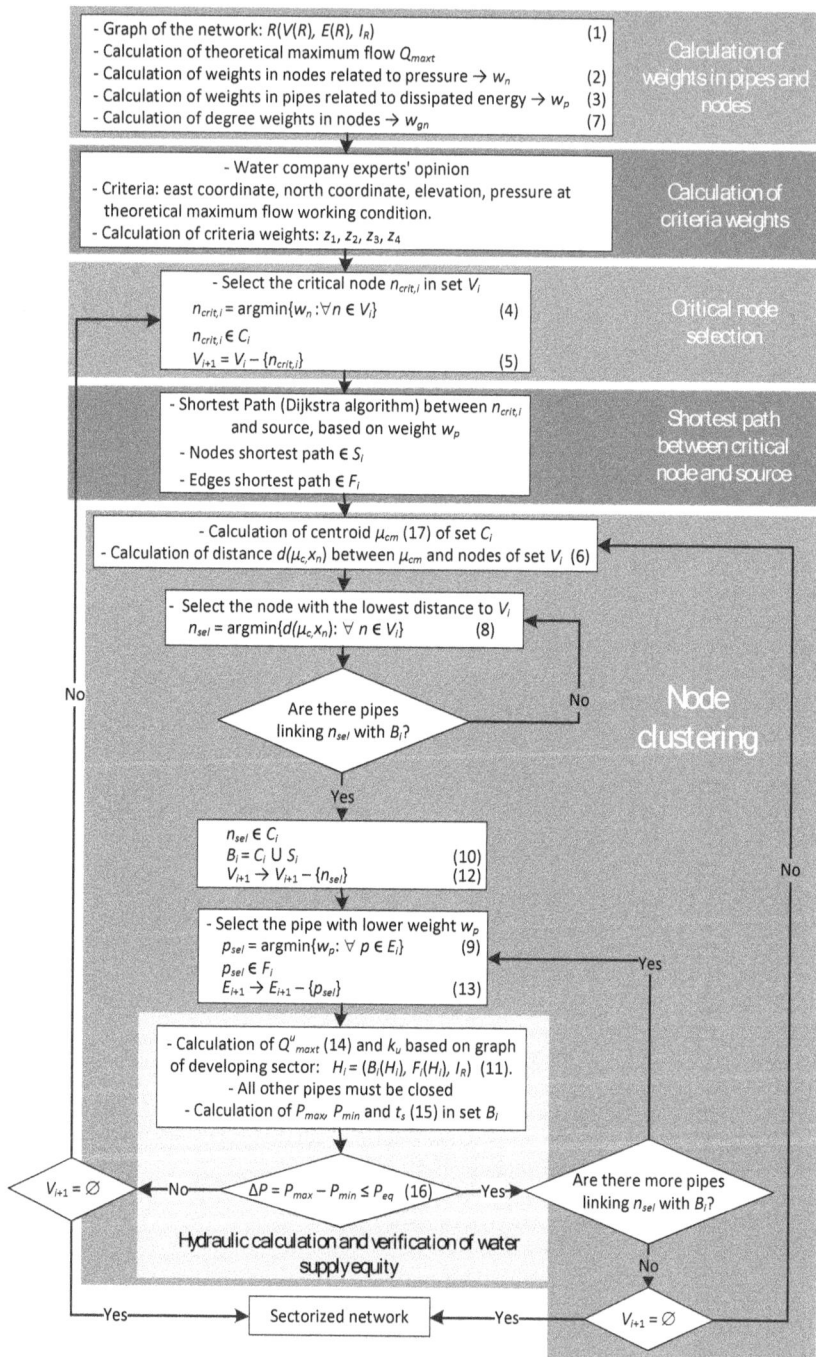

Figure 2. Flowchart, including high-level pseudocode and conceptual description, for implementation of sectors in intermittent water supply networks.

2.4.1. Calculation of Weights in Pipes and Nodes

The network is represented by a graph R, which consists of a triplet, namely, the network node set $V(R)$, the set of network pipes $E(R)$, and the incidence relation I_R, which relates each element (edge) in $E(R)$ to a unique non-ordered pair of nodes (vertices) in $V(R)$:

$$R = (V(R), E(R), I_R). \tag{1}$$

Based on this network, we can determine the initial theoretical maximum flow, Q_{maxt}, as described above (see [30] for specific details). Pipes are subjected to their maximum to fulfill the minimum pressure requirements. The calculated pressure, P_n^{Qmaxt}, at each node n; and the obtained flow, Q_p^{Qmaxt}, and head loss, h_p^{Qmaxt} for each pipe p, are used in weight calculation as follows.

Node weights, w_n, are directly related to the pressure at each node n:

$$w_n = P_n^{Qmaxt}. \tag{2}$$

Under the same working condition, a pipe weight, w_p, is determined by the inverse of the power dissipation [42], a function of the water specific weight, γ, and the calculated flow and head loss on pipe p, as in (3).

$$w_p = \frac{1}{\gamma \cdot \left| Q_p^{Qmaxt} \right| \cdot h_p^{Qmaxt}}. \tag{3}$$

With the pipe weights, the network becomes an undirected weighted graph, in which it is possible to recognize least-loss-energy pipes.

At this stage, using the graph of the entire network, we also calculate the degree of each node, g_n, which is later used to determine the degree weight, w_{gn}, and the similarity distance (see below).

2.4.2. Calculation of Criteria Weights

In the process, for node selection and subsequent sector development, various node-related criteria are considered, namely: east coordinate; north coordinate; elevation; pressure at theoretical maximum flow working condition; and connection degree.

Criteria weights, z_m, ($m = 1$ for east coordinate, $m = 2$ for north coordinate, $m = 3$ for elevation, and $m = 4$ for service pressure) are derived from the opinion of the water company experts, since they are fully acquainted with the network characteristics and performance. To derive those weights we use pairwise comparison matrices and their Perron eigenvectors to transform opinions into weights or priorities, as in the analytic hierarchy process (AHP) [43,44]. A different treatment is given to the connection degree, as explained below.

2.4.3. Critical Node Selection

This is an iteration process starting after completing the initialization stages 2.4.1 and 2.4.2. In each iteration step, first an individual (a node) for building the next cluster is identified. Some individuals are then grouped around it, and clusters (sectors or DMAs) are thus defined iteratively.

To build the i-th sector, we first identify the critical network node, $n_{crit,i}$ in the set of remaining nodes, V_i (initially, all the nodes of the entire network belong to this set). This critical node is selected to be the least-supply-pressure node during the maximal theoretical flow working condition, according to (2):

$$n_{crit,i} = \text{argmin}\{w_n : \forall n \in V_i\}. \tag{4}$$

This node is also the seed element of cluster C_i under development. Thus, it is the first element in cluster C_i, and must be included in this set: $n_{crit,i} \in C_i$.

Moreover, to avoid further selecting the critical node from the next set, V_{i+1}, it must be removed from the previous set, V_i:

$$V_{i+1} = V_i - \{n_{crit,i}\}. \tag{5}$$

2.4.4. Shortest Path between Critical Node and Source

With the dissipation energy weight, w_p, of every pipe, the critical node as a start, and the supply source as a destination, we determine the shortest path between both using the Dijkstra algorithm [45]. If there is more than one supply source, the shortest path must be determined for all sources. This step is essential to identify sectors, since each sector will have its own starting shortest path. Due to pipe weight characteristics, the shortest path will usually be made up of larger diameter pipes.

This path is used for hydraulic calculations as a sector entrance. A second node subset, S_i, which groups shortest path nodes, is also defined, as well as a pipe subset, F_i, which groups the shortest path pipes.

2.4.5. Node Clustering

The critical node becomes the cluster initial centroid, μ_c, (see (17) below for an exemption) and the next node is selected from subset V_i. This selection is determined by using the similarity distance,

$$d(\mu_c, x_n) = w_{gn} \cdot \sqrt{\sum_{m=1}^{4} z_m \cdot (\mu_{cm} - x_{nm})^2}, \tag{6}$$

between centroid μ_{cm} and normalized value x_{nm} for each node n, depending on the m criteria, and on the cluster connection through an edge (pipe). Before stating the selection mechanism, we first explain (6) further.

The weight w_{gn} is described below. Note that normalization for each criterion is performed by dividing each value by the sum of the criterion values.

Using east and north coordinates, we determine an equivalent value to the horizontal distance between the centroid and every network node. Closer nodes to the centroid are more likely to be grouped in the forming cluster. Normalization of these coordinates must refer to a common value to avoid modifying scales of the reference plane axes. This common value may be the greatest value of the east or north coordinates sum.

Node elevation and pressure criteria are particularly useful to achieve equity in sector supply. In this way, clusters are integrated by nodes with similar pressure and elevation.

This selection process may leave isolated nodes that connect with a sector through a single pipe and are unable to form a new sector. For this reason, similarity distance (6) is calculated using a weight, w_{gn}, which depends on the degree, g_n, of node n in the network. Nodes with a low connection degree are prioritized in the selection by means of

$$w_{gn} = 1 + \frac{g_n}{M}. \tag{7}$$

M is a constant that depends on the importance of the node degree. Assuming low values for M (1 to 10) implies giving more importance to the node connection degree in the network. Low values are recommended for branched networks, in which branches are in an unfavorable location (distant nodes or nodes with differing elevations or pressures). In the case of looped networks with uniform characteristics, M may be greater (50 to 100), and a value $w_{gn} = 1$ may be adopted. Prioritizing low connection degree nodes may increase pressure differences between the highest and lowest pressure nodes in the cluster. Consequently, smaller sectors are created.

To select the next node to belong to the cluster, we consider the graph used in the hydraulic calculation. The selected node, n_{sel}, is the graph node minimizing (6), that is to say, the graph node with the smallest similarity distance:

$$n_{sel} = \mathrm{argmin}\{d(\mu_c, x_n) : \forall n \in V_i\}. \tag{8}$$

However, we also need to guarantee the existence of an edge between the previously selected nodes, and the newly selected node (Figure 3). As a result, to select a node we need more than one iteration.

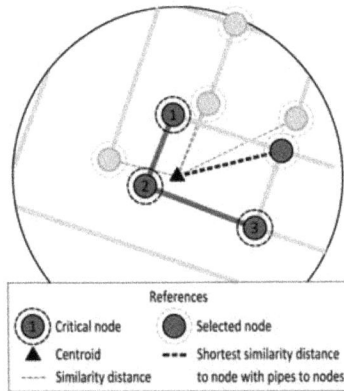

Figure 3. Selection of the node with shortest similarity distance with the pipe to be clustered.

If a new selected node has many pipes connecting with the sector (Figure 4), then the water has many options to flow and sector capacity is likely to increase.

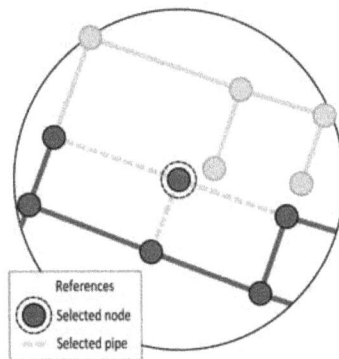

Figure 4. Selected node with many connection pipes to open.

Conversely, if the selected node has few links with the new sector, reducing pressure loss by changing the number of available circulation routes is less likely to succeed. Increasing or decreasing the network capacity and achieving the desired equity depends on the elevation and pressure of the selected node. If a node has a comparatively low elevation in the sector, it has a high pressure. Thus, this node may become the highest pressure node, which reduces equity and defines the further selection process. If a node has a higher elevation than the elevation node average, it may become a new critical node due to its minimum pressure, whose effect tends to reduce sector capacity.

By selecting the smaller diameter pipes first, the selected pipe, p_{sel}, from the set of current available pipes, E_i, is the pipe with lowest weight w_p:

$$p_{sel} = \text{argmin}\{w_p : \forall p \in E_i\}. \tag{9}$$

Every selected node, n_{sel}, and pipe, p_{sel}, must be included in the developing cluster, C_i ($n_{sel} \in C_i$) and in the shortest path pipe subset, F_i, ($p_{sel} \in F_i$), respectively. Moreover, the subset of the critical path nodes, S_i, must also join the cluster node subset, C_i, to obtain node subset B_i, as in (10), which is the base for the new graph, H_i, as specified in (11). This graph is used for hydraulic calculations.

$$B_i = C_i \cup S_i, \tag{10}$$

$$H_i = (B_i(H_i), F_i(H_i), I_R). \tag{11}$$

To avoid picking more than once any nodes and pipes previously selected for other sectors, each must be removed respectively from the new vertex set, V_{i+1}, and edge set, E_{i+1}, used in the next iteration:

$$V_{i+1} \rightarrow V_{i+1} - \{n_{sel}\}, \tag{12}$$

$$E_{i+1} \rightarrow E_{i+1} - \{p_{sel}\}. \tag{13}$$

Now it is time for hydraulic calculations with the current sector B_i.

2.4.6. Hydraulic Calculation and Verification of Water Supply Equity

At the beginning of the hydraulic calculations, only pipes in subset F_i are considered open, while the remaining pipes are considered closed until a node that connects them to the developing sector is selected. This situation may have a huge influence in the sector capacity calculation and, consequently, in equity and supply times.

We now calculate the theoretical maximum flow, Q^u_{maxt}, with the graph of the developing sector H_i for a working condition u. We also determine the maximum, P_{max}, and the minimum, P_{min}, pressures for the selected set of nodes, B_i. Thus, we are able to determine the peak factor, k_u, and, using the average demand, Q_j, for any selected node j, $j = 1, \ldots, ns$, we obtain for this working condition

$$Q^u_{maxt} = k_u \cdot \sum_{j=1}^{ns} Q_j. \tag{14}$$

To determine the supply time, t_s, we assume that the consumed water volume in continuous supply equals that of intermittent supply, $V_s = \sum_{j=1}^{ns} Q_j \cdot 24 = Q^h_{maxt} \cdot t_s$. Furthermore, we consider that the average flow is distributed 24 h a day, and the network capacity [30] is high enough to supply a high flow, Q^u_{maxt}, in a short supply time. Thus,

$$t_s = \frac{24}{k_u}. \tag{15}$$

Usually, the greater the number of grouped nodes, the lower the peak factor k_u value, so supply periods tend to 24 h. If the number of nodes is low, the peak factor increases, and thus supply time is shorter. In this case, having fewer supply hours is useful for avoiding supply schedule overlap.

The configuration and number of nodes in a hydraulic sector limits its theoretical maximum flow and, thus, its peak factor as well. Consequently, there is an intrinsic relation between a sector size and its supply time to guarantee an appropriate pressure.

The process of cluster selection of nodes comes to an end when the pressure difference, ΔP, of the hydraulic calculations surpasses a limit value, P_{eq}, which assures water supply equity:

$$\Delta P = P_{max} - P_{min} > P_{eq}. \tag{16}$$

If the pressure difference, ΔP, still guarantees equitable supply, perhaps new elements (nodes and pipes) can be incorporated in the current sector. To this end, if there are still unassigned nodes (V_i is not empty) a new centroid, μ_{cm}, is determined for each criterion m. We use the normalized values x_{qm} for each node q, which makes up the developing sector B_i, N_c being its total number of nodes:

$$\mu_{cm} = \frac{1}{N_c} \sum_{q=1}^{N_c} x_{qm}. \tag{17}$$

If, on the other hand, P_{eq} is effectively surpassed and there are still unassigned nodes (V_i is not empty), we start the iteration process in Section 2.4.3 again, and use the next critical network node, which is selected from all the excluded nodes (already grouped in previous clusters). From this new critical node, a new sector is built. Each network sector is built this way until all the nodes are assigned to a sector. This ends the sectorization process.

3. Case Study Description

The case-study network, shown in Figure 5 and summarized in Table 2, corresponds to a subsystem of the water supply network of Oruro (Bolivia). This network is supplied for 4 h a day, its demand flow during this period is 12.64 L/s, and its minimum pressure is 5.30 m. The minimum water level at its source is 3737 masl (meters above sea level), and the network average elevation is 3718 masl.

To achieve an equitable supply and build large sectors, we adopt a minimum pressure of 10 m and a pressure difference of 5 m, which is the maximum value recommended by CPHEEO.

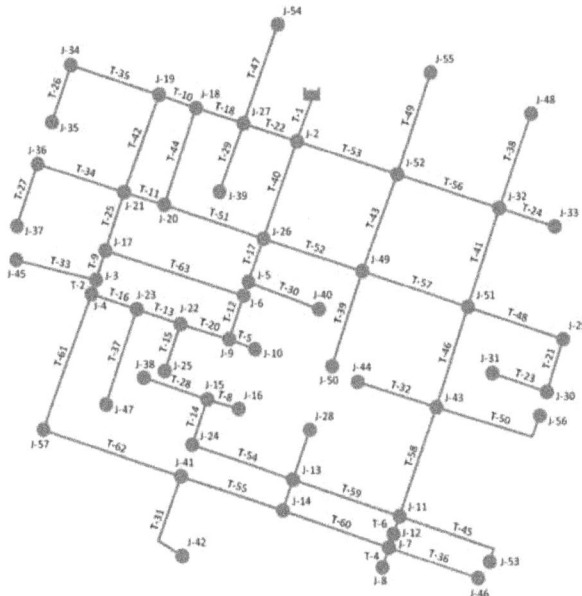

Figure 5. Network model with intermittent water supply to sectorize.

Table 2. Main network characteristics of the case study.

Description	Value
Number of network nodes	56 nodes
Number of network pipes	61 pipes
Average demand flow in intermittent water supply	12.64 L/s
Current supply time	4 h
Minimum pressure	5.30 m
Maximum pressure	17.20 m

Preliminary Evaluation of Water Supply Equity

Before applying our process, we evaluate some modifications in the current network management to try to achieve equitable supply. First, we determine the setting curve and network maximum theoretical flow [30] that satisfies the minimum service pressure of 10 m (Figure 6). The theoretical maximum flow or network capacity is 10.04 L/s, which does not fulfill the population demand in intermittent supply (12.64 L/s). One way to satisfy this requirement is by reducing the minimum service pressure to P_{min} = 5.30 m, which rearranges the setting curve to a fulfillment of the demanded flow (Figure 6).

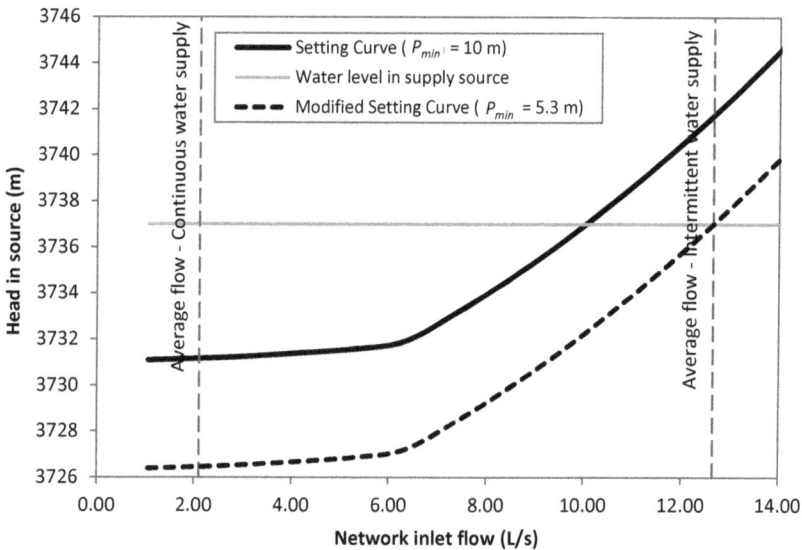

Figure 6. Reduction of the minimum pressure to reach the demand flow in intermittent supply.

As a result, the minimum pressure (10 m) is not met. Moreover, the difference in pressure ΔP = 11.99 m between the maximum pressure 17.29 m and the minimum pressure 5.30 m far exceeds the desired pressure of 5 m. Thus, other solutions must be evaluated.

A second alternative is increasing the number of supply hours (18). In this way, we reduce the average flow in intermittent supply (Q_{int} = 12.64 L/s) to a value that equals the current network capacity (Q_{maxt} = 10.04 L/s), using

$$t_{min} = \frac{t_s \cdot Q_{int}}{Q_{maxt}}. \tag{18}$$

If we modify the initial supply time, t_s = 4 h, to a minimum supply time, t_{min} = 5.04 h, the demand is satisfied by the network capacity, 10.04 L/s, and the pressure at each node is over 10 m.

Nevertheless, we must also evaluate pressure differences. We determined the pressure difference $\Delta P = 7.88$ m between the maximum pressure 17.88 m and the minimum pressure 10.00 m, which clearly exceeds 5 m and thus equity is not guaranteed.

As a consequence, a sectorization alternative needs to be studied. In the next section, we apply the process developed in this paper.

4. Results and Discussion

As shown, for the sectorization process, we use the following criteria: east coordinate; north coordinate; elevation; pressure; and node degree (Table 3). All except the node degree, need to be normalized (Table 4).

Table 3. Criteria for clustering process.

Node	East Coordinate (m)	North Coordinate (m)	Elevation (m)	Pressure (m)	Degree
J-2	698,074.22	8,010,604.23	3719.00	17.87	4
J-3	697,855.66	8,010,454.61	3719.00	16.05	3
J-4	697,853.41	8,010,448.53	3718.98	16.06	3
⋮	⋮	⋮	⋮	⋮	⋮
J-57	697,801.55	8,010,310.70	3718.60	13.14	2
Sum	-	448,583,649.73	208,217.36	808.56	-

Table 4. Normalized values.

Node	x_{n1}	x_{n2}	x_{n3}	x_{n4}
J-2	0.00155617	0.01785755	0.01786114	0.02210271
J-3	0.00155569	0.01785721	0.01786115	0.01984983
J-4	0.00155568	0.01785720	0.01786103	0.01985699
⋮	⋮	⋮	⋮	⋮
J-57	0.00155557	0.01785689	0.01785921	0.01625678

Criteria weights are determined based on interviews with water company experts. In this case, study, three company experts were interviewed. Thus, we set pairwise comparison matrices [44] that influence every criterion (except for node degree, which, as explained above, receives a different treatment). Perron eigenvectors represent the criteria weights that were defined by the company experts. Table 5 shows the pairwise comparison matrix of expert 1 as well as its Perron eigenvector. These values have a consistency ratio (*CR*) of 5.1%, which is suitable for the criteria [44]. The final weights are obtained through the component geometric from the Perron eigenvectors for the experts (Table 6), which also had acceptable *CR* values.

Table 5. Pairwise comparison matrix, expert 1.

Criterion	East and North Coordinates	Elevation	Pressure	Eigenvector
East and north coordinates	1	3	1/2	0.333
Elevation	1/3	1	1/3	0.140
Pressure	2	3	1	0.528

Table 6. Normalized weight of each criterion.

Criterion	Expert 1	Expert 2	Expert 3	Geometric Mean	Normalized Weight
East and north coordinates	0.333	0.333	0.200	0.281	$z_1 = z_2 = 0.291$
Elevation	0.140	0.333	0.200	0.210	$z_3 = 0.218$
Pressure	0.528	0.333	0.600	0.473	$z_4 = 0.490$
Total	1	1	1	0.964	1

Due to the network characteristics, it is less likely that disconnected nodes are left during sector building. Therefore, as discussed above, taking (7) into account, we assume a weight $w_g = 1$ for each node.

To start building the first sector, we identify node J-38 as the most critical node in the network. Starting from this node, we determine the shortest path to its supply source (see Figure 11, which compiles the final results), and thus we set the first sector. Later, we group nodes according to their similarity distance. Every step is evaluated until each node has a pressure difference that assures the desired equity (Figure 7).

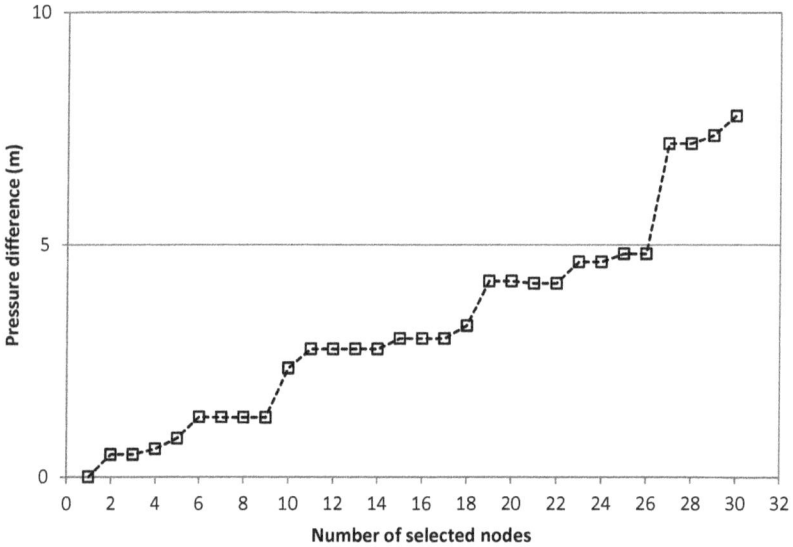

Figure 7. Variation of the pressure difference as a function of the selected nodes.

The clustering process produces evident jumps in pressure differences (Figure 7), due to selection of nodes that enable either raising the pressure, or reducing the minimum pressure. After surpassing the pressure difference of 5 m, DMA implementation stops, and according to this condition, the first 26 nodes selected make up the first sector, without considering the first nodes of the identified shortest path (see Figure 11).

The sudden increase in the developing sector capacity is caused by selecting nodes that have a high degree of connection. This situation causes a reduction in supply time, because the greater the capacity, the shorter the supply time (Figure 8).

Let us continue with the process. The network critical node has already been selected for the first sector. Consequently, there is a new critical node among the unselected nodes. Since pressure difference between this node and the supply source is large, an equitable supply is difficult to achieve. Consequently, we consider reducing the head in the source, or creating more sectors. For better network performance, in terms of supply equity, and to reduce leaks, it is better to reduce the head at the supply source.

We now evaluate situations in which the head at supply source, H_s, is reduced. Moreover, we increase the pressure difference value to analyze the sector configuration behavior at high pressures (Figure 9).

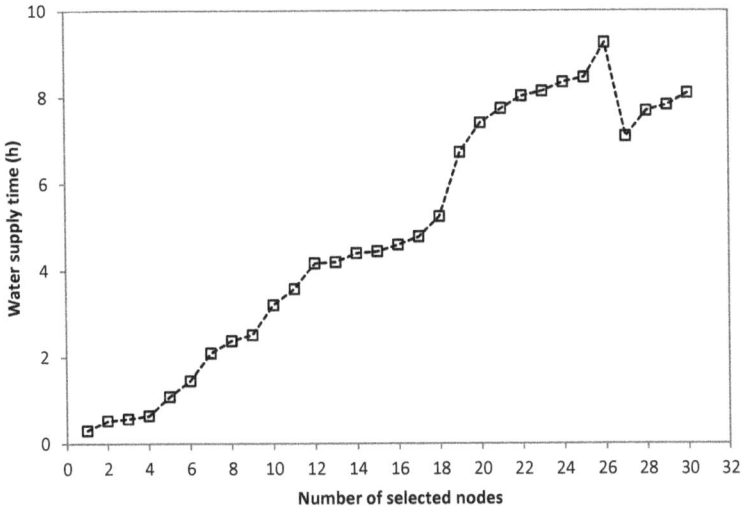

Figure 8. Variation in water supply time as a function of the selected nodes.

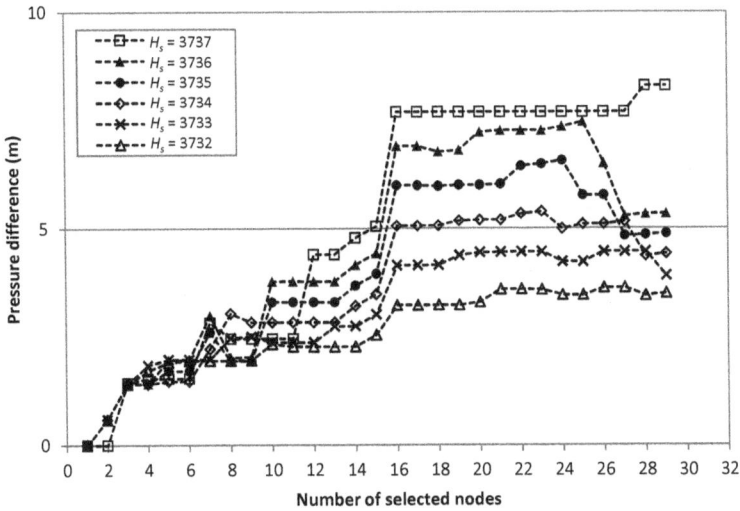

Figure 9. Variation of the pressure difference as a function of the selected nodes and the head in the water supply source.

If the head in the source is 3737 m or 3736 m, for which the pressure difference surpasses 5 m (Figure 9), we would need to create more additional sectors. This becomes necessary because pressures must be adjusted to the pressure difference between the pressure at the nodes near the supply source and the lowest pressure node.

If the head in the source is reduced, we obtain pressure differences lower than 5 m starting from values lower than 3735 m. The lower the head in the source, the lower the pressure difference between the supply source and the critical node, with which we achieve a greater supply equity. Nevertheless, pressure difference reduction means increasing supply service time.

As for the first sector, sudden reductions in supply hours (Figure 10) are caused by increasing the network capacity, which, in turn, is due to the selection of high connectivity degree nodes.

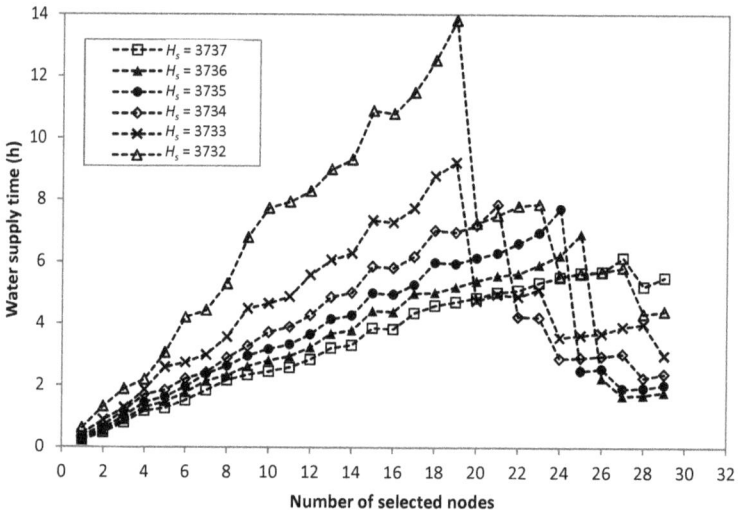

Figure 10. Variation of the water supply time as a function of the selected nodes and the head in the water supply source.

It is not recommendable to reduce the current number of supply hours, because users may complain. Thus, to have a 4 h supply, we define a pressure head of $H_s = 3732$ m (Figure 10). Under these conditions, we create the second sector (Figure 11) and guarantee the desired equity.

The sectorization process produces two sectors with intermittent supply (Table 7). The pressure difference is lower than 5 m, which assures equitable supply. We also determine the supply time based on the hydraulic characteristics of each sector.

Table 7. Characteristics of the sectors after the sectorization process.

Sector	P_{max} (m)	P_{min} (m)	ΔP (m)	Q_{maxt} (L/s)	Supply Time t_s (h)	H_s (mca)	Clustering Nodes C_i
Sector 1	14.81 (J-30)	10.00 (J-57)	4.81 < 5	2.76	8.46	3737	26
Sector 2	13.54 (J-10)	10.00 (J-34)	3.54 < 5	6.18	4.41	3732	30

Sector delimitation is achieved by installing sectioning valves at pipes T-56, T-22, T-53, T-51, T-17, T-39, T-43 and T-61. Pipe T-57 controls the incoming water flow to sector 1.

Due to the network characteristics, initial nodes of first shortest path, namely J-2, J-26, and J-49, work in both sectors, and supply time is longer (12.87 h). This situation could be avoided, for example, by installing a direct connection pipe between the source and sector 1.

Sector 1 includes the network critical node, which reduces its capacity and conditions the sector to have a longer supply time. Conversely, sector 2 may have greater capacity because the critical node does not belong to it, and the new critical node favors capacity increases.

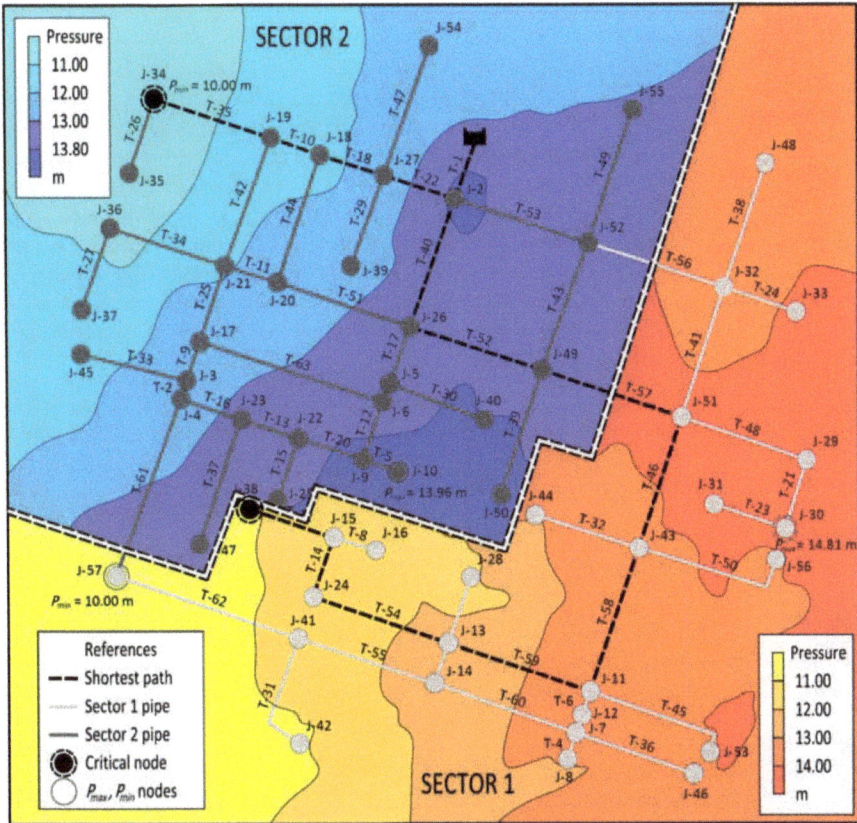

Figure 11. Development of sectors 1 and 2.

5. Conclusions

In this paper, we have considered a procedure to define sectors in a water distribution network with intermittent supply. We develop sectors based on equity criteria, and using water company expert opinions. Moreover, we determine the supply time using the sector hydraulic conditions. The authors claim that these characteristics are innovative in methodologies of this kind. The developed methodology uses soft computing elements of graph theory and clustering.

The sectorization process is a very useful technical management tool for those intermittent supply systems that are unable to evolve to continuous supply, and for systems that could evolve to continuous supply.

Sector construction based on equity criteria may also be useful for a future intermittent-to-continuous-supply transition, because sectors help define areas for pressure management.

A sectorized network by itself does not guarantee equitable and predictable intermittent water supply. It is also necessary to manage the supply schedules for all sectors to avoid schedule overlaps with consequent pressure reduction [17].

In our case study, due to the network characteristics, some shortest path nodes were selected for more than one sector. This could be avoided, for example, by setting up a shortcut pipe between source and sector. However, let us note that, although these nodes are supplied for a longer period of time, they satisfy the pressure difference condition in their sector.

In larger networks with more than one supply source, we need greater computing capacity and suitable process supervision.

There are few tools for the management of intermittent supply networks. It is necessary to develop more sectorization techniques for this type of network, in which sectors are intrinsic elements. Future sectorization network research must aim at reducing the number of pipes with isolation valves, developing efficient equity indicators, and evaluating the resilience of created sectors to assure constant equity in supply.

Related to this last issue, despite the absence of explicit resilience reference values (such as for the pressure difference as recommended by CPHEEO [2]) for studies including equity as a criterion, we mention here that the resilience index [42] for the entire network is 0.894, which is, as expected, clearly improved after sectorization. In effect, the new resilience values are 0.969 for the first sector and 0.997 for the second. From our point of view, this improvement clearly backs our sectorization proposal, which we consider to be promising.

Acknowledgments: The authors are grateful to SeLA (water company in Oruro (Bolivia)) for providing information. The use of English has been supervised by John Rawlins, a qualified member of the UK Institute of Translation and Interpreting.

Author Contributions: Amilkar E. Ilaya-Ayza, Carlos Martins and Joaquín Izquierdo conceived and designed the study. Amilkar E. Ilaya-Ayza wrote the manuscript, which was fully revised and discussed by the authors. All the authors approved the final version of the manuscript.

Conflicts of Interest: The authors declare no conflicts of interest.

References

1. World Health Organization. Constraints Affecting the Development of the Water Supply and Sanitation Sector. 2003. Available online: http://www.who.int/docstore/water_sanitation_health/wss/constraints.html (accessed on 20 July 2016).
2. Central Public Health and Environmental Engineering Organisation. *Manual on Operation and Maintenance of Water Supply Systems*; Ministry of Urban Development; World Health Organisation: New Delhi, India, 2005.
3. Dahasahasra, S.V. A model for transforming an intermittent into a 24 × 7 water supply system. *Geospat. Today* **2007**, *8*, 34–39.
4. Faure, F.; Pandit, M.M. Intermittent Water Distribution. 2010. Available online: http://www.sswm.info/category/implementation-tools/water-distribution/hardware/water-distribution-networks/intermittent-w (accessed on 15 October 2017).
5. Charalambous, B. The Effects of Intermittent Supply on Water Distribution Networks. Water Loss. 2012. Available online: http://www.leakssuite.com/wp-content/uploads/2012/09/2011_Charalambous.pdf (accessed on 15 October 2015).
6. Knobelsdorf, J.; Mujeriego, R. Crecimiento bacteriano en las redes de distribución de agua potable: Una revisión bibliográfica. *Ingeniería Del Agua* **1997**, *4*, 17–28. [CrossRef]
7. Semenza, J.C.; Roberts, L.; Henderson, A.; Bogan, J.; Rubin, C.H. Water distribution system and diarrheal disease transmission: A case study in Uzbekistan. *Am. J. Trop. Med. Hyg.* **1998**, *59*, 941–946. [CrossRef] [PubMed]
8. Mermin, J.H.; Villar, R.; Carpenter, J.; Roberts, L.; Gasanova, L.; Lomakina, S.; Hutwagner, L.; Mead, P.; Ross, B.; Mintz, E. A massive epidemic of multidrug-resistant typhoid fever in Tajikistan associated with consumption of municipal water. *J. Infect. Dis.* **1999**, *179*, 1416–1422. [CrossRef] [PubMed]
9. Tokajian, S.; Hashwa, F. Water quality problems associated with intermittent water supply. *Water Sci. Technol.* **2003**, *47*, 229–234. [PubMed]
10. Tokajian, S.; Hashwa, F. Phenotypic and genotypic identification of *Aeromonas* spp. isolated from a chlorinated intermittent water distribution system in Lebanon. *J. Water Health* **2004**, *2*, 115–122. [PubMed]
11. Lee, E.; Schwab, K. Deficiencies in drinking water distribution systems in developing countries. *J. Water Health* **2005**, *3*, 109–127. [PubMed]
12. Kumpel, E.; Nelson, K.L. Comparing microbial water quality in an intermittent and continuous piped water supply. *Water Res.* **2013**, *47*, 5176–5188. [CrossRef] [PubMed]

13. Kumpel, E.; Nelson, K.L. Mechanisms affecting water quality in an intermittent piped water supply. *Environ. Sci. Technol.* **2014**, *48*, 2766–2775. [CrossRef] [PubMed]

14. Ercumen, A.; Arnold, B.F.; Kumpel, E.; Burt, Z.; Ray, I.; Nelson, K.; Colford, J.M., Jr. Upgrading a Piped Water Supply from Intermittent to Continuous Delivery and Association with Waterborne Illness: A Matched Cohort Study in Urban India. *PLoS Med.* **2015**, *12*, e1001892. [CrossRef] [PubMed]

15. Vairavamoorthy, K.; Akinpelu, E.; Lin, Z.; Ali, M. Design of sustainable water distribution systems in developing countries. In Proceedings of the American Society of Civil Engineers (ASCE) Conference, Houston, TX, USA, 10–13 October 2001.

16. Galaitsi, S.E.; Russell, R.; Bishara, A.; Durant, J.L.; Bogle, J.; Huber-Lee, A. Intermittent domestic water supply: A critical review and analysis of causal-consequential pathways. *Water* **2016**, *8*, 274. [CrossRef]

17. Ilaya-Ayza, A.E.; Martins, C.; Campbell, E.; Izquierdo, J. Gradual transition from intermittent to continuous water supply based on multi-criteria optimization for network sector selection. *J. Comput. Appl. Math.* **2017**, *330*, 1016–1029. [CrossRef]

18. Ilaya-Ayza, A.E.; Benitez, J.; Izquierdo, J.; Pérez-García, R. Multi-criteria optimization of supply schedules in intermittent water supply systems. *J. Comput. Appl. Math.* **2017**, *309*, 695–703. [CrossRef]

19. Di Nardo, A.; Di Natale, M.; Santonastaso, G.; Tzatchkov, V.; Alcocer-Yamanaka, V. Water network sectorization based on a genetic algorithm and minimum dissipated power paths. *Water Sci. Technol.* **2013**, *13*, 951–957. [CrossRef]

20. Ziegler, D.; Sorg, F.; Fallis, P.; Hübschen, K.; Happich, L.; Baader, J.; Trujillo, R.; Mutz, D.; Oertlé, E.; Klingel, P.; et al. *Guidelines for Water Loss Reduction, A Focus on Pressure Management*; Deutsche Gesellschaft für Internationale Zusammenarbeit (GIZ) GmbH and VAG Armaturen GmbH: Eschborn, Germany, 2012.

21. Hajebi, S.; Temate, S.; Barrett, S.; Clarke, A.; Clarke, S. Water Distribution Network Sectorisation Using Structural Graph Partitioning and Multi-objective Optimization. *Procedia Eng.* **2014**, *89*, 1144–1151. [CrossRef]

22. Farley, M. Are there alternatives to DMA? *Asian Water* **2010**, *26*, 10–16.

23. Herrera, A.M. Improving Water Network Management by Efficient Division into Supply Clusters. Available online: https://riunet.upv.es/bitstream/handle/10251/11233/tesisupv3599.pdf?sequence=1 (accessed on 15 January 2017).

24. Izquierdo, J.; Herrera, M.; Montalvo, I.; Pérez-García, R. Division of water supply systems into district metered areas using a multi-agent based approach. In *Software and Data Technologies*; Springer: Berlin/Heidelberg, Germany, 2011; pp. 167–180.

25. Campbell, E.; Izquierdo, J.; Ilaya-Ayza, A.E.; Pérez-García, R.; Tavera, M. A flexible methodology to sectorize water supply networks based on social network theory concepts and on multi-objective optimization. *J. Hydroinform.* **2016**, *18*, 62–76. [CrossRef]

26. Alvisi, S.; Franchini, M. A heuristic procedure for the automatic creation of district metered areas in water distribution systems. *Urban Water J.* **2014**, *11*, 137–159. [CrossRef]

27. Tzatchkov, V.; Alcocer-Yamanaka, V.; Bourguett-Ortiz, V. Sectorización de redes de distribución de agua potable a través de algoritmos basados en la teoría de grafos. *Tlaloc-AMH* **2008**, *40*, 14–22.

28. Di Nardo, A.; Di Natale, M.; Santonastaso, G.; Tzatchkov, V.; Alcocer-Yamanaka, V. Water network sectorization based on graph theory and energy performance indices. *J. Water Resour. Plan. Manag.* **2014**, *140*, 620–629. [CrossRef]

29. McIntosh, A.C. *Asian Water Supplies Reaching the Urban Poor*; Asian Development Bank: Metro Manila, Philippines, 2003.

30. Ilaya-Ayza, A.E.; Campbell, E.; Pérez-García, R.; Izquierdo, J. Network capacity assessment and increase in systems with intermittent water supply. *Water* **2016**, *8*, 126. [CrossRef]

31. Vairavamoorthy, K.; Gorantiwar, S.D.; Pathirana, A. Managing urban water supplies in developing countries—Climate change and water scarcity scenarios. *Phys. Chem. Earth A B C* **2008**, *33*, 330–339. [CrossRef]

32. Soltanjalili, M.-J.; Bozorg Haddad, O.; Mariño, M.A. Operating water distribution networks during water shortage conditions using hedging and intermittent water supply concepts. *J. Water Resour. Plan. Manag.* **2013**, *139*, 644–659. [CrossRef]

33. Shrinking Glaciers Cause State-of-Emergency Drought in Bolivia. Available online: https://www.theguardian.com/environment/2016/nov/28/shrinking-glaciers-state-of-emergency-drought-bolivia (accessed on 18 January 2017).

34.	Jain, A.K.; Dubes, R.C. *Algorithms for Clustering Data*; Prentice Hall: Englewood Cliffs, NJ, USA, 1988.
35.	Rokach, L.; Maimon, O. *Data Mining and Knowledge Discovery Handbook*; Springer: New York, NY, USA, 2005; pp. 321–352.
36.	Fontanazza, C.M.; Freni, G.; La Loggia, G.; Notaro, V.; Puleo, V. Evaluation of the Water Scarcity Energy Cost for Users. *Energies* **2013**, *6*, 220–234. [CrossRef]
37.	Gottipati, P.V.; Nanduri, U.V. Equity in water supply in intermittent water distribution networks. *Water Environ. J.* **2014**, *28*, 509–515. [CrossRef]
38.	Chandapillai, J.; Sudheer, K.P.; Saseendran, S. Design of water distribution network for equitable supply. *Water Resour. Manag.* **2012**, *26*, 391–406. [CrossRef]
39.	Manohar, U.; Kumar, M.M. Modeling Equitable Distribution of Water: Dynamic Inversion-Based Controller Approach. *J. Water Resour. Plan. Manag.* **2013**, *140*, 607–619. [CrossRef]
40.	De Marchis, M.; Fontanazza, C.M.; Freni, G.; La Loggia, G.; Napoli, E.; Notaro, V. A model of the filling process of an intermittent distribution network. *Urban Water J.* **2010**, *7*, 321–333. [CrossRef]
41.	Rossman, L.A. *EPANET 2: Users Manual*; EPA: Cincinnati, OH, USA, 2000.
42.	Todini, E. Looped water distribution networks design using a resilience index based heuristic approach. *Urban Water* **2000**, *2*, 115–122. [CrossRef]
43.	Saaty, T.L. A scaling method for priorities in hierarchical structures. *J. Math. Psychol.* **1977**, *15*, 234–281. [CrossRef]
44.	Saaty, T.L.; Vargas, L. *Models, Methods, Concepts & Applications of the Analytic Hierarchy Process*; Springer: New York, NY, USA, 2012.
45.	Dijkstra, E.W. A note on two problems in connexion with graphs. *Numer. Math.* **1959**, *1*, 269–271. [CrossRef]

water

MDPI

Article

Applications of Graph Spectral Techniques to Water Distribution Network Management

Armando di Nardo [1,2,3], Carlo Giudicianni [1], Roberto Greco [1,2] (iD), Manuel Herrera [4,*] (iD) and Giovanni F. Santonastaso [1,2]

1   Dipartimento di Ingegneria Civile, Design, Edilizia, e Ambiente, Università degli Studi della Campania 'L. Vanvitelli', via Roma 29, 81031 Aversa, Italy; armando.dinardo@gmail.com (A.d.N.); carlo.giudicianni@gmail.com (C.G.); roberto.greco@unicampania.it (R.G.); giovannifrancesco.santonastaso@gmail.com (G.F.S.)
2   Action Group CTRL + SWAN of the European Innovation Partnership on Water EU, via Roma 29, 81031 Aversa, Italy
3   Institute for Complex Systems (Consiglio Nazionale delle Ricerche), via dei Taurini 19, 00185 Roma, Italy
4   Centre for Energy and the Design of Environments (EDEn)—Department of Architecture & Civil Engineering, University of Bath, Claverton Down, Bath BA2 7AZ, UK
*   Correspondence: A.M.Herrera.Fernandez@bath.ac.uk; Tel.: +44-1225-38-4440

Received: 18 December 2017; Accepted: 5 January 2018; Published: 9 January 2018

Abstract: Cities depend on multiple heterogeneous, interconnected infrastructures to provide safe water to consumers. Given this complexity, efficient numerical techniques are needed to support optimal control and management of a water distribution network (WDN). This paper introduces a holistic analysis framework to support water utilities on the decision making process for an efficient supply management. The proposal is based on graph spectral techniques that take advantage of eigenvalues and eigenvectors properties of matrices that are associated with graphs. Instances of these matrices are the adjacency matrix and the Laplacian, among others. The interest for this application is to work on a graph that specifically represents a WDN. This is a complex network that is made by nodes corresponding to water sources and consumption points and links corresponding to pipes and valves. The aim is to face new challenges on urban water supply, ranging from computing approximations for network performance assessment to setting device positioning for efficient and automatic WDN division into district metered areas. It is consequently created a novel tool-set of graph spectral techniques adapted to improve main water management tasks and to simplify the identification of water losses through the definition of an optimal network partitioning. Two WDNs are used to analyze the proposed methodology. Firstly, the well-known network of C-Town is investigated for benchmarking of the proposed graph spectral framework. This allows for comparing the obtained results with others coming from previously proposed approaches in literature. The second case-study corresponds to an operational network. It shows the usefulness and optimality of the proposal to effectively manage a WDN.

Keywords: water distribution system management; spectral analysis; complex networks

1. Introduction

Starting from 19th Century, Water Distribution Networks (WDN) were designed using a traditional approach based on mathematical models to find their optimal system layout in terms of water demand and pressure level satisfaction in each node. Nowadays, new challenges come from network management of an old water system designed more than 50–70 years ago. For instance, significant water losses in the WDN can usually be spotted, raising some cases up to 70% [1]. The issue often leads to having nodal pressures that are lower than a minimum service level. On top of this, there is a bigger problem regarding WDNs delay in terms of management and innovations when compared to other network public services (electricity, transport, gas, etc.). This fact is noticeable nowadays when there still is a bias on a lack of development of urban water issues with respect to smart cities research [2,3]. It is necessary to propose new paradigms, creating a novel framework analysis in research and development for urban water management.

The complexity of WDN management depends on different peculiar aspects, such as network connectivity or asset location (e.g., pipes, pumps, valves). In addition, any WDN performance shows a strong dependency on the complex network geometry produced by traditional design criteria, i.e., placing looped pipes under every street. These complex geometries and topologies require innovative approaches for the analysis and management of a WDN with a densely layout of up to tens of thousands of nodes and hundreds of looped paths that can be considered as complex networks [4]. Recently, there have flourished algorithms and mathematical tools in graph and complex network theory to better analyse the behaviour and evolution of complex systems [5–7]. All of these tools are focused on how "structure affects function" [5] as key aspect for their development. Among the most important methodologies handling complex networks are the Graph Spectral Techniques (GSTs) [8]. GSTs analyze network topologies by exploiting the properties of some graph matrices, providing useful information about the global and local performance and evolution of network systems.

A number of GSTs have been applied to WDNs over the last years. These shown to be useful to define an optimal clustering layout through spectral clustering [9–11]. GSTs also supported approaching preliminary assessments of the global network robustness through graph matrices eigenvalues [12–14], providing surrogate robustness metrics. However, these studies only use some GSTs properties and do not provide an overall framework regarding the opportunities offered by the study of network eigenvalues and eigenvectors.

This paper proposes a GST tool-set based on two graph matrices and their relative spectra for supporting several applications on WDNs management. The aim is to present a complete outline on the capabilities provided by graph spectral techniques applied to WDNs and assemble them into a unique framework. The paper highlights how GST metrics and their algorithms aid to face some crucial tasks of WDN management by just using topological and geometric information. In literature exist several approaches enhancing graph theoretic approaches for WDN management with hydraulic information. There are addressed this way the problem of network failures quantified both with respect to physical connectivity and water supply service level [15–18], resilience analysis [19], ranking pipes [20], and vulnerability analysis [21]. However, there are a series of advantages of focusing the analysis only on the network topology. The GST tool-set provides a solution in the frequent case of not having available hydraulic information, fosters real-time response for WDN management, makes it easier to deal with large-scale WDNs, provides an initial solution to further applications (e.g., specific algorithms for sensor location), presents a surrogate solution for WDN management in all of the cases, even for disruption scenarios (such as single or multiple component removal), and can be easily extended to contain hydraulic information by weighting the graph, but using similar methodologies to those proposed in this paper.

This paper approaches several issues. Firstly, it is done a robustness analysis by computing the strength of the network connectivity using a number of spectral metrics. This is of high interest to assess the impact of any network perturbation (single or multiple component removal) resulting from random network failures or targeted attacks [22]. The paper also undertakes through GSTs a water network clustering to define the optimal dimension and shape of a District Meter Area (DMA) [23,24]. In addition, there are also tackled both the problem of an optimal sensor placement [25–27] and the identification of the most sensitive nodes to malicious attacks [28,29]. Besides providing a unique GST framework for urban water management, this work also presents novelty elements such as the application of spectral tools for several WDN tasks: approaching connectivity and continuity analysis, finding an optimal number of clusters for the water network partitioning, and selecting the most "influential" nodes for locating quality sensors and metering stations. The GST framework is especially useful for aiding the decision making process for real-time WDN management and in the frequent case of not having available hydraulic information.

Last but not least, another two important aspects supporting the use of graph spectral techniques are the following: (a) dealing with easy to implement metrics that can be efficiently solved by standard linear algebra methods; and (b) providing mathematical elegance to the proposed procedures, as they are supported by mathematical theorems. The outline of the paper is the following. First, it provides a brief survey of the principal graph spectral techniques, independently of the application field in which they are used. The main graph matrices and some important eigenvalues and their eigenvectors are defined and explained. In order to better show the meaning and efficiency of spectral tools, a simple Example Network is analyzed. Finally, the GST tool-set is tested on two case studies, a real small-size and an artificial medium-size water system. The conclusions section includes a comparison and analysis of the results.

2. Spectral Graph Theory

Spectral graph theory is a mathematical approach combining both linear algebra and graph theory [30] in order to exploit eigenvalue and eigenvector properties. This way, the main benefit of spectral graph theory is its simplicity, as any system can be successfully analyzed just through the spectrum of its associated graph matrix, M. Spectral graph parameters contain a lot of information on both local and global graph structure. The computational complexity to compute eigenvalues and eigenvectors of graph matrices is $O(n^3)$, where n is the number of vertices/nodes (it is usual to name the elements of a graph as vertices and edges and the elements of a network as nodes and links; we make this distinction throughout the paper.) in the associated graph/network. From the 1990s, graph spectra have been used for several important applications in many fields [31]; such as expanders and combinatorial optimization, complex networks and the internet topology, data mining, computer vision and pattern recognition, internet search, load balancing and multiprocessor interconnection networks, anti-virus protection, knowledge spread, statistical databases and social networks, quantum computing, bioinformatics, coding theory, control theory, and computer sciences.

2.1. Graph Matrices

The Adjacency matrix, A, and the Laplacian matrix, L, are widely used in graph analysis. Another matrices such as the Modularity matrix, the Similarity matrix, and the sign-less Laplacian are omitted from the current GST tool-set. Using them will make a wider GST mathematical framework but require a further investigation that falls out of the scope of this proposal. The following items synthetically describe a number of graph matrices that are related to A and L, whose properties are introduced and developed in this paper.

- Adjacency Matrix A: let $G = (V, E)$ be an undirected graph with n-vertices set V and m-edges set E. A common way to represent a graph is to define its Adjacency matrix A, whose elements $a_{ij} = a_{ji} = 1$ if nodes i and j are directly connected and $a_{ij} = a_{ji} = 0$ otherwise. The degree of node i of A is defined as $k_i = \sum_{j=1}^{n} a_{ij}$;

- Weighted Adjacency Matrix W: it is possible to express the weighted Adjacency matrix W, in case to be available information about the connection strength between vertices of the graph G. Edge weights are expressed in terms of proximity and/or similarity between vertices. Thus, all of the weights are non-negative. That is, $w_{ij} = w_{ji} \geq 0$ if i and j are connected, $w_{ij} = w_{ji} = 0$ otherwise. The degree of a node i of W is defined as $k_i = \sum_{j=1}^{n} w_{ij}$;

- Un-normalized Laplacian Matrix L: one of the main utilities of spectral graph theory is the Laplacian matrix [32] and both its un-normalized and normalized version [8]. Let $D_k = diag(k_i)$ be the diagonal matrix of the vertex connectivity degrees, the Laplacian matrix is defined as the difference between D_k and the Adjacency matrix A (or the weighted Adjacency matrix W if it is

considered a weighted graph). The un-normalized Laplacian matrix is defined by $L = D_k - A$
($L = D_k - W$);

- Random Walk Normalized Laplacian Matrix L_{rw}: it is closely related to a random walk
  representation. Its definition comes from the Laplacian matrix L being multiplied by the inverse
  of the diagonal matrix of the vertex connectivity degrees, D_k. Then, $L_{rw} = D_k^{-1}L$ [33].

It is worth to highlight that the above described Laplacian matrices are positive semi-definite
and have n non-negative real-valued eigenvalues $0 = \lambda_1 \leq \ldots \leq \lambda_n$. These properties are of main
importance in the graph spectral theory.

2.2. Network Eigenvalues

This section provides a quick survey of some graph eigenvalues properties. It is not exhaustive.
However, there are enounced the most important properties for further mathematical reference.
These are about eigenvalues that are used in the paper regarding WDN applications.

- The Largest eigenvalue (Spectral radius or Index) λ_1: it refers to the Adjacency graph matrix
  A and it plays an important role in modelling a moving substance propagation in a network.
  It takes into account not only immediate neighbours of vertices, but also the neighbours of the
  neighbours [34]. Spectral radius concept is often introduced by using the example of how a virus
  spread in a network. The smaller the Spectral radius the larger the robustness of a network against
  the spread of any virus in it. In this regard, the epidemic threshold is proportional to the Inverse
  of Spectral radius $1/\lambda_1$ [35]. This fact can be explained as the number of walks in a connected
  graph is proportional to λ_1. The greater the number of walks of a network, the more intensive is
  the spread of the moving substance in it. The other way round, the higher the Spectral radius,
  the better is the communication into a network.
- The Spectral gap $\Delta\lambda$: it represents the difference between the first and second eigenvalue of an
  Adjacency matrix, A. It is a measure of network connectivity strength. In particular, it quantifies
  the robustness of network connections and the presence of bottlenecks, articulation points,
  or bridges. This is of significant importance, as the removal of a bridge splits the network
  in two or more parts. The larger the Spectral gap the more robust is the network [36].
- The Multiplicity of zero eigenvalue m_0: the multiplicity of the eigenvalue 0 of L is equal to
  the number of connected components $A_1, \ldots, A_k$ in the graph; thus, the matrix L has as many
  eigenvalues 0 as connected components [37].
- The Eigengap $\lambda_{k+1} - \lambda_k$: it is a spectral utility specifically designed for network clustering.
  A suitable number of clusters k may be chosen such that all eigenvalues $\lambda_1, \ldots, \lambda_k$ of Laplacian
  matrix L are very small, but λ_{k+1} is relatively large [38]. The more significant the difference for
  a-priori proposing the number of clusters the better is the further clustering configuration.
- The Second smallest eigenvalue (Algebraic connectivity) λ_2: it refers to the Laplacian matrix.
  λ_2 plays a special role in many graph theories related problems [39]. It quantifies the strength of
  network connections and its robustness to link failures. The larger the Algebraic connectivity is
  the more difficult to cut a graph into independent components. It is also related to the min-cut
  problem of a data set for spectral clustering [37].

A simple Example Network with $n = 18$ nodes and a varying number of links m (from 27 to 30)
is illustrated in the Figure 1 by its different possible layouts. Example Network will be useful as an
instance for spectral metrics computation. This will also show the possible applications for water
distribution network management. The first Example Network layout, A), is composed by two
separated network subregions. Layout B) comes from adding a single link to A) to obtain a connected
network. An additional link is added to B) to obtain C). Table 1 and Figures 2 and 3 show the spectral
metrics computed on the previous described network layouts (Figure 1).

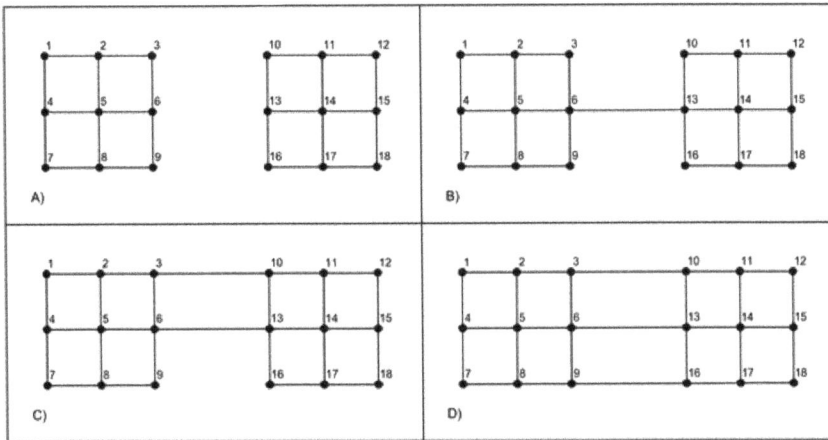

Figure 1. Four layouts of the Example Network with the same number of nodes and a different number of links. **A)** two separated subregions; **B)** a single edge links the two subregions; **C)** two edges link the two subregions; **D)** three edges link the two subregions.

Table 1 reports how the Spectral radius, the Spectral gap, and the Algebraic connectivity increase with the number of links between the subregions. The same result is also shown in Figure 1, where it is clear that the general connectivity and robustness increase from A) to D). Algebraic connectivity and Spectral gap start from zero for the separated layout A). Both measures significantly increase in the other layouts, A) to D). This show how these two metrics may be used as a measure of the network connectivity strength [40].

The measures for Spectral radius (Table 1) start from values greater than zero for layout A). Then, these values decrease as the number of connections increase. In this regard, Spectral radius can be used as a parameter to quantify the communication rate or the connectivity level of the network. It is also noticed how Spectral radius hardly varies for the four analyzed Example Network layouts. This result is explained as the measure ranges from the average node degree *kmean* and the maximum node degree of the network *kmax* [41] that in Example Network ranges between *kmean* = 2.67 to *kmax* = 4.00 (for layout A) and *kmean* = 3.00 to *kmax* = 4.00 (for layout D).

Table 1. Spectral metrics for the four cases of the example network.

Metric	Layout A	Layout B	Layout C	Layout D
Inverse of Spectral radius $1/\lambda_1$	0.354	0.332	0.320	0.311
Spectral gap $\Delta\lambda$	0.000	0.275	0.422	0.555
Eigengap $\lambda_{k+1} - \lambda_k$	1.000	0.875	0.806	0.732
Multiplicity of zero m_0	2	1	1	1
Algebraic connectivity λ_2	0.000	0.125	0.194	0.268

Figure 2 shows the top five eigenvalues $\lambda_1, \ldots, \lambda_5$ of the Laplacian matrix for the four layout configurations of Example Network. It is noticeable that some eigenvalues are equal for all of the layouts. The first eigenvalue λ_1 is always equal to zero because the graph Laplacian matrix is positive semi-definite [37].

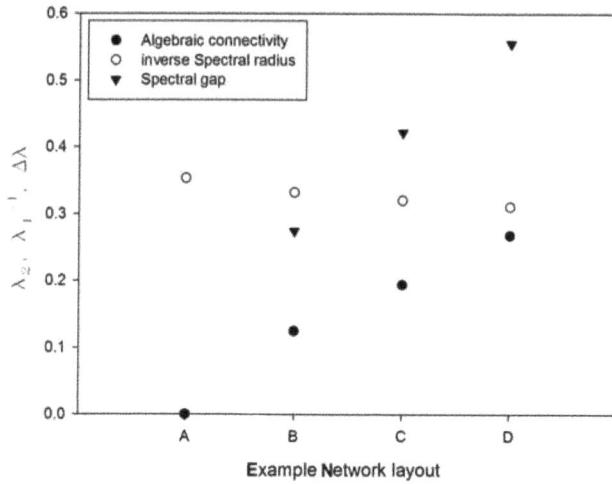

Figure 2. Algebraic connectivity, Inverse Spectral radius and Spectral radius for the layout A, B, C, and D of Example Network.

In layout A) the Multiplicity of zero, m_0, is equal to 2. Consequently, also the second eigenvalue λ_2 (the Algebraic connectivity) is equal to zero (Table 1). This means that there are two separated subregions in the network, as the number of multiplicity of zero, m_0, is equal to the number of the disconnected subregions. In all four layouts, the maximum eigengap occurs between the third eigenvalue λ_3 and the second eigenvalue λ_2. This indicates that, from a topological point of view, the optimal number of clusters to split the network is two. These results match with those naturally expected by the Example Network construction and also by its visualization. It also important to highlight that the value of the eigengap decreases as the number of links between the two A) regions increases. This suggests that the eigengap criterion works better when the clusters in the network can be well defined (not overlapping).

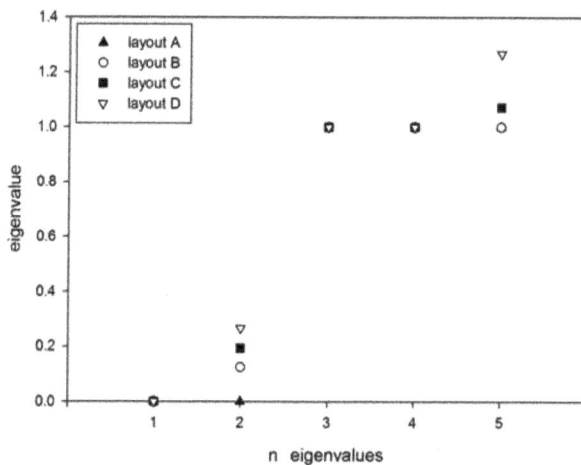

Figure 3. First five eigenvalues for the cases A, B, C, and D of the Example Network.

2.3. Network Eigenvectors

Graph eigenvectors contain a lot of information about the graph structure. The above described matrices are based on eigenvalue spectra and have been proposed into several applications [34,42,43]. It is worth highlighting that graph eigenvectors are not graph invariants since they depend on the labelling of graphs [30]. This characteristic can become into an advantage at some cases. This is shown in the following subsection where there are introduced the principal eigenvector, the Fiedler eigenvector, and problems that are related to simultaneous usage of several eigenvectors.

- Principal eigenvector: it corresponds to the largest A-eigenvalue, v_1, of a connected graph. It gives the possibility to rank graph vertices by its coordinates with respect to the number of paths passing through them to connect two nodes in the network [44]. The number of paths can be seen as the "importance" (also called the centrality) of node i. In this regard, the eigenvector centrality attributes a score to each node equals to the corresponding coordinate of the principal eigenvector. Groups of highly interconnected nodes are more "important" for the communication in comparison to equally high connected nodes do not form groups, that is, whose neighbours are less connected than them (according to the social principle that "I am influential if I have influential friends"). An important Principal eigenvector application is on Web search engines as Google's PageRank algorithm [45];
- The Fiedler eigenvector: it corresponds to the second smallest Laplacian (or normalized Laplacian) eigenvalue of a connected graph. Fiedler [39] first demonstrated that the eigenvector v_2 associated to the second smallest eigenvalue λ_2 provides an approximate solution to the graph bi-partitioning problem. This is approached according to the signs of the components of v_2. A subgraph is encompassed by nodes with positive components in the Fiedler eigenvector. The other subgraph contains nodes that are related to negative Fiedler eigenvector components. The v_2 values closer to 0 correspond to "better" splits. In this regard, if a number of clusters $k \geq 2$ is needed, then it is useful to resort to the Recursive spectral bisection [46,47]. According to this, the Fiedler eigenvector is used to bi-divide the vertices of the graph by the sign of its coordinates and the process is iterated then for each defined sub-part until reach the targeted number k of clusters.
- Other Eigenvector: an alternative to obtain a good graph partitioning for $k \geq 2$ clusters is related to the first k smallest eigenvector of the Laplacian matrix (or normalized Laplacian). The approach is based on solving the relaxed versions of the *RCut* problem (*NCut* problem) to define the so-called spectral clustering (normalized spectral clustering). It has been demonstrated in literature [33] that the normalized spectral clustering, based on the Random Walk Normalized Laplacian Matrix L_{rw}, shows a superior performance to other spectra alternatives to find a clustering configuration. The solution is simultaneously characterized by both a minimum number of cuts and a well-balanced clusters size. According to [33], the minimization of the *NCut* problem is equal to the minimization of the Rayleigh quotient.

$$\min(NCut(x)) = \min \frac{y^T (D_k - A) y}{y^T D y} \tag{1}$$

The expression of Equation (1) is minimized by the smallest eigenvalue of the $(D - A)$ matrix that is in correspondence to its smallest eigenvector. In this regard, the minimization of the *NCut* problem is related to the solution of the generalized eigenvalues system.

$$(D_k - A) y = \lambda D_k y. \tag{2}$$

According to the expression of $L = D_k - A$, and pre-multiplying by D_k^{-1}, the problem is reduced to the classical eigenvalues system.

$$L_{rw} y = \lambda y. \tag{3}$$

Finally, the spectral clustering consists of the following steps:

1. definition of Adjacency matrix A (or weighted Adjacency matrix W);
2. computation of the Laplacian L;
3. computation of the first k eigenvectors of normalized Laplacian L_{rw} matrix;
4. definition of the matrix U_{nxk} containing the first k eigenvectors as columns; and,
5. clustering the nodes of the network into clusters $C_1, \ldots, C_k$ using the k-means algorithm applied to the rows of the U_{nxk} matrix.

It is important to clarify that the boundary links, *Nec*, are those for which each of the connected nodes belong to different clusters C_k. An important aspect according of the spectral algorithm is to change the representation of the nodes from Euclidean space to points in the U_{nxk} matrix. This new data space enhances important cluster-properties and the final configuration has an easier detection [37]. Successful applications for the water distribution networks can be found in [11,14].

Figures 4 and 5 show the outcome from applying eigenvector techniques to Example Network. Regarding the Principal eigenvector, the eigenvector centrality $v_{1,i}$ is evaluated for layout D). Table 2 shows that the two most important nodes are the node 6 and the node 13 (marked in Table 2), as those nodes correspond the maximum value of the eigenvector. The connectivity degree for these nodes is $k_i = 4$, and they are connected to other nodes with a connectivity degree $k_i = 4$ (that is node 5 and node 13 are connected to node 6; node 14 and node 6 are connected to node 13). So, the two most important nodes, identified with the eigenvector centrality, are those nodes that have highly connected adjacent neighbour. These nodes 6 and 13 can consequently be considered "central" nodes for the communication of the network (from a topological point of view). Similar results are obtained also for the other Example Network layouts.

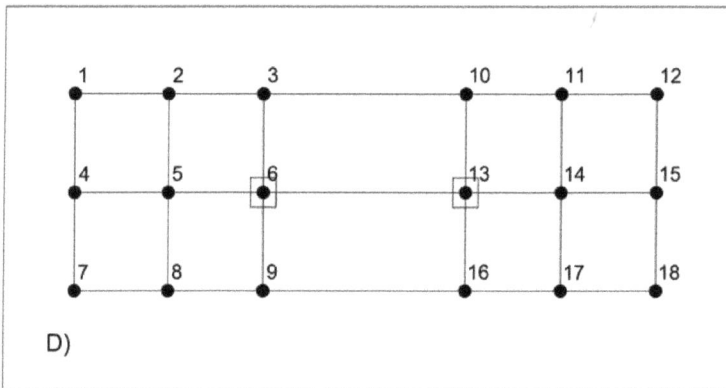

Figure 4. Two most important nodes, computed by the eigenvector centrality, for the layout D of Example Network.

Regarding the Fiedler eigenvector, the coordinates of v_2 for the four layouts of Example Network are shown in Figure 5. The Fiedler eigenvector has a number of components (coordinates) equal to the number of nodes. It is clear that the coordinates have positive and negative values for the four layouts. In particular, it is possible to define two well separated groups. The first ranges from node 1 to node 9 (negative values), while the second is made by node 10 up to node 18 (positive values). By splitting the nodes of the network according to their coordinates for v_2, it is possible to define a bisection of them.

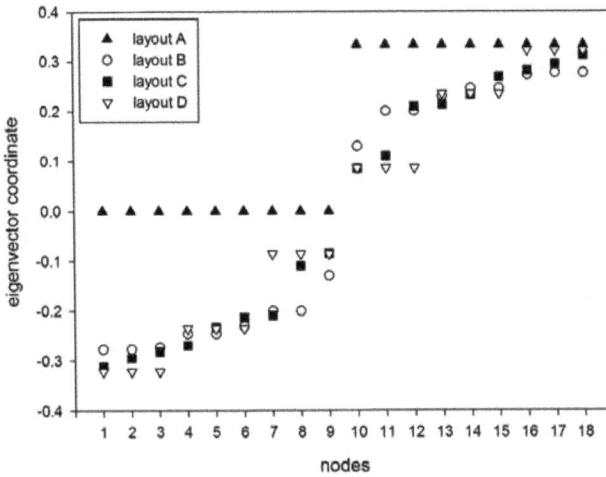

Figure 5. Fiedler eigenvector coordinates for the layout A, B, C, and D.

Table 2. Eigenvector centrality for all the nodes in Example Network, layout D.

n	1	2	3	4	5	6	7	8	9	10	11	12	13	14	15	16	17	18
$v_{1,i}$	0.12	0.21	0.26	0.16	0.30	0.37	0.12	0.21	0.26	0.26	0.21	0.12	0.37	0.30	0.16	0.26	0.21	0.12

Analysing layout A (two separated groups), it is straightforward to see how the two groups of coordinates are well defined, having a constant value for each group. In the other layouts, the difference between two groups is less clear, as the number of connected links increases. However, the bisection of the nodes of the network can still be defined for these networks because the sign is preserved. In all of the layouts, the two clusters are defined having the same number of nodes (Figure 5).

Regarding to the clustering problem via the *NCut* minimization problem, the optimal clustering layout for Example Network proposes to take two clusters ($k = 2$), in compliance with the eigengap property (Figure 3). The Fiedler bipartition, according to the second eigenvector of the Laplacian matrix, provides the same clustering configuration than *NCut* algorithm. This is an expected result, as only the second eigenvector is considered in the definition of the matrix U_{nxk} for $k = 2$.

3. Case Study

All of the metrics and algorithms based on the Graph Spectral Techniques described above can be considered as an operational GST tool-set that is able to solve key management issues of water distribution networks. GSTs are tested on the real small-size water system of Parete (a town with 10,800 inhabitants located in a densely populated area near Caserta, Italy) and on the synthetic medium-size water system of C-Town [48]. The main characteristics of both WDNs are reported in Table 3.

Table 3. Main characteristics of water distribution network of C-Town and Parete. The symbol in brackets "-" indicates that the parameter is dimensionless.

Network	n (-)	m (-)	nr (-)	L_{TOT} (km)
C-Town	396	444	1	56.7
Parete	184	282	2	34.7

The Eigenvalues significance, explained in the previous section, is described for the two case studies. The Adjacency and the Laplacian matrices of these two networks are defined and the principal eigenvalues computed. It is important to note that the graphs are considered unweighted to better show the efficiency of the proposed management framework. This is based only on the topological knowledge of WDNs, as it is frequent to do not have available any hydraulic information about the network. Then, a novel GST tool-set is proposed that provides global and local network information key to develop operational algorithms and procedures to face complex tasks in WDNs management. It is however possible to attribute some weights to the network by taking into account the "strength" of the link between nodes [7]. In the WDNs case, the weights could represent background knowledge on geometric and hydraulic characteristics of the pipes (diameter, length, conductivity, flow, and velocity, among others).

Table 4 shows the network eigenvalues for the two case studies. The multiplicity of the 0-eigenvalue from the Adjacency matrix is, for both of the case studies, equal to $m_0 = 1$. This means that in both WDNs, there is only one connected component. It is interesting to note that also for complex network models (made by thousands of components) it is still easy to check if any anomaly observed in the water supply is caused by the decomposition of the original network in several subregions (as it is the case of unexpected pipe disruptions or valve malfunctions).

Table 4. Principal Eigenvalues of the Adjacency and Laplacian matrices of water distribution network of C-Town and Parete.

Network	m_0	$\Delta\lambda$	λ_2	$1/\lambda_1$	$\lambda_{k+1} - \lambda_k$
C-Town	1	0.0303	0.0006	0.358	5
Parete	1	0.0685	0.0212	0.303	4

GSTs also provide support to compute a surrogate index for the topological WDNs robustness regarding the following two features: (a) The presence of "bottlenecks" or articulation points. These are subregions that are connected to others through a single link. Removing any node or link at the bottleneck causes network disconnection. Bottlenecks are computed through the value of the Spectral gap $\Delta\lambda$, as calculated on the Adjacency matrix; (b) The network "strength" to get split into subregions, computed through the value of the Algebraic connectivity λ_2 calculated on the Laplacian matrix. The values of the Spectral gap and the Algebraic connectivity aid and simplify the assessment of robustness of a WDN, as it was preliminary proposed by [12–14]. In the current case studies, it is clear that the corresponding values of the two spectral measures are small and near to zero, $\Delta\lambda = 0.0685$ and $\lambda_2 = 0.0212$ for Parete, while $\Delta\lambda = 0.0303$ and $\lambda_2 = 0.0006$ for C-Town. These small values are justified by the fact that WDNs are sparser than other networks as Internet or social networks. This is due to both geographical embedding and economic constraints [7,11].

The larger Spectral gap for Parete than for C-Town suggests that Parete has a smaller number of bottlenecks. When considering the Algebraic connectivity, Parete shows greater tolerance to the efforts to be split into isolated parts with respect to C-Town. Comparing the two case studies, Parete evidently is more robust against node and link failure than C-Town (as we can expect from comparing a real utility network design as it has Parete to a synthetic WDN). The smaller value of the Spectral radius inverse shows that Parete have a more efficient layout than C-Town in terms of communication and degree connectivity. In this regard, the inverse of the Spectral radius can be used as a global measure of the reachability of network elements and the paths multiplicity. These first results obtained with spectral metrics support a preliminary visual analysis of the two WDNs, through which it is possible to observe a more cohesive shape (and so a more robust structure) for Parete than C-Town. These GSTs measures aid hydraulic experts to quantify several intuitive aspects of WDNs performance. In addition, GSTs make it possible to approach a structure analysis of large networks for which just a visual analysis does not provide enough information.

The three Eigenvectors techniques explained in the previous section are tested on Parete and C-Town WDNS. These are the Fiedler eigenvector, *Ncut* methods based on the other eigenvectors and the principal eigenvector. Through the Fielder Eigenvector and *Ncut* methods, it is possible to face the important and arduous task associated with permanent water network partitioning (WNP) [23]. WNP consists into define optimal discrete network areas, District Meter Area (DMA), aimed to improve the water network management (i.e., water budget, pressure management, or water losses localization). This should be done avoiding to negatively affect the hydraulic performance of the system that could be significantly deteriorated by shutting-off some pipes [23,49]. Choosing a suitable number of subregions and their respective layouts by a clustering algorithm is essential to design a WDN partition into DMAs. The definition of the number of clusters attempts to take into account some peculiarities of the system (i.e., water demand, pressure distribution, or elevation), which often are not available for the entire water network. A clustering method based on GSTs only considers network topological characteristics and is able to capture inherent cluster-properties of the system.

While the second smallest eigenvalue (Algebraic connectivity) is interpreted as a measure of the strength to split the network in sub-graphs, the eigengap $\lambda_{k+1} - \lambda_k$ could be interpreted as a measure of the surplus of the strength needed to split the network from $k + 1$ to k clusters. Once defining the maximum eigengap $\lambda_{k+1} - \lambda_k$, it is clear that, from a topological point of view, it is better to split the network at most up to k clusters, since a greater surplus of strength is needed to split the network in $k + 1$ and more clusters. For this reason, the maximum eigengap can be used to define the optimal number of clusters from a topological and connectivity point of view. Figure 6 shows the first ten eigenvalues of the Laplacian matrix for the graph of C-Town and Parete. It is clear that the first largest eigengap for C-Town, occurs between the sixth and the fifth eigenvalue ($\lambda_6 - \lambda_5 = 0.002$), while for Parete occurs between the fifth and the fourth eigenvalue ($\lambda_5 - \lambda_4 = 0.042$). This metric suggests that, an optimal number of clusters on which subdivided the water distribution networks of C-Town and Parete is, respectively, $k = 5$ and $k = 4$.

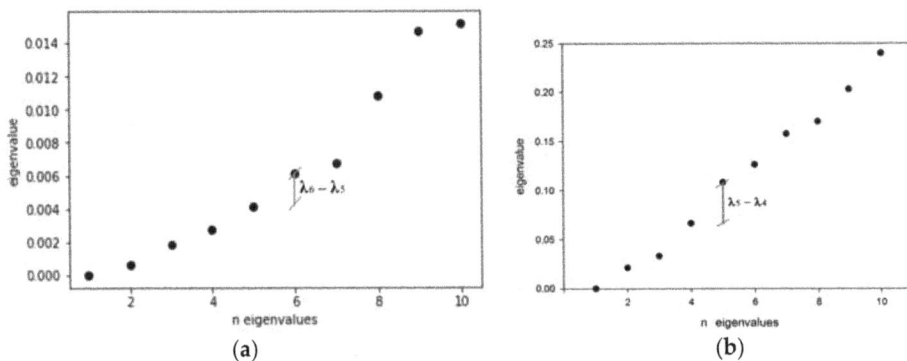

Figure 6. First 10 eigenvalues for the two case studies: (**a**) C-Town network; and, (**b**) Parete network.

Once it is defined a suitable number of clusters for a WDN, it is necessary to set the optimal layout at each sub-region in which the WDN is subdivided (clustering phase) to approach a complete water network partitioning [23]. The clustering phase focuses on identify clusters shape, aiming both to balance the number of the nodes and to minimize the number of boundary pipes between clusters. Approaching an appropriate network clustering is essential. This constitutes the starting point for the subsequent division phase that consists on choosing the boundary pipes in which to insert gate valves and flow meters, as it is widely described in [50].

Spectral clustering offers a valid and powerful tool to exploit the properties of the Laplacian matrix spectrum. Figure 7 reports the Fiedler eigenvectors, v_2, for C-Town and Parete WDNs. It is clear, as it was shown on Example Network, that the coordinates of the second eigenvector, v_2, easily define an optimal bipartition layout for the network. These divide the network nodes according to the signs (positive or negative) for the corresponding value of the Fiedler eigenvector. It is worth highlighting that this procedure ensures the continuity of each defined cluster, as each node of a cluster is linked at least to another node of the same cluster.

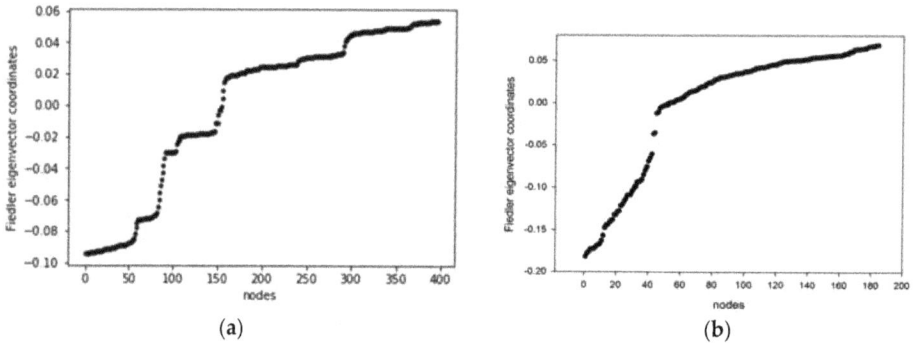

(a)   (b)

Figure 7. Fiedler eigenvector v_2 coordinates for the two case studies: (**a**) C-Town network; and, (**b**) Parete network.

In case of the optimal number of clusters (defined by the maximum value of the eigengap) is higher than two, then the first clustering configuration obtained as outcome of the Fiedler eigenvector v_2, can be used as input for a recursive bisection process. That is, for each cluster, the Fiedler eigenvector v_2 can be computed for the next clustering up to reach the targeted number of clusters. This network bisection can also represent a starting layout for other recursive algorithms that require an initial random choice of the clustering layout. Another GST based powerful tool for the optimal clustering layout of a water distribution network, is the *Ncut* spectral clustering [33], already explained in the Eigenvector techniques section, based on the use of other eigenvectors further than v_2.

Figure 8 shows the optimal clustering layout through *Ncut* spectral clustering. The results are given for a number of $k = 4$ clusters for Parete and $k = 5$ clusters for C-Town, according to the optimal number of clusters defined through the eigengap for both of the case studies. It is worth to point out that the clusters are well balanced in terms of number of nodes (a standard deviation $dst = 2.7\%$ for Parete and $dst = 8.1\%$ for C-Town). The number of boundary pipes is small with respect to the total number of pipes (about $Nec = 16$ for Parete and $Nec = 4$ for C-Town, corresponding to 5.7% and 1%, respectively).

GSTs propose a solution for ranking WDN nodes and then select the most important points. The WDNs of Parete and C-Town are ranked according to the score attributed by the corresponding coordinates to the first eigenvector, v_1, of the Adjacency matrix. Ranking WDN nodes is useful for locating optimal nodes in which locate devices (i.e., chlorine stations, pressure regulation valves, quality sensors, flow meters, etc.). The identification of the most important nodes can also contribute as initial guess for further development of specific device location algorithms. The applications range, for instance, from detecting accidental or intentional contamination to control pipe flows and node pressures. These challenging tasks can be approached through GSTs, even when no other information is available rather than the network topology. As it is explained in the previous section, the eigenvector centrality can spot the most "influential" nodes, according to the number of neighbours of the adjacent nodes. The idea behind the network centrality concept is to identify which points are traversed by the greatest number of connections. Central nodes are thus considered as essential nodes for network connectivity and have influence over large network areas. Figure 8 points out also the most important nodes based on the eigenvector centrality criterion. The results show the highest centrality node per each DMA of the C-Town and Parete partitioned WDNs. After WDNs clustering, the process is focus on every single Adjacency matrix related to water distribution sub-networks. The eigenvector centrality provides most the important nodes per cluster or DMA, from a topological and connectivity point of view.

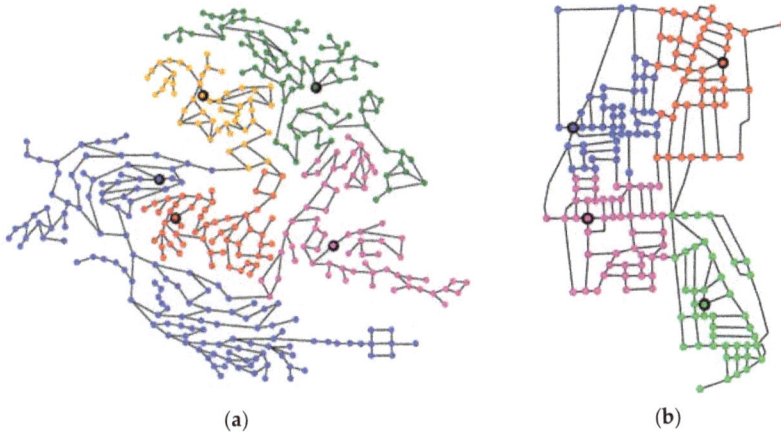

(a)   (b)

Figure 8. Optimal clustering layout for the two case studies with different colors for each clusters and highlighting the most important nodes of each cluster according to the eigenvector centrality of the partitioned networks: (**a**) C-Town network ($k = 5$); and (**b**) Parete network ($k = 4$).

4. Conclusions

This paper proposes a survey of the possibilities offered by graph spectral techniques. There is provided a complete tool-set of several metrics and algorithms, borrowed from graph spectral techniques (GSTs), and applied to water network operations and management. The tool-set is based on topological and geometric information of the water network layout. No hydraulic data (such as diameter, roughness, pressure, etc.) is required. This made the proposal particularly attractive, as it is a common situation that often face water utilities. Another advantage of the proposal lies on the huge GST tool-set applicability to any water distribution network. It also is straightforward its adaptation to deal with near real-time challenges, as avoiding any hydraulic simulation that often stall having a suitable speed on having network performance results.

The application of the proposed GST tool-set has shown to provide useful metrics for continuity check, testing if there is any unconnected part of the water network. GSTs also made it possible to approach topological robustness analysis, aiding to develop water system design or to network resilience assessments. Another challenges in water management have been also addressed, such as partitioning the water distribution network into district metered areas through a spectral clustering process. Ranking nodes importance in a water distribution network is useful for approaching valve or sensor location. The most "influential" or important nodes have also been obtained thanks to the GST tool-set framework.

Further work will lead to investigate new opportunities coming from GSTs for water distribution management. These will be towards using meaningful weights on pipes and nodes. The aim will be to add partial or complete hydraulics knowledge to the purely topological based solutions provided by GSTs.

Acknowledgments: The authors have no funding to report.

Author Contributions: Each of the authors contributed to the design, analysis and writing of the manuscript.

Conflicts of Interest: The authors declare no conflict of interest.

References

1. Farley, M.; World Health Organization. *Leakage Management and Control: A Best Practice Training Manual*; World Health Organization: Geneva, Switzerland, 2001.
2. Neirotti, P.; De Marco, A.; Cagliano, A.C.; Mangano, G.; Scorrano, F. Current trends in Smart City initiatives: Some stylised facts. *Cities* **2014**, *38*, 25–36. [CrossRef]
3. Albino, V.; Berardi, U.; Dangelico, R.M. Smart Cities: Definitions, dimensions, performance, and initiatives. *J. Urban Technol.* **2015**, *22*, 3–21. [CrossRef]
4. Mays, L.W. *Water Distribution System Handbook*; McGraw-Hill Education: New York, NY, USA, 1999; ISBN 978-0-07-134213-1.
5. Watts, D.J.; Strogatz, S.H. Collective dynamics of 'small-world' networks. *Nature* **1998**, *393*, 440–442. [CrossRef] [PubMed]
6. Barabási, A.-L.; Albert, R. Emergence of scaling in random networks. *Science* **1999**, *286*, 509–512. [CrossRef] [PubMed]
7. Boccaletti, S.; Latora, V.; Moreno, Y.; Chavez, M.; Hwang, D.-U. Complex networks: Structure and dynamics. *Phys. Rep.* **2006**, *424*, 175–308. [CrossRef]
8. Chung, F.R.K. *Spectral Graph Theory*; American Mathematical Society: Providence, RI, USA, 1996; ISBN 978-0-8218-0315-8.
9. Herrera, M.; Canu, S.; Karatzoglou, A.; Pérez-García, R.; Izquierdo, J. An approach to water supply clusters by semi-supervised learning. In Proceedings of the 9th International Congress on Environmental Modelling and Software, Ottawa, ON, Canada, 1 July 2010.
10. Gutiérrez-Pérez, J.A.; Herrera, M.; Pérez-García, R.; Ramos-Martínez, E. Application of graph-spectral methods in the vulnerability assessment of water supply networks. *Math. Comput. Model.* **2013**, *57*, 1853–1859. [CrossRef]
11. Di Nardo, A.; Di Natale, M.; Giudicianni, C.; Greco, R.; Santonastaso, G.F. Complex network and fractal theory for the assessment of water distribution network resilience to pipe failures. *Water Sci. Technol. Water Supply* **2017**, *17*, ws2017124. [CrossRef]
12. Yazdani, A.; Jeffrey, P. Robustness and vulnerability analysis of water distribution networks using graph theoretic and complex network principles. In Proceedings of the 2010 Water Distribution System Analysis, Tucson, Arizona, 12–15 September 2010; pp. 933–945. [CrossRef]
13. Yazdani, A.; Jeffrey, P. Complex network analysis of water distribution systems. *Chaos Interdiscip. J. Nonlinear Sci.* **2011**, *21*, 016111. [CrossRef] [PubMed]
14. Di Nardo, A.; Di Natale, M.; Giudicianni, C.; Musmarra, D.; Varela, J.M.R.; Santonastaso, G.F.; Simone, A.; Tzatchkov, V. Redundancy features of water distribution systems. *Procedia Eng.* **2017**, *186*, 412–419. [CrossRef]
15. Diao, K.; Sweetapple, C.; Farmani, R.; Fu, G.; Ward, S.; Butler, D. Global resilience analysis of water distribution systems. *Water Res.* **2016**, *106*, 383–393. [CrossRef] [PubMed]
16. Shuang, Q.; Zhang, M.; Yuan, Y. Performance and reliability analysis of water distribution systems under cascading failures and the identification of crucial pipes. *PLoS ONE* **2014**, *9*, e88445. [CrossRef] [PubMed]
17. Torres, J.M.; Duenas-Osorio, L.; Li, Q.; Yazdani, A. Exploring topological effects on water distribution system performance using graph theory and statistical models. *J. Water Resour. Plan. Manag.* **2017**, *143*, 04016068. [CrossRef]
18. Candelieri, A.; Soldi, D.; Archetti, F. Network analysis for resilience evaluation in water distribution networks. *Environ. Eng. Manag. J.* **2015**, *14*, 1261–1270.
19. Soldi, D.; Candelieri, A.; Archetti, F. Resilience and vulnerability in urban water distribution networks through network theory and hydraulic simulation. *Procedia Eng.* **2015**, *119*, 1259–1268. [CrossRef]
20. Candelieri, A.; Giordani, I.; Archetti, F. Supporting resilience management of water distribution networks through network analysis and hydraulic simulation. In Proceedings of the 2017 21st International Conference on Control Systems and Computer Science (CSCS), Bucharest, Romania, 29–31 May 2017; pp. 599–605.
21. Agathokleous, A.; Christodoulou, C.; Christodoulou, S.E. Topological robustness and vulnerability assessment of water distribution networks. *Water Resour. Manag.* **2017**, *31*, 4007–4021. [CrossRef]
22. Greco, R.; Di Nardo, A.; Santonastaso, G. Resilience and entropy as indices of robustness of water distribution networks. *J. Hydroinform.* **2012**, *14*, 761–771. [CrossRef]

23. Di Nardo, A.; Di Natale, M.; Santonastaso, G.F.; Venticinque, S. An automated tool for smart water network partitioning. *Water Resour. Manag.* **2013**, *27*, 4493–4508. [CrossRef]
24. Alvisi, S.; Franchini, M. A procedure for the design of district metered areas in water distribution systems. *Procedia Eng.* **2014**, *70*, 41–50. [CrossRef]
25. Pérez, R.; Puig, V.; Pascual, J.; Peralta, A.; Landeros, E.; Jordanas, L. Pressure sensor distribution for leak detection in Barcelona water distribution network. *Water Sci. Technol. Water Supply* **2009**, *9*, 715–721. [CrossRef]
26. Antunes, C.H.; Dolores, M. Sensor location in water distribution networks to detect contamination events—A multiobjective approach based on NSGA-II. In Proceedings of the 2016 IEEE Congress on Evolutionary Computation (CEC), Vancouver, BC, Canada, 24–29 July 2016; pp. 1093–1099.
27. Tinelli, S.; Creaco, E.; Ciaponi, C. Sampling significant contamination events for optimal sensor placement in water distribution systems. *J. Water Resour. Plan. Manag.* **2017**, *143*, 04017058. [CrossRef]
28. Gomes, R.; Marques, A.S.; Sousa, J. Decision support system to divide a large network into suitable District Metered Areas. *Water Sci. Technol.* **2012**, *65*, 1667–1675. [CrossRef] [PubMed]
29. Di Nardo, A.; Di Natale, M.; Musmarra, D.; Santonastaso, G.F.; Tzatchkov, V.; Alcocer-Yamanaka, V.H. Dual-use value of network partitioning for water system management and protection from malicious contamination. *J. Hydroinform.* **2015**, *17*, 361–376. [CrossRef]
30. Arsić, B.; Cvetković, D.; Simić, S.K.; Škarić, M. Graph spectral techniques in computer sciences. *Appl. Anal. Discrete Math.* **2012**, *6*, 1–30. [CrossRef]
31. Cvetković, D.; Simić, S. Graph spectra in computer science. *Linear Algebra Appl.* **2011**, *434*, 1545–1562. [CrossRef]
32. Mohar, B. The Laplacian spectrum of graphs. In *Graph Theory, Combinatorics, and Applications*; Wiley: Hoboken, NJ, USA, 1991; pp. 871–898.
33. Shi, J.; Malik, J. Normalized cuts and image segmentation. *IEEE Trans. Pattern Anal. Mach. Intell.* **2000**, *22*, 888–905. [CrossRef]
34. Bonacich, P. Power and centrality: A family of measures. *Am. J. Sociol.* **1987**, *92*, 1170–1182. [CrossRef]
35. Wang, Y.; Chakrabarti, D.; Wang, C.; Faloutsos, C. Epidemic spreading in real networks: An eigenvalue viewpoint. In Proceedings of the 2003 22nd International Symposium on Reliable Distributed Systems, Florence, Italy, 6–8 October 2003; pp. 25–34.
36. Donetti, L.; Neri, F.; Muñoz, M.A. Optimal network topologies: Expanders, cages, Ramanujan graphs, entangled networks and all that. *J. Stat. Mech. Theory Exp.* **2006**, *2006*, P08007. [CrossRef]
37. Von Luxburg, U. A tutorial on spectral clustering. *Stat. Comput.* **2007**, *17*, 395–416. [CrossRef]
38. Mohar, B. Some applications of laplace eigenvalues of graphs. In *Graph Symmetry*; NATO ASI Series; Springer: Dordrecht, The Netherlands, 1997; pp. 225–275; ISBN 978-90-481-4885-1.
39. Fiedler, M. Algebraic connectivity of graphs. *Czechoslov. Math. J.* **1973**, *23*, 298–305.
40. Estrada, E. Network robustness to targeted attacks. The interplay of expansibility and degree distribution. *Eur. Phys. J. B Condens. Matter Complex Syst.* **2006**, *52*, 563–574. [CrossRef]
41. Cvetkovic, D.M.; Doob, M.; Sachs, H. *Spectra of Graphs: Theory and Application*; Academic Press: Berlin, Germany, 1980; ISBN 978-0-12-195150-4.
42. Dorogovtsev, S.N.; Mendes, J.F.F. *Evolution of Networks: From Biological Nets to the Internet and WWW*; Oxford University Press: Oxford, MS, USA; New York, NY, USA, 2014; ISBN 978-0-19-968671-1.
43. Wang, Z.; Thomas, R.J.; Scaglione, A. Generating random topology power grids. In Proceedings of the 41st Annual Hawaii International Conference on System Sciences (HICSS 2008), Waikoloa, HI, USA, 7–10 January 2008; p. 183.
44. Wei, T.-H. Algebraic Foundations of Ranking Theory. Ph.D. Thesis, University of Cambridge, Cambridge, UK, 1952.
45. Brin, S.; Page, L. The anatomy of a large-scale hypertextual web search engine. In Proceedings of the Seventh International World-Wide Web Conference (WWW 1998), Brisbane, Australia, 14–18 April 1998.
46. Barnard, S.T.; Simon, H.D. Fast multilevel implementation of recursive spectral bisection for partitioning unstructured problems. *Concurr. Comput. Pract. Exp.* **1994**, *6*, 101–117. [CrossRef]
47. Pothen, A.; Simon, H.; Liou, K. Partitioning sparse matrices with eigenvectors of graphs. *SIAM J. Matrix Anal. Appl.* **1990**, *11*, 430–452. [CrossRef]

48. Ostfeld, A.; Salomons, E.; Ormsbee, L.; Uber, J.G.; Bros, C.M.; Kalungi, P.; Burd, R.; Zazula-Coetzee, B.; Belrain, T.; Kang, D.; et al. Battle of the water calibration networks. *J. Water Resour. Plan. Manag.* **2012**, *138*, 523–532. [CrossRef]
49. Herrera, M.; Abraham, E.; Stoianov, I. A graph-theoretic framework for assessing the resilience of sectorised water distribution networks. *Water Resour. Manag.* **2016**, *30*, 1685–1699. [CrossRef]
50. Sela Perelman, L.; Allen, M.; Preis, A.; Iqbal, M.; Whittle, A.J. Automated sub-zoning of water distribution systems. *Environ. Model. Softw.* **2015**, *65*, 1–14. [CrossRef]

water

MDPI

Article

Transient Wave Scattering and Its Influence on Transient Analysis and Leak Detection in Urban Water Supply Systems: Theoretical Analysis and Numerical Validation

Huan-Feng Duan [ID]

Department of Civil and Environmental Engineering, The Hong Kong Polytechnic University, Hung Hom, Kowloon, Hong Kong 999077, China; hf.duan@polyu.edu.hk; Tel.: +852-3400-8449

Received: 8 September 2017; Accepted: 11 October 2017; Published: 13 October 2017

Abstract: This paper investigates the impacts of non-uniformities of pipe diameter (i.e., an inhomogeneous cross-sectional area along pipelines) on transient wave behavior and propagation in water supply pipelines. The multi-scale wave perturbation method is firstly used to derive analytical solutions for the amplitude evolution of transient pressure wave propagation in pipelines, considering regular and random variations of cross-sectional area, respectively. The analytical analysis is based on the one-dimensional (1D) transient wave equation for pipe flow. Both derived results show that transient waves can be attenuated and scattered significantly along the longitudinal direction of the pipeline due to the regular and random non-uniformities of pipe diameter. The obtained analytical results are then validated by extensive 1D numerical simulations under different incident wave and non-uniform pipe conditions. The comparative results indicate that the derived analytical solutions are applicable and useful to describe the wave scattering effect in complex pipeline systems. Finally, the practical implications and influence of wave scattering effects on transient flow analysis and transient-based leak detection in urban water supply systems are discussed in the paper.

Keywords: water supply pipeline; transient wave; non-uniformities; wave scattering; transient modelling; leak detection

1. Introduction

Wave scattering has been commonly studied in shallow water fields where the water waves propagate through the channel bottom with randomly varying bars, as depicted in Figure 1 [1]. Experimental results for such cases together with some theoretical considerations are investigated and discussed in [2,3]. The results in these studies showed that, with the existence of random non-uniformities (inhomogeneities) of channel bottom elevation, eventually the amplitude of the generated wave decreases along the longitudinal direction and it tends to zero if the longitudinal distance is large enough.

Similar random non-uniformities can be found in pipelines (closed conduits) such as random variations in the pipe cross-sectional area (pipe diameter), with respect to length, developing with age (refer to Figure 2). In practical systems, many factors can attribute to the random non-uniformities of pipe cross-sectional area, for example, bio-film build up, corrosion, and deposition in water supply pipes, drainage pipes, crude oil pipes, and arterial systems. In particular, in water pipelines, non-uniformities of pipe diameter may be induced by various different factors as shown in Figure 3, including corrosion, sediment, junctions and complex connections.

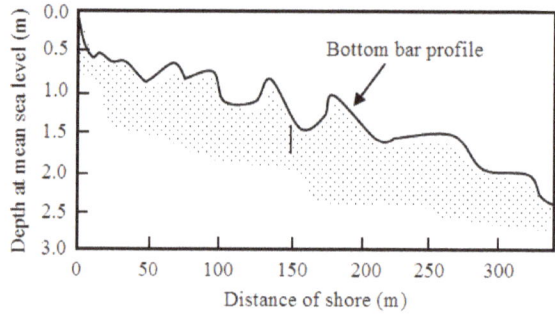

Figure 1. Bottom bar profile in Chesapeake Bay (adapted from [1]).

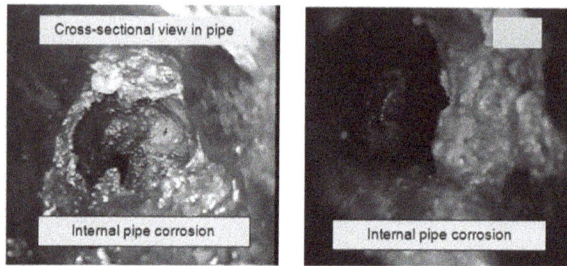

Figure 2. Cross-sectional views of aged water pipelines (adapted from [4]).

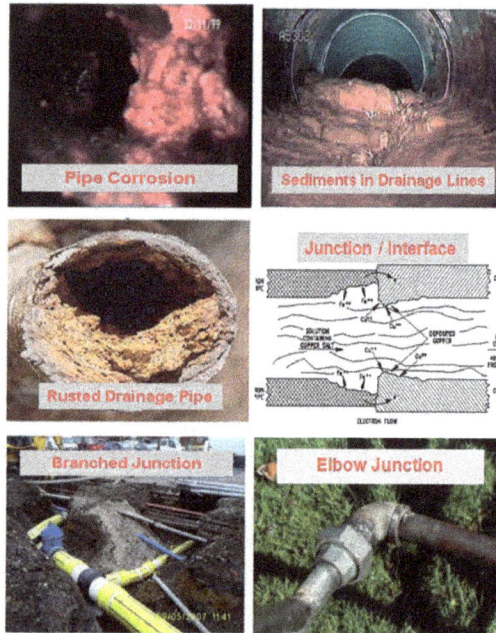

Figure 3. Different factors attributed to non-uniformities of pipe diameter (pictures adapted from online public sources on google websites).

From the perspective of steady flow in water piping systems, pipe diameter non-uniformities may cause additional energy losses and, thus, require more pumping capacity. On the other hand, from the perspective of unsteady flow, random variations in pipe diameters can result in random reflections and damping of waterhammer waves. So far, the phenomenon of wave scattering in water piping systems and its implications for the structural integrity and robustness of such systems are still not well understood. Recent numerical and experimental studies [4,5] have demonstrated the significant influence of irregular pipe diameters (e.g., roughness and blockage) on the transient wave propagation behaviors, and also on the transient-based defect detection methods (e.g., leakage and blockage). Moreover, their results also indicate that wave scattering by pipe diameter non-uniformities is more important and influential than the corresponding friction effect induced by roughness/irregularities to the wave damping and reflection in both the time and frequency domains. Therefore, an in-depth understanding of such wave scattering effects due to pipe diameter non-uniformities is necessary and critical to transient modelling, analysis and application in water supply pipe systems.

In fact, the wave scattering phenomenon and its relevant influence has been widely studied in many different application fields in the literature such as condensed matter, electromagnetism, seismology and fluid mechanics. For example, a classic paper [6] showed that random impurities have an important consequence on the propagation of electrons in a dirty crystal: the diffusive motion of electrons is terminated and all electrons become localized (this phenomenon is known as Anderson localization). Anderson's idea was used later in the analysis of the propagation of surface waves over a random seabed. Experimental results from previous studies [1,2] demonstrated the localization of water waves over the rough bottom. The analogy between water wave dynamics and Anderson localization is pointed out in [3]. Furthermore, the behavior and propagation of slowly modulated waves in random media has been studied in [7]. However, so far there is no such systematic analysis and theoretical investigation of the transient wave scattering phenomenon in urban water supply pipelines, although it is widely observed from laboratory experiments and field tests in this research area [4,5,8].

To investigate the potential wave-scattering phenomenon induced by random pipe diameter non-uniformities (i.e., pipe cross-sectional area) and to understand its impact on the transient wave propagation in water piping systems, in the present paper the method of multi-scale perturbation from the literature is firstly applied to one-dimensional (1D) waterhammer equations, which describe the flow dynamics in a pipe under the additional assumption that the pipe cross-sectional area (diameter) varies in a random manner along the longitudinal coordinate. Two cases—regular and random pipe non-uniformities—are considered for the analytical derivations. The obtained analytical results are then compared and validated by numerical simulations, which are achieved by the step-discretization approximation for different pipe non-uniformities. Thereafter, further discussion of the practical implications of the results and findings in this study to transient system modelling and pipe leak detection is performed in the paper. Finally, relevant conclusions are drawn at the end of this study.

2. Problem Statement and Study Framework

In realistic water supply pipelines, the non-uniformities of pipe diameters could be formed by various different reasons, as shown in Figure 3, resulting in relatively random geometries and distributions of such non-uniformities, as sketched in Figure 4a. In many theoretical studies (e.g., [4,5]), these random non-uniformities are usually treated approximately as different regular shapes or their combinations, in order to conduct mathematical operations and numerical computations. For example, Figure 4b with a relatively smooth variation (e.g., sinusoidal shape) and Figure 4c with relatively sharp variation (e.g., step shape) are two commonly used approximations. From the perspective of mathematics, the complicated random situation in Figure 4a could be a superposition of different numbers of simplified cases in Figure 4b,c. Therefore, it is a good start to investigate and understand the simplified cases, which can provide insights and a basis to explore and explain more complicated situations, such as the random case in Figure 4a.

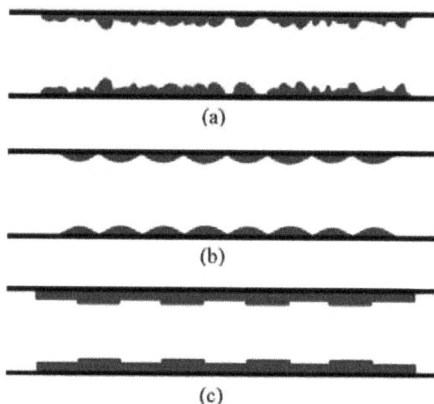

Figure 4. Sketch of different types of pipe diameter non-uniformities (side-sectional profile): (**a**) realistic and random situation; (**b**) regular approximation by sinusoidal variation; (**c**) regular approximation by step variation.

To this end, in this study, the regular case of pipe diameter non-uniformities in Figure 4b is used for preliminary analytical analysis for transient wave propagation in pipelines, while the random case in Figure 4a is applied for further analytical derivations so as to obtain the complete characteristics of transient flows in realistic pipelines. Thereafter, the other regular case in Figure 4c is adopted as a discrete approximation of a random case in order to achieve numerical simulations for the validation of the derived analytical results in this study. The detailed settings for such numerical simulations are provided later in the part covering numerical applications. The obtained analytical and numerical results are finally discussed for an in-depth understanding of the transient wave behavior and propagation in non-uniform pipelines in urban water supply systems.

3. Models and Methods

For clarity, the main models and analysis methods used in this study for investigating transient wave scattering effect in water supply pipelines are summarized as follows.

3.1. One-Dimentional (1D) Transient Model

The continuity and momentum equations of the 1D waterhammer model for compressible pipe flow with pipe diameter non-uniformities (i.e., varying pipe cross-sectional area) by neglecting the friction and visco-elastic effects are considered herein [4,9],

$$\frac{\partial(\rho A)}{\partial t} + \frac{\partial(\rho Q)}{\partial x} = 0, \tag{1}$$

$$\frac{\partial(\rho Q)}{\partial t} + A\frac{\partial P}{\partial x} + \tau_w \pi D = 0, \tag{2}$$

where ρ is fluid density; $A = A(x)$ is pipe cross-sectional area; $D = D(x)$ is pipe diameter; $Q = Q(x, t)$ is pipe discharge; $P = P(x, t)$ is pressure; τ_w is wall shear stress; x is spatial coordinate; and t is temporal coordinate. In the numerical simulations, the wall shear stress is modelled by the Darcy–Weisbach formula, where only the steady state friction is included. The method of characteristics (MOC) is used for the 1D numerical simulations in this study, and the details for implementing this numerical scheme into above transient model can refer to the classic textbooks and references in this field [9,10]. While in the analytical analysis, the friction effect (wall shear stress term in the equation) is excluded due to the mathematical complexity and difficulty of analytical derivation, and so as to highlight the effect of wave scattering during transient flow process.

For analytical derivation, the continuity and momentum Equations (1) and (2) can be further lumped into wave equation form through following transformation,

$$\frac{\partial^2 P}{\partial t^2} = -\frac{\rho a^2}{A} \frac{\partial^2 Q}{\partial x \partial t},$$ (3)

$$\rho \frac{\partial^2 Q}{\partial x \partial t} = -\frac{\partial}{\partial x}\left(A \frac{\partial P}{\partial x}\right).$$ (4)

where a is acoustic wave speed. After the mathematical elimination operation, the result becomes,

$$A \frac{\partial^2 P}{\partial t^2} = a^2 \frac{\partial}{\partial x}\left(A \frac{\partial P}{\partial x}\right),$$ (5)

with the pipe cross-sectional area $A(x)$ varying with x. Furthermore, Equation (5) can also be rewritten as,

$$\underbrace{\frac{\partial^2 P}{\partial t^2} = a^2 \frac{\partial}{\partial x}\left(A \frac{\partial P}{\partial x}\right)}_{(a)} + \underbrace{(1-A)\frac{\partial^2 P}{\partial t^2}}_{(b)},$$ (6)

where part (a) in Equation (6) has a similar form solved in previous studies for shallow surface wave problems [7], while the other part (b) of Equation (6) is an additional term originated from the case of pressurized wave propagation in water supply pipelines that is focused and dealt with in this study. Similarly, the multi-scale perturbation method from previous studies is further adapted and applied to solve this transient wave equation for pressurized water pipelines [7,11], which is elaborated in the following section.

3.2. Multi-Scale Perturbation Method

The method of multi-scale perturbation used in this study follows the previous studies [7,11], with three sets of coordinate scales in both spatial and time introduced as follows:

$$\text{spatial domain: } x, x_1 = \varepsilon x, x_2 = \varepsilon^2 x,$$ (7)

$$\text{time domain: } t, t_1 = \varepsilon t, t_2 = \varepsilon^2 t,$$ (8)

where the three scales (x, x_1, x_2 and t, t_1, t_2) correspond to wave oscillations, initial wave modulation, and the modulation by randomness when waves propagate along the pipeline, respectively; ε characterizes the ratio of different scales and $\varepsilon \ll 1$. The derivatives with respect to x and t are transformed in accordance with the chain rule as [11],

$$x = x; x_1 = \varepsilon x; x_2 = \varepsilon^2 x; t = t; t_1 = \varepsilon t; t_2 = \varepsilon^2 t;$$

$$\frac{\partial}{\partial x} \rightarrow \frac{\partial}{\partial x} + \varepsilon \frac{\partial}{\partial x_1} + \varepsilon^2 \frac{\partial}{\partial x_2} + \cdots; \frac{\partial}{\partial t} \rightarrow \frac{\partial}{\partial t} + \varepsilon \frac{\partial}{\partial t_1} + \varepsilon^2 \frac{\partial}{\partial t_2} + \cdots.$$ (9)

As a result, the solution is represented in the form of a perturbation series such as,

$$P = P_0(x, x_1, x_2, t, t_1, t_2) + \varepsilon P_1(x, x_1, x_2, t, t_1, t_2) + \varepsilon^2 P_2(x, x_1, x_2, t, t_1, t_2) + \cdots,$$ (10)

where P_0, P_1 and P_2 correspond to the above three scales of wave propagation and modification, respectively. It is important to note that high order terms (>2) with regard to ε from Equations (9) and (10) are neglected in the following analytical analysis under the assumption of a relatively small extent of non-uniformities of pipe diameters or cross-sectional areas. This assumption will be validated and discussed through numerical applications later in this study.

4. Analytical Results and Analysis

By applying the multi-scale perturbation method in Equation (9) to the transient wave equation in Equation (6), the analytical results of wave scattering with regard to the pressure wave envelopment and evolution can be obtained for both the regular case in Figure 4b and the random case in Figure 4a of pipe diameter variations. For clarity and due to the page space limit, only the key steps and results are presented as follows, while the detailed derivations are neglected in this paper.

4.1. Results of Regular Non-Uniformities

For the analysis of the regular case of pipe diameter non-uniformities, it is assumed that the disordered pipe section has a regular variation magnitude of a pipe cross-section area as shown in Figure 4b, which is defined by the relatively disordered cross-sectional area (i.e., $\delta A = \Delta A / A_0$) following a periodic co-sinusoidal variation along the pipeline, as follows:

$$A(x) = A_0(1 + \varepsilon \delta A \cos(\lambda_b x)) = 1 + \varepsilon \delta A \cos(\lambda_b x), \tag{11}$$

where A_0 is the mean value of the pipe cross-sectional area, assuming $A_0 = 1.0 \text{ m}^2$ in this study for simplification; and λ_b is the periodic length of pipe diameter disorders.

Based on the multi-scale wave perturbation technique, the wave scattering results in the regularly disordered pipeline can be obtained as three following cases:

1. Subcritical detuning: $0 < \Omega < \Omega_0$,

$$\begin{cases} T(x_1) = \frac{\Omega \sinh K(L-x_1) + iaK \cosh K(L-x_1)}{\Omega \sinh(KL) + iaK \cosh(KL)} \\ R(x_1) = \frac{\Omega_0 \sinh K(L-x_1)}{\Omega \sinh(KL) + iaK \cosh(KL)} \end{cases}, \tag{12}$$

where T and R are transmission and reflection coefficients respectively; Ω_0 and Ω represent the incident wave frequency and pipe disorder variation frequency, respectively, and $\Omega_0 = \delta A \omega / 2$, $\Omega = a\lambda_b$; x_1 is the distance along the disordered section in the pipeline; i is the imaginary unit, and $i^2 = -1$; L is the total length of disordered section along the pipeline; and K is the detuning (group) wave number and, $K = \sqrt{\left|\Omega_0^2 - \Omega^2\right|}/a$.

2. Supercritical detuning: $\Omega > \Omega_0$,

$$\begin{cases} T(x_1) = \frac{\Omega \sin K(L-x_1) + iaP \cos K(L-x_1)}{\Omega \sin(KL) + iaK \cos(KL)} \\ R(x_1) = \frac{\Omega_0 \sin K(L-x_1)}{\Omega \sin(KL) + iaK \cos(KL)} \end{cases}. \tag{13}$$

3. Bragg resonance: $\Omega = \Omega_0$,

$$\begin{cases} T(x_1) = \frac{\cosh \Omega_0(L-x_1)/a}{\cosh \Omega_0 L/a} \\ R(x_1) = -i\frac{\sinh \Omega_0(L-x_1)/a}{\cosh \Omega_0 L/a} \end{cases}. \tag{14}$$

Particularly, the Bragg resonance of the regularly disordered pipe in Equation (14) indicates that all waves are reflected completely by the disordered section along the pipeline. The analytical reflection coefficient (R) along the disorder distance ($X = x_1$) can be obtained according to Equation (14) and shown in Figure 5. The results of Figure 5 show clearly that, under a fixed disorder magnitude, the reflection coefficient (R) is decreasing along the pipeline, which indicates that the wave perturbation energy is decayed gradually by the disorder section. As expected, the reflection coefficient (R) at a fixed location of the pipe disorder section increases with the disorder magnitude

(δA) due to more wave energy reflecting back at the initial disorder section in the larger disorder magnitude (δA) case. For example, the reflection coefficient at the starting location of the disorder section (i.e., $X/L_0 = 0$) could attain 0.9 when the disorder magnitude is about 10% of the mean value (i.e., $\varepsilon \sim 0.1$). Under this situation, there would be very little wave energy (perturbations) remaining at the end of pipe disorder section (since the total energy in the entire pipeline is conserved), resulting in a relatively high decrease-gradient of the reflection coefficient curve for larger δA case as shown in Figure 5. More numerical validations to corroborate the analytical result are conducted later in this study.

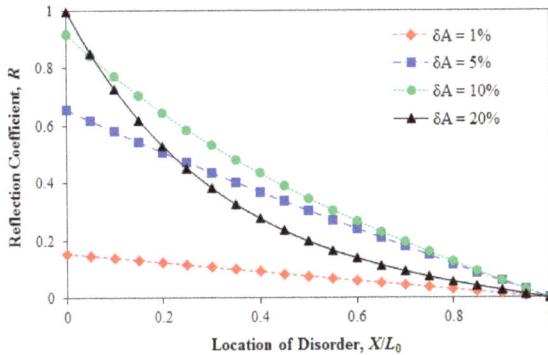

Figure 5. Analytical results of the reflection coefficient for the regular non-uniformities case.

4.2. Results of Random Non-Uniformities

For the case of random pipe diameter non-uniformities, as shown in Figure 4a, it is assumed that the pipe cross-sectional area follows the random variation function as,

$$A(x) = A_0(1 + \varepsilon\zeta(x)) = (1 + \varepsilon\zeta(x)), \tag{15}$$

where $\zeta(x)$ is a function that represents the random variation characteristics of the pipe cross-sectional area and is assumed to be of zero mean and be of standard deviation of $\sigma(x)$; other symbols are as defined above.

After applying the multi-scale perturbation analysis, the solution of the pressure wave envelopment to Equation (6) has the following form [4],

$$B = B_0 e^{-\lambda x} = B_0 e^{-\lambda a t}, \tag{16}$$

where $B = B(x) =$ the amplitude of the wave envelope with distance or with the equivalent time $t = x/a$ with $a =$ wave speed along the pipe disorder section; $B_0 =$ amplitude of the incident wave; $\lambda = \lambda_r - i\lambda_i$ is complex wave number, with λ_r and $\lambda_i =$ wave damping factor and wave phase change (frequency shift) factor, respectively, and

$$\lambda_r = \frac{\alpha k^2 \sigma^2}{a^2 + 4k^2}, \lambda_i = -\frac{k\alpha^2 \sigma^2}{2(a^2 + 4k^2)}. \tag{17}$$

where $k =$ incident wave number and $k = \omega/a$, with $\omega =$ wave frequency; $\alpha =$ spatial correlation coefficient of the blockage and $\alpha \sim 1/\lambda_b$ with $\lambda_b =$ correlation length which describes the spatial variability of pipe diameter non-uniformities.

The result of Equation (17) indicates that the wave amplitude exponentially decreases with longitudinal distance (x). In other words, the wave is localized by the random non-uniformities of the pipe diameter or cross-sectional area along the pipeline. The localization distance can be defined and used for characterizing the wave scattering by the random diameter non-uniformities as

$$L_{loc} = \frac{1}{\varepsilon^2 \lambda_r} = \frac{\alpha^2 + 4k^2}{\varepsilon^2 \sigma^2 \alpha k^2},$$ (18)

which represents the spatial distance for the wave amplitude decreased by an exponential factor of e^{-1} as shown in Figure 6. This parameter (L_{loc}) is used later in this study for the evaluation of the wave scattering effect due to different pipe diameter non-uniformities.

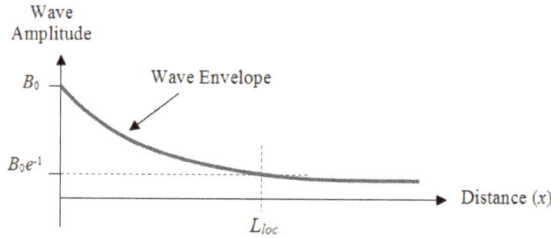

Figure 6. Wave localization by random pipe diameter non-uniformities.

Particularly, for a specified water supply pipeline under investigation, the correlation length of random pipe diameter non-uniformities (i.e., $L_{cor} \sim 1/\alpha$) is usually determinate (but maybe known or unknown for the analyst), and therefore, a dimensionless parameter (termed as wave scattering factor) can be further defined for better characterizing the wave scattering effect in that system as,

$$\varphi = \alpha L_{loc} = \alpha \frac{1}{\varepsilon^2 \lambda_r} = \frac{\alpha^2 + 4k^2}{\varepsilon^2 \sigma^2 k^2} = \frac{1}{\sigma_A^2}\left(4 + \left(\frac{\alpha}{k}\right)^2\right).$$ (19)

Specifically, a smaller φ value (shorter localized distance) means a relatively more significant wave scattering effect, and vice versa. Based on this result, the typical dependence relationship of the wave scattering factor on the properties of incident waves and pipe diameter non-uniformities can be shown in Figure 7. It is clearly shown in Figure 7 that when the incident wave length is around twice as long as the correlation length of the random pipe diameter non-uniformities (i.e., $k/\alpha = 1/2$), the wave scattering effect would attain to maximum (i.e., minimum localized distance), and thus the incident wave can be strongly attenuated (scattered) along the pipe disorder section.

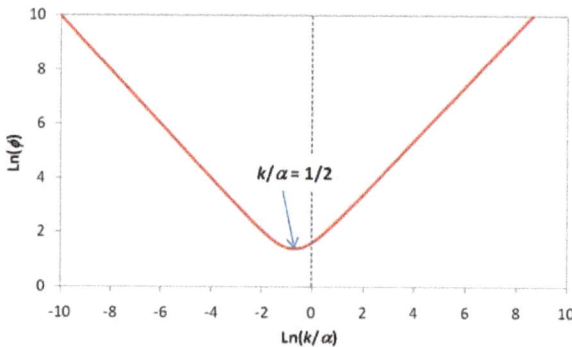

Figure 7. Dependence relationship of wave scattering factor on the incident waves and pipe diameter non-uniformities.

It is necessary to point out that the obtained results of Equation (14) for the regular pipe disorder case and Equation (19) for the random pipe disorder case are obtained under the assumptions of linearization with relatively small extent of non-uniformities and no-reflection boundary conditions in the water pipeline system. The validity and accuracy of these results and assumptions are to be validated by the numerical simulations conducted later in this study.

5. Numerical Validation

5.1. Settings of Numerical Tests

To validate the derived analytical results presented above and evaluate the importance of the wave-scattering effect on transient wave propagation, the hypothetic pipeline system shown in Figure 8a is used for a numerical simulation, which consists of three pipe sections: upstream uniform pipe section, middle disordered pipe section (for testing), and downstream uniform pipe section. The length of each section is 2000 m, and the no-reflection boundary condition from the two ends of the whole pipeline is imposed for the numerical simulation. For simplicity, all pipes are assumed to have a constant steady-state friction factor and wave speed (e.g., $f = 0.01$ and $a = 1000$ m/s). The step approximation illustrated in Figure 4c is applied for the numerical simulation for both regular and random non-uniformities, which can be shown in Figure 8b.

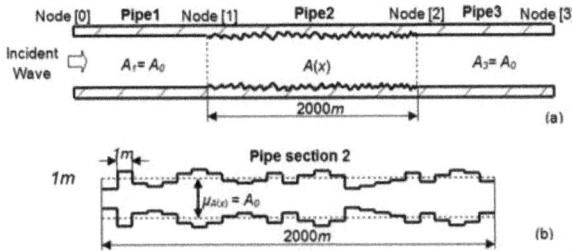

Figure 8. Schematic of numerical pipeline system: (**a**) three-section pipeline; (**b**) middle disorder section for testing.

A total of nine test cases, listed in Table 1, were conducted for numerical analysis, with cases 1~3 applied for a regular pipe disorder situation and cases 4~9 for a random pipe disorder situation. Moreover, for each test, three different relationships between the incident wave length ($\lambda_w = 1/k$) and the characteristic/correction length (distance) of pipe diameter non-uniformities ($\lambda_b = 1/\alpha$) were considered for the evaluation. It is assumed that both types of non-uniformities (represented by pipe cross-sectional area) have a zero mean relative to the original nominal value. Note that λ_b represents the periodic length of disordered diameters for the regular disorder case, while it represents the correlation length of disordered diameters in the pipeline for the random disordered cases (i.e., $1/\alpha$).

Table 1. Settings for the numerical test cases.

Type	Case No.	$\lambda_w/2\lambda_b$	A_0 (m²)	δA	Distribution Function	Correlation Function
Regular	1	>1			Degenerate	0 for $\zeta \neq 0$
	2	=1	1.0	$\sigma A/A_0 = 0.20$	(deterministic)	1 for $\zeta = 0$
	3	<1				
Random	4	>1				
	5	=1	1.0	$\sigma_A/A_0 = 0.23$	Uniform	$e^{-\alpha\|\zeta\|}$
	6	<1				
	7	>1				
	8	=1	1.0	$\sigma_A/A_0 = 0.23$	Upper triangular	$e^{-\alpha\|\zeta\|}$
	9	<1				

Initially, the pipeline system was considered to be under a steady state. Note that for comparison, the results for the completely uniform pipeline without non-uniformities (termed as the "intact" case hereafter) were also obtained for each test case. For transient generation, the incident wave at the upstream entrance of the pipeline was assumed to be a sinusoidal perturbation of pressure head signal as:

$$H(t) = H_0\left[1 + R_f \sin(\omega t)\right], \tag{20}$$

where H = instant pressure head; H_0 = initial steady pressure head level; R_f = amplitude factor of incident wave and $R_f = 0.2$ in this study; and ω = incident wave frequency.

For test cases of the random pipe diameter non-uniformities, two kinds of probability distributions were considered for the randomness of the non-uniformities: one followed uniform distribution (for cases 4~6) and the other was upper triangular distribution (for cases 7~9). It was assumed that the random variables of pipe diameter disorder were correlated with an exponential function along the pipeline in the spatial domain. Other numerical settings for different cases are listed in Table 1.

5.2. Validation for Regular Case

In the regular disordered tests, the total disordered distance was assumed to be 2000 m (L_0) and there were a total of 20 uniformly spaced reaches with each 100 m (λ_b = 200 m). A continuously sinusoidal incident wave defined by Equation (20) was used in this study and the incident wave frequency was adjusted to achieve the three cases: $\lambda_w/2\lambda_b > 1$, $\lambda_w/2\lambda_b = 1$ and $\lambda_w/2\lambda_b < 1$. The envelope of the maximum and minimum pressure head profiles along the disordered pipe section was extracted from the numerical results and plotted in Figure 9a–c.

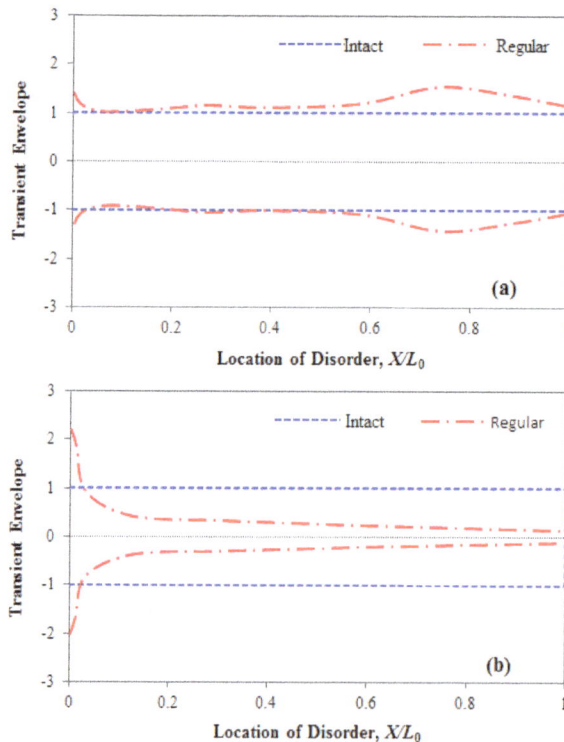

Figure 9. *Cont.*

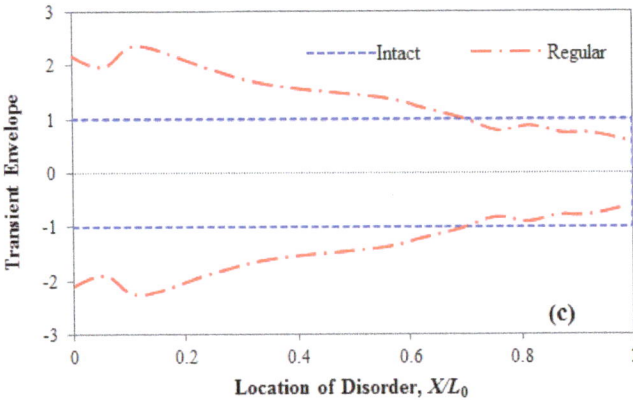

Figure 9. Results of the pressure wave envelope for regular disorder cases: (**a**) $\lambda_w/2\lambda_b = 1$ and $\delta A = 0.05$; (**b**) $\lambda_w/2\lambda_b = 1$ and $\delta A = 0.10$; and (**c**) $\lambda_w/2\lambda_b = 1$ and $\delta A = 0.20$.

The results comparison in Figure 9 indicates that the attenuation of the pressure wave along the pipeline due to the wave-scattering effect behaves much more significantly for case two, with $\lambda_w/2\lambda_b = 1$ in Figure 9b, than other two cases, ($\lambda_w/2\lambda_b > 1$ in Figure 9a and $\lambda_w/2\lambda_b < 1$ in Figure 9c, which is consistent with the analytical results of Equations (12)–(14) and Figure 7. Meanwhile, for the cases of $\lambda_w/2\lambda_b > 1$ and $\lambda_w/2\lambda_b < 1$, the pressure wave envelopes were larger than that of the intact case because of the superposition of the scattered waves. To further validate the analytical solution, the reflection coefficients (R) of perfect resonance for case no. two were calculated and plotted in Figure 10. As shown in Figure 10, obvious discrepancies were observed between the analytical and numerical results, which were actually increasing with the disorder magnitude (δA). This result implies that the linearized analytical solution can provide good estimations for the wave scattering effect of a relatively small pipe disorder situation, which is due to the linearization assumption imposed by the analytical analysis.

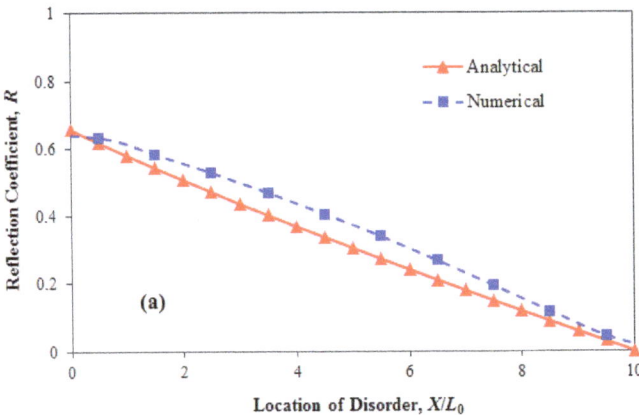

Figure 10. *Cont.*

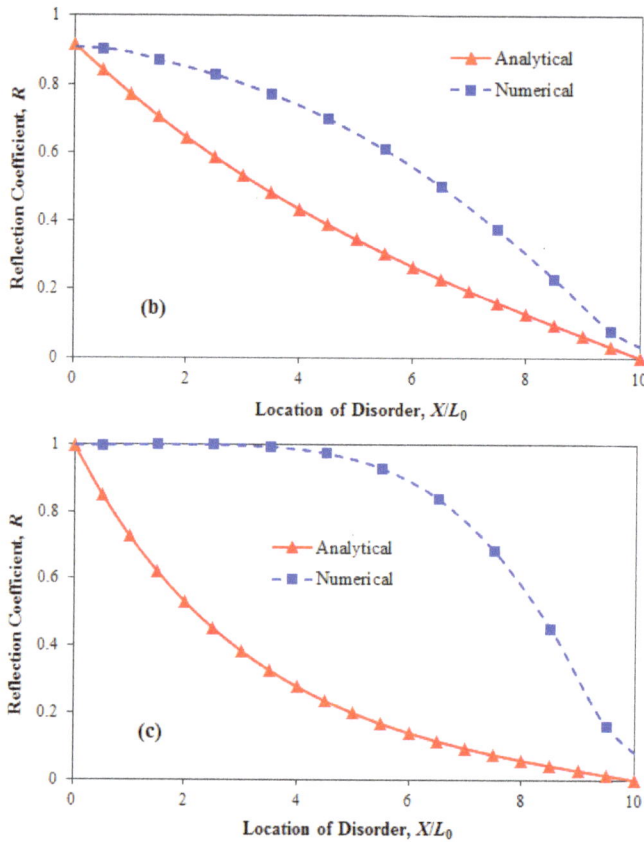

Figure 10. Comparisons of the numerical and analytical results of the reflection coefficients for the case of $\lambda_w/2\lambda_b = 1$: (**a**) $\delta A = 0.05$; (**b**) $\delta A = 0.10$; and (**c**) $\delta A = 0.20$.

5.3. Validation for the Random Case

In the numerical tests involving randomly disordered pipes, the disordered diameters were assumed to be spatially correlated along the pipeline. For simulations, it was assumed that the continuously-correlated random diameters could be discretized into many small reaches, with each reach 1 m representing one spatial random point of the original continuous disordered section. In this study, the generation of samples of randomly correlated diameters was based on the "NORTA" (normal to anything) theorem, which was developed by Ghosh [12]. Thereafter, a Monte-Carlo simulation (MCS) with 500 samples was conducted and the statistical results were retrieved for the analysis [13].

With the MCS-based numerical simulations, the pressure wave profiles were obtained and shown in Figures 11–13 for the cases of $\lambda_w/2\lambda_b > 1$, $\lambda_w/2\lambda_b < 1$ and $\lambda_w/2\lambda_b = 1$, respectively. It is clear from these results that the pressure wave amplitude decays exponentially with distance along the pipe with random diameter non-uniformities. Moreover, the results for both uniform and triangular distributions of random non-uniformities indicate that the wave scattering effect behaves most significantly when $\lambda_w/2\lambda_b = 1$ (see Figure 12), which is similar to the results of the regular disorder cases analyzed above. The results also imply that the different probability distributions (uniform or triangular) for random non-uniformities along the pipeline have little impact on the wave-scattering effect, as long as the other parameters remain the same, e.g., mean, standard deviation and correlation.

For validation, the analytical and numerical results of dimensionless localization length (φ) in Equation (19) for different cases and their relative errors were calculated and are listed in Table 2. The results show that the maximum relative error is less than 5%, which implies good prediction by the linear analytical solutions of Equation (19) for wave scattering in random pipe disorder cases. On this point, the analytical results of Equation (19) have been validated for describing the qualitative influence and importance trend of wave scattering induced by random diameter non-uniformities in the pipe.

Table 2. Analytical and numerical results of the wave scattering factor.

Case No.		Uniform Distribution			Upper Triangular Distribution		
		4	5	6	7	8	9
Wave scattering factor (φ)	Analytical	125.0	75.0	159.4	125.0	75.0	159.4
	Numerical	127.5	74.5	157.0	125.8	76.3	152.5
	Relative error (%)	2.0	0.7	1.5	0.7	1.7	4.5

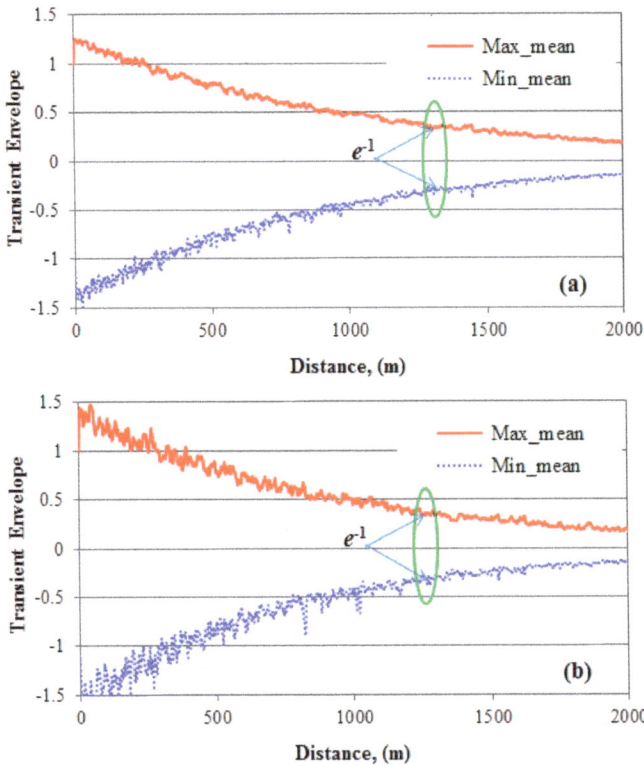

Figure 11. Results of the random disorder case with $\lambda_w/2\lambda_b > 1$: (**a**) uniform distribution and $\alpha = 0.6$ m^{-1}; (**b**) triangular distribution and $\alpha = 0.6$ m^{-1}.

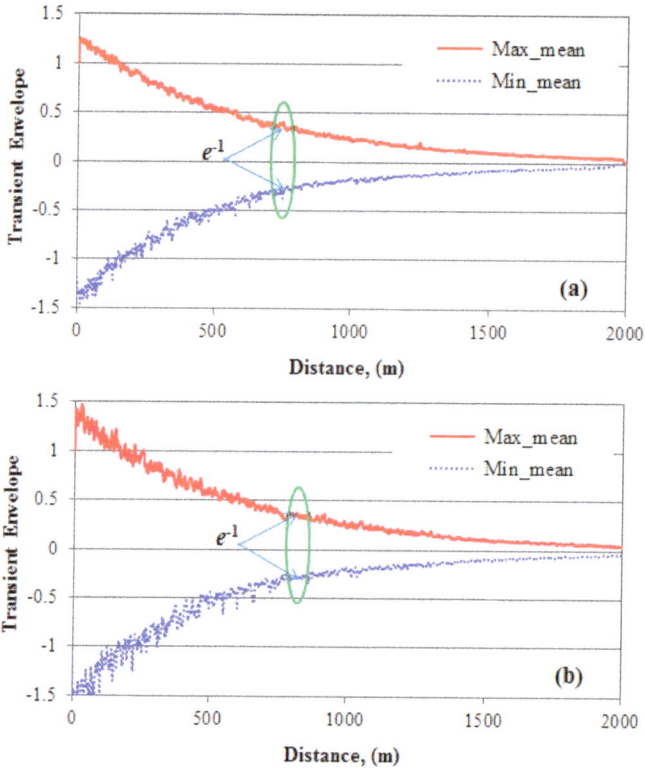

Figure 12. Results of the random disorder case with $\lambda_w/2\lambda_b$ =1: (a) uniform distribution and $\alpha = 0.2 \text{ m}^{-1}$; (b) triangular distribution and $\alpha = 0.2 \text{ m}^{-1}$.

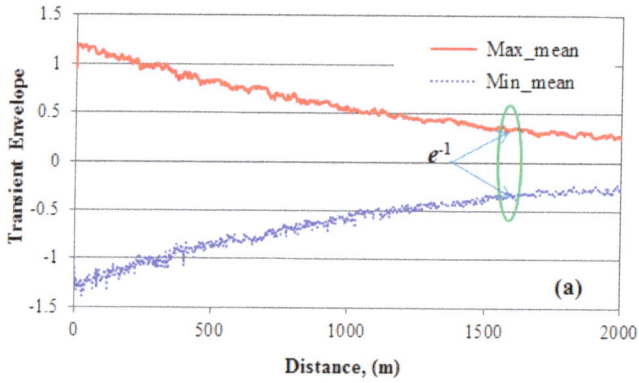

Figure 13. *Cont.*

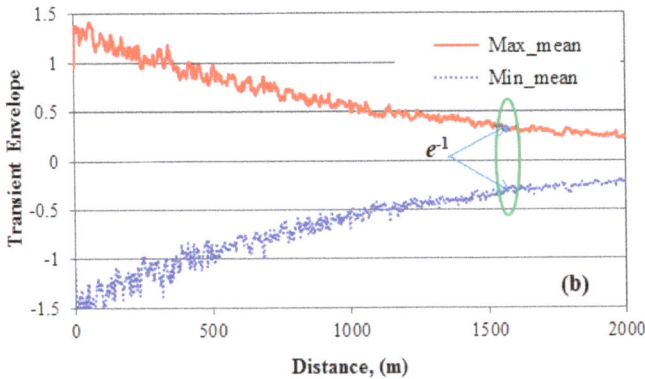

Figure 13. Results of the random disorder case with $\lambda_w/2\lambda_b < 1$: (**a**) uniform distribution and $\alpha = 0.05$ m^{-1}; (**b**) triangular distribution and $\alpha = 0.05$ m^{-1}.

6. Results Discussion and Implications

6.1. Energy Analysis of Transient Wave Scattering

It is clear in Equation (19) that the localization distance (by wave scattering factor φ) decreases with an increase of the amplitude of pipe random non-uniformities (σ_A) because of more serious reflections by these non-uniformities in the pipeline. To further explain and understand the wave scattering effect, an energy analysis is performed based on the energy formulations in previous studies [14,15]. The results of case no. four in Table 2 are retrieved from the model and plotted in Figure 14. It is clearly shown in Figure 14 that the total energy in the pipeline system with random variation in diameters is always conserved, although each form of the energy (kinetic or internal) changes significantly with time. In other words, as a result of the wave scattering effect, the total energy has been re-distributed in the system due to the pipe diameter non-uniformities such that pressure wave envelopment is scattered significantly along the pipeline as indicated in the analytical solutions.

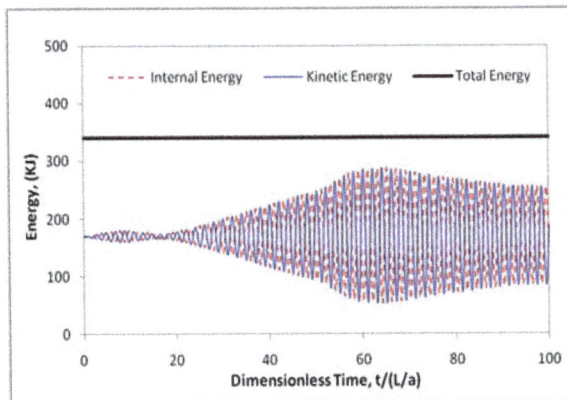

Figure 14. Energy results of transient wave scattering along the pipeline.

6.2. Impacts on Transient Modelling and Analysis

The pressure wave envelopment attenuation in the present waterhammer models is usually attributed to friction damping (and viscoelasticity damping if pipes are plastic) [9,15]. With the existence of wave scattering effects, especially in the aged pipeline system, the actual damping of pressure waves might not be fully represented by only the friction (and pipe-wall viscoelasticity if appropriate), while the wave scattering effect with energy re-distribution in the system may provide a great contribution to the total attenuation. Therefore, it is necessary to inspect the relative importance of wave scattering and friction and their roles in pressure wave envelopment attenuation, such that the transient flow behaviors in these disordered pipes can be well understood and accurately simulated by waterhammer models.

In this regard, the analytical results derived above in this study may provide useful guidance and estimation of the influence of wave scattering. Specifically, the analytical results of Equation (19) in this study indicates an increasing localization distance (φ) and thus a decreasing wave scattering effect with an increase of the ratio of α/k. That is, when the incident wave length is shorter in comparison with the correlation length of random diameter non-uniformities (i.e., $\alpha << k$), the wave scattering effect becomes more significant and the waves are localized more seriously in the pipeline. As a result, high frequency waves propagating in the randomly disordered pipeline can be greatly scattered such that most of the waves are reflected back upstream (i.e., the incident part) and hardly transmitted downstream (i.e., the outgoing part), if only the pipeline is long enough.

In practical transient (waterhammer) systems, however, the incident waves are usually "fast and sharp" signals, for example caused by a sudden closure or opening of valves, the starting or stopping of pumps, etc., where the operation time duration is rather short. From this perspective, in pipe systems with potential random diameter non-uniformities and under transient conditions, the wave scattering induced wave envelopment attenuation is generally dominant in comparison with friction damping. In this regard, many typical examples have been shown in previous studies [4,5,8], where significant discrepancies were commonly observed between the real data (from both laboratory and field tests) and the numerical model results (e.g., MOC-based simulation with steady and/or unsteady friction components).

Another important implication for transient system analysis is that the random non-uniformities can cause an increase in the pressure head (also energy) in certain regions of the pipeline system since most of the waves are reflected or trapped by the disordered section of the pipeline. This is clearly worrisome for aged pipes since such a pressure increase was likely not accounted for when pipes were designed for the water supply system. Therefore, the wave behavior in aged pipelines might become very complicated due to the potential wave scattering effect, such that the design schemes of system strength from their initial new states may become overestimated or underestimated for some sections of the pipe system.

6.3. Impacts on Transient-Based Leak Detection

Transient-based defect detection techniques are being developed by various researchers [16–26]. The idea is to intentionally inject a wave, typically a pressure variation by, for example, changing a valve setting, and then measure the subsequent pressure response of the system. The key is to find the signature of the defect in the measured signal and use it to identify the nature, location and size of the defect.

A previous study by the author [27] has demonstrated that the current transient-based pipe defect detection methods are mainly dependent on wave damping and reflections and that the "fast and sharp" input wave signals are preferable to these methods. Clearly, such approaches could become intractable in the presence of random non-uniformities in the pipeline system. The non-uniformities in pipes may be due to pipe diameter, material and thickness as well as fluid properties such the case of pumped sewerage. Consequently, these pipe defect detection methods are particularly difficult to be applied in the aged pipes and/or sewage drainage systems where potential non-uniformities commonly exist. For illustration, two test cases (T1 and T2) listed in Table 3 are examined in which the disordered pipe diameter shown in Figure 8 is present. Four types of leak detection methods—transient reflection method (TRM), transient damping method (TDM), system response function method (SRFM), and inverse transient method (ITM)—were used in this investigation [27].

Table 3. Results of leak detection under the presence of wave scattering in the pipeline.

Case	Real Leak Information, x_L^* & A_L^*	Predicted Leak Information, x_p^* & A_p^*				Max. Error, $|x_L^* - x_p^*|$ & $|A_L^* - A_p^*|$
		TRM	TDM	SRFM	ITM	
T1	No Leak	0.50 & 0.01	0.50 & 0.038	0.25 & 0.024	0.46 & 0.031	— & —
T2	0.1 & 0.002	0.49 & 0.012	0.44 & 0.042	0.17 & 0.019	0.34 & 0.035	39% & 40%

In Table 3, leak location and size (x_L^* & A_L^*) were normalized by the total pipe length and average pipe cross-sectional area, respectively. The relative errors of predicted dimensionless leak location by using these four methods are also listed in the table. The results of case T1 show that the additional pseudo leak is detected by all four methods, while actually there is no leak along the pipeline. In case T2, the maximum predicted error using the four methods can reach 39% and 40% for the leak location and size, respectively. This also indicates that the four leakage detection methods are invalid or inaccurate when wave scattering induced reflections and "damping" exist.

Actually, many recent studies have evidenced the wave scattering phenomenon in water supply distribution systems, where the water supply demand (and thus the pressure head) was observed to pulse frequently and continuously, although transient oscillations were relatively small [28–33]. Consequently, these preliminary results and analysis indicated that the wave scattering effect could have a great influence on both transient system analysis and transient-based utilization in urban water pipeline systems. More attention needs to be paid to the impact of the wave scattering effect on transient wave behavior and propagation (reflection and damping) so that present models and techniques can be applied with confidence to practical pipeline systems.

7. Conclusions

The analytical expressions for transient wave evolution in water pipelines with different non-uniformities were derived in this paper by using the multi-scale wave perturbation method, which was validated and examined through extensive 1D numerical simulations. The analytical and numerical results showed the fact that pressure waves are attenuated significantly by both the regular and random pipe diameter non-uniformities along the longitudinal direction, which has been widely observed in the numerical and experimental results in the literature. Meanwhile, the derived results imply that the importance and influence of the wave scattering effect in the pipeline is dependent on the relationship between the incident wave frequency and non-uniform pipe diameter variation frequency. Particularly, the wave scattering induced wave localization length becomes smaller and thus the attenuation of wave envelope is more significant when the ratio of incident wave length and the correlation length of the non-uniformities becomes smaller. As a result, for the specific pipeline system with the existence of pipe diameter non-uniformities, the wave scattering effect becomes critical for the high frequency incident waves, which is, however, common in water hammer flows.

The preliminary results and findings of this study are useful and implicative to both transient theory (transient modelling and analysis) and practice (transient utilization). Firstly, the wave behavior in the aged pipelines might become very complicated due to the unavoidable wave scattering effect such that the design schemes of the system strength from their initial new states may become invalid/inaccurate (overestimated or underestimated) for the regional or global pipe system. Secondly, the complicated wave reflections and amplitude attenuation induced by the wave scattering effect may result in inaccurate predictions or even the invalidity of current transient-based pipe defect detection methods. Finally, but not least importantly, the transient (waterhammer) flow theories, such as friction and viscoelasticity models, which are usually validated and calibrated through the measured data of pressure wave attenuation and reflections from practical systems, may be wrongly represented and explained if the potential wave scattering effect has not been considered or not been well included in the analysis.

It is important to note that the assumption of a relative small extent of pipe diameter non-uniformities has been used in the analytical analysis of this study, where high-order (>2) terms were ignored in the derivation process. With this assumption, clear discrepancies, especially for the regular case of pipe non-uniformities, were observed between the analytical and numerical results obtained in this study. From this perspective, more future work is required to further validate and verify the accuracy and applicability of the derived analytical results in this paper.

Acknowledgments: This research work was supported by (1) the Hong Kong Research Grants Council (RGC) under the projects No. 25200616, No. 15201017 and No. 3-RBAB; (2) the Hong Kong Polytechnic University (PolyU) under projects No. 1-ZVCD and No. 1-ZVGF.

Conflicts of Interest: The author declares no conflict of interest.

References

1. Dolan, M.; Dean, R.G. Multiple longshore sand bars in the Upper Chesapeake Bay. *Estuar. Coast. Shelf Sci.* **1985**, *21*, 721–743. [CrossRef]
2. Belzons, M.; Guazzelli, E.; Parodi, O. Gravity waves on a rough bottom: Experimental evidence of one-dimensional localization. *J. Fluid Mech.* **1988**, *186*, 539–558. [CrossRef]
3. Devillard, P.; Dunlop, F.; Souvillard, B. Localization of gravity waves on a channel with random bottom. *J. Fluid Mech.* **1988**, *186*, 521–538. [CrossRef]
4. Duan, H.F.; Lee, P.J.; Che, T.C.; Ghidaoui, M.S.; Karney, B.W.; Kolyshkin, A.A. The influence of non-uniform blockages on transient wave behavior and blockage detection in pressurized water pipelines. *J. Hydro-Environ. Res.* **2017**, *17*, 1–7. [CrossRef]
5. Duan, H.F.; Lee, P.J.; Tuck, J. Experimental investigation of wave scattering effect of pipe blockages on transient analysis. *Procedia Eng.* **2014**, *89*, 1314–1320. [CrossRef]
6. Anderson, P.A. Absence of diffusion in certain random lattices. *Phys. Rev.* **1958**, *109*, 1492–1505. [CrossRef]
7. Mei, C.C.; Stiassnie, M.; Yue, D.K.P. *Theory and Applications of Ocean Surface Waves, Part 1: Linear Aspects*; World Scientific: Singapore, 2005.
8. McInnis, D.; Karney, B.W. Transients in distribution networks: Field tests and demand models. *J. Hydraul. Eng.* **1995**, *121*, 218–231. [CrossRef]
9. Ghidaoui, M.S.; Zhao, M.; McInnis, D.A.; Axworthy, D.H. A review of waterhammer theory and practice. *Appl. Mech. Rev.* **2005**, *58*, 49–76. [CrossRef]
10. Wylie, E.B.; Streeter, V.L.; Suo, L. *Fluid Transients in Systems*; Prentice Hall, Inc.: Englewood Cliffs, NJ, USA, 1993.
11. Nayfeh, A.H. *Introduction to Perturbation Techniques*; John Wiley & Sons: New York, NY, USA, 1981.
12. Ghosh, S. Dependence in Stochastic Simulation Models. Ph.D. Thesis, Cornell University, New York, NY, USA, 2004.
13. Tung, Y.K.; Yen, B.C.; Melching, C.S. *Hydrosystems Engineering Reliability Assessment and Risk Analysis*; McGraw-Hill Company, Inc.: New York, NY, USA, 2006.
14. Karney, B.W. Energy relations in transient closed-conduit flow. *J. Hydraul. Eng.* **1990**, *116*, 1180–1196. [CrossRef]
15. Duan, H.F.; Ghidaoui, M.S.; Lee, P.J.; Tung, Y.K. Unsteady friction and visco-elasticity in pipe fluid transients. *J. Hydraul. Res.* **2010**, *48*, 354–362. [CrossRef]
16. Brunone, B. Transient test-based technique for leak detection in outfall pipes. *J. Water Resour. Plan. Manag.* **1999**, *125*, 302–306. [CrossRef]
17. Sun, J.L.; Wang, R.; Duan, H.F. Multiple-fault detection in water pipelines using transient time-frequency analysis. *J. Hydroinform.* **2016**, *18*, 975–989. [CrossRef]
18. Wang, X.J.; Lambert, M.F.; Simpson, A.R.; Liggett, J.A.; Vítkovský, J.P. Leak detection in pipeline systems using the damping of fluid transients. *J. Hydraul. Eng.* **2002**, *128*, 697–711. [CrossRef]
19. Duan, H.F.; Lee, P.J.; Ghidaoui, M.S.; Tung, Y.K. Extended blockage detection in pipelines by using the system frequency response analysis. *J. Water Resour. Plan. Manag.* **2011**, *138*, 55–62. [CrossRef]

20. Duan, H.F.; Lee, P.J.; Ghidaoui, M.S.; Tung, Y.K. Leak detection in complex series pipelines by using system frequency response method. *J. Hydraul. Res.* **2011**, *49*, 213–221. [CrossRef]
21. Duan, H.F.; Lee, P.J. Transient-based frequency domain method for dead-end side branch detection in reservoir-pipeline-valve systems. *J. Hydraul. Eng.* **2016**, *142*, 04015042. [CrossRef]
22. Ferrante, M.; Brunone, B. Pipe system diagnosis and leak detection by unsteady-state tests-1: Harmonic analysis. *Adv. Water Resour.* **2003**, *26*, 95–105. [CrossRef]
23. Lee, P.J.; Lambert, M.F.; Simpson, A.R.; Vítkovský, J.P.; Liggett, J. Experimental verification of the frequency response method for pipeline leak detection. *J. Hydraul. Res.* **2006**, *44*, 693–707. [CrossRef]
24. Meniconi, S.; Duan, H.F.; Lee, P.J.; Brunone, B.; Ghidaoui, M.S.; Ferrante, M. Experimental investigation of coupled frequency and time-domain transient test-based techniques for partial blockage detection in pipes. *J. Hydraul. Eng.* **2013**, *139*, 1033–1040. [CrossRef]
25. Kim, S. Impedance method for abnormality detection of a branched pipeline system. *Water Resour. Manag.* **2016**, *30*, 1101–1115. [CrossRef]
26. Sattar, A.M.; Chaudhry, M.H. Leak detection in pipelines by frequency response method. *J. Hydraul. Res.* **2008**, *46*, 138–151. [CrossRef]
27. Duan, H.F.; Lee, P.J.; Ghidaoui, M.S.; Tung, Y.K. Essential system response information for transient-based leak detection methods. *J. Hydraul. Res.* **2010**, *48*, 650–657. [CrossRef]
28. Stephens, M.L. Transient Response Analysis for Fault Detection and Pipeline Wall Condition Assessment in Field Water Transmission and Distribution Pipelines and Networks. Ph.D. Thesis, The University of Adelaide, Adelaide, Australia, 2008.
29. Duan, H.F. Investigation of Factors Affecting Transient Pressure Wave Propagation and Implications to Transient Based Leak Detection Methods in Pipeline Systems. Ph.D. Thesis, The Hong Kong University of Science and Technology, Hong Kong, China, 2011.
30. Blokker, E.J.M.; Vreeburg, J.H.G.; van Dijk, J.C. Simulating residential water demand with a stochastic end-use model. *J. Water Resour. Plan. Manag.* **2010**, *136*, 19–26. [CrossRef]
31. Creaco, E.; Campisano, A.; Franchini, M.; Modica, C. Unsteady flow modeling of pressure real-time control in water distribution networks. *J. Water Resour. Plan. Manag.* **2017**, *143*, 04017056. [CrossRef]
32. Creaco, E.; Pezzinga, G.; Savic, D. On the choice of the demand and hydraulic modeling approach to WDN real-time simulation. *Water Resour. Res.* **2017**, *53*, 6159–6177. [CrossRef]
33. Buchberger, S.G.; Carter, J.T.; Lee, Y.H.; Schade, T.G. *Random Demands, Travel Times and Water Quality in Dead-Ends, Prepared for American Water Works Association Research Foundation*; Report No. 294; American Water Works Association Research Foundation: Denver, CO, USA, 2003.

water

MDPI

Article

A Hybrid Heuristic Optimization Approach for Leak Detection in Pipe Networks Using Ordinal Optimization Approach and the Symbiotic Organism Search

Chao-Chih Lin [ID]

Institute of Environmental Engineering, National Chiao Tung University, Hsinchu 30010, Taiwan;
tom.r1000000@gmail.com; Tel.: +886-3-571-2121 (ext. 55526)

Received: 22 September 2017; Accepted: 22 October 2017; Published: 24 October 2017

Abstract: A new transient-based hybrid heuristic approach is developed to optimize a transient generation process and to detect leaks in pipe networks. The approach couples the ordinal optimization approach (OOA) and the symbiotic organism search (SOS) to solve the optimization problem by means of iterations. A pipe network analysis model (PNSOS) is first used to determine steady-state head distribution and pipe flow rates. The best transient generation point and its relevant valve operation parameters are optimized by maximizing the objective function of transient energy. The transient event is created at the chosen point, and the method of characteristics (MOC) is used to analyze the transient flow. The OOA is applied to sift through the candidate pipes and the initial organisms with leak information. The SOS is employed to determine the leaks by minimizing the sum of differences between simulated and computed head at the observation points. Two synthetic leaking scenarios, a simple pipe network and a water distribution network (WDN), are chosen to test the performance of leak detection ordinal symbiotic organism search (LDOSOS). Leak information can be accurately identified by the proposed approach for both of the scenarios. The presented technique makes a remarkable contribution to the success of leak detection in the pipe networks.

Keywords: leak detection; pipe network; inverse transient analysis (ITA); water distribution networks (WDNs); ordinal optimization approach (OOA); symbiotic organism search (SOS)

1. Introduction

The issue of potable water shortages has become more and more critical in many parts of the world. Water loss is considered as a serious problem in both developed and developing countries [1], and is attracting more and more public concern. Non-revenue water or unaccounted for water is estimated at between 20 to 40% for most countries investigated [2,3]. The inverse transient analysis (IWA) method separates water losses in distribution systems into real and apparent losses. Real losses (leakage) from pipelines or pipe networks may not only cause large economic loss, but can also affect environmental hygiene [4,5]. Leaks may create serious water quality problems, resulting in equipment failure, problematic operations management, errors in pipeline design, and significant costs [4,6–10]. Among the various reasons for water losses, leaks in water distribution networks (WDNs) are considered to be one of the major problems to be solved.

Basically, the leak detection and location techniques can be generally divided into two main categories, steady-state based and transient based. Many steady-state based methods, such as Acoustic Emissions [11,12] and vibration monitoring [13–16], were developed for leak detection in the pressurized pipeline. These steady-state vibro-acoustic based methods have been proven to be quite effective in previous studies. The most important advantage is that the steady-state based methods can provide the high precision results without altering the operating conditions of the system. However, these methods were generally developed based on the specialized hardware, which lead to high costs [17]. On the other hand, transient-based methods (e.g., pressure wave propagation) have been widely used to detect leaks in WDN. A transient event can be generated by a simple operation, such as valve closure and opening [18]. In the transient condition, the leak location, leak orifice size, and frictional losses will affect the head changes in the pipe network when compared with those of the system in steady-state. The major advantage of transient-based methods is that information about such leaks can be efficiently and cost-effectively obtained, because transient waves travel along fluid-filled pipes at high speed [4,5,19]. A large amount of data can be collected by a simple operation in a very short period of time. However, the transient operation may cause some undesired damage or failure in the system if the operation is not properly handled.

Many previous studies have focused on the development of numerical simulations with optimization algorithms for leak detection in a WDN. Moreover, heuristic optimization approaches have also been widely discussed in the community of water supply engineers. Several earlier researchers have used heuristic algorithms with a synthetic WDN to test their ability to detect leaks. Examples are genetic algorithm (GA) [20], hybrid genetic algorithm (HGA) [21], particle swarm optimization (PSO) [22], sequential quadratic programming (SQP) [23], central force optimization (CFO) [24], and simulating annealing (SA) [25]. Vítkovský et al. [20] demonstrated the ability of the GA in lieu of the Levenberg-Marquardt (LM) method used in [26] to identify leaking nodes and pipe friction factors in the same WDN. Kapelan et al. [21] combined the GA and LM method as a hybrid inverse transient model (HGA) to exploit the advantages of combining two methods. The HGA is more stable than LM model, and it is more accurate and faster than the GA model. Jung and Karney [22] showed that both GA and PSO are capable of solving the ITA problem. It was found that PSO is more suitable than GA not only in convergence but also in accuracy. Haghighi and Ramos [24] exploited an ITA-based optimal algorithm, termed CFO, as an inverse problem solver for leak detection in a reference leaking pipe network. CFO exhibited good accuracy for identifying the friction factor and leak location. Recently, Huang et al. [25] presented an ITA-type optimization approach, called LDSA, based on the combination of the transient flow simulation and SA to detect leaks in a laboratory pipeline and a synthetic pipe network. The SA was used to solve the inverse problem with a least-squares criterion objective function.

On the other hand, some studies applied their leak detection techniques to a laboratory pipe system or real WDNs. Ferrante et al. [27] coupled the wavelet analysis with a Lagrangian model to identify the leaks in a laboratory branched pipe system in the Water Engineering Laboratory at the University of Perugia. Their approach memorized the amplitude of each wave and the moment at which it passed the leak, and then identified the leak. Nazif et al. [28] introduced a heuristic method combined the artificial neural network (ANN) model and GA to find the optimal hourly water level variation in a water distribution storage tank for different seasons. Ferrante et al. [29] investigated the relationship between leak geometry and detectability within steady-state and transient based techniques. They used the experimental tests to demonstrate the effect of higher system pressure for leak detectability in steady-state and transient conditions. Casillas et al. [30] introduced a sensor placement approach based on either GA or PSO to detect leaks in WDNs in Hanoi and in Limassol. The results showed that PSO obtains results faster than GA, and it is very effective for smaller WDN or with fewer sensors. However, the GA provided better placement solutions with higher efficiency for larger networks or more sensors. Meniconi et al. [31] used a Lagrangian model simulating pressure wave propagation to evaluate the pipe pressure wave speed and to locate the possible anomalies by coupling GA and wavelet analysis, respectively. Their approach was further executed in a part of the WDN of Milan for providing the diagnosis of the pipe system. Lee et al. [32] integrated the advantage of the methods of cumulative sum and wavelet transform to effectively detect the sudden pressure changes of WDN. The pressure data obtained from the real burst accident were used to verify their burst detection and location algorithm. Moreover, many other studies were devoted to the development of optimization approaches for leak detection in real cases [19,33,34].

This paper focuses on solving the leak detection problem, as well as the number of leaks, their location, the value of C_dA (discharge coefficient multiplies area opening of the orifice) in the pipe network, with optimum transient perturbations. A hybrid heuristic approach, called leak detection ordinal symbiotic organism search (LDOSOS), is developed based on the ordinal optimization algorithm (OOA) and symbiotic organism search (SOS) for automatically determining leak information in a leaking pipe network. In order to examine the performance of the proposed approach, two synthetic leaking scenarios with different pipe network configurations are considered. The ability of convergence compared with different optimization algorithms pertaining to the detection results is addressed in first scenario. Furthermore, the use of the optimum transient generation is demonstrated in the second scenario.

2. Methodology

This section deals with the mathematical background of forward flow simulation models, and includes steady-state and transient flow simulations in the pipeline network. The procedure of transient generation point selection is introduced in this section. Brief descriptions of OOA and SOS, and the combination of a forward flow model with OOA and SOS for the leak detection are also described.

2.1. Flow Simulation Model

Yeh and Lin [35] developed a numerical approach, termed PNSA, for estimating the steady-state nodal head and flow rate for any given pipe network. This approach uses the SOS in lieu of the SA in PNSA to solve an optimal water head distribution and the nodal flow rates in a network. The flowchart of PNSOS is shown in Figure 1.

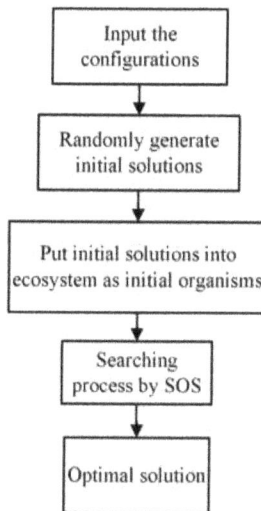

Figure 1. Flowchart of pipe network analysis model (PNSOS).

The Hazen-Williams equation is used to express the relationship between the flow rate and head loss for each pipe [36,37]. The loss coefficient (K_{ij}) in the Hazen-Williams equation for a pipe is defined as [38]:

$$K_{ij} = \frac{10.66667 \cdot L_{ij}}{C_{ij}^{1.851852} \cdot D_{ij}^{4.870370}} \tag{1}$$

where ij is the variable defined from node i to node j, L_{ij} is the length (m) of the pipe, C_{ij} is the Hazen-Williams friction coefficient depending on the pipe material [36], and D_{ij} is the pipe diameter (m). Based on Equation (1), the flow rate Q_{ij} in each pipe can be expressed as

$$Q_{ij} = \left(\frac{\Delta H_{ij}}{K_{ij}}\right)^{0.54} \tag{2}$$

where ΔH_{ij} is the frictional head loss in a pipe. The equation of mass conservation at node i can be written as

$$MC_i = \sum_{j=1}^{nn} Q_{ij} + QI_i \tag{3}$$

where nn is the number of total neighbor nodes to node i and QI_i is the demand or the source at node i. The flow rate is positive for flowing out of node i and negative for flowing into node i, while QI_i is positive for inflow and negative for outflow. Therefore, the objective function used in the PNSOS is defined as

$$Minimize \sum_{i}^{nd} (MC_i)^2 \tag{4}$$

where nd is the total number of nodes needed to estimate the nodal heads and flows in a network system.

2.2. Hydraulic Transient Model

Since the steady-state water head and flow rate for each node in the network has been estimated, the hydraulic transient can then be evaluated from the following momentum equation and continuity equation [25,38]:

$$momentum : gA\frac{\partial H}{\partial x} + \frac{\partial Q}{\partial t} + \frac{f}{2DA}Q|Q| = 0 \tag{5}$$

$$continuity : \frac{\partial H}{\partial t} + \frac{a^2}{gA}\frac{\partial Q}{\partial x} = 0 \tag{6}$$

where g is gravity acceleration, A is pipe cross-sectional area, H is hydraulic head, x is distance along the pipe, Q is volume rate of flow, t is time, D is diameter of pipe, a is the wave speed, and f is the friction factor, which can be described by steady, quasi-steady, or unsteady state conditions. Many approaches have been presented for unsteady friction modeling, such as weighting function-based model [39,40] and the instantaneous acceleration-based model [41,42]. Since this work is not experimentally oriented, the friction factor is considered to be steady. Moreover, the elastic behavior is assumed for pipe material and then the effects of viscoelasticity are not considered. The method of characteristics (MOC) is a common technique for solving hydraulic transient equations [43]. Equations (5) and (6) are then transferred to the following two sets of ordinary differential equations along the characteristic lines $(dx/dt = \pm a)$:

$$C^+ : \begin{cases} \frac{dH}{dt} + \frac{a}{gA}\frac{dQ}{dt} + \frac{fa}{2gDA^2}Q|Q| = 0 \\ \frac{dx}{dt} = +a \end{cases} \tag{7}$$

$$C^- : \begin{cases} -\frac{dH}{dt} + \frac{a}{gA}\frac{dQ}{dt} + \frac{fa}{2gDA^2}Q|Q| = 0 \\ \frac{dx}{dt} = -a \end{cases} \tag{8}$$

where dx is the distance differential and dt is the time differential. With appropriate initial and boundary conditions, the finite difference method is applied to approximate Equations (7) and (8), and simultaneously calculate the transient head and flow rate in the grid points in the network. Boundary conditions often consist of known pressures and flow rates at the pipe network, which are generally reservoirs, connections, valves, pumps, and leakages. The leaks in the networks may be represented by orifices and expressed as [44]:

$$Q_L = C_d A \sqrt{2g\Delta H} \tag{9}$$

where Q_L is the volumetric flow rate of the leak, $C_d A$ is the discharge coefficient times the effective area of orifice, and ΔH is the head loss caused by the orifice.

2.3. Excitation Procedure for Transient Generation

For leak detection in pipe networks using ITA, the problem of the generation of pressure waves is a very important issue of transient test-based techniques. The worst transient fluctuations are usually required because the severity of transient fluctuations is related to the efficacy of applying ITA. The possibility of the use on hydraulic transient for leak detection have been experimentally validated in laboratory conditions by using the Portable Pressure Wave Maker (PPWM) device [45,46], and further applied in field tests [47–49]. Generally, valve operation is a common way to create a change in the outflow discharge for obtaining transient fluctuations [18]. Vítkovský et al. [19] suggested that a valve, as a transient generator, should be closed very quickly with a small discharge magnitude. Thus, the optimal transient generation location and operation parameters for generating the optimal transient perturbations should be determined while using ITA. Valve operation is also termed excitation, as shown in Figure 2. Note that Figure 2 are modified from [7]. The two major valve operation parameters are the duration of change y and the amount of change z for controlling the intensity of transient fluctuations. On the other hand, another parameter is the side curves of the excitation, which are usually non-linear and depend on valve type and its operation. The side curve is considered to be linear for convenience in numerical verification. Haghighi and Shamloo [50] noted that the intensity index, E, is the accumulative energy of transient heads evaluated from function of parameters y and z for each candidate point (i.e., node in pipe network). To obtain the optimum transient perturbations in the pipe network, the intensity index E should be maximized while using each candidate node as a transient generation point with corresponding operation parameters. The objective function for determining the optimal parameters y and z for each candidate node with maximum accumulative energy is described as [50]:

$$Maximize : E(y,z) = \sum_{t_s=1}^{nt} \sum_{i=1}^{nd} \left| \frac{H_{it_s} - H_{i0}}{H_r} \right| \times \prod_{t_s=1}^{nt} \prod_{i=1}^{nd} Pe_{it_s} \tag{10}$$

$$Pe_{it_s} = \begin{cases} \frac{H_{it_s}}{H_{min}} & H_{min} > H_{it_s} \\ 1 & H_{min} \le H_{it_s} \le H_{max} \\ \frac{H_{max}}{H_{it_s}} & H_{max} < H \end{cases} \tag{11}$$

where E is the overall energy of heads, y is the duration of outflow change, z is the consumption of the change, H_{i0} is the initial steady head, H_{it_s} is the piezometric head at node i at time step t_s, nt is number of transient modelling time steps, H_r is the reservoir head, Pe_{it_s} is the penalty factor to impose pressure constraints, H_{max} and H_{min} are, respectively, the maximum and the minimum permissible heads in system.

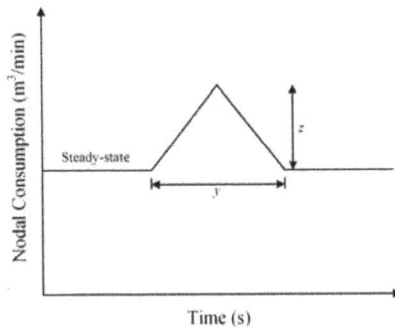

Figure 2. Transient excitation at operation point.

2.4. Ordinal Optimization Approach

The key idea behind an OOA is to rank the values of the objective function in ascending order. In reality, an OOA is much easier than the process of searching for a global optimal solution [51]. Two major procedures are used in an OOA: ordinal comparison and goal softening procedures. The first procedure looks for the relative relationship between each solution in order to find better solutions; the second is to determine a reliable and satisfactory solutions instead of directly evaluating the optimal solution in a complex optimization model. Thus, the optimum solution could be obtained from a feasible solution space. The searching process in the procedure would be reduced. To obtain the top optimum proportion solutions, Lau and Ho [52] noted that the top 5% of solutions can be treated as a reliable criterion with a very high probability ($\geq$0.95) of obtaining satisfactory solutions.

2.5. Symbiotic Organism Search (SOS)

Cheng and Prayogo [53] developed a new meta-heuristic algorithm, termed the symbiotic organism search (SOS) algorithm, which was inspired by actual biological interactions. The biological interaction between two organisms or species in a symbiotic system can be generally categorized into three types: mutualism, commensalism, and parasitism. Mutualism denotes that the relationship is beneficial each organism. Commensalism is the relationship in which one organism benefits from the other without affecting it, while parasitism represents a non-mutual relationship in which one species benefits at the expense of the other. The characteristics of SOS are similar to other population-based meta-heuristic algorithms, such as GA and PSO. SOS shares the same following four features: (1) the control parameters, such as the initial population size and the maximum number of iterations should be appropriately settled; (2) the population of organisms which contains candidate solutions are used to determine the global optimal solution in the search space during the searching process; (3) the candidate solutions are used to guide the searching process; and (4) a selection mechanism to retain better good solutions and to abandon poor solutions. Furthermore, the SOS algorithm is a parameter-free technique, and only the control parameters, such as initial population size and maximum number of iteration are required. Algorithm-specific parameters for other competing algorithms might increase computational time and produce local optimal solutions. Hence, SOS has been successfully applied to various types of problem such as the construction management [54], work shift [55], and optimal reservoir operation [56].

Figure 3 shows the flowchart of an SOS algorithm. SOS starts with an initial population named the ecosystem. A group of organisms is randomly generated in the feasible solution domain and then added into the initial ecosystem. Each organism is considered as a candidate solution (CAS) for the corresponding problem with a certain objective function value (OFV). The search procedure begins when the initial ecosystem is set up. In the searching procedure, each organism will get to benefit or be harmed from continuously interacting with another organism in three different types/forms of symbiosis explained above.

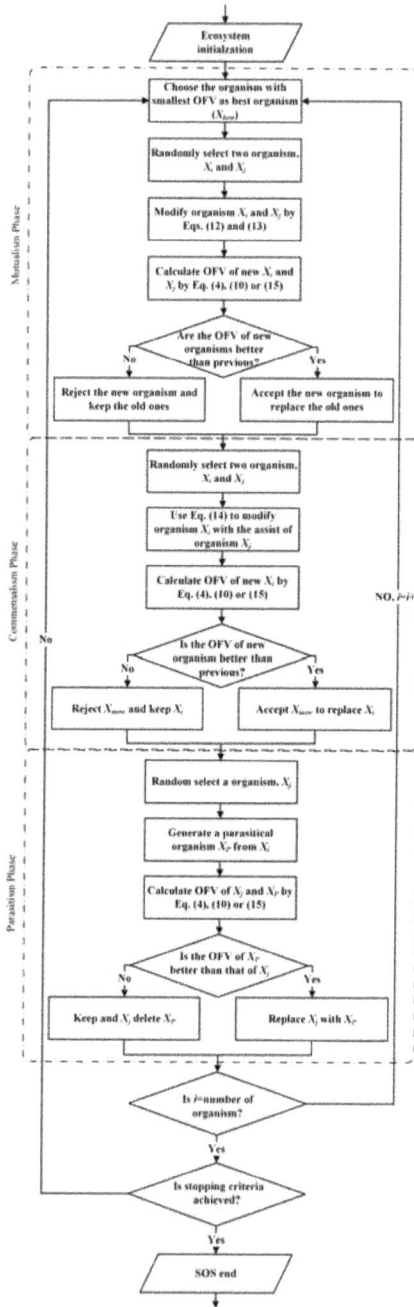

Figure 3. Flowchart of symbiotic organism search (SOS).

2.5.1. Mutualism State

In mutualism, organism X_i and X_j are randomly selected from an ecosystem for interaction. Both organisms interact to mutual benefit in order to increase chances of survival in the ecosystem. Hence, the new candidate solutions for X_i and X_j are modified based on the mutualistic mechanism between X_i and X_j, and is illustrated as:

$$X_{inew} = X_i + RD_M \times (X_{best} - MV \times BF_1) \tag{12}$$

$$X_{jnew} = X_j + RD_M \times (X_{best} - MV \times BF_2) \tag{13}$$

where RD_M is a vector of random numbers range from 0 to 1, X_{best} is the current best organism with the best OFV in the ecosystem, MV is the mutual vector defined as $MV = (X_i + X_j)/2$, and BF_1, and BF_2 are the benefit factors randomly as either 1 or 2.

2.5.2. Commensalism State

Similar to mutualism, the organism X_j is randomly chosen from the ecosystem to interact with another random organism X_i. X_i gains benefit from X_j, but X_j is not affected by this relationship. The new X_i can be modified as:

$$X_{inew} = X_i + RD_C \times (X_{best} - X_j) \tag{14}$$

where RD_C is the vector of random numbers range from -1 to 1.

2.5.3. Parasitism State

In parasitism, a parasitical organism X_P is generated by cloning and mutating it from X_i in random dimensions, using a random number with a range between given lower and upper bounds. A parasite X_P tries to replace the random host organism X_j. Both X_P and X_j are then evaluated for their fitness (OFVs). If the parasite has a better OFV, the host organism will be immediately replaced by the parasite. If the OFV of X_j is better, then X_j will survive and kill the parasite X_P.

2.6. Development of LDOSOS

Abhulimen et al. [57] recorded that pressure measurements are more sensitive than volume measurements for leak detection. Hence the objective function of ITA is defined in terms of pressure head as:

$$Minimize \sum_{j=1}^{m} \sum_{i=1}^{n} (H_{oij} - H_{sij})^2 \tag{15}$$

where m is the total number of observation points in the network, n is the number of observations made at an observation point, and H_{oij} and H_{sij} are ith observed, and simulated heads at the observation point j, respectively. The LDOSOS can automatically determine the leak information based on the minimization of Equation (15). The procedures of LDOSOS are summarized in Figure 4. The LDOSOS can be used to determine the optimal leak location, leak size, and the number of leaking pipes simultaneously. The procedure for detecting the leaks using LDOSOS is given below:

1. Import the network configurations.
2. Use SOS to determine the optimal transient generating point with its corresponding operating parameters (i.e., y: duration of outflow change, z: amount of nodal consumption variation) by maximizing Equation (10). The optimum solution is obtained when the OFV of Equation (10) does not change within four iterations.
3. For the pipe sifting procedure in OOA, successively generate a temporary leak which is located at the middle of each pipe; the location and C_dA of the orifices are treated as temporary solutions.
4. Since the temporary leak solutions are available, PNSOS is then used to calculate the steady-state nodal heads and flow rates in the network.

5.   Generate a hydraulic transient event at the optimal generation point and apply Equations (7) and (8) to simulate the head distribution in the network.
6.   Apply Equation (15) to calculate the temporary OFV for the temporary solution of each pipe.
7.   Arrange all of the pipes according to the values of temporary OFVs. A quarter of pipes with smaller OFVs are chosen as candidate pipes (CAPs). Only the CAPs will be used in the further steps.
8.   Randomly generate 200 CASs with the information of a leaking pipe, leak location and C_dAs of the orifices, and calculate their OFVs. The top 5% CASs would then be selected for the next step.
9.   Consider the best 5% CASs as the initial organisms for the ecosystem.
10. Execute a searching process using SOS. In general, mutualism and commensalism states are used to guide the organisms toward the current best organism, and the parasitism state is applied to avoid the organism becoming stuck in a local optimal solution.
11. Check whether the optimization process satisfies the stopping criterion or not. If so, the LDOSOS is then terminated; otherwise, the searching process goes on and back to the tenth step.

Note that the first stopping criterion is defined as the absolute value of the difference between the two successive optimal OFVs of X_{best} which are all less than 10^{-4} within four iterations. The second criterion is that the iteration reaches 10,000 times.

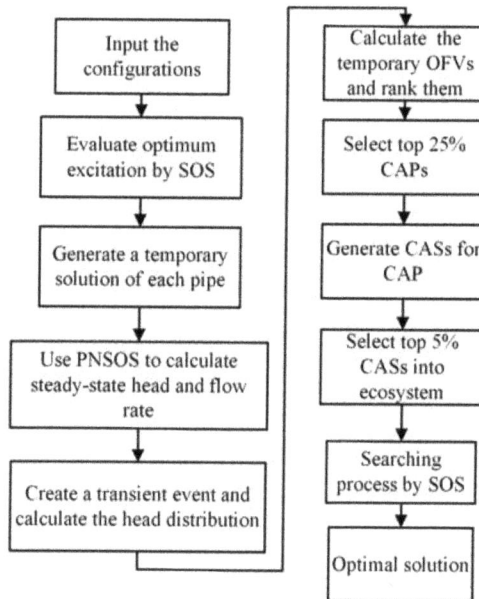

Figure 4. Flowchart of leak detection ordinal symbiotic organism search (LDOSOS).

3. Results and Discussion

3.1. Pipe Networks Setting

To test the applicability of LDOSOS in leak detection, two scenarios with different synthetic pipe network systems adopted from the literature are used. These two pipe networks are designed based on the concept of district metering areas (DMA), in which inflow and outflow are monitored. User demands and leaks are assumed as constants and can be separated through continuous observation of mass conservation of flow measurements.

The first scenario (S1) is the one presented in [25], with a layout referred to as pipe network A, as shown in Figure 5. The configuration of this network consists of eight pipes, six nodes, two potential leaks, and one downstream valve, and the characters "N", "P", and "L" represent the node, pipe, and leak, respectively. The pipe diameter varies from 250 to 500 mm, while the pipe length varies from 1000 to 1250 m. Nodal consumptions are considered as 0 for all nodes. N1 is considered as a reservoir with 120 m constant-head, and downstream of P8 is a valve with a discharge flow rate of 5 cubic m per min (m^3/min). Two potential leaks, denoted as L1 and L2, are both in P6 and are at 300 and 310 m from N5, respectively, with the same C_dA = 0.00025 m^2 and same Q_L = 0.5 m^3/min. The downstream valve is the only outflow of pipe network A. Hence, the optimum operation point is located at the valve. In order to compare the proposed approach with the work of [25], the operation parameters y and z are fixed to 1 s and -5 m^3/min, respectively, for the simulation of a sudden closure of the valve. The characteristics of the nodes and pipes of the pipe network A are listed in Table 1.

In order to solve the leak detection problems in a real WDN, scenario 2 (S2), with a synthetic pipe network B presented in [35] is considered in this study. The pipe properties are listed in Table 2. Figure 6 shows the configuration of the network which consists of 11 pipes, nine nodes, seven continually outflow points, and two potential leaks. The pipe length ranges from 400 m to 1250 m, and the diameter ranges 200 to 405 mm. The N1 is the water supply node with a constant inflow rate of 25 m^3/min and constant head of 120 m. Seven outflow nodes N2, N3, N4, N5, N6, N8, and N9 continuously discharge 5, 2.5, 2.2, 2.2, 2.5, 5 and 5 m^3/min, respectively. The leak L1 is in P11 and 300 m away from N3, while the leak L2 is at the middle of P7 and 250 m away from N6. The C_dA and Q_L of L1 are 0.00025 m^2 and 0.5 m^3/min and C_dA and Q_L of L2 are 0.0001 m^2 and 0.1 m^3/min. In pipe network B, N2, N8, and N9 are available as candidate transient generation points with larger discharges. The optimal operation point (i.e., N2, N8 and N9) and the optimal parameters, y and z, must be determined.

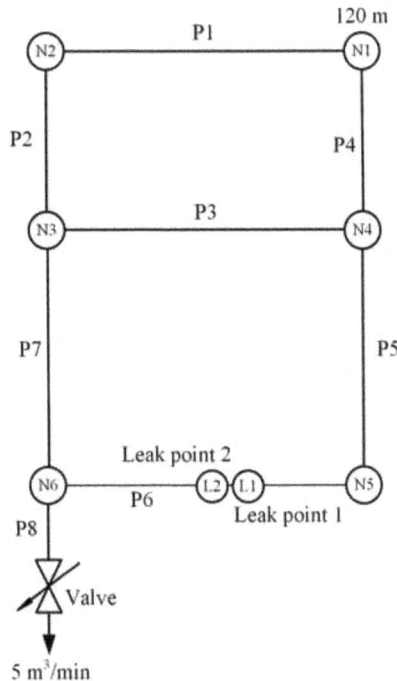

Figure 5. Configuration of synthetic pipe network A.

Table 1. The characteristics of the synthetic pipe network A.

Pipe	Node		Diameter (mm)	Length (m)
Number	From	To		
P1	N1	N2	305.0	1000.0
P2	N2	N3	305.0	1000.0
P3	N3	N4	250.0	1100.0
P4	N1	N4	405.0	1250.0
P5	N4	N5	355.0	1000.0
P6	N5	N6	305.0	1100.0
P7	N3	N6	305.0	1250.0
P8	N6	Valve	500.0	1000.0

Table 2. The characteristics of the synthetic pipe network B.

Pipe	Node		Diameter (mm)	Length (m)
Number	From	To		
P1	N1	N2	305.0	1000.0
P2	N2	N3	305.0	1000.0
P3	N3	N4	250.0	1100.0
P4	N1	N4	405.0	1250.0
P5	N4	N5	200.0	500.0
P6	N5	N6	400.0	400.0
P7	N7	N6	200.0	500.0
P8	N4	N7	355.0	400.0
P9	N7	N8	355.0	600.0
P10	N8	N9	305.0	1100.0
P11	N3	N9	305.0	1250.0

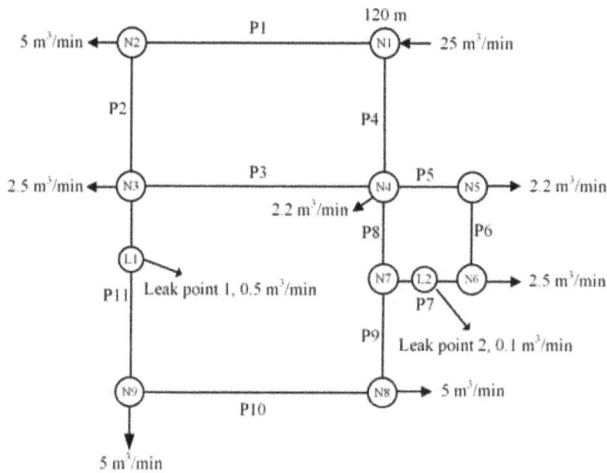

Figure 6. Configuration of synthetic pipe network B.

In both of the scenarios, all of the pipes are old cast-iron pipes and the Hazen-Williams friction coefficient is chosen as 100 [18]. The wave speed a in pipe is thus assumed to be 1000 m/s [18]. The total simulation time is 30 s. The time interval (dt) is chosen as 0.005 s for Sections 3.2 and 3.3 On the other hand, dt is chosen as 0.005 and 0.05 s for various cases for Section 3.4. In the following sections, the application of LDOSOS to the leak detection in the simple pipe network and WDN is assessed. Moreover, the error analysis is also addressed. The simulations for the leak detection are performed on a personal computer with Intel 3.3 G E3-1230v2 CPU and 32 GB RAM.

3.2. Applicability of LDOSOS to Leak Detection

In this section, the proposed method is used to demonstrate the convergence and efficiency of leak detection in S1. The results are then compared with leak detection symbiotic organism search (LDSOS) and LDSA approaches. LDSOS is a simplified optimization approach similar to LDOSOS, but without the elimination procedure OOA. All of the pipes are treated as candidate pipes in LDSOS. Furthermore, 10 initial solutions are randomly generated from a feasible solution domain as the initial organisms for the ecosystem of LDSOS. On the other hand, LDSA is a heuristic technique based on SA for solving leak detection problems in pipe network A. In the leak detection process, LDSA randomly generated a trial solution in the network then adopted the least squares method with the minimal OFV to find the possible leak information, including the location and the value of C_dA. However, the searching space in SA may be enormous and required a large computing time to find the optimal solution. By contrast, OOA is adopted in the LDOSOS to sift through the searching space. Based on the search procedure, P6 and P7 in pipe network A are first sifted and ranked as the CAPs; the top 5% CASs from CAPs with different leak information are sifted by calculating all CASs' OFV (Equation (15)). More specifically, the top 10 best CASs (200 × 0.05 = 10) are sifted as the initial organisms for LDOSOS.

Figure 7 shows the predicted heads versus time at the valve, based on the intact network and the network with two leaks. The predicted head distributions by LDSA, LDSOS, and LDOSOS display very similar transient patterns; however, their efficiencies are quite different in obtaining optimal results. Table 3 gives the results of leak detection for S1 from LDSA, LDSOS, and LDOSOS. The results show that LDOSOS is capable to detect leaks that are close to each other. The LSDA takes 372 min and 9815 iterations, while LDSOS required 120 min and 3481 iterations to obtain the results. The efficiency of LDSOS is better than LDSA. Moreover, LDOSOS takes 50 min and 1469 iterations to complete the detection process and to obtain the optimal locations and C_dAs. The convergence and efficiency of LDOSOS is greatly enhanced as a result of the sifting procedure OOA. The computing efficiency of LDOSOS is approximate 86 and 58% better than that of LDSA and LDSOS. The LDOSOS obtained the optimum solution after about 1500 iterations which is significantly less than the other two approaches. Obviously, the performance of LDOSOS is much more efficient than the other two algorithms.

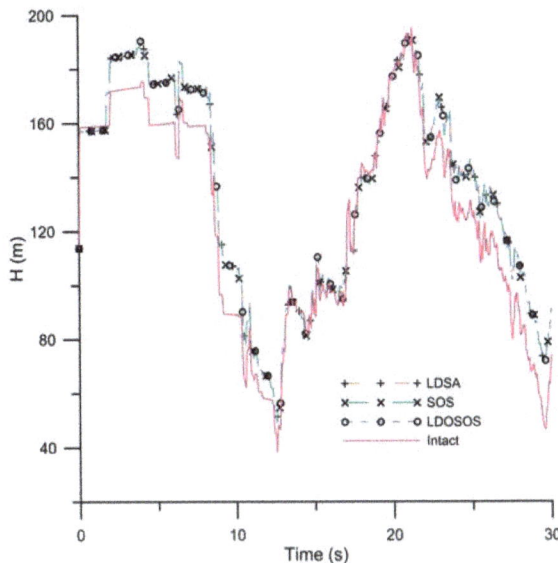

Figure 7. The simulated head distributions at valve for leak detection using various approaches.

Table 3. Actual leak information and predicted results from three approaches for S1.

Header	L1			L2			CPU Time (min)	Iterations
	Pipe No.	Location (m)	$C_dA \times 10^{-4}$ (m^2)	Pipe No.	Location (m)	$C_dA \times 10^{-4}$ (m^2)		
Actual	6	300	2.5	6	310	2.5	-	-
LDSA	6	300	2.5	6	310	2.5	372	9815
LDSOS	6	300	2.5	6	310	2.5	120	3481
LDOSOS	6	300	2.5	6	310	2.5	50	1469

3.3. Leak Determination in WDN with Optimal Transient Operation

The proposed approach is further used to demonstrate the performance of LDOSOS on leak detection in the WDN. The nodes with several continuity nodal consumptions in the WDN system are considered as the discharge outlet for supplying water to users. To obtain the optimal transient fluctuations in the pipe network, a best generation point is first selected from the feasible locations (i.e., N2, N8, and N9). The initial outflow of the feasible locations is then changed to the triangular form, as shown in Figure 2. The optimal operation parameters, y and z are optimized to generate the worst transient fluctuations, while $1 < y < 10$ s and $-5 < z < 5$ m^3/min, with all of the nodal heads being greater than 0 and less than 160 m. The maximum value of Equation (10) with the optimal parameters for each candidate node is listed in Table 4. The overall maximum value reaches to 1978, while using N8 as the transient generation point. The optimum operation duration time y is determined as 2.7 s, while the discharge change z is estimated as -2.58 m^3/min. The maximum transient energy of N8 is higher than that of N2 and N9. Thus, the N8 with its relevant parameters is the best point to generate the transient fluctuations for S2. Furthermore, N8 and N9 are considered as the observation and generation point in LDOSOS to compare the leak detection results using different operation points with its relevant parameters. The transient pressures are sampled for 30 s after the excitation.

Table 4. Optimum operation parameters for candidate nodes.

Node	y (s)	z (m^3/min)	E_{Max}
N2	7.2	-5.0	1239
N8	2.7	-2.58	1978
N9	3.6	-3.22	1843

Figure 8a,b, respectively, show the optimal temporal head distributions observed at N8 and N9 for pipe network B with two potential leaks. The optimal simulated temporal heads at N8 and N9 are precisely reconstructed as compared to the observations. Apparently, the oscillation of transient fluctuation at N8 is heavier than that of N9. Table 5 shows the optimal solutions of the estimated leak information predicted by using N8 or N9 as the transient generation point in LDOSOS. The predicted results clearly demonstrates the ascendancy of the optimal location (N8) as the best transient generation point. Although the leak sizes are quite different to each other, the leaks are successfully detected by the proposed approach in the short time. It takes about 60 min and 1843 iterations to obtain the optimal solution by using N8 to generate transient fluctuations. On the other hand, it took about 64 min and 1967 iterations for using N9 as the generation point. The efficiency between two generation points is not obvious. However, the predicted results of L2 yields about a 3.6% relative error in leak area of L2 while using N9 as the generation point. The accuracy is better when using the optimal generation point N8 with its relevant parameters, which indicates that LDOSOS could be used for leak detection, even though the node continues to supply water to other purposed/use. The results show that the accuracy of the proposed approach increased when using the optimal transient generation point and the optimal operation parameters.

Table 5. Estimated leak information for S2 using N8 or N9.

Header		N8						N9				
		L1			L2			L1			L2	
	Pipe No.	Location (m)	$C_dA \times 10^{-4}$ (m^2)	Pipe No.	Location (m)	$C_dA \times 10^{-4}$ (m^2)	Pipe No.	Location (m)	$C_dA \times 10^{-4}$ (m^2)	Pipe No.	Location (m)	$C_dA \times 10^{-4}$ (m^2)
Actual	11	300	2.5	7	250	1	11	300	2.5	7	250	1
LDOSOS	11	300	2.499	7	250	1	11	300	2.479	7	250	0.964

Note that the computation time are 60 min and 64 min for using N8 and N9, respectively.

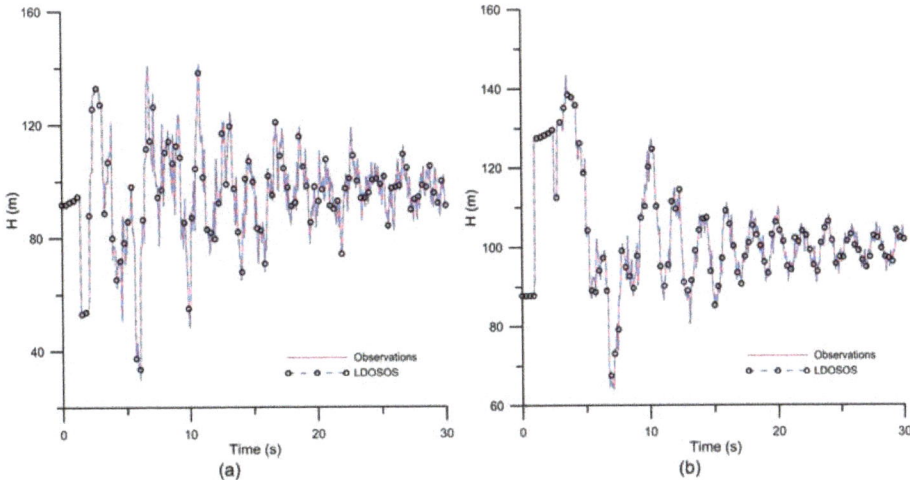

Figure 8. Simulated head distributions observed at (a) N8 and (b) N9.

3.4. Measurement Error Analysis

To evaluate the accuracy of the results predicted by proposed approach, two error criteria, standard error of the estimate (SEE) and mean error (ME), are used to assess the influence of measurement error to LDOSOS for leak detection. SEE is the square root of the sum of squared errors between the simulated and computed heads divided by the number of degrees of freedom, which equals the number of observed data points minus the number of unknowns. On the other hand, ME is the average of the sum of errors between the simulated and computed heads. More details on the use of ME and SEE are provided in [58].

Two cases for each scenario are considered in Analysis: 1. 6001 simulated data with high frequency of 0.005 s are used; 2. 601 data with low frequency of 0.05 s with measurement errors are applied. The white noise with zero mean and 0.01 standard deviation are generated and added to those low frequency data to represent the measurement errors. Note that N8 is chosen as the transient generation point for the mentioned cases of S2. Table 6 shows the results of the error analysis for leak detection in those two scenarios with two cases. The error analyses demonstrate slight differences among those two cases in each scenario. The MEs in each Case 1 of those two scenarios are one order smaller than those of Case 2. Furthermore, SEEs in each Case 2 are two orders larger than those of Case 1. Such a result is consistent with the magnitude of noise added to the original simulated data. The results of Case 2 of both scenarios demonstrate that the influence of measurement error is insignificant, implying that LDOSOS can precisely predict the leak information even the observations contained measurement errors.

Table 6. The prediction errors of leak detection in two scenarios.

	Prediction Errors	
Scenario 1	ME (m)	SEE (m)
Case 1	-1.76×10^{-6}	4.11×10^{-4}
Case 2	1.35×10^{-5}	5.63×10^{-2}
Scenario 2	ME (m)	SEE (m)
Case 1	6.13×10^{-6}	8.13×10^{-4}
Case 2	7.41×10^{-5}	6.58×10^{-2}

4. Conclusions

This study presents a hybrid heuristic optimization approach, termed LDOSOS, to detect leaks in pipe networks based on the combination of hydraulic transient flow simulations, OOA and SOS. Leaks in the pipes are represented by orifices. The PNSOS is used to determine the head at each node and the flow rate at each pipe before transient generation. To enhance the efficiency and accuracy of LDOSOS, a procedure is used to estimate the optimal transient generation point and its operation parameters. Moreover, the OOA is applied to sift the searching space. The top 25% CAPs with smaller OFVs are sifted, and the top 5% CASs are selected as the initial organisms for the ecosystem in SOS algorithm. After obtaining the OFVs of the initial organisms, LDOSOS identifies the leaks using three symbiotic states to guide the organisms forward and move toward the variable best organism (X_{best}) with the smallest OFV.

A pipe network with two leaks in the same pipe is first used to verify the ability of the proposed approach for leak detection. The temporal head distribution, leak locations, and C_dAs are accurately predicted and agreed well with those from the other two algorithms. When these three algorithms are compared, LDOSOS had an approximately 86 and 58% better computing efficiency than LDSA and LDSOS. Moreover, the LDOSOS only takes about 50 min and 1469 iterations to obtain the optimal solution, implying that the searching space is largely reduced by the elimination procedure OOA, and the solutions quickly converged to the optimal solution during iterations. The simulation results show that LDOSOS not only greatly enhances the computation efficiency but also increases the convergence ability. On the other word, LDOSOS significantly outperforms LDSA and LDSOS.

Using ITA for leak detection, the worst transient fluctuations with drastic changes are theoretically essential for good performance of ITA. The optimal transient energy estimation is used in the proposed approach to obtain the optimal transient generation point and its relevant parameters. On the basis of DMA, a WDN is used in this study consisting of 11 pipes, nine nodes, with several continuously outflows, and two potential leaks in two different pipes. The SOS is first used to optimize the operation parameter of three feasible transient generation points. N8 and N9 are then selected to compare the performance of LDOSOS using different transient generation points. The leak information is accurately predicted by LDOSOS with fairly high efficiency by using the best generation point N8 and relevant parameters. The detected results adopting the suboptimal generation point N9 yield a relative error of about 3.6% for the predicted leak area of L2. The results show that the optimal generation point and operation parameters provide more reliable results than other candidate generation points.

In summary, it has been demonstrated from the simulations that LDOSOS has the ability to detect the number of leaking pipe, location of the leak, and their size in a synthetic simple pipe network and a synthetic WDN. The proposed approach speeds up the ITA convergence and improves the reliability of the results. Moreover, the head at measurement point can be precisely computed by LDOSOS even the observations contained measurement errors. However, real drink water systems usually have more complicated pipe and system properties. Pipe characteristics in real situations will also be different if different pipe materials are used. We expect that we will be able improve the capability of LDOSOS to apply to real situations and provide a sound method to detect leaks in pipes in future work.

Acknowledgments: The author appreciates the three anonymous reviewers for comments and thanks Hund-Der Yeh for providing the research resource used in this study. This study was partly supported by the grants from Taiwan Ministry of Science and Technology under the contract number MOST106-2221-E009-066.

Conflicts of Interest: The author declares no conflict of interest.

References

1. Thornton, J.; Sturm, R.; Kunkel, G. *Water Loss Control*, 2nd ed.; Mcgraw-Hill: New York, NY, USA, 2008.
2. Lambert, A.O. International report: Water losses management and techniques. *Water Sci. Technol. Water Supply* **2002**, *2*, 1–20.
3. Kanakoudis, V.; Tsitsifli, S. Urban water services public infrastructure projects: Turning the high level of the NRW into an attractive financing opportunity using the PBSC tool. *Desalin. Water Treat.* **2012**, *39*, 323–335. [CrossRef]
4. Colombo, A.F.; Lee, P.; Karney, B.W. A selective literature review of transient-based leak detection methods. *J. Hydro-Environ. Res.* **2009**, *2*, 212–227. [CrossRef]
5. Puust, R.; Kapelan, Z.; Savic, D.A.; Koppel, T. A review of methods for leakage management in pipe networks. *Urban Water J.* **2010**, *7*, 25–45. [CrossRef]
6. Al-Khomairi, A. Leak detection in long pipelines using the least squares method. *J. Hydraul. Res.* **2008**, *46*, 392–401. [CrossRef]
7. Shamloo, H.; Haghighi, A. Optimum leak detection and calibration of pipe networks by inverse transient analysis. *J. Hydraul. Res.* **2010**, *48*, 371–376. [CrossRef]
8. Covelli, C.; Cozzolino, L.; Cimorelli, L.; Della Morte, R.; Pianese, D. A model to simulate leakage through joints in water distribution systems. *Water Sci. Technol. Water Supply* **2015**, *15*, 852–863. [CrossRef]
9. Covelli, C.; Cimorelli, L.; Cozzolino, L.; Della Morte, R.; Pianese, D. Reduction in water losses in water distribution systems using pressure reduction valves. *Water Sci. Technol. Water Supply* **2016**, *16*, 1033. [CrossRef]
10. Covelli, C.; Cozzolino, L.; Cimorelli, L.; Della Morte, R.; Pianese, D. Optimal location and setting of prvs in wds for leakage minimization. *Water Resour. Manag.* **2016**, *30*, 1803–1817. [CrossRef]
11. Juliano, T.M.; Meegoda, J.N.; Watts, D.J. Acoustic emission leak detection on a metal pipeline buried in sandy soil. *J. Pipeline Syst. Eng. Pract.* **2013**, *4*, 149–155. [CrossRef]
12. Martini, A.; Troncossi, M.; Rivola, A. Leak detection in water-filled small-diameter polyethylene pipes by means of acoustic emission measurements. *Appl. Sci.* **2017**, *7*, 2. [CrossRef]
13. Martini, A.; Troncossi, M.; Rivola, A. Automatic leak detection in buried plastic pipes of water supply networks by means of vibration measurements. *Shock Vib.* **2015**, *2015*. [CrossRef]
14. Martini, A.; Troncossi, M.; Rivola, A. Vibroacoustic measurements for detecting water leaks in buried small-diameter plastic pipes. *J. Pipeline Syst. Eng. Pract.* **2017**, *8*. [CrossRef]
15. Yazdekhasti, S.; Piratla, K.R.; Atamturktur, S.; Khan, A. Experimental evaluation of a vibration-based leak detection technique for water pipelines. *Struct. Infrastruct. Eng.* **2017**, 1–10. [CrossRef]
16. Zahab, S.E.; Mosleh, F.; Zayed, T. *An Accelerometer-Based Real-Time Monitoring and Leak Detection System for Pressurized Water Pipelines*; Pipelines: Kansas City, MO, USA, 2016; pp. 257–268.
17. Li, R.; Huang, H.; Xin, K.; Tao, T. A review of methods for burst/leakage detection and location in water distribution systems. *Water Sci. Technol. Water Supply* **2015**, *15*, 429–441. [CrossRef]
18. Chaudhry, M.H. *Applied Hydraulic Transients*, 3rd ed.; Springer: New York, NY, USA, 2014.
19. Vítkovský John, P.; Lambert Martin, F.; Simpson Angus, R.; Liggett James, A. Experimental observation and analysis of inverse transients for pipeline leak detection. *J. Water Resour. Plan. Manag.* **2007**, *133*, 519–530. [CrossRef]
20. Vítkovský John, P.; Simpson Angus, R.; Lambert Martin, F. Leak detection and calibration using transients and genetic algorithms. *J. Water Resour. Plan. Manag.* **2000**, *126*, 262–265. [CrossRef]
21. Kapelan, Z.S.; Savic, D.A.; Walters, G.A. A hybrid inverse transient model for leakage detection and roughness calibration in pipe networks. *J. Hydraul. Res.* **2003**, *41*, 481–492. [CrossRef]
22. Jung, B.S.; Karney, B.W. Systematic exploration of pipeline network calibration using transients. *J. Hydraul. Res.* **2008**, *46*, 129–137. [CrossRef]

23. Shamloo, H.; Haghighi, A. Leak detection in pipelines by inverse backward transient analysis. *J. Hydraul. Res.* **2009**, *47*, 311–318. [CrossRef]

24. Haghighi, A.; Ramos, H.M. Detection of leakage freshwater and friction factor calibration in drinking networks using central force optimization. *Water Resour. Manag.* **2012**, *26*, 2347–2363. [CrossRef]

25. Huang, Y.-C.; Lin, C.-C.; Yeh, H.-D. An optimization approach to leak detection in pipe networks using simulated annealing. *Water Resour. Manag.* **2015**, *29*, 4185–4201. [CrossRef]

26. Liggett, J.A.; Chen, L.C. Inverse transient analysis in pipe networks. *J. Hydraul. Eng.* **1994**, *120*, 934–955. [CrossRef]

27. Ferrante, M.; Brunone, B.; Meniconi, S. Leak detection in branched pipe systems coupling wavelet analysis and a lagrangian model. *J. Water Supply Res. Technol. AQUA* **2009**, *58*, 95. [CrossRef]

28. Nazif, S.; Karamouz, M.; Tabesh, M.; Moridi, A. Pressure management model for urban water distribution networks. *Water Resour. Manag.* **2010**, *24*, 437–458. [CrossRef]

29. Ferrante, M.; Brunone, B.; Meniconi, S.; Karney, B.W.; Massari, C. Leak size, detectability and test conditions in pressurized pipe systems. *Water Resour. Manag.* **2014**, *28*, 4583–4598. [CrossRef]

30. Casillas, V.M.; Garza-Castañón, E.L.; Puig, V. Optimal sensor placement for leak location in water distribution networks using evolutionary algorithms. *Water* **2015**, *7*, 6496–6515. [CrossRef]

31. Meniconi, S.; Brunone, B.; Ferrante, M.; Capponi, C.; Carrettini, C.A.; Chiesa, C.; Segalini, D.; Lanfranchi, E.A. Anomaly pre-localization in distribution–transmission mains by pump trip: Preliminary field tests in the milan pipe system. *J. Hydroinform.* **2015**, *17*, 377–389. [CrossRef]

32. Lee, S.J.; Lee, G.; Suh, J.C.; Lee, J.M. Online burst detection and location of water distribution systems and its practical applications. *J. Water Resour. Plan. Manag.* **2016**, *142*, 04015033. [CrossRef]

33. Soares, A.K.; Covas, D.I.C.; Reis, L.F.R. Leak detection by inverse transient analysis in an experimental pvc pipe system. *J. Hydroinform.* **2011**, *13*, 153–166. [CrossRef]

34. Casillas Ponce, M.V.; Garza Castañón, L.E.; Cayuela, V.P. Model-based leak detection and location in water distribution networks considering an extended-horizon analysis of pressure sensitivities. *J. Hydroinform.* **2014**, *16*, 649–670. [CrossRef]

35. Yeh, H.-D.; Lin, Y.-C. Pipe network system analysis using simulated annealing. *J. Water Supply Res. Technol. AQUA* **2008**, *57*, 317–327. [CrossRef]

36. Mays, L.W. *Water Supply Systems Security*; McGraw-Hill: New York, NY, USA, 2004.

37. Savić, D.A.; Banyard, J.K. *Water Distribution Systems*, 2nd ed.; ICE: London, UK, 2011.

38. Larock, B.E.; Jeppson, R.W.; Watters, G.Z. *Hydraulics of Pipeline Systems*; CRC Press: Boca Raton, FL, USA, 2000.

39. Vítkovský, J.; Stephens, M.; Bergant, A.; Simpson, A.; Lambert, M. Numerical error in weighting function-based unsteady friction models for pipe transients. *J. Hydraul. Eng.* **2006**, *132*, 709–721. [CrossRef]

40. Duan, H.-F.; Ghidaoui, M.; Lee, P.J.; Tung, Y.-K. Unsteady friction and visco-elasticity in pipe fluid transients. *J. Hydraul. Res.* **2010**, *48*, 354–362. [CrossRef]

41. Vítkovský, J.P.; Bergant, A.; Simpson, A.R.; Lambert, M.F. Systematic evaluation of one-dimensional unsteady friction models in simple pipelines. *J. Hydraul. Eng.* **2006**, *132*, 696–708. [CrossRef]

42. Reddy, H.P.; Silva-Araya, W.F.; Chaudhry, M.H. Estimation of decay coefficients for unsteady friction for instantaneous, acceleration-based models. *J. Hydraul. Eng.* **2012**, *138*, 260–271. [CrossRef]

43. Boulos, P.F.; Lansey, K.E.; Karney, B.W. *Comprehensive Water Distribution Systems Analysis Handbook for Engineers and Planners*, 2nd ed.; MWH Soft, Incorporated: Pasadena, CA, USA, 2006.

44. Brunone, B. Transient test-based technique for leak detection in outfall pipes. *J. Water Resour. Plan. Manag.* **1999**, *125*, 302–306. [CrossRef]

45. Meniconi, S.; Brunone, B.; Ferrante, M.; Massari, C. Small amplitude sharp pressure waves to diagnose pipe systems. *Water Resour. Manag.* **2011**, *25*, 79–96. [CrossRef]

46. Brunone, B.; Ferrante, M.; Meniconi, S. Portable pressure wave-maker for leak detection and pipe system characterization. *J. Am. Water Works Assoc.* **2008**, *100*, 108–116.

47. Shucksmith, J.D.; Boxall, J.B.; Staszewski, W.J.; Seth, A.; Beck, S.B.M. Onsite leak location in a pipe network by cepstrum analysis of pressure transients. *J. Am. Water Works Assoc.* **2012**, *104*, E457–E465. [CrossRef]

48. Stephens, M.L.; Lambert, M.F.; Simpson, A.R. Determining the internal wall condition of a water pipeline in the field using an inverse transient. *J. Hydraul. Eng.* **2013**, *139*, 310–324. [CrossRef]

49. Taghvaei, M.; Beck, S.B.M.; Boxall, J. Leak detection in pipes using induced water hammer pulses and cepstrum analysis. *Int. J. COMADEM* **2010**, *13*, 19–25.

50. Haghighi, A.; Shamloo, H. Transient generation in pipe networks for leak detection. *Proc. Inst. Civ. Eng. Water Manag.* **2011**, *164*, 311–318. [CrossRef]

51. Ho, Y.C.; Cassandras, C.G.; Chen, C.H.; Dai, L. Ordinal optimisation and simulation. *J. Oper. Res. Soc.* **2000**, *51*, 490–500. [CrossRef]

52. Lau, T.W.E.; Ho, Y.C. Universal alignment probabilities and subset selection for ordinal optimization. *J. Optim. Theory Appl.* **1997**, *93*, 455–489. [CrossRef]

53. Cheng, M.-Y.; Prayogo, D. Symbiotic organisms search: A new metaheuristic optimization algorithm. *Comput. Struct.* **2014**, *139*, 98–112. [CrossRef]

54. Cheng, M.-Y.; Prayogo, D.; Tran, D.-H. Optimizing multiple-resources leveling in multiple projects using discrete symbiotic organisms search. *J. Comput. Civ. Eng.* **2016**, *30*, 04015036. [CrossRef]

55. Tran, D.-H.; Cheng, M.-Y.; Prayogo, D. A novel multiple objective symbiotic organisms search (mosos) for time–cost–labor utilization tradeoff problem. *Knowl-Based. Syst.* **2016**, *94*, 132–145. [CrossRef]

56. Bozorg-Haddad, O.; Azarnivand, A.; Hosseini-Moghari, S.-M.; Loáiciga, H.A. Optimal operation of reservoir systems with the symbiotic organisms search (SOS) algorithm. *J. Hydroinform.* **2017**, *19*, 507–521. [CrossRef]

57. Abhulimen, K.E.; Susu, A.A. Liquid pipeline leak detection system: Model development and numerical simulation. *Chem. Eng. J.* **2004**, *97*, 47–67. [CrossRef]

58. Yeh, H.-D. Theis' solution by nonlinear least-squares and finite-difference newton's method. *Ground Water* **1987**, *25*, 710–715. [CrossRef]

![water logo] *water*

MDPI

Article

On the Role of Minor Branches, Energy Dissipation, and Small Defects in the Transient Response of Transmission Mains

Silvia Meniconi [1,*,†] ![ORCID], **Bruno Brunone** [1,†] ![ORCID] and **Matteo Frisinghelli** [2,†]

1   Department of Civil and Environmental Engineering, University of Perugia, 06125 Perugia, Italy;
    bruno.brunone@unipg.it
2   Novareti SpA, 38068 Rovereto, Italy; M.Frisinghelli@novareti.eu
*   Correspondence: silvia.meniconi@unipg.it; Tel.: +39-075-585-3893
†   These authors contributed equally to this work.

Received: 30 December 2017; Accepted: 8 February 2018; Published: 11 February 2018

Abstract: In the last decades several reliable technologies have been proposed for fault detection in water distribution networks (DNs), whereas there are some limitations for transmission mains (TMs). For TM inspection, the most common fault detection technologies are of inline types—with sensors inserted into the pipelines—and then more expensive with respect to those used in DNs. An alternative to in-line sensors is given by transient test-based techniques (TTBTs), where pressure waves are injected in pipes "to explore" them. On the basis of the results of some tests, this paper analyses the relevance of the system configuration, energy dissipation phenomena, and pipe material characteristics in the transient behavior of a real TM. With this aim, a numerical model has been progressively refined not only in terms of the governing equations but also by including a more and more realistic representation of the system layout and taking into account the actual functioning conditions. As a result, the unexpected role of the minor branches—i.e., pipes with a length smaller than the 1% of the length of the main pipe—is pointed out and a preliminary criterion for the system skeletonization is offered. Moreover, the importance of both unsteady friction and viscoelasticity is evaluated as well as the remarkable effects of small defects is highlighted.

Keywords: transient simulation; fault detection; transmission main; branches; transient test-based techniques

1. Introduction

Because of the many differences, it is common ground that fault detection in transmission mains (TMs) is a completely different matter with respect to distribution networks (DNs). On one side, as an example, the accessibility is in favor of DNs. In fact, TMs have less appurtenances and are buried more deeply and in less accessible locations with respect to DNs. This implies the need of using in TMs inspection technologies of an inline type, more expensive than the traditional fixed probes and devices installed in DNs [1]. On the other side, the topology is in favor of TMs. In fact, if transient test-based techniques (TTBTs) were used in DNs, the numerous branches and users would absorb the pressure waves injected in the system before they can interact with the anomalies. This said, two points are very well established: (i) fault detection is a straightforward matter nor for TMs nor for DNs, and (ii) TMs and DNs require different techniques to achieve positive results. Two examples—but several ones could be given—to confirm the validity of these arguments. The first concerns TTBTs: adequate results have been obtained in the leak survey of the Milan (Italy) DN by transient tests only when some parts of the system were deliberately disconnected and then the topology of the system was simplified significantly [2] (it is remarkable to note that the pioneering paper by Liggett and Chen [3], where the Inverse Transient Analysis (ITA) was proposed, concerns looped pipe systems). The second relates the vibroacoustic measurements which are very reliable when short pipes (i.e., with a length of few dozens of meters as those connecting the users with the main pipe in DNs) are checked [4,5] but which cannot be used for long pipes because of the large signal attenuation.

A review of the existing methods for fault detection in pipe systems is beyond the aims of this paper where the attention is focused on TMs and, specifically, on the use of TTBTs in these systems. In this context, a possible procedure to follow is outlined in Figure 1 where the most important phases are pointed out. The procedure may be initiated by a failure alarm (phase #0) which in many cases derives simply from leakage clearly visible on a road or an abrupt pressure decrease, pointed out by the monitoring system. The successive phase #1 ("System survey") requires collecting information about the topology, pipe characteristics, boundary conditions, and appurtenances, if they are not already available. As it will be discussed later on, it should be emphasized that the more accurate the data from the pipe survey the faster the field test campaign and the more reliable the TTBT results.

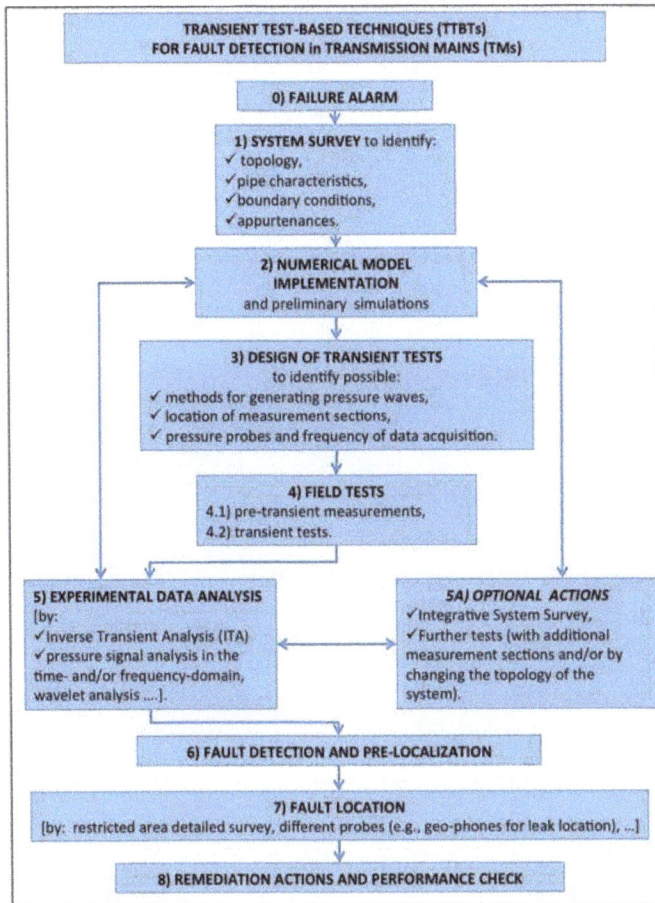

Figure 1. Transient test-based techniques (TTBTs) for fault detection in Transmission Mains (TMs): main phases of the procedure with highlighted the interactions between them.

Based on the findings of the pipe survey, phase #2 ("Numerical model implementation") can start. Intentionally the word "models" has been chosen to stress that within the procedure several types of models can be used not only with regard to the followed approach—e.g., in the time or frequency domain—but also with respect to their complexity (e.g., Lagrangian models vs. Method of Characteristics models). It is worthy of noting that within the procedure the role of the numerical models is not limited to the analysis of the experimental data (phase #5) but it is also of a crucial importance to design properly and safely the field transients (phase #3). In fact, the preliminary numerical simulations may give an idea of the pressure extreme values achieved during the tests—which is very important for the managers of the pipe system—as well as they are a valuable tool to indicate the measurement sections, to be checked during the successive phase.

Within phase #3 ("Design of transient tests"), the most appropriate technique for generating effective pressure waves is identified. It is important to note that this is a very important point within TTBTs, since in most cases transients generated by closing the installed valves or by pump shutdown are too slow to point out clearly the existing anomalies. In fact, particularly in TMs, valves are often too large to be operated by hand quickly and when they are motorized the prescribed closure speed is set as small as it does not generate unsafe overpressures. In the case of the pump shutdown, the time the pump takes to stop depends on its inertia which, of course, cannot be changed. As a consequence, there is a need for reliable techniques to generate proper pressure waves (i.e., sharp and compatible with the mechanical characteristics of the pipes). With this aim, the closure of a side discharge valve [6,7], the use of the Portable Pressure Wave Maker (PPWM) device [8–10], and the underwater explosion of a cavitating bubble [11,12] have been proposed in literature. Simultaneously, according to the available appurtenances, the measurement sections must be chosen. As mentioned above, TMs are often in less accessible locations but, beyond this, severe limits for the selection of proper measurement sections may also derive from design characteristics. In this respect, Figure 2 reports two examples of TMs (very frequent indeed!) in which the branch, where a probe should be installed, enters an inaccessible tunnel, and then no further measurements can be executed downstream. With regard to the pressure probes, the choice must fall on those with a quite large frequency response (typically of the order of few milliseconds)—to capture rapidly varying pressure signals—and a full scale not much larger than the expected pressure extreme values (from phase #2)—to ensure the best accuracy.

Figure 2. Typical TM branches where the inaccessibility is evident: the only appurtenance is just upstream of an inaccessible tunnel.

In the successive phase #4 ("Field tests"), preliminary measurements are executed with regard to pre-transient conditions. This may allow understanding the behavior of the system and setting more appropriate initial conditions in transient simulations. Afterwards the transient tests can be executed by using the identified technique. In this respect, according to the functioning conditions and the needs of the water company, it would be important to carry out the same test several times to check its repeatability, as well as the stationarity of the system behavior.

Within phase #5 ("Experimental data analysis"), the acquired pressure signals are examined to extract all the information (i.e., location, type and severity) about possible defects (e.g., leaks [13], partial blockages [14,15], unwanted partially closed in-line valves [16,17], illegal branches [18], and pipe wall deterioration [6]). A review of the existing techniques for the optimal analysis of the experimental data is beyond the aims of this paper which is focused, as shown below, on the crucial role played by the topology and functioning conditions of the system with respect to transient data analysis. Ambiguity and uncertainties occurring within such a phase may suggest refining the system survey (phase #5A) in order to detect possible components (e.g., very short branches, malfunctioning valves) neglected during phase #1. Moreover, further tests can be executed possibly after having simplified the topology of the system (e.g., by closing some branches) in order to improve the effectiveness of the transient tests in terms of the propagation of the pressure waves (i.e., to limit their absorption by secondary parts of the system). Such optional actions improve the performance of the pressure signal analysis and allow detecting and pre-locating the defects (phase #6). It is worthy of pointing out the great importance of the impact of phases #5 and #5A on the numerical models built within phase #2. In fact, the analysis of the experimental data, the execution of further tests, and the availability of a larger number of measurement sections may suggest improvements not only in the model parameters but also in the governing equations.

In the successive phase #7, faults can be located more precisely by means of proper probes (e.g., geo-phones for leak detection, and transients with high frequency waves [19,20]), after having executed a detailed survey of the part of the system highlighted by the previous phase as a possible fault location. Then the remediation actions and the check of the performance of the restored system complete the procedure (phase #8).

This paper presents clear evidence—based on transient tests executed on a real TM—of the unexpected relevance of some components and functioning conditions that, at a first glance, one could be authorized to neglect. Specifically, the crucial role played by some short branches in the transient behavior of the investigated pipe system is pointed out, as well it is confirmed the importance of the malfunctioning [21] of some installed valves that, presumed as totally closed, actually allow leakage, even if quite small. Moreover, the effect on transient pressure signals of the unsteady friction (UF) and difference in pipe materials (elastic and polymeric) is discussed. The method used in this paper for analyzing the experimental data from field tests is based on the use of a numerical model simulating transients in a pipe system. According to the quality of the numerical results, the model complexity is progressively increased by including more realistic representations of the topology as well as more refined governing equations. However, the aim of this paper is not to simulate at the best the experimental results by calibrating the model parameters. On the contrary, its very aim is to point out the improvement in the efficiency of the numerical simulations that can be achieved by including more and more appropriate representation of both the topology and the physical phenomena.

2. The Investigated Transmission Main and Transient Tests

Transient tests have been executed on the Trento steel TM, managed by Novareti SpA, connecting the Spini well-field to the "10,000" reservoir; it supplies the city of Trento, in the northeast of Italy. The original aim of these tests was to increase the number of the available experimental data concerning transients in elastic pipes (e.g., [22–24]) with a large value of the initial Reynolds number, Re_0 (= $V_0 D / \nu$, with V_0 = pre-transient mean flow velocity, D = pipe internal diameter, and ν = kinematic viscosity). Such a TM (hereafter referred to as the main pipe), buried at a depth of about 2 m in a porphyry sand, has a total length L = 1322 m, nominal diameter DN500, D = 506.6 mm, and wall thickness equal to 4.19 mm; it was selected since it is classified as a single pipe. In fact, the few minor branches are quite short and were certified by the system manager as inactive (i.e., connecting the main pipe to a dead end or with a closed valve at about the inlet). The diameter of such minor branches, D_b, ranges between 80 mm and 506.6 mm, whereas their length, L_b, between 0.7 m and 6.8 m and then between the 0.05 % and 0.5 % of L (Figure 3, and Table 1). All branches are in steel with the exception of the E one, which is a high-density polyethylene (HDPE) pipe and consists of two reaches: the first between nodes 12 and 13 (where there is a valve, certified as fully closed) with $L_{b,E'}$ = 3.0 m, and the second between nodes 13 and 14 (where there is an inactive well) with $L_{b,E''}$ =15.5 m. During the tests, the initial supplied discharge has been measured at the well-field just upstream of the check valve by means of an electromagnetic flow meter. The pressure signal, H, has been acquired at section M (Figure 3), just downstream of the check valve, by means of a piezoresistive pressure transducer with a full scale (fs) of 10 bar, accuracy of 0.25% $\times fs$, and response time of 0.5 ms; the level of the end reservoir has been provided by the data system acquisition of Novareti SpA. Steady-state flow measurements ($Re_0 \approx 10^5$) indicate that a fully rough pipe flow regime happens and provide an estimate of the roughness height equal to 0.8 mm.

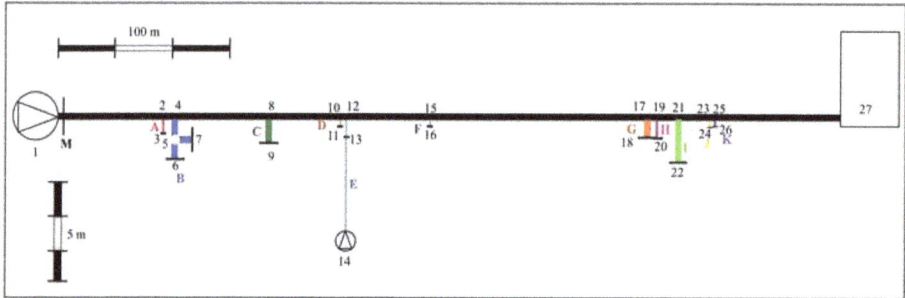

Figure 3. Trento TM layout (note that letters indicate the branches, whereas numbers indicate the nodes of the system; in particular: 1-well-field; M-measurement section; 27-downstream end reservoir; a different length scale has been used for the main pipe and minor branches).

Table 1. Characteristics and relevance of the branches.

Branch	Initial Node—End Node	L_b (m)	D_b (mm)	Material (—)	E_f (—)
A	2–3	1.7	150	steel	−0.71
B	4–5	3.1	506	steel	
	5–6	1.8	506	steel	−0.45
	5–7	2.7	200	steel	
C	8–9	3.5	506	steel	−0.64
D	10–11	0.7	80	steel	−0.73
E′	12–13	3	247	PEAD	−0.47
F	15–16	1.1	100	steel	−0.72
G	17–18	3	506	steel	−0.67
H	19–20	3	200	steel	−0.74
I	21–22	6.8	506	steel	−0.66
J	23–24	1	150	steel	−0.72
K	25–26	0.76	200	steel	−0.72

Transients have been generated by pump shutdown at the well-field, by stopping abruptly the electricity supply, and the repeatability of the tests has been checked (Figure 4). On the basis of the experimental pressure signals, the value of the pressure wave speed, a (= 1030 m/s), has been obtained.

In the below analysis, the pressure signal $H_{e,1}$ (hereafter referred to simply as H_e), with $Re_0 = 1.6 \times 10^5$, has been considered as representative of all the executed transients (the subscript e indicates the experimental values). It is worthy of noting that, as it will be discussed in the next section, at a first glance the transient response of the examined TM is very different from the one expected if it behaved actually as a single pipe (SP).

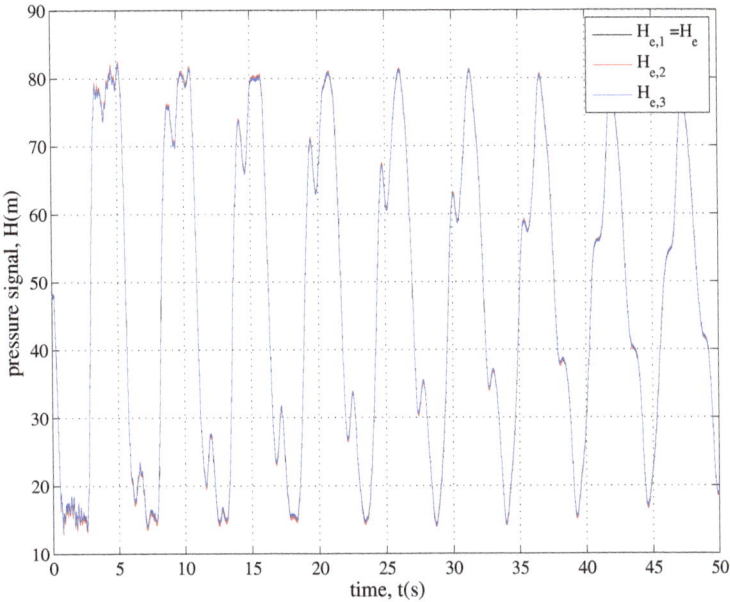

Figure 4. Trento TM: transient tests generated by pump shutdown. Note that the behavior of the pressure signals testifies the repeatability of tests.

3. Numerical Tests for Transient Simulation

Notwithstanding the above mentioned clear perception, in the first step of the pressure signal analysis, the TM has been considered as a single pipe, and the classical water-hammer equations [25] have been used (numerical test for the case of a single pipe, NTSP):

$$\frac{\partial H}{\partial s} + \frac{V}{g}\frac{\partial V}{\partial s} + \frac{1}{g}\frac{\partial V}{\partial t} + J = 0, \tag{1}$$

being the momentum equation, with s = spatial co-ordinate, t = time (elapsed since the beginning of the transient), g = acceleration due to gravity, and the friction term, J, assumed as equal to the steady-state component, J_s, given by the Darcy-Weisbach friction formula:

$$J = J_s = \lambda \frac{V^2}{2gD}, \tag{2}$$

with λ = friction factor, and

$$\frac{\partial H}{\partial t} + \frac{a^2}{g}\frac{\partial V}{\partial s} = 0, \tag{3}$$

being the continuity equation. As a result, the pressure signal, H_n, of Figure 5 is obtained (the subscript n indicates the numerical values), with a value of the Nash-Sutcliffe efficiency coefficient,

$$E_f = 1 - \sum_{i=1}^{M} \frac{(H_{e,i} - H_{n,i})^2}{(H_{e,i} - \tilde{H}_e)^2}, \tag{4}$$

E_f equal to -0.67, with M = number of samples, and $\tilde{H}_e$ = experimental mean value. Such a poor performance of the classical water hammer equations implies—as it was quite easy to predict—the need of substantial improvements in the model. However, an in-depth analysis of the experimental and numerical pressure signals indicates that a large part of the differences is due to the presence in H_e of some further sharp pressure changes—both rises and decreases—other those caused by the pump shutdown and the reflection at the check valve and the downstream reservoir. According to literature [18], the shape of such further pressure changes suggests to explore firstly the role played by the system configuration and specifically that of the minor branches neglected in the SP scheme. Thereafter, the relevance of the unsteady energy dissipation mechanisms (i.e., unsteady friction, UF), the viscoelastic effects (VE) in branch E, and possible small defects will be checked. In Table 2 the main characteristics of the simulated systems and model assumptions and performance are reported.

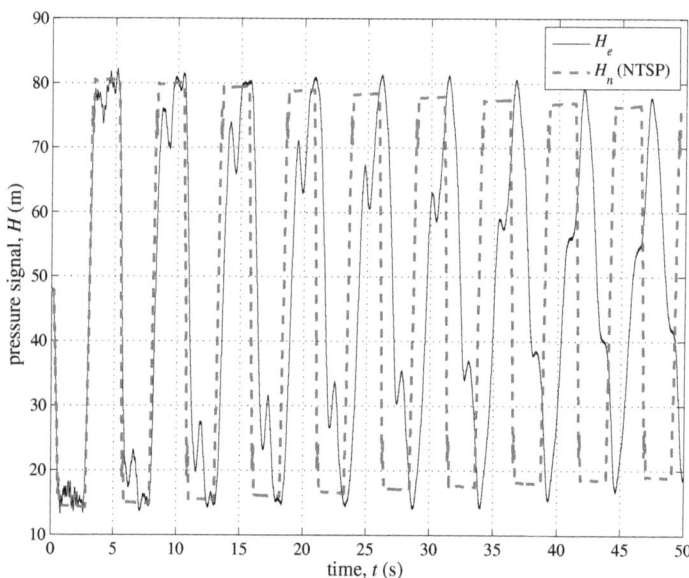

Figure 5. Trento TM: transient generated by pump shutdown at the well-field. Experimental pressure signal, H_e, at section M vs. model simulation, H_n, for the single pipe scheme (NTSP).

Table 2. Numerical tests: simulated systems, model assumptions, and performances.

Numerical Test (NT#)	Simulated Topology and Functioning Conditions	Model Assumptions	E_f (–)
SP	single pipe (i.e., no branches)	Equations (1) and (2)	−0.67
5b	only branches B, C, E, G, and I inactive and valve 13 fully closed	" "	−0.07
UF	as 5b	UF included	0.30
VE	as 5b	VE included	−0.07
UF + VE	as 5b	UF + VE included	0.30
MV13	as 5b but with a malfunctioning valve at node 13	" "	0.80
L14	as MV13 but with E branch with a small leak Q_L	" "	0.83

3.1. The Role of the Minor Branches

The large complexity of many pipe systems—particularly DNs—has always encouraged the attempt to simplify them but preserving the behavior in steady-state conditions of the real system with respect to a given feature of interest. Emblematic is the case of a branch with numerous users which can be transformed in a pipe with a fictitious constant discharge with the two pipes—the real and the simplified one—having in common only the total head loss (which is the retained feature indeed). In literature, the procedure to simplify a pipe system by excluding the less important branches is defined as skeletonization [26]. In transient conditions, of course, the point is completely different and there is not a clear rule to decide which components of a pipe system must be regarded as crucial. As a consequence, to analyze the relevance of the minor branches of the considered TM, numerical tests have been executed by considering the branches one at a time. Then the obtained values of E_f (Table 1) have been compared in Figure 6 with the one (=-0.67) pertaining the single pipe (SP). This plot and data reported in Table 1 show that in principle the relevance of a given branch decreases with the distance from the section where the pressure wave is injected into the system. This not only in terms of the mere distance from the injection section, but also with respect to the number and characteristics (e.g, diameter and pressure wave speed) of the branches with which the pressure wave interacts along its path. This result is confirmed by the transient response (Figure 7) of the systems with only branch C (NTbC) or branch G (NTbG), which have almost the same characteristics, but with branch G being at a larger distance from the injection section; Figure 7 shows that the performance of NTbC is quite better than NTbG ($E_f = -0.64$ and -0.67, respectively).

A scrupulous analysis of the effect of a given series of branches on the transient behavior of a TM is beyond the scope of this paper. However, as a preliminary criterion to simplify the system, the branches with a value of E_f smaller than the one of SP (i.e., branches A, D, F, H, J, and K) are excluded. As a consequence, in the successive phases of the analysis of the experimental pressure signal only branches B, C, E, G, and I will be retained (NT5b). In Figure 8 the numerical test with such a simplified system is reported: a clear improvement of the performance of the model can be observed ($E_f = -0.07$) with respect to the SP approach ($E_f = -0.67$). Such a significant increase of E_f (about the 857% with respect to NTSP) clearly highlights the crucial importance of the branches even if their length is very small with respect to the one of the main pipe.

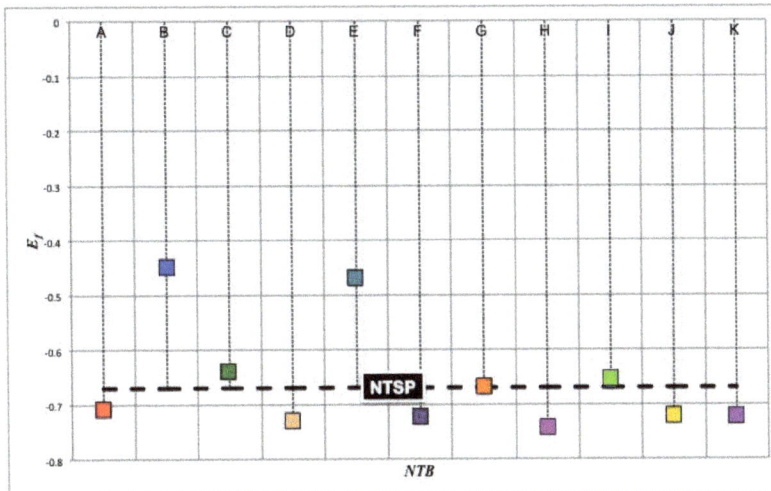

Figure 6. Trento TM: transient tests generated by pump shutdown. Performance of the numerical model by considering the branches one at a time.

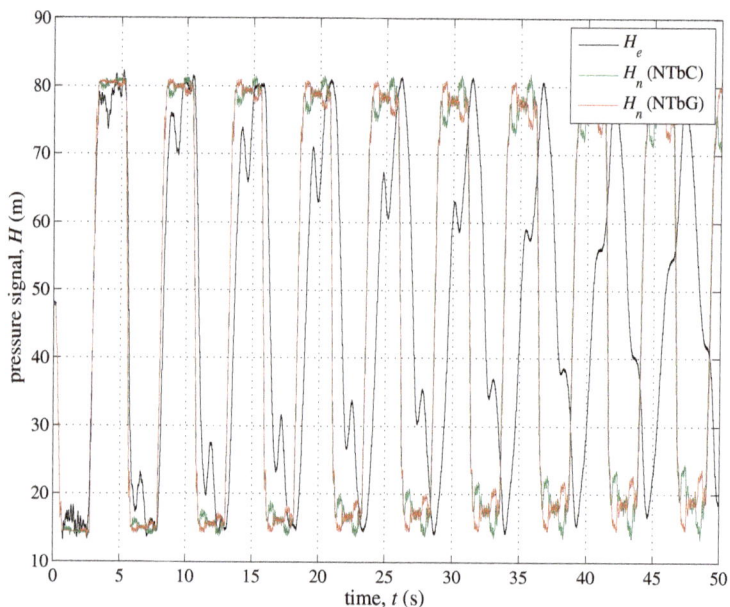

Figure 7. Trento TM: transient generated by pump shutdown at the well-field. Experimental pressure signal, H_e, at section M vs. model simulation, H_n, for the system with only branch C (NTbC), and only branch G (NTbG).

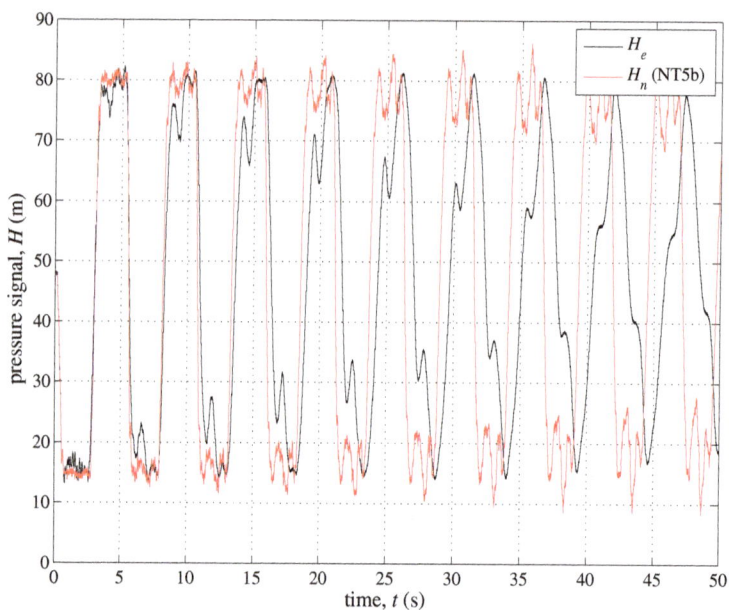

Figure 8. Trento TM: transient generated by pump shutdown at the well-field. Experimental pressure signal, H_e, at section M vs. model simulation, H_n, for the system with 5 branches (B, C, E, G and I) selected by means of the value of E_f (NT5b).

3.2. The Role of the Unsteady Friction and Viscoelasticity

To take into account the effect of the unsteadiness on the energy dissipation, in Equation (1) the additional unsteady friction (UF) term, J_u, evaluated by means of an Instantaneous Acceleration Based (IAB) model [27–30]:

$$J_u = \frac{k_{UF}}{g} \left(\frac{\partial V}{\partial t} + \text{sign} \left(V \frac{\partial V}{\partial s} \right) a \frac{\partial V}{\partial s} \right), \tag{5}$$

has been included ($J = J_s + J_u$), with k_{UF} = unsteady friction coefficient, and $\text{sign}(V \partial V / \partial s) = (+1$ for $V \partial V / \partial s \geq 0$ or -1 for $V \partial V / \partial s < 0$). In the used UF model, the coefficient k_{UF} is the only parameter to evaluate. In line with the aims of this paper, the same value of k_{UF} (i.e., the one pertaining to the main pipe) has been considered for all pipes. According to literature—which points out the importance of the initial conditions and relative roughness [31]—the value $k_{UF} = 8 \times 10^{-3}$ has been chosen.

As pressure traces of Figure 9 clearly show, the performance of the model including UF (NTUF) improves significantly with the simulated damping of the pressure peaks quite closer to the experimental one. As a consequence, the efficiency coefficient of NTUF (Table 2) is equal to 0.30, with an increase of about the 528% with respect to NT5b.

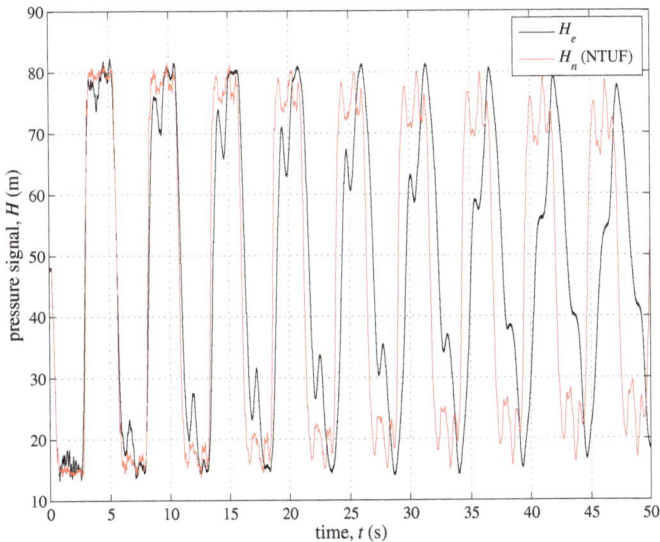

Figure 9. Trento TM: transient generated by pump shutdown at the well-field. Experimental pressure signal, H_e, at section M vs. model simulation, H_n, for the system with the 5 branches (B, C, E, G and I) and the unsteady friction included in the model (NTUF).

To evaluate singly the role played by the mechanical characteristics of pipe material, for the E branch (with $L_{b,E'}$) the modified continuity equation (see Appendix A):

$$\frac{\partial H}{\partial t} + \frac{a^2}{g} \frac{\partial V}{\partial s} + \frac{2a^2}{g} \frac{d\epsilon_r}{dt} = 0, \tag{6}$$

has been considered instead of Equation (3), to simulate the viscoelastic (VE) effects, with ϵ_r = retarded strain. In order to evaluate ϵ_r, the viscoelastic parameters of the Kelvin-Voigt element, T_r (= retardation time) and E_r (= dynamic modulus of elasticity) have been chosen according to literature [32]. The resulting pressure signal (Figure 10) indicates clearly that, because of the very small percentage of polymeric pipes (= 0.22%) with respect to the elastic main pipe, the role of viscoelasticity is negligible. As a consequence, the results of NTVE are virtually the same of NT5b ($E_f = -0.07$) as well as the results of NTUF + VE (Figure 11) replicate the ones of NTUF ($E_f = 0.30$).

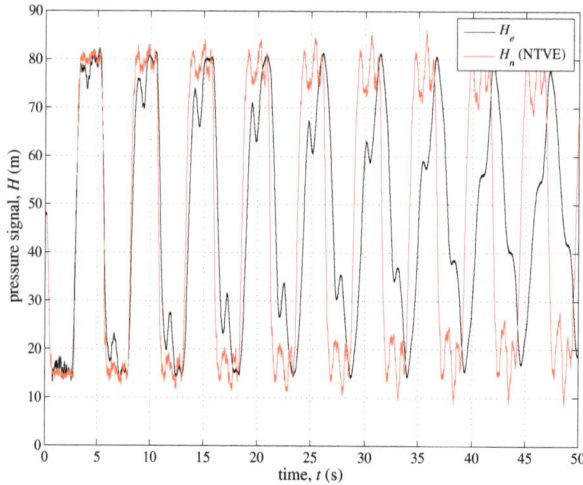

Figure 10. Trento TM: transient generated by pump shutdown at the well-field. Experimental pressure signal, H_e, at section M vs. model simulation, H_n, for the system with the 5 branches (B, C, E, G and I) and the viscoelasticity (for branch E, with $L_{b,E} = L_{b,E'}$) included in the model (NTVE).

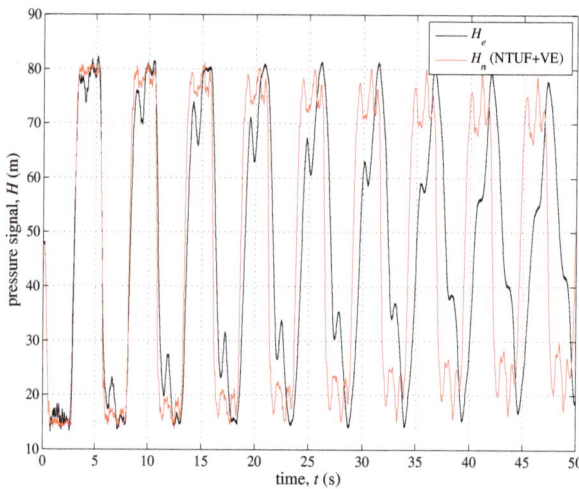

Figure 11. Trento TM: transient generated by pump shutdown at the well-field. Experimental pressure signal, H_e, at section M vs. model simulation, H_n, for the system with the 5 branches (B, C, E, G and I) and both the unsteady friction and the viscoelasticity (for branch E, with $L_{b,E} = L_{b,E'}$) included in the model (NTUF + VE).

3.3. The Role of Small Defects

According to literature, from the point of view of transient simulation, the used model contains the most important mechanisms. As a consequence, the not satisfactory values of E_f could be ascribed to a not reliable conformity of the assumed layout to the reality. Then, according to phase #5A (Figure 1), an integrative system survey has been executed which revealed the malfunctioning of valves at nodes 13 and 14. This outcome allows hypothesizing different functioning conditions of the E branch: (i) the length is 18.5 m, by assuming $L_{b,E} = L_{b,E'} + L_{b,E''}$ (NTMV13), because of the malfunctioning of valve 13; (ii) a small discharge, Q_L, of about 2 L/s happens towards the unused well located at node 14 (NTL14), because of the malfunctioning of valve at node 14 . Such a value of Q_L is compatible with the difference between the discharge supplied at the well-field and the outflow at the end reservoir. Both these scenarios (Figures 12 and 13) exhibit a clear improvement, with E_f being equal to 0.80 (NTMV13) and 0.83 (NTL14), and then with an increase of about the 167%, and 177% with respect to NTUF + VE, respectively. In fact, most of the discontinuities of the experimental pressure signal are now captured reasonably well.

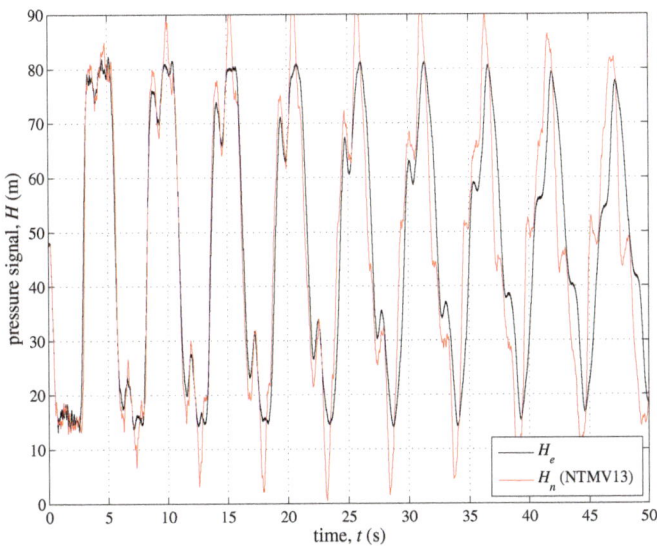

Figure 12. Trento TM: transient generated by pump shutdown at the well-field. Experimental pressure signal, H_e, at section M vs. model simulation, H_n, for the system with the 5 branches B, C, E, G, and I, the unsteady friction and the viscoelasticity (for branch E, with $L_{b,E} = L_{b,E'} + L_{b,E''}$) included in the model, and a malfunctioning valve at node 13 (NTMV13).

Moreover, it is worthy of noting that the reasons of the improvement of E_f for NTMV13 are two, both linked to the increase of $L_{b,E}$, due to the malfunctioning of the valve at node #13. In fact, the larger $L_{b,E}$, the most significant the role of the branch itself, and the more valuable the effect of the viscoelasticity (e.g., [33,34]).

Notwithstanding the valuable improvement in terms of E_f from NTSP to NTL14, a quite remarkable difference between the experimental and numerical pressure signals still remains. Such a partial failure of the numerical model can be ascribed to several reasons. First of all the deliberately omitted parameter calibration must be mentioned. In fact, in the paper the values of the unsteady friction coefficient and viscoelastic parameters of literature have been used, even if reliable criteria for evaluating such quantities is still an open problem. Secondly, a possible inaccuracy in the description of the system topology which, as shown in this paper, has a significant effect on the numerical simulations, even if quite small. Finally, possible undetected localized phenomena (e.g., water column separation) might have caused further pressure waves.

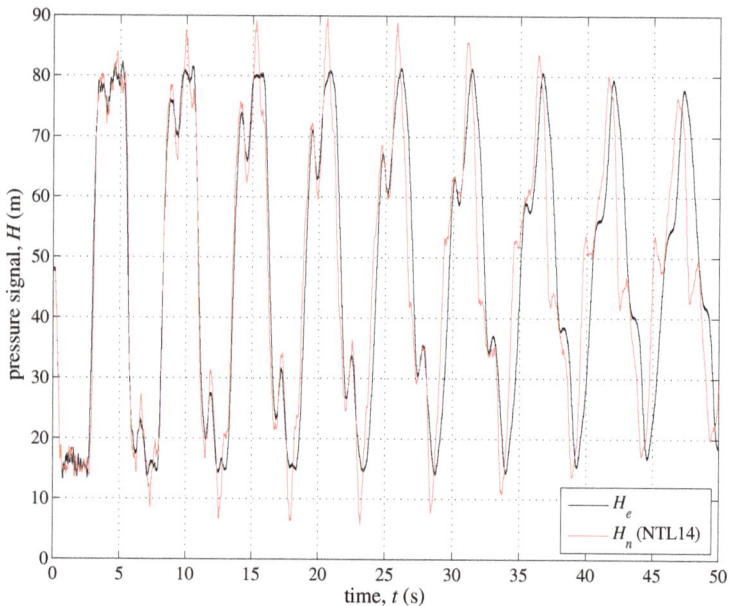

Figure 13. Trento TM: transient generated by pump shutdown at the well-field. Experimental pressure signal, H_e, at section M vs. model simulation, H_n, for the system with the 5 branches B, C, E, G, and I, the unsteady friction and the viscoelasticity (for branch E, with $L_{b,E} = L_{b,E'} + L_{b,E''}$) included in the model, and a small leak, Q_L, at node 14 (NTL14).

4. Conclusions

Unsteady-state tests executed on the Trento TM by a pump shutdown have disclosed the unexpected remarkable effect of the short minor branches—essentially inactive—on the transient response of the investigated pipe system. In fact, the experimental pressure signal shows clear sharp changes beyond those due to the pressure wave reflection at the upstream and downstream boundaries (i.e., the check valve at the well-field and the downstream end reservoir). By means of the numerical model, the relevance of the topology, pipe material characteristics, transient energy dissipation, and defects has been explored. The performance of the numerical model has been evaluated on the basis of the Nash Sutcliffe efficiency coefficient. A preliminary criterion for the skeletonization of the TM has been proposed.

In Figure 14a, the progressively refinement of the model, with the relative values of the efficiency (Figure 14b), is clarified, as a succession of more and more complex numerical model and topology. From Figure 14b, it emerges the very crucial role played by secondary branches, particularly the five most important. Specifically, the largest increase ($\simeq$857%) in the numerical performance is achieved when these branches are included in the system topology. As a consequence, a more in depth analysis for the skeletonization of pipe systems with respect to the unsteady-state flow is an urgent need, in order to use TTBT reliably for fault detection in complex pipe systems. A less important but still significant improvement is obtained when the unsteady friction is taken into account ($\simeq$528%). On the contrary, the role of the viscoelasticity becomes relevant only when the length of the polymeric branch is appreciable. Finally, an important contribution for the simulation of sharp pressure changes is given by the inclusion of small defects (i.e., malfunctioning valves).

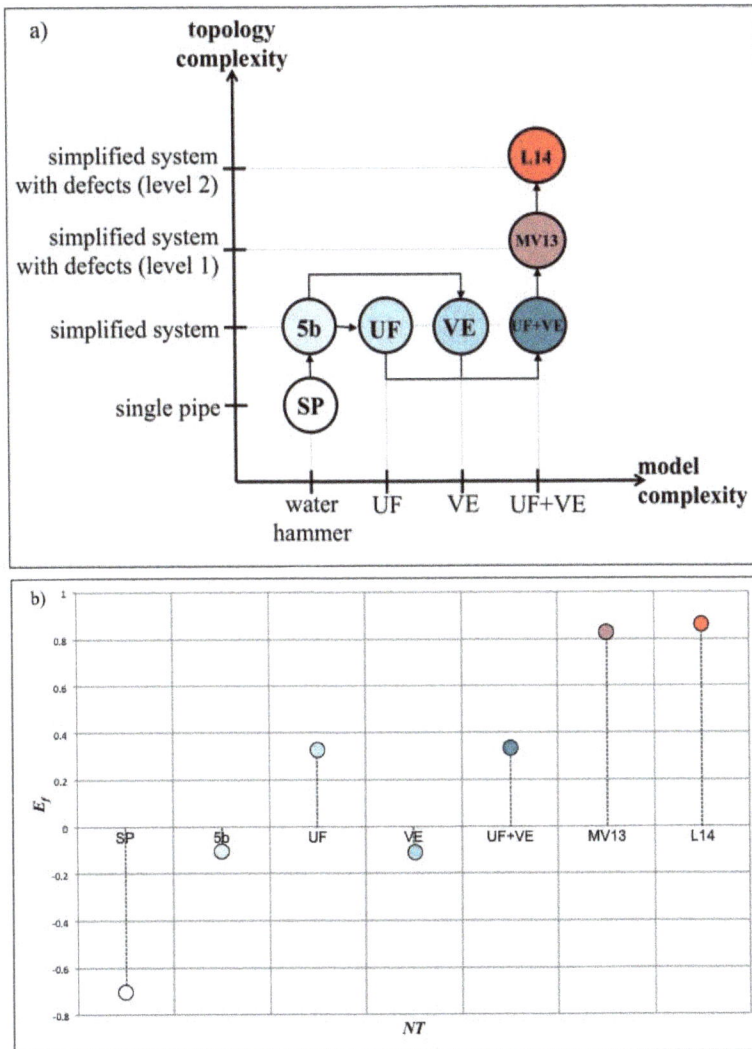

Figure 14. Procedure within the numerical model implementation for evaluating the relevance of the topology simulation (**a**) and model complexity in the transient response of the TM (**b**).

Acknowledgments: This research has been funded by the Hong Kong (HK) Research Grant Council Theme-Based Research Scheme and the HK University of Science and Technology (HKUST) under the project "Smart Urban Water Supply System (Smart UWSS)", University of Perugia, and Fondazione Cassa Risparmio Perugia (project No. 2017.0234.021). Tests have been executed within a research project in co-operation with Novareti SpA, Trento. The support in the preparation and execution of the field tests of E. Mazzetti, C. Capponi, and C. Del Principe of the University of Perugia, and M. Larentis, and C. Costisella of Novareti SpA is highly appreciated.

Author Contributions: S. Meniconi, B. Brunone and M. Frisinghelli conceived, designed and performed the experiments; S. Meniconi, and B. Brunone analyzed the data; M. Frisinghelli contributed materials tools; S. Meniconi, and B. Brunone wrote the paper.

Conflicts of Interest: The authors declare no conflict of interest.

Appendix A. Single Element Kelvin-Voigt (1 K-V) Models

With respect to elastic materials, when a circumferential stress, σ, is applied to a viscoelastic material, the total strain, ϵ, is given by the sum of the instantaneous elastic, ϵ_{el}, and retarded component, ϵ_r [35–42]:

$$\epsilon = \epsilon_{el} + \epsilon_r. \tag{A1}$$

Such a behavior can be simulated by means of a single element Kelvin-Voigt (1 K-V) model where a viscous damper and an elastic spring, connected in parallel, are jointed to a simple elastic spring in series. Within 1 K-V models, the following relationship links σ and ϵ_r:

$$\sigma = E_r \epsilon_r + \frac{E_r}{T_r} \frac{d\epsilon_r}{dt}, \tag{A2}$$

where E_r = dynamic modulus of elasticity, and T_r = retardation time of the KV element. According to the Hooke's law, the elastic strain, ϵ_{el}, of the spring is given by:

$$\epsilon_{el} = \frac{\sigma}{E_{el}}, \tag{A3}$$

where the elastic Young's modulus of elasticity, E_{el}, is linked to a [43] by:

$$a = \sqrt{\frac{\frac{k}{\rho}}{1 + \psi \frac{kD}{eE_{el}}}}, \tag{A4}$$

with k = bulk modulus of elasticity, and ψ = dimensionless parameter accounting for longitudinal support situation [44,45].

The above difference between elastic and viscoelastic materials reflects in the continuity equation and then the following term must be added in Equation (3):

$$\frac{2a^2}{g} \frac{d\epsilon_r}{dt}. \tag{A5}$$

References

1. Laven, L.; Lambert, A.O. What do we know about real losses on transmission mains? In Proceedings of the IWA Specialised Conference "Water Loss 2012", Manila, Philippines, 26–29 February 2012.
2. Meniconi, S.; Brunone, B.; Ferrante, M.; Capponi, C.; Carrettini, C.A.; Chiesa, C.; Segalini, D.; Lanfranchi, E.A. Anomaly pre-localization in distribution-transmission mains by pump trip: Preliminary field tests in the Milan pipe system. *J. Hydroinform.* **2015**, *17*, 377–389.
3. Liggett, J.A.; Chen, L.C. Inverse transient analysis in pipe networks. *J. Hydraul. Eng.* **1994**, *120*, 934–955.
4. Martini, A.; Troncossi, M.; Rivola, A. Vibroacoustic measurements for detecting water leaks in buried small-diameter plastic pipes. *J. Pipeline Syst. Eng. Pract.* **2017**, *8*, 04017022.
5. Martini, A.; Troncossi, M.; Rivola, A. Leak detection in water-filled small-diameter polyethylene pipes by means of acoustic emission measurements. *Appl. Sci.* **2017**, *7*, doi:10.3390/app7010002.
6. Stephens, M.; Lambert, M.F.; Simpson, A.R. Determining the internal wall condition of a water pipeline in the field using an inverse transient. *J. Hydraul. Eng.* **2013**, *139*, 310–324.
7. Taghvaei, M.; Beck, S.B.M.; Boxall, J.B. Leak detection in pipes using induced water hammer pulses and cepstrum analysis. *Int. J. COMADEM* **2010**, *13*, 19–25.
8. Brunone, B.; Ferrante, M.; Meniconi, S. Portable pressure wave-maker for leak detection and pipe system characterization. *J. Am. Water Works Assoc.* **2008**, *100*, 108–116.
9. Meniconi, S.; Brunone, B.; Ferrante, M.; Massari, C. Small amplitude sharp pressure waves to diagnose pipe systems. *Water Resour. Manag.* **2011**, *25*, 79–96.

10. Meniconi, S.; Brunone, B.; Frisinghelli, M.; Mazzetti, E.; Larentis, M.; Costisella, C. Safe transients for pipe survey in a real transmission main by means of a portable device: The case study of the Trento (I) supply system. *Procedia Eng.* **2017**, *186*, 228–235.

11. Mazzocchi, E.; Pachoud, A.J.; Farhat, M.; Hachem, F.E.; Cesare, G.D. Signal analysis of an actively cavitation bubble in pressurized pipes for detection of wall stiffness drops. *J. Fluids Struct.* **2016**, *65*, 60–75.

12. Gong, J.; Lambert, M.F.; Nguyen, S.T.N.; Zecchin, A.C.; Simpson, A.R. Detecting thinner-walled pipe sections using a spark transient pressure wave generator. *J. Hydraul. Eng.* **2018**, *144*, 06017027.

13. Covas, D.; Ramos, H. Case studies of leak detection and location in water pipe systems by inverse transient analysis. *J. Wat. Res. Plan. Man.* **2010**, *136*, 248–257.

14. Duan, H.; Lee, P.; Ghidaoui, M.; Tuck, J. Transient wave-blockage interaction and extended blockage detection in elastic water pipelines. *J. Fluids Struct.* **2014**, *46*, 2–16.

15. Meniconi, S.; Brunone, B.; Ferrante, M.; Capponi, C. Mechanism of interaction of pressure waves at a discrete partial blockage. *J. Fluids Struct.* **2016**, *62*, 33–45.

16. Contractor, D.N. The reflection of waterhammer pressure waves from minor losses. *J. Basic Eng.* **1965**, *87*, 445–451.

17. Meniconi, S.; Brunone, B.; Ferrante, M. In-line pipe device checking by short period analysis of transient tests. *J. Hydraul. Eng.* **2011**, *137*, 713–722.

18. Meniconi, S.; Brunone, B.; Ferrante, M.; Massari, C. Transient tests for locating and sizing illegal branches in pipe systems. *J. Hydroinform.* **2011**, *13*, 334–345.

19. Louati, M.; Ghidaoui, M.S. High frequency acoustic wave properties in a water-filled pipe. Part 1: Dispersion and multi-path behavior. *J. Hydraul. Res.* **2017**, *55*, 613–631.

20. Louati, M.; Ghidaoui, M.S. High frequency acoustic wave properties in a water-filled pipe. Part 2: Range of propagation. *J. Hydraul. Res.* **2017**, *55*, 632–646.

21. Meniconi, S.; Brunone, B.; Ferrante, M.; Massari, C. Potential of transient tests to diagnose real supply pipe systems: What can be done with a single extemporary test. *J. Water Resour. Plan. Manag.* **2010**, *137*, 238–241.

22. Meniconi, S.; Duan, H.; Brunone, B.; Ghidaoui, M.; Lee, P.; Ferrante, M. Further developments in rapidly decelerating turbulent pipe flow modeling. *J. Hydraul. Eng.* **2014**, *140*, 04014028.

23. Adamkowski, A.; Lewandowski, M. Experimental examination of unsteady friction models for transient pipe flow simulation. *J. Fluids Eng. Trans. ASME* **2006**, *128*, 1351–1363.

24. Liou, J.C.P. Understanding line packing in frictional water hammer. *J. Fluids Eng. Trans. ASME* **2016**, *138*, 081303.

25. Wylie, E.; Streeter, V. *Fluid Transients in Systems*; Prentice-Hall Inc.: Upper Saddle River, NJ, USA, 1993; p. 463.

26. Larock, B.E.; Jeppson, R.W.; Watters, G.Z. *Hydraulics of Pipeline Systems*; CRC Press: Boca Raton, FL, USA, 1999; p. 552.

27. Bergant, A.; Simpson, A.R.; Vitkovsky, J. Developments in unsteady pipe flow friction modelling. *J. Hydraul. Res.* **2001**, *39*, 249–257.

28. Brunone, B.; Morelli, L. Automatic control valve induced transients in an operative pipe system. *J. Hydraul. Eng.* **1999**, *125*, 534–542.

29. Brunone, B.; Golia, U.M. Discussion of "Systematic evaluation of one-dimensional unsteady friction models in simple pipelines" by J.P. Vitkovsky, A. Bergant, A.R. Simpson, and M. F. Lambert. *J. Hydraul. Eng.* **2008**, *134*, 282–284.

30. Ghidaoui, M.S.; Zhao, M.; McInnis, D.A.; Axworthy, D.H. A review of water hammer theory and practice. *Appl. Mech. Rev.* **2005**, *58*, 49–76.

31. Pezzinga, G. Evaluation of unsteady flow resistances by quasi-2D or 1D models. *J. Hydraul. Eng.* **2000**, *126*, 778–785.

32. Pezzinga, G.; Brunone, B.; Meniconi, S. Relevance of pipe period on Kelvin-Voigt viscoelastic parameters: 1D and 2D inverse transient analysis. *J. Hydraul. Eng.* **2016**, *142*, 04016063.

33. Pezzinga, G.; Scandura, P. Unsteady flow in installations with polymeric additional pipe. *J. Hydraul. Eng.* **1995**, *121*, 802–811.

34. Pezzinga, G. Unsteady flow in hydraulic networks with polymeric additional pipe. *J. Hydraul. Eng.* **2002**, *128*, 238–244.

35. Covas, D.; Stoianov, I.; Mano, J.F.; Ramos, H.; Graham, N.; Maksimovic, C. The dynamic effect of pipe-wall viscoelasticity in hydraulic transients. Part II—Model development, calibration and verification. *J. Hydraul. Res.* **2005**, *43*, 56–70.

36. Franke, P.G.; Seyler, F. Computation of unsteady pipe flow with respect to visco-elastic material properties. *J. Hydraul. Res.* **1983**, *21*, 345–353.

37. Keramat, A.; Tijsseling, A.S.; Hou, Q.; Ahmadi, A. Fluid-structure interaction with pipe-wall viscoelasticity during water hammer. *J. Fluids Struct.* **2012**, *28*, 434–455.

38. Keramat, A.; Kolahi, A.G.; Ahmadi, A. Water hammer modeling of viscoelastic pipes with a time-dependent Poisson's ratio. *J. Fluids Struct.* **2013**, *43*, 164–178.

39. Soares, A.K.; Covas, D.; Reis, L.F. Analysis of PVC pipe-wall viscoelasticity during water hammer. *J. Hydraul. Eng.* **2008**, *134*, 1389–1395.

40. Weinerowska-Bords, K. Accuracy and parameter estimation of elastic and viscoelastic models of the water hammer. *Task Q.* **2007**, *11*, 383–395.

41. Meniconi, S.; Brunone, B.; Ferrante, M.; Massari, C. Energy dissipation and pressure decay during transients in viscoelastic pipes with an in-line valve. *J. Fluid Struct.* **2014**, *45*, 235–249.

42. Pezzinga, G.; Brunone, B.; Cannizzaro, D.; Ferrante, M.; Meniconi, S.; Berni, A. Two-dimensional features of viscoelastic models of pipe transients. *J. Hydraul. Eng.* **2014**, *140*, 04014036.

43. Hachem, F.E.; Schleiss, A.J. A review of wave celerity in frictionless and axisymmetrical steel-lined pressure tunnels. *J. Fluids Struct.* **2011**, *27*, 311–328.

44. Parmakian, J. *Waterhammer Analysis*; Dover: New York, NY, USA, 1963.

45. Montuori, C. Colpo d'ariete in presenza di resistenze in condotte di notevole spessore. *L'Energia Elettrica* **1966**, *XLIII*, 1–18. (In Italian)

water

MDPI

Article

Data-Driven Study of Discolouration Material Mobilisation in Trunk Mains

Gregory Meyers * [iD], Zoran Kapelan and Edward Keedwell

College of Engineering, Mathematics and Physical Sciences, University of Exeter, North Park Road,
Exeter EX4 4QF, UK; Z.Kapelan@exeter.ac.uk (Z.K.); E.C.Keedwell@exeter.ac.uk (E.K.)
* Correspondence: gmm206@exeter.ac.uk

Received: 14 September 2017; Accepted: 20 October 2017; Published: 24 October 2017

Abstract: It has been shown that sufficiently high velocities can cause the mobilisation of discolouration material in water distribution systems. However, how much typical hydraulic conditions affect the mobilisation of discolouration material has yet to be thoroughly investigated. In this paper, results are presented from real turbidity and flow observations collected from three U.K. trunk main networks over a period of two years and 11 months. A methodology is presented that determines whether discolouration material has been mobilised by hydraulic forces and the origin of that material. The methodology found that the majority of turbidity observations over 1 Nephelometric Turbidity Units (NTU) could be linked to a preceding hydraulic force that exceeded an upstream pipe's hydraulically preconditioned state. The findings presented in this paper show the potential in proactively managing the hydraulic profile to reduce discolouration risk and improve customer service.

Keywords: water distribution systems; velocity; discolouration; modelling; turbidity; hydraulic events; water quality; mains conditioning

1. Introduction

Historically, water supply systems and thus water companies have been primarily focused on the sufficient delivery of safe drinking water to customers. In recent years, higher customer expectations of water quality standards have been reflected by regulatory bodies through the implementation of fines and penalties for a number of discolouration contacts [1]. Discoloured water has long been the largest cause of water quality customer contacts in the U.K. water industry [2,3]. Even putting aside the validity of public health concerns, discoloured water can still undermine consumer confidence and negatively impact a water utility's reputation.

Reducing discolouration risk is especially challenging due to the complex chemically, biologically and hydraulically dependent nature of discolouration material accumulation and mobilisation not being fully understood [4–6]. While the hydraulic mobilisation of iron and manganese deposits has been long known to result in discoloured water, the presence of discoloured water can also be due to other processes such as biofilm mobilisation or chemical interactions between pipe materials and water acidity [7–11]. Discolouration has been shown to significantly vary even between different parts of the same water distribution network and yet is still similarly experienced throughout different countries regardless of widely-varying factors between their Water Distribution Systems (WDS) [12–14].

Water companies primarily deal with discolouration by cleaning, i.e., flushing WDS mains. Once a sufficient number of discolouration complaints have been reported in the area, the company may decide to reline (or replace) old mains believed to be the cause of significant discolouration [6,11], particularly if this is going to help address additional issues (e.g., leakage). However, cleaning and especially rehabilitating WDS mains is expensive and can still potentially only have limited effect if the discolouration material was mobilised from a different section of the network [15]. Thus, determining where the significant causes of discolouration are in a WDS is important to efficiently reduce discolouration risk.

Trunk mains have long been considered high discolouration risks due to their large potential to act as a form of a reservoir for discolouration material. Trunk mains can passively send low concentrations of material downstream to accumulate in distribution pipes and actively cause widespread discolouration. However, only recently has research indicated that a significant number of discolouration events in downstream distribution networks can actually be attributed to upstream trunk mains [14,15].

Due to the potential consequences associated with trunk mains, research on trunk mains has been primarily limited to areas where the benefits are clearly evident to water companies. In particular, significant research has been carried out on developing methods to intermittently clean trunk mains with minimal cost and required downtime [16–18].

The process of incrementally increasing the flow in a trunk main to remove discolouration material from the trunk main, also known as flushing, has slowly gained more popularity due to its relatively low capital cost and ease of implementation. Increases in the applied hydraulic force on the pipe walls have been shown to mobilise discolouration material in pipes [19,20], and it is now sufficiently well understood that the resulting turbidity response from the flushing can to various degrees be modelled and predicted [16,21,22].

Unfortunately, a significant limitation of using many of these previous studies to investigate typical discolouration events in water distribution networks is that the discolouration was manually induced. As a result, atypical flow patterns are created for the express purpose of inducing discolouration mobilisation, and changes in the configuration of the network are also sometimes made to ensure that customers are not negatively impacted during the flushing event or works. Equally as important however is that data are only gathered for a short time around a single flushing event, usually on the scale of hours or days. Even when repeated flushes are carried out on the same WDS, the time between flushes is not monitored.

Therefore, while it has been proven that discolouration mobilisation can be caused by hydraulic forces, little and even conflicting evidence has been shown on the scale and frequency of hydraulically mobilised discolouration events under usual WDS operating conditions. Gaffney and Boult [23] showed no turbidity events in a District Metered Area (DMA) under two Formazin Nephelometric Units could be attributed to a change in pressure. Cook et al. [15] showed a number of turbidity events in DMAs could be associated with increased flows at inlet meters; however, the percentage of turbidity events associated with increased flows significantly varied between the five analysed DMAs.

To the authors' best knowledge, no long-term study with continuous turbidity and flow data on trunk mains exists. Likewise, no studies could be found assessing whether discolouration in trunk mains under typical operating conditions is primarily caused by hydraulic events.

This paper presents a long-term continuous study on discolouration mobilisation and a methodology to determine the amount of turbidity that can be attributed to changes in the hydraulic profile in trunk mains. This methodology additionally aims to identify the origin of turbidity in the network to aid in targeted proactive cleaning strategies.

2. Methodology

The methodology presented here evaluates the percentage of turbidity observations that can be linked to preceding hydraulic events in an upstream pipe and thus identifying where discolouration material is more likely to be accumulating in the WDS. This in turn can enable targeted trunk main rehabilitation and cleaning operations.

The methodology is formed from three principles: (a) a hydraulic force that mobilised the discolouration material resulting in the high turbidity observation occurring just prior to the high turbidity observation; (b) a stronger hydraulic force would result in more discolouration material being mobilised, provided that there is available material to mobilise [19,24,25]; (c) discolouration material is constantly being regenerated/built up in all pipes [18,26]. Based on these three principles, a turbidity observation is thought to be the result of a hydraulically-based mobilisation process if a hydraulic force in an upstream pipe preceding the turbidity observation exceeds the recent prior hydraulic forces experienced in that pipe.

The percentage of selected turbidity observations that can be linked to preceding hydraulic events in an upstream pipe is given by the Hydraulically Mobilised Turbidity Percentage (HMTP) shown below:

$$\text{HMTP}(\varepsilon, T, x, y) = \frac{\sum_{\tau \in T} \left[\beta^{\varepsilon}_{y,\tau} > \alpha^{\varepsilon}_{x,\tau} \right]}{|T|} \times 100 \tag{1}$$

where ε is the upstream pipe being assessed, T is the set of turbidity observations τ given in Nephelometric Turbidity Units (NTU), x and y are periods of time, $\beta^{\varepsilon}_{y,\tau}$ is the recent peak velocity (m/s) in pipe ε during a period of time of y duration preceding the turbidity observation τ and $\alpha^{\varepsilon}_{x,\tau}$ is the peak velocity (m/s) in pipe ε during a period of time of x length that precedes the period of time for $\beta^{\varepsilon}_{y,\tau}$.

From the perspective of a turbidity observation, the recent preceding peak velocity $\beta^{\varepsilon}_{y,\tau}$ of pipe ε is only assumed to have caused that turbidity observation if it has exceeded the prior peak velocity $\alpha^{\varepsilon}_{x,\tau}$ of that pipe. This is because the prior velocities in that pipe should have mobilised all the discolouration material that they could, and only a higher velocity should be able to mobilise significantly more material. The prior peak velocity $\alpha^{\varepsilon}_{x,\tau}$ will be called the preconditioned velocity threshold, and the recent preceding peak velocity $\beta^{\varepsilon}_{y,\tau}$ will be called the peak mobilising velocity. Thus, the y parameter determines how far back in time the HMTP should look for hydraulic mobilisation, and the x parameter determines the minimum size of hydraulic events being considered.

An example of the methodology is shown in Figure 1 where velocity and turbidity observations from a real trunk main system are displayed. For the sake of brevity, the methodology is visualised for a single turbidity observation and for which the peak turbidity observation is chosen. The length of time set for y is 24 h because it was sufficiently long enough for all potentially mobilised material from the furthest upstream point to reach the downstream turbidity meter. The length of time set for x is 7 days and was chosen solely for the ease of visualizing this example.

The velocity profile 24 h preceding the turbidity observation (highlighted green) is where the peak mobilising velocity is calculated. The 7 days prior (highlighted yellow) is where the preconditioned velocity threshold is calculated. The peak mobilising velocity (i.e., $\beta^{\varepsilon}_{y,\tau}$) of this upstream pipe is greater than its preconditioned velocity threshold (i.e., $\alpha^{\varepsilon}_{x,\tau}$), and thus, the turbidity observation is determined to have resulted from the hydraulic mobilisation of discolouration material in this pipe.

From the velocity and turbidity measurements shown in Figure 1, it can also be seen that the velocity just before the start of Day 10 also exceeds the peak velocity indicated on Day 3. However, no subsequent turbidity response is seen on Day 10 because that pipe is now reconditioned to the new preconditioned velocity threshold at the end of Day 8 (i.e., all discolouration material that could have been mobilised by this new velocity was already mobilised by the peak velocity at the end of Day 8).

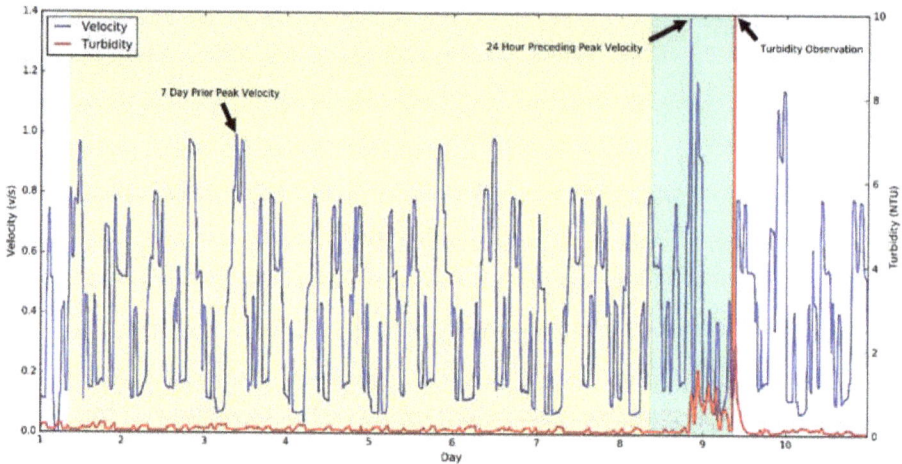

Figure 1. An example of the methodology showing that the turbidity observation could be linked to the preceding upstream hydraulic mobilisation of discolouration material. The 24 h preceding the turbidity observation are highlighted green, and the 7 prior days are highlighted yellow. The preceding 24-h peak velocity is shown to be the cause of the high turbidity observation as it exceeds the prior 7-day peak velocity. NTU: Nephelometric Turbidity Units.

2.1. Chi-Square Test for Independence

While the high turbidity observation examined in Figure 1 is determined to have been caused by the hydraulic mobilisation of discolouration material, it is possible that the velocity profile and preceding turbidity response were coincidental. Thus, the chi-square test for independence will be used to determine the statistical significance of the results.

All turbidity observations will be divided into two turbidity sets of over 1 NTU observations and under 1 NTU observation, and then, each set of turbidity observations will be examined separately. $HMTP_{>1NTU}$ will show the percentage of turbidity observations above 1 NTU that are deemed to be caused by hydraulic mobilisation, and likewise, $HMTP_{<1NTU}$ will show the percentage of turbidity observations below 1 NTU that are deemed to be caused by hydraulic mobilisation. The turbidity threshold of 1 NTU was chosen as it is a clear quantifiable response above background turbidity levels and is the U.K. regulatory limit for water leaving water treatment works [27]. Therefore, a turbidity observation over 1 NTU can be considered as part of a turbidity event, and turbidity observations under 1 NTU can be considered as the absence of a turbidity event.

The proposed null hypothesis is that the turbidity level (i.e., over 1 NTU or under 1 NTU) is independent of an upstream pipe's preceding peak velocity that exceeds the preconditioned velocity threshold. The proposed alternative hypothesis is that higher turbidity levels (i.e., over 1 NTU) are dependent on an upstream pipe's preceding peak velocity that exceeds the preconditioned velocity threshold. Thus, for a statistically-significant result where the null hypothesis can be rejected, a $HMTP_{>1NTU}$ significantly greater than a corresponding $HMTP_{<1NTU}$ is expected to be seen.

The significance level chosen is 0.01, and the chi-square test statistic with 1 degree of freedom is used to calculate the statistical significance.

2.2. Pipes and Pipes in Series

The methodology examines each pipe upstream of the turbidity meter, where a pipe is determined here by stretches of piping where the velocity remains the same. This means an import and export branch or change in diameter determines the boundaries of a pipe.

While each pipe can be examined individually to estimate the amount of discolouration material linked to that pipe, the preconditioned velocity threshold of multiple pipes can be simultaneously exceeded and discolouration material mobilised from multiple pipe simultaneously. This would mean that some turbidity observations are counted as originating from more than one pipe.

Thus, to accurately assess the total amount of turbidity observations that can be linked to hydraulic mobilisation, all pipes upstream of the turbidity meter are also jointly assessed. This is done by separately assessing if any pipe upstream of the turbidity meter experienced a velocity that exceeded their preconditioning velocity threshold. The multiple pipes that are jointly assessed will be called pipe sets.

3. Case Studies

3.1. Description of Sites

Flow and turbidity measurements were taken over two years and 11 months from three hydraulically distinct parts of a real Water Resource Zone (WRZ) in the U.K., starting from 1 September 2013 until 1 August 2016. The three sites range from 6 km to 23 km in network length, 300 mm to 700 mm in pipe diameter size and are each primarily comprised of Ductile Iron (DI). While Site 1 has two turbidity meters, Sites 2 and 3 both have only a single turbidity meter. All turbidity meters were placed at the downstream end of each site, just upstream of a flow meter so that each turbidity measurement has an associated flow measurement. Except for a few insignificantly small water consumptions taken directly off some trunk mains, every inlet and outlet of each site was hydraulically metered by a flow meter.

Site 1 is a trunk main network with one import and six exports. Aside from a flow meter placed directly after the upstream service reservoir, which is the sole inlet for the trunk main network, the other six flow meters were each placed at an exporting branch. Site 1 can be broken down into six pipes and denoted as A, B, C, D, E, F. Pipes A, B, C, D are located upstream of Turbidity Meter (TM) A, and pipes A, B, E, F are located upstream of TM B.

Site 2 is a 6.5 km trunk main with one import and two exports. The flow velocity in this trunk main is primarily determined by two pumps at the downstream end of the main. Site 2 can be broken down into two pipes that are upstream of TM C and will be denoted as Pipes G and H.

Unlike Sites 1 and 2, Site 3 has two flow imports from two different water sources. Site 3 can be broken down into three pipes that are upstream of TM D and will be denoted as I, J and K. As the water from the further downstream import between Pipes J and K is less expensive, only a small flow is typically seen across the almost 20 km length of Pipes I and J. However, when the downstream reservoir is low on water, the upstream pumps engage to supply additional water. A schematic of Site 3 is shown in Figure 2.

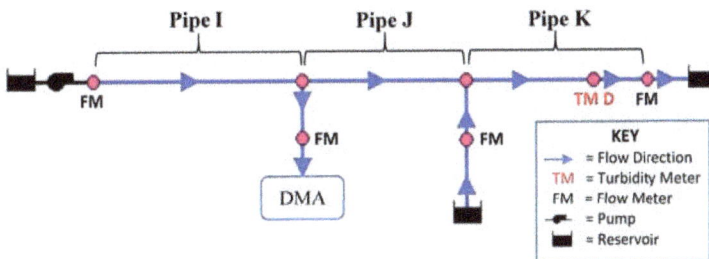

Figure 2. Schematic of Site 3. DMA, District Metered Area.

A summary of each pipe upstream of each turbidity meter is shown in Table 1.

Table 1. Pipe characteristics and the 99th velocity percentile over all observed data for each site.

Site	Pipe	Length	Diameter	Upstream of Turbidity Meter	99th Velocity Percentile
1	A	1.8 km	700 mm	A, B	0.94 m/s
	B	1.6 km	700 mm	A, B	0.92 m/s
	C	1.9 km	600 mm	A	0.17 m/s
	D	5.1 km	300 mm	A	0.65 m/s
	E	1.8 km	400 mm	B	0.86 m/s
	F	4.4 km	400 mm	B	0.80 m/s
2	G	1.9 km	450 mm	C	0.83 m/s
	H	4.6 km	450 mm	C	0.80 m/s
3	I	11 km	400 mm	D	0.79 m/s
	J	8.5 km	400 mm	D	0.70 m/s
	K	3.6 km	400 mm	D	0.73 m/s

3.2. Flow and Turbidity Data

The flow and turbidity observations were logged at 15-min intervals with flow recorded as the sum of water through the meter during that interval and turbidity observations recorded as the current turbidity value at the interval. Flow was originally recorded in cubic meters per 15 min ($m^3/15$ min) and turbidity in NTU. A summary of the turbidity data for each turbidity meter is shown in Table 2.

Table 2. Summary of turbidity observations for each site.

Turbidity Meter (TM)	Duration Monitored	99th Percentile (NTU)	Observations >1 NTU
TM A (Site 1)	2 years, 11 months	0.41	265
TM B (Site 1)	2 years, 11 months	0.42	328
TM C (Site 2)	2 years, 11 months	0.36	290
TM D (Site 3)	2 years, 5 months	0.46	204

While all turbidity meters captured data over the same time period, TM D in Site 3 was offline for a total of six months, from July 2014 to November 2014 and then from June 2016 to August 2016. This is a considerable factor in why TM D has fewer turbidity observations over 1 NTU than the other turbidity meters. As shown by the 99th percentiles in Table 2, the vast majority of turbidity observations are significantly less than the 1 NTU threshold chosen in the methodology.

4. Results

4.1. Hydraulically Mobilised Turbidity Percentage

The results of the HMTP using an x of 1 day and y of 1 day applied to each turbidity meter and its corresponding jointly assessed pipe set are shown in Table 3.

Table 3. Hydraulically Mobilised Turbidity Percentage (HMTP) carried out on each pipe in series between the upstream sources and downstream service reservoirs to assess the amount of turbidity observations that can be linked to hydraulic mobilisation. The x and y parameters of HMTP were set to 1 day.

Turbidity Meter	Pipes in Set	HMTP$_{<1NTU}$	HMTP$_{>1NTU}$	*p*-Value
TM A (Site 1)	A, B, C, D	81%	100%	$p < 10^{-9}$
TM B (Site 1)	A, B, E, F	77%	91%	$p \approx 10^{-8}$
TM C (Site 2)	G, H	53%	93%	$p < 10^{-9}$
TM D (Site 3)	I, J, K	66%	84%	$p \approx 10^{-6}$

A length of 1 day was chosen for the y parameter because it was sufficiently long enough for material mobilised from the furthest upstream points of each site to reach their respective downstream turbidity meters. The x parameter was set to 1 day to show the maximum amount of turbidity in each site that can be linked to preceding upstream hydraulic events.

The $HMTP_{>1NTU}$ results in Table 3 range from 84% to 100%, thus showing that the majority of turbidity can be linked to preceding hydraulic events. However, because the requirements for the methodology to determine if there were a hydraulic event preceding a turbidity observation are quite low with only an x of 1 day (i.e., the peak velocity in the previous 24 h exceeds the prior 24-h peak velocity), the $HMTP_{<1NTU}$ results in Table 3 are also substantially high. While the p-values show that the null hypothesis can be rejected at a 0.01 level of significance, a significantly larger gap between the $HMTP_{<1NTU}$ and $HMTP_{>1NTU}$ results would indicate greater confidence in the methodology and results.

The effect of increasing the x parameter for $HMTP_{<1NTU}$ and $HMTP_{>1NTU}$ can be seen as plotted in Figure 3a. Figure 3a shows that the $HMTP_{<1NTU}$ of each TM exponentially decays while the $HMTP_{>1NTU}$ decreases at a substantially slower rate. An objective function calculating the trade-off between the $HMTP_{<1NTU}$ and $HMTP_{>1NTU}$ is given by the formula shown below:

$$\varphi = \frac{HMTP_{>1NTU} + (1 - HMTP_{<1NTU})}{2} \tag{2}$$

Figure 3b shows φ plotted for each TM over increasing values of x. A higher φ indicates a better trade-off between a low $HMTP_{<1NTU}$ and a high $HMTP_{>1NTU}$.

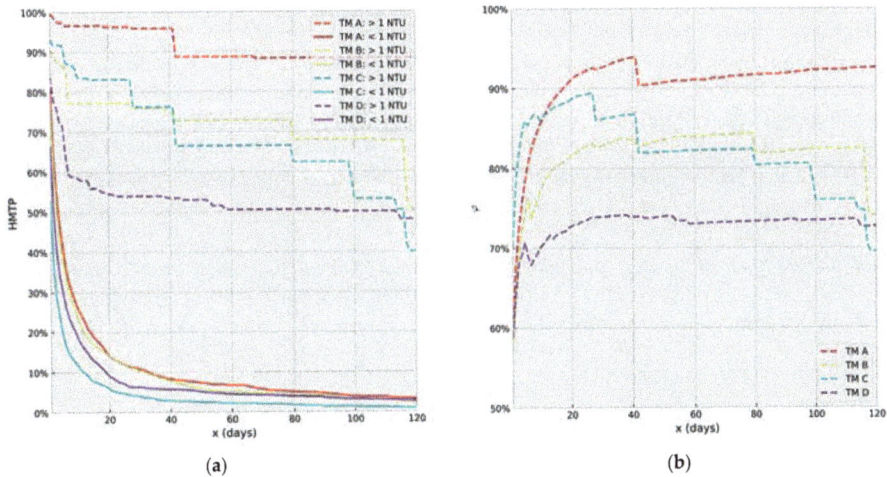

(a)   (b)

Figure 3. (a) The $HMTP_{<1NTU}$ and $HMTP_{>1NTU}$ are shown for each TM over increasing x parameters; (b) the objective formula φ is shown for each TM over increasing x parameters.

From the results shown in Figure 3b, a 30-day length was chosen for x, and the corresponding results for each jointly assessed pipe set followed by the results for each individual pipe that makes up that pipe set are shown in Table 4. The percentage of turbidity observations under 1 NTU that the methodology deemed to be hydraulically mobilised ranges from 3% to 5% for individual pipes and 4% to 11% for the grouped pipe sets; while the percentage of turbidity observations over 1 NTU ranges from 0% to 84% for individual pipes and 54% to 96% for the grouped pipe sets.

Table 4. Hydraulically Mobilised Turbidity Percentage (HMTP) results with the x parameter set to 30 days and the y parameter set to 1 day.

Turbidity Meter	Pipes	HMTP$_{<1NTU}$	HMTP$_{>1NTU}$	p-Value
TM A (Site 1)	A, B, C, D	11%	96%	$p < 10^{-9}$
	A	4%	26%	$p < 10^{-9}$
	B	4%	33%	$p < 10^{-9}$
	C	5%	84%	$p < 10^{-9}$
	D	5%	81%	$p < 10^{-9}$
TM B (Site 1)	A, B, E, F	10%	76%	$p < 10^{-9}$
	A	4%	75%	$p < 10^{-9}$
	B	4%	75%	$p < 10^{-9}$
	E	5%	1%	$p = 1$
	F	5%	0%	$p = 1$
TM C (Site 2)	G, H	4%	76%	$p < 10^{-9}$
	G	4%	76%	$p < 10^{-9}$
	H	3%	76%	$p < 10^{-9}$
TM D (Site 3)	I, J, K	6%	54%	$p < 10^{-9}$
	I	3%	41%	$p < 10^{-9}$
	J	3%	39%	$p < 10^{-9}$
	K	4%	52%	$p < 10^{-9}$

Note that the sum of HMTP$_{>1NTU}$ for individual pipes belonging to each turbidity meter exceeds 100%. This was expected because, as mentioned above, a turbidity observation can be linked to multiple pipes if a hydraulic event occurs in both pipes simultaneously.

For the results of TM A, a high HMTP$_{>1NTU}$ of 84% is given for Pipe C. This is important to note because as can be seen from Table 1, Pipe C had a low 99th velocity percentile of 0.17 m/s, which indicates a high potential for material build up. The 12% difference between the HMTP$_{>1NTU}$ of 84% for Pipe C and the HMTP$_{>1NTU}$ of 96% for the pipe set is assumed to come from Pipes A and B and not Pipe D because Pipes C and D have very similar velocity profiles (when compared to Pipes A and B).

Comparing the TM A and TM B cases shows two very different sets of HMTP$_{>1NTU}$ results for Pipes A and B, even though they are both located on the same site. This indicates that significant discolouration material is being mobilised from Pipes A and B; and the material that travels towards TM B reaches it, but a portion of the material that travels towards TM A ends up settling/attaching in Pipe C and then remobilises at a later time.

4.2. Turbidity and Velocity Relationships

Figure 4 shows the peak velocity in the 24 h preceding a turbidity observation plotted against the peak velocity in the 30 days prior to the 24 h preceding the turbidity observation for two pipes that had the highest HMTP$_{>1NTU}$ for their respective turbidity meters as shown in Table 4. The size of each data point in a plot is relative to the turbidity measurement where a higher turbidity value results in a bigger data point. Because all turbidity observations from the same turbidity event should have the same 24-h preceding peak velocity and prior 30-day peak velocity, each visible data point is actually an individual turbidity event with the size of the data point representing the biggest single turbidity observation seen during that turbidity event.

The dashed line, which is also the identity line, shows where the preconditioned velocity threshold for data points along the y axis is. Hence, data points above the identity line are considered to have exceeded their preconditioned velocity thresholds for that pipe because they have experienced a velocity in the preceding 24 h that is higher than all velocities experienced in the prior 30 days. Although data points below the identity line of a specific pipe cannot be linked to hydraulic mobilisation from that pipe, it does not mean that they cannot be linked to hydraulic mobilisation from a different pipe upstream of the turbidity meter.

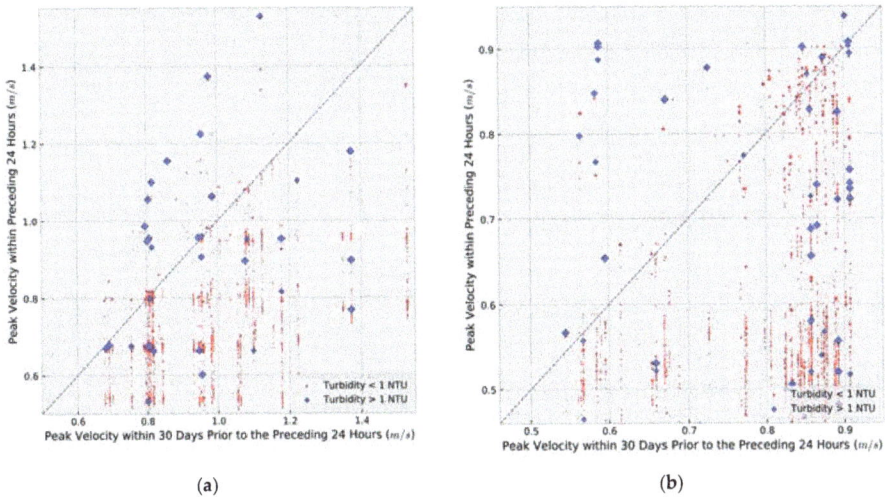

Figure 4. For each turbidity observation, the 24-h preceding peak velocity was plotted against the peak velocity of the 30 days prior to the 24 h preceding the turbidity observation. (**a**) Site 2, Turbidity Meter C, Pipe H; (**b**) Site 3, Turbidity Meter D, Pipe K.

If a cleaning velocity (i.e., velocity at which all material is removed from pipe) existed in the any pipes, a significant number of low and only low turbidity observations would be seen above the identity line after a sufficiently high prior peak velocity (i.e., the *x* axis of Figure 4). However, this is not observed in any pipes, and thus, no clear mobilisation limit is seen. Similarly, by looking at only the peak velocities in pipes during the 24 h preceding the arrival of turbidity (i.e., *y* axis of Figure 4), a clear minimum mobilising velocity was not seen in any pipes. Instead, it can be clearly seen that turbidity observations under 1 NTU are ubiquitously present below the identity line, but not above.

Table 5 shows Spearman's rank correlation coefficients for three different sets of velocity calculations correlated with their associated turbidity sets. Spearman's rank correlation coefficient measures the dependence of two parameters as described as a monotonic function. Linearity between the two parameters is not assumed in Spearman correlation, which is not the case for Pearson's correlation coefficient. A Spearman correlation coefficient of 1 or -1 indicates a perfect monotonic relationship. For each correlation coefficient, an associated *p*-value is derived from a statistical *t*-test, which indicates the probability of an uncorrelated system generating datasets that have a correlation at least as extreme.

The first correlation is the *24h Peak Velocity* which is the 24-h preceding peak velocities of each turbidity observation, regardless of turbidity value, correlated with those turbidity observations. The very weak to weak positive correlation across all pipes shows that higher preceding velocities alone rarely indicate the appearance of higher turbidity concentrations downstream. However, these correlations are predominantly driven by the many low bulk flow turbidity observations and tell little about the correlation between peak velocities preceding a turbidity event and the amount of turbidity mobilised in that event. Thus, the second correlation set shown in Table 5 is the *24-h Peak Event Velocity* where the 24-h preceding peak velocities of turbidity events are correlated with the downstream turbidity observations of those turbidity events. Interestingly, there are a few negative correlations, but because they are very weak correlations, not much can be inferred.

Table 5. Spearman's rank correlation coefficients and associated *p*-values for three sets of correlations: (**a**) 24-h preceding peak velocities of each turbidity observation correlated with those turbidity observations; (**b**) 24-h preceding peak velocities of each turbidity event correlated with the turbidity observations of that event; (**c**) the difference between the 24-h preceding peak velocity and the 30-day preconditioned threshold of each turbidity event correlated with the turbidity observations of that event.

Turbidity Meter	Pipe	(a) 24-h Peak Vel.	*p*-Value	(b) 24-h Peak Event Vel.	*p*-Value	(c) Exceeded Vel. Difference	*p*-Value
TM A (Site 1)	A	0.30	$p < 10^{-40}$	−0.16	$p \approx 10^{-23}$	0.55	$p < 10^{-40}$
	B	0.28	$p < 10^{-40}$	−0.13	$p \approx 10^{-15}$	0.38	$p \approx 10^{-21}$
	C	0.29	$p < 10^{-40}$	0.40	$p < 10^{-40}$	0.13	$p \approx 10^{-08}$
	D	0.29	$p < 10^{-40}$	0.41	$p < 10^{-40}$	0.16	$p \approx 10^{-10}$
TM B (Site 1)	A	0.38	$p < 10^{-40}$	0.22	$p < 10^{-40}$	0.42	$p < 10^{-40}$
	B	0.38	$p < 10^{-40}$	0.22	$p < 10^{-40}$	0.48	$p < 10^{-40}$
	E	0.06	$p < 10^{-40}$	−0.18	$p \approx 10^{-32}$	−0.30	$p \approx 10^{-14}$
	F	0.02	$p \approx 10^{-10}$	−0.17	$p \approx 10^{-31}$	−0.26	$p \approx 10^{-08}$
TM C (Site 2)	G	0.32	$p < 10^{-40}$	0.40	$p < 10^{-40}$	0.41	$p < 10^{-40}$
	H	0.30	$p < 10^{-40}$	0.40	$p < 10^{-40}$	0.41	$p < 10^{-40}$
TM D (Site 3)	I	0.13	$p < 10^{-40}$	0.15	$p \approx 10^{-22}$	0.44	$p < 10^{-40}$
	J	0.27	$p < 10^{-40}$	0.14	$p \approx 10^{-20}$	0.52	$p < 10^{-40}$
	K	0.31	$p < 10^{-40}$	0.22	$p < 10^{-40}$	0.24	$p \approx 10^{-20}$

The third correlation set is the *Exceeded Velocity Difference* where the difference between the 24-h preceding peak velocity and the 30-day preconditioned velocity threshold (i.e., the identity lines shown in Figure 4) of each turbidity event is correlated with the turbidity observations of those events. While the largest correlation is only 0.55, the majority of correlations are moderately positive, which is significantly stronger compared to the *24-h Peak Velocity* and *24-h Peak Event Velocity* correlations. This shows that the exceeded velocity difference is a better indicator of the resulting turbidity event size than a preceding increase in velocity alone.

5. Discussion

A sum total of 1087 turbidity observations of over 1 NTU were recorded from just under three years' worth of turbidity measurements at four turbidity meters. As each observation is taken at a 15-min interval, this is equivalent to using 54 turbidity events where the turbidity level of each event is at least 1 NTU for over 5 h straight. When considering that turbidity is akin to a concentration and is significantly diluted by the high flow rates typical of trunk mains, then even relatively low turbidity observations in trunk mains should be of somewhat concern. This is because the discoloration material that does not directly reach a customer's tap can still resettle in a downstream network. Then, that same discolouration material when remobilised in a smaller distribution pipe with a fraction of the flow rate could result in a significantly higher turbidity reading.

Discolouration material clearly does build up in the trunk mains observed in this paper, this is despite the relatively high velocity percentiles shown in Table 1 that vastly exceed the 'self-cleaning' velocity ranges (0.25 m/s–0.4 m/s) associated with smaller distribution pipes [3,28]. This agrees with the findings of other authors conducted on large diameter pipes (i.e., over 200 mm), which find no evidence for self-cleaning velocities or shear stresses [18,24,29].

The majority of *p*-values in Tables 3 and 4 show the null hypothesis being overwhelmingly rejected at a 0.01 level of significance and thus conclude that higher turbidity levels (i.e., over 1 NTU) can be explained by an upstream pipe's preceding peak velocity exceeding its preconditioned velocity threshold. These *p*-values here are particularly low due to the high number of turbidity observations considered (i.e., over 100,000), thus making it very unlikely that these values would be seen in uncorrelated results.

The only exceptions in results seen in Tables 4 and 5 are Pipes E and F for TM B, which are distinctly different from all other results as, conversely, only a few high turbidity observations can be linked to preceding hydraulic events in these pipes. This may be indicative of an underlying process that is either preventing discolouration material from sufficiently accumulating in these pipes or the methodology from accurately modelling them.

Important to note is that the velocity peaks preceding turbidity observations are a relatively small increase in comparison to the average daily peak velocity, typically being less than 110% of the average daily peak velocity. This shows how sensitive discolouration material can be to mobilising velocities and indicates why discolouration is so often attributed to scheduled works that alter velocities in WDS.

While Table 3 shows that a maximum amount of 84% to 100% of turbidity observations over 1 NTU in the trunk mains examined here can be linked to hydraulic mobilisation, this also conversely shows that between 0% and 16% of turbidity observations cannot be linked to the hydraulically-driven mobilisation process outlined in this paper. This leaves a few possibilities about the mobilisation of the remaining turbidity: (a) some of the discolouration material was mobilised from further upstream (i.e., reservoirs or treatment works); (b) a hydraulic process not accounting for such as a transient event or flow reversal caused some mobilisation; (c) a non-hydraulic process caused some mobilisation (e.g., biofilm detachment/sloughing that can sometimes occur without an increase in hydraulic force).

The methodology presented here does not make any assumptions about what the discolouration material consists of (e.g., manganese, biofilms), what form the discolouration material takes inside pipes (e.g., sediment, cohesive layers), nor does it assume a rate (e.g., linear, exponential) at which discolouration material is mobilised. Additionally, because the mobilisation condition has been reduced to a simple "greater than prior" condition, as long as the hydraulic force has a monotonic relationship to the flow rate, it also does not matter what the hydraulic force is (e.g., velocity, shear stress, laminar boundary layer size). This means, in theory, that the methodology could be applied to almost any WDS regardless of the material composition, layout and range of flow rates of the WDS.

As flow meters are already ubiquitous in WDS, this methodology also shows the potential information gain from installing even a single turbidity meter. As the accuracy of the methodology to identify the primary sources of discolouration increases with more data, installing a turbidity meter at a downstream service reservoir where regular maintenance is easily achievable is advised. If possible, further turbidity meters should be installed at the downstream ends of different network branches. This would enable the correlation of methodology results to further identify high discolouration risk pipes, as was shown done for Site 1.

Regarding the frequency of flow and turbidity observations, while a 15 min sampling frequency was deemed sufficient for the sites examined here, a higher frequency may be required for WDSs that can experience sharp, but short-lived velocity spikes. This is because a significant, but short-lived velocity spike that could cause discolouration may only present as a minor increase in the cumulative flow over a 15 min period.

6. Conclusions

This paper presents a long-term continuous study of discolouration mobilisation and a methodology to determine the approximate amount and origin of hydraulically mobilised turbidity in trunk mains. The methodology is validated on three real sites in the U.K. The following conclusions are made based on the case studies results obtained:

(a) The methodology shows that for the four turbidity meters used in this study, a maximum of 84%, 91%, 93% and 100% of turbidity observations over 1 NTU could be linked to preceding hydraulic forces that exceeded an upstream pipe's hydraulically preconditioned state. This shows that the mobilisation of discolouration material is predominantly determined by hydraulic forces, which, in turn, indicates significant potential for modelling and predicting discolouration events.

(b) The methodology showed that even without a calibrated hydraulic model, it is possible to determine the approximate origin of discolouration material that had been hydraulically mobilised within each site analysed. This can be used as an aid in the prioritisation of cleaning trunk mains and targeted mains rehabilitation.

(c) The level of turbidity is shown to be significantly dependent on preceding upstream velocities that exceed a pipe's preconditioned state. Furthermore, discolouration material is shown to accumulate regardless of the velocity magnitude, thus indicating that controlling the shape of the hydraulic profile is vital in effectively managing discolouration risk.

Acknowledgments: The authors are grateful to the Engineering and Physical Sciences Research Council (EPSRC) for providing the financial support as part of the STREAM project and to Julian Collingbourne of South West Water for supplying the data used in this paper.

Author Contributions: Each of the authors contributed to the design, analysis and writing of the study.

Conflicts of Interest: The authors declare no conflict of interest.

References

1. OFWAT Serviceability Outputs for PR09 Final Determinations. Available online: https://www.ofwat.gov.uk/wp-content/uploads/2015/11/det_pr09_finalfull.pdf (accessed on 22 August 2017).
2. DWI Drinking Water Quality Events in 2013. Available online: http://dwi.defra.gov.uk/about/annual-report/2013/dw-events.pdf (accessed on 22 August 2017).
3. Vreeburg, I.J.H.G.; Boxall, D.J.B. Discolouration in potable water distribution systems: A review. *Water Res.* **2007**, *41*, 519–529. [CrossRef] [PubMed]
4. Armand, H.; Stoianov, I.I.; Graham, N.J.D. A holistic assessment of discolouration processes in water distribution networks. *Urban Water J.* **2017**, *14*, 263–277. [CrossRef]
5. Douterelo, I.; Boxall, J.B.; Deines, P.; Sekar, R.; Fish, K.E.; Biggs, C.A. Methodological approaches for studying the microbial ecology of drinking water distribution systems. *Water Res.* **2014**, *65*, 134–156. [CrossRef] [PubMed]
6. Vreeburg, J.H.G.; Schippers, D.; Verberk, J.Q.J.C.; van Dijk, J.C. Impact of particles on sediment accumulation in a drinking water distribution system. *Water Res.* **2008**, *42*, 4233–4242. [CrossRef] [PubMed]
7. Liu, G.; Zhang, Y.; Knibbe, W.-J.; Feng, C.; Liu, W.; Medema, G.; van der Meer, W. Potential impacts of changing supply-water quality on drinking water distribution: A review. *Water Res.* **2017**, *116*, 135–148. [CrossRef] [PubMed]
8. Makris, K.C.; Andra, S.S.; Botsaris, G. Pipe scales and biofilms in drinking-water distribution systems: Undermining finished water quality. *Crit. Rev. Environ. Sci. Technol.* **2014**, *44*, 1477–1523. [CrossRef]
9. Fish, K.; Osborn, A.M.; Boxall, J.B. Biofilm structures (EPS and bacterial communities) in drinking water distribution systems are conditioned by hydraulics and influence discolouration. *Sci. Total Environ.* **2017**, *593–594*, 571–580. [CrossRef] [PubMed]
10. Husband, S.; Boxall, J.B. Field studies of discoloration in water distribution systems: Model verification and practical implications. *J. Environ. Eng.* **2009**, *136*, 86–94. [CrossRef]
11. Husband, P.S.; Whitehead, J.; Boxall, J.B. The role of trunk mains in discolouration. *Proc. Inst. Civ. Eng. Water Manag.* **2010**, *163*, 397–406. [CrossRef]
12. Armand, H.; Stoianov, I.; Graham, N. Investigating the impact of sectorized networks on discoloration. *Procedia Eng.* **2015**, *119*, 407–415. [CrossRef]
13. Blokker, E.J.M.; Schaap, P.G. Temperature influences discolouration risk. *Procedia Eng.* **2015**, *119*, 280–289. [CrossRef]
14. Cook, D.M.; Boxall, J.B. Discoloration material accumulation in water distribution systems. *J. Pipeline Syst. Eng. Pract.* **2011**, *2*, 113–122. [CrossRef]
15. Cook, D.M.; Husband, P.S.; Boxall, J.B. Operational management of trunk main discolouration risk. *Urban Water J.* **2015**, *13*, 382–395. [CrossRef]
16. Husband, S.; Boxall, J. Understanding and managing discolouration risk in trunk mains. *Water Res.* **2016**, *107*, 127–140. [CrossRef] [PubMed]
17. Sunny, I.; Husband, S.; Drake, N.; Mckenzie, K.; Boxall, J. Quantity and quality benefits of in-service invasive cleaning of trunk mains. *Drink. Water Eng. Sci. Discuss.* **2017**, 1–9. [CrossRef]
18. Vreeburg, J.H.G.; Beverloo, H. Sediment accumulation in drinking water trunk mains. In Proceedings of the 11th International Conference on Computing and Control for the Water Industry, Urban Water Management: Challenges and Opportunities, Exeter, UK, 5–7 September 2011.
19. Boxall, J.B.; Saul, A.J. Modeling discoloration in potable water distribution systems. *J. Environ. Eng.* **2005**, *131*, 716–725. [CrossRef]

20. Verberk, J.; Hamilton, L.A.; O'Halloran, K.J.; Van Der Horst, W.; Vreeburg, J. Analysis of particle numbers, size and composition in drinking water transportation pipelines: Results of online measurements. *Water Sci. Technol. Water Supply* **2006**, *6*, 35–43. [CrossRef]

21. Meyers, G.; Kapelan, Z.; Keedwell, E.; Randall-Smith, M. Short-term forecasting of turbidity in a UK water distribution system. *Procedia Eng.* **2016**, *154*, 1140–1147. [CrossRef]

22. Meyers, G.; Kapelan, Z.; Keedwell, E. Short-term forecasting of turbidity in trunk main networks. *Water Res.* **2017**, *124*, 67–76. [CrossRef] [PubMed]

23. Gaffney, J.W.; Boult, S. Need for and use of high-resolution turbidity monitoring in managing discoloration in distribution. *J. Environ. Eng.* **2012**, *138*, 637–644. [CrossRef]

24. Husband, P.S.; Jackson, M.; Boxall, J.B. Trunk main discolouration trials and strategic planning. In Proceedings of the 11th International Conference on Computing and Control for the Water Industry, Urban Water Management: Challenges and Opportunities, Exeter, UK, 5–7 September 2011.

25. Slaats, P.G.G.; Rosenthal, L.P.M.; Siegers, W.G.; van den Boomen, M.; Beuken, R.H.S.; Vreeburg, J.H.G. *Processes Involved in the Generation of Discolored Water*; AWWA Research Foundation: Denver, CO, USA, 2003.

26. Furnass, W.R.; Collins, R.P.; Husband, P.S.; Sharpe, R.L.; Mounce, S.R.; Boxall, J.B. Modelling both the continual erosion and regeneration of discolouration material in drinking water distribution systems. *Water Sci. Technol. Water Supply* **2014**, *14*, 81. [CrossRef]

27. DWI Public Water Supplies in the Western Region of England. Available online: http://dwi.defra.gov.uk/about/annual-report/2014/ (accessed on 13 March 2016).

28. Blokker, E.J.M.; Vreeburg, J.H.G.; Schaap, P.G.; van Dijk, J.C. The self-cleaning velocity in practice. In Proceedings of the Water Distribution System Analysis Conference, Tucson, AZ, USA, 12–15 September 2010.

29. Seth, A.D.; Husband, P.S.; Boxall, J.B. Rivelin trunk main flow test. In Proceedings of the 10th International on Computing and Control for the Water Industry, Sheffield, UK, 1–3 September 2009; Maksimovic, B., Ed.; Taylor Francis: Oxfordshire, UK, 2009; pp. 431–434.

water

MDPI

Article

Influence of Extreme Strength in Water Quality of the Jucazinho Reservoir, Northeastern Brazil, PE

Rafael Roney Camara de Melo [1,*], **Ioná Maria Beltrão Rameh Barbosa** [2], **Aida Araújo Ferreira** [2], **Alessandra Lee Barbosa Firmo** [2], **Simone Rosa da Silva** [3], **José Almir Cirilo** [4] and **Ronaldo Ribeiro Barbosa de Aquino** [5]

[1]   Federal Institute of Pernambuco—Campus Pesqueira, Pesqueira 55200-000, Brazil
[2]   Federal Institute of Pernambuco—Campus Recife, Recife 50740-540, Brazil; ionarameh@recife.ifpe.edu.br (I.M.B.R.B.); aidaferreira@recife.ifpe.edu.br (A.A.F.); alessandrafirmo@recife.ifpe.edu.br (A.L.B.F.)
[3]   Polytechnic School of Pernambuco, University of Pernambuco, Recife 50720-00, Brazil; simonerosa@poli.br
[4]   Agreste Academic Campus, Federal University of Pernambuco, Caruaru 55002-970, Brazil; almir.cirilo@gmail.com
[5]   Department of Electrical Engineering, Federal University of Pernambuco, Recife 50740-530, Brazil; rrba@ufpe.br
*   Correspondence: rafael.roney@yahoo.com.br

Received: 28 August 2017; Accepted: 27 November 2017; Published: 7 December 2017

Abstract: The Jucazinho reservoir was built in the State of Pernambuco, Northeastern Brazil, to water supply in a great part of the population that live in the semi-arid of Pernambuco. This reservoir controls the high part of Capibaribe river basin, area affected several actions that can compromise the reservoir water quality such as disposal of domestic sewage, industrial wastewater and agriculture with use of fertilizers. This study aimed to identify the factors that lead to water quality of the Jucazinho reservoir using a database containing information of nine years of reservoir water quality monitoring in line with a multivariate statistical technique known as Principal Component Analysis (PCA). To use this technique, it was selected two components which determine the quality of the reservoir water. The first principal component, ranging from an annual basis, explained the relationship between the development of cyanobacteria, the concentration of dissolved solids and electrical conductivity, comparing it with the variation in the dam volume, total phosphorus levels and turbidity. The second principal component, ranging from a mensal basis, explained the photosynthetic activity performed by cyanobacteria confronting with the variation in the dam volume. It observed the relationship between water quality parameters with rainfall, featuring an annual and seasonal pattern that can be used as reference to behaviour studies of this reservoir.

Keywords: water supply; principal component analysis; Jucazinho reservoir

1. Introduction

The Jucazinho reservoir, located in the Capibaribe River Basin in the state of Pernambuco, presents deteriorated water quality as a result of nutrient insertion due to the use and occupation of the contributing hydrographic basin. This reservoir is responsible for the water supply of approximately 800 thousand inhabitants in the Agreste region of Pernambuco [1].

The deterioration of water quality over the years and the consequent investments for the improvement of treatment systems to make it drinkable preoccupy water resource managers and technicians of COMPESA—the Sanitation Company of Pernambuco State—responsible for water supply and sanitation services to the majority of the population of Pernambuco.

Due to the Jucazinho reservoir's economic and social importance for the region, the quality of its water must be frequently monitored, noting its seasonal variation and investigating the possible causes of the changes that occur. The reservoirs of the Brazilian semi-arid region present seasonal variations, in which, generally, there is a greater concentration of nutrients impacting the trophic state, during the periods of drought [2–5]. In the semi-arid region of Brazil, reservoir levels and their variations impact the quality of the stored water, that is, in a period of little rainfall, there is a reduction in the level of the reservoir, and there are thus consequent changes in water quality [3].

The excessive concentration of nutrients in many lakes and reservoirs is pointed out as anthropogenic [6–9]. This subject has been extensively discussed by public authorities, considering that low water quality implies a greater expenditure of time and financial resources to make the water suitable [10]. In addition, there is a risk of a presence of toxins and their impacts on drinkable water when the eutrophication process and high concentration of algae occurs [11].

Water quality is measured by parameters related to the physical, chemical, and biological characteristics of the water [12]. Such parameters are obtained through the collection of water samples from the reservoir for subsequent physical, chemical, and biological analyses.

A continuous collection of data that portray water quality, meteorological records, and conditions of use and occupation of the soil in the basin leads to a set of information that are often not trivial and that can be used to carry out analyses on water quality. For a large data set that involves many variables, the solution for a better understanding of the relationships among variables is not simple, so there is a need to apply a statistical technique of size reduction, that is, this technique replaces the set of existing variables by another, smaller, set that is a combination of the original variables. These applied statistical techniques shows the most important relationships among the variables and provide results that allow for a temporal analysis of the behavior of the data in study area [13,14]. Factorial analysis (FA), cluster analysis, and principal component analysis (PCA) are some of the statistical analyses applied to size reduction [15,16].

With the evolution of computers and software, there is a growing application of multivariate statistics, leading to many studies with applications in the area of water resources. Several authors have used techniques of multivariate statistical analysis to explain the variables that govern the water quality of various sources, mainly for the verification of the spatial and temporal behavior of these variables, since the amount of data involved in the study sometimes hinders the interpretation of results [17,18].

The present study had as an objective the verification of the behavior of water quality of the Jucazinho reservoir during a period of extreme drought. For this, one of the above-mentioned multivariate statistical techniques was used, the PCA.

2. Materials and Methods

This section presents information about the study area, where the water samples were collected to determine the physical, chemical, and biological parameters. The method by which the data of the water level of the dam was collected, the treatment of the data, and the statistical analysis is also described.

2.1. Spring Studied and Its Area of Hydric Contribution

The Jucazinho reservoir (location shows in Figure 1) was built in 1998 by the National Department of Works against Droughts (DNOCS), aiming, among other objectives, to ensure the water supply of the region and increase flood control along the Capibaribe river downstream from the dam until Recife, the capital of Pernambuco [19].

This reservoir has a flow contribution area greater than half of the drainage area of the Capibaribe basin, responsible for the water supply of about 800 thousand inhabitants in the Agreste of Pernambuco. It is located between the municipalities of Surubim and Cumaru and has an accumulation capacity of 327 million cubic meters. The flood control system is also integrated by three other reservoirs: Carpina, Goitá, and Tapacurá. Together, the reservoirs protect 3 million people against floods [20].

The part of the river basin controlled by the Jucazinho reservoir is fully situated within the semi-arid region of Northeastern Brazil. While there were seasons with high rainfall during the first decade of this century, this region suffered the worst drought of the last 60 years from 2011 to 2016. This drought was characterized by irregular and low rainfall concentrated within a short period of each year. This phenomenon, associated with high evaporation rates (2500–3000 mm per year), is causing collapse in almost all reservoirs, including Jucazinho.

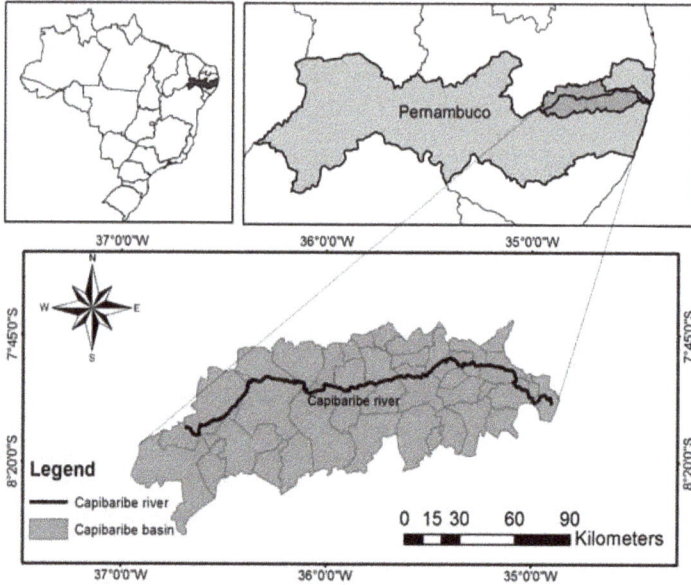

Figure 1. Location of the Capibaribe river basin and the area of water contribution of the Jucazinho reservoir.

Figure 2 shows the variation in the accumulated volume in Jucazinho (2005–2013). Since then, the volume reduction trend continued until the collapse in 2016 [19].

Upstream of this reservoir, the Capibaribe river stretch is the destination of both domestic sewage from urban areas of the cities Santa Cruz do Capibaribe and Toritama, and industrial effluent of jeans laundries in the latter municipality [21].

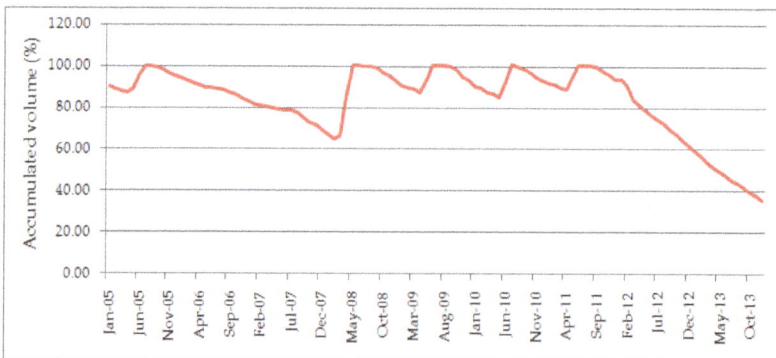

Figure 2. The variation in accumulated volume in Jucazinho in the period from 2005 to 2011. Source: [8].

2.2. Acquisition of Data

Currently, the Water and Climate Agency of Pernambuco (APAC) and the State Environment Agency (CPRH) work in the fulfillment of the task of monitoring the water quality of state water bodies, with a view to the same attribution to both agencies.

The systematic monitoring of the rivers and reservoirs has been performed more frequently since 2005, when there was a resizing of the mesh of the sampling stations. Specifically for this year, monitoring was performed monthly. Currently, most of the water bodies, where, basically, surface water samples are collected, are monitored quarterly.

In this study, a series of data collected from January 2005 to August 2013, ceded by the APAC/CPRH, was analyzed. The variables studied were temperature (T), electrical conductivity (EC), hydrogen potential (pH), dissolved oxygen (DO), biochemical demand for oxygen (BDO), total phosphorus (P), turbidity (TUR), total solids (TS), cyanobacteria (CB), and the volume of the dam % (VD). These variables were selected because they are part of the main parameters monitored by CPRH.

It is worth mentioning that the series of data for the volume of the dam was obtained in the historical database of DNOCS. Specifically for these data, filling omissions and gaps in the information was necessary, which was carried out by linear interpolation, because the decrease of the level of the reservoir presents linear behavior.

2.3. Statistical Analyses

The statistical analysis used aimed to reduce the size of 10 water quality parameters in some components to facilitate interpretation of the original data. The multivariate statistical technique of PCA was applied from the data available. The database consisted of 10 variables and 45 cases, totalizing a record set of 450 data points. In matrix form, the original data were expressed by $X = (x_{i,j})$, where $i = 1 \ldots n$ samplings (450), and $j = 1 \ldots p$ variables (10). In the PCA technique, the first step is to transform the array of original data into a correlation matrix $[R]$ $(p \times p)$, for p equal to 10 water quality parameters analyzed in this study.

The data were then subjected to a statistical standardization procedure, wherein the original value is subtracted from the mean and divided by the standard deviation. The standardization is a statistical procedure with the original data to facilitate the comparison between the variables of different magnitudes, since all values of the mean and standard deviation after standardization present 0 and 1, respectively. These data were standardized in an electronic spreadsheet using Equation (1). Y_{ij}, X_{ij}, $S(\overline{X}_j)$, and $\overline{X}_j$ represent, respectively, the standardized value of the variable, the value of the original variable, the standard deviation, and the mean of the record set.

$$Y_{ij} = \frac{X_{ij} - \overline{X}_j}{S(\overline{X}_j)} a = 1. \tag{1}$$

The overall consistency of the data was measured by the Kayser–Mayer–Olkim method (KMO) [16]. The KMO method uses a criterion to identify whether a factorial analysis model that is being used is properly adjusted to data, testing the overall consistency of the data. This procedure checks whether the inverse correlation matrix is next to the diagonal matrix, consisting of comparing the values of the coefficients of linear correlation observed with the values of the partial coefficients of correlation. The number of components removed was defined, considering only components with an eigenvalue exceeding 1 [14].

3. Results and Discussion

3.1. Descriptive Statistics of the Original Variables and Correlations between Variables

Table 1 presents the descriptive statistics of the variables originals (non-standard). For each variable, the mean, standard deviation, variance, and coefficient of variation are presented.

The coefficient of variation is a measure of dispersion useful for the comparison of different distributions. The standard deviation is also a measure of dispersion, but it is relative to the mean. As two distributions may have different means, the deviation of these two distributions are not comparable. The solution is to use the coefficient of variation, which is equal to the standard deviation divided by the mean [15].

Analyzing the variation coefficients of the variables presented in Table 1, it is verified that the variable cyanobacteria (CB) presented a high degree of dispersion, with coefficient of variation equal to 3.62, whereas the variables turbidity (TUR) and biochemical demand for oxygen (BDO) showed values that were lower, yet still significant, 0.80 and 0.62 respectively. The other variables showed low dispersion with coefficients of variation below 0.50.

Table 1. Descriptive statistical values of the variables.

Variable	Mean	Standard Deviation	Variance	Coefficient of Variation
CB (cell mL^{-1})	3.81×10^7	1.38×10^7	1.90×10^7	3.62
TUR (UNT)	5.45	4.36	19.01	0.80
BDO (mg L^{-1})	3.26	2.03	4.11	0.62
P (mg L^{-1})	0.25	0.09	0.01	0.38
DO (mg L^{-1})	6.85	2.05	4.21	0.30
TS (mg L^{-1})	1084.16	172.80	29,859.41	0.16
VD (%)	88.13	13.52	182.74	0.15
EC (dSm^{-1})	1.680	0.232	0.054	0.14
pH	8.37	0.62	0.38	0.07
T (°C)	27.66	1.23	1.52	0.04

Table 2 shows the correlation matrix that was prepared to check the parameters of the greatest correlation and help with the interpretation of data.

Higher values were observed for correlations between dissolved oxygen (DO) and hydrogen potential (pH) (R = 0.700), indicating the elevation of pH and DO in function of algal photosynthetic activity (production of oxygen), promoted by the contribution of nutrients [18,22], and between total solids (TS) and electrical conductivity (EC) (R = 0.839), indicating that the greater part of total solids is in a dissolved state.

Table 2. Correlation matrix of the variables.

	T	pH	EC	DO	BDO	P	TUR	TS	CB	VD
T	1.000									
pH	0.362	1.000								
EC	−0.078	0.055	1.000							
DO	0.383	0.700	−0.155	1.000						
BDO	0.279	0.500	−0.262	0.427	1.000					
P	−0.038	0.081	−0.537	0.099	0.449	1.000				
TUR	0.309	0.088	−0.510	0.156	0.099	0.395	1.000			
TS	−0.114	0.024	0.839	−0.082	−0.293	−0.455	−0.435	1.000		
CB	0.000	−0.036	0.421	−0.141	−0.195	−0.375	−0.220	0.390	1.000	
VD	−0.137	−0.107	−0.537	0.009	−0.011	0.097	−0.061	−0.483	−0.105	1.000

Note: Temperature (T) in °C hydrogen potential (pH), electrical conductivity (EC) in dS m^{-1}, dissolved oxygen (DO) in mg·L^{-1}, Biochemical Demand for Oxygen (BDO) in mg·L^{-1}, total phosphorus (P) in mg·L^{-1}, turbidity (TUR) in UNT, total solids (TS) in mg·L^{-1}, cyanobacteria (CB) in cel·mL^{-1}, and volume of the dam (VD) in %.

There is a proportional relation between the content of salts dissolved and the electric conductivity and that the content of salts can be estimated by measuring the conductivity of the water [4]. In water sources in the state of Ceará, in Northeast Brazil, they observed a strong correlation of electrical conductivity with sodium, magnesium, calcium, hardness and chloride, that is, dissolved solids [23]. Further strengthening the strong correlation between electrical conductivity and dissolved solids, the authors of [24] found a correlation exceeding 0.9 for the electrical conductivity and the chlorates.

The justification of this result consisted in the increase of chlorides as a function of the high rate of evaporation during the dry season in the semi-arid region of Ceará-Brazil.

3.2. Principal Component Analysis

The application of the technique of PCA resulted in the extraction of several components, whose first four explained approximately 77% of the total variance (Table 3). It was verified that the first (PC1), second (PC2), third (PC3), and fourth (PC4) components explained, respectively, 34%, 22%, 11%, and 10% of the total variance.

Table 3. Eigenvalues and explained variance of the variables.

Components	Eigenvalues	Explained Variance (%)	Accumulated Eingenvalues	Accumulated Explained Variance (%)
PC1	3.35	33.52	3.35	33.52
PC2	2.22	22.18	5.57	55.70
PC3	1.12	11.23	6.69	66.93
PC4	1.02	10.22	7.72	77.15
PC5	0.73	7.32	8.45	84.47
PC6	0.59	5.88	9.04	90.35
PC7	0.34	3.38	9.37	93.73
PC8	0.29	2.86	9.66	96.59
PC9	0.23	2.26	9.89	98.86
PC10	0.11	1.14	10.00	100.00

After that, only PC1 and PC2 were selected, because the two together add up to more than 55% of the variance studied, explaining the greater part of existing correlations in the data set. According to [25], coefficients of correlation greater than 0.5 express a strong relationship between the variables of water quality.

Table 4 shows the weights of the variables that most contributed to these principal components. The parameters that most contributed positively to PC1 were phosphorus and turbidity, while the parameters electrical conductivity, total solids, and cyanobacteria contributed negatively to it.

The parameters that most contributed positively to PC2 were pH and OD, while the parameter associated with the volume of water accumulated contributed negatively to it.

Table 4. Weights of the variables for PC1 and PC2.

Components	PC 1	PC 2
T	0.299	0.561
pH	0.306	0.81
EC	−0.853	0.389
DO	0.432	0.696
BDO	0.576	0.464
P	0.688	−0.133
TUR	0.596	−0.012
TS	−0.812	0.364
CB	−0.548	0.155
VD	0.366	−0.472

With the graph of projections for the weights of the variables plotted on the PC1 × PC2 plane (Figure 3), it is suggested that PC1 explains the increase in the concentration of dissolved solids and cyanobacterial proliferation as a function of the drought period (absence of rainfall), and, as a consequence of the lack of rainfall, turbidity and the levels of total phosphorus in the spring were reduced.

Turbidity results from the presence of colloidal particles in suspension, divided organic matter, plankton, and other microscopic organisms [26]. Turbidity can also be related to the inflow of effluents [27], and these are rich in phosphorus [6,28]. Thus, these two variables are associated with, and contribute positively to, this component, while the total dissolved solids, electrical conductivity, and cyanobacteria contribute negatively.

On the other hand, the authors of [24] stated that the total dissolved salts in the waters may be from both natural sources (mineralization and marine aerosols) and anthropic sources (domestic sewage). Some researchers studying water bodies of the state of Ceará made conclusions about the origin of the salts in some of these streams. The authors of [23], studying the waters of the Acaraú river basin, verified that the common origin of these minerals was the weathering of rocks and subsequent runoff from drained areas. Studies in the Jaibaras River, also in Ceará, correlated the salts with domestic sewage and spa waste [29]. Studies about the Trussu river basin, identified the washing of clothes and domestic sewage as being the source of chloride and sodium [30].

The water contribution area of the Jucazinho reservoir have a predominance of soil types Planosol, Solonetz, and Bruno but not Calcic or Regosol [20], which may be contributing to the salinization of water [31]. It is important to emphasize that the salts used by laundries/dyers in Toritama for fixing colors onto fabric confer a high salinity of the effluent. These wastes, when discarded without being properly treated, may be another source of salts to the river and consequently to the reservoir [1].

PC2 indicates the intensification of the process of photosynthesis performed by cyanobacteria, which can be justified by the contributions of the weights of pH, DO (positively), and the percentage of the volume of the dam (negatively).

According to [32], the northeastern region of Brazil presents more propitious conditions for cyanobacterial blooms, because the climate is always hot, reservoirs with low levels, caused by recurrent periods of drought, and a lack of sanitation services, among other factors that favor the excessive increase of biomass in these bodies. According to [33], the maximum rate of growth of the cyanobacteria is present at temperatures above 25 °C and the range of optimal growth happens at a pH level from 7.5 to 10, inhibited below 5. The authors of [34] affirm that the aquatic communities can interfere with the values of pH, mainly through the metabolism of CO_2. During the process of photosynthesis, in which there is consumption of this gas by phytoplankton, an increase in the pH values of the medium occurs. Thus, PC2 shows the relationship between photosynthetic activity and the production of oxygen, when the accumulated volume remains at lower levels.

Reseachers performed a statistical evaluation in rivers in the state of Minas Gerais through PCA in the period from 2007 to 2011 [35]. The results showed a positive correlation of density of cyanobacteria with pH, chlorophylla, temperature, and nitrate, and an inverse relationship with turbidity, color, and solids in suspension. The researcher highlights that high loads of nitrate and phosphorus, alkaline pH values, high water temperatures, and prevalence of long periods of drought is the necessary combination for the increase in densities of cyanobacteria.

In statistical study performed in the Teles Pires and Cristalino rivers in the Alto Tapajós basin, located in the state of Mato Grosso, observed, both for the DO and for the pH, during the study period, a negative correlation with the rainfall [36].

Figure 3 displays the graph of weights for the first two principal components. Geometrically, the weights correspond to the cosines of the angles that the principal components make with the original variables. The weights of the original variables in linear combination define each principal component. The relation between the variables can be observed in the graph of the weights. Based on these relations, it is possible initially to infer a physical interpretation for the principal components. Still in the same figure, it is interesting to note the provision of variables along PC1, which shapes 33.52% of the variance of the data matrix. The volume of the dam (VD) has the opposite sign of cyanobacteria (CB), electrical conductivity (EC), and total solids (TS).

Figure 4 shows the score graph with the objective of analyzing the seasonal behavior of the data series. It was possible to group together the years 2005, 2006, 2007, 2012, and 2013 (Group 1) with more positive contributions to the explanation of PC1, while the years 2008, 2009, 2010, and 2011 (Group 2) formed a group that contributed negatively to it.

The scores of the PCA, presented in Figure 4, revealed differences resulting from the seasonal influence and annual accumulated volume, which in turn is directly related to rainfall in the water contribution area of the reservoir.

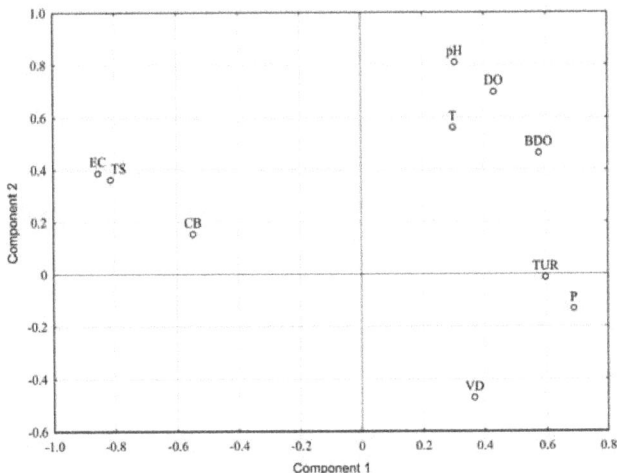

Figure 3. Weights of principal components plotted on the PC1 × PC2 plane.

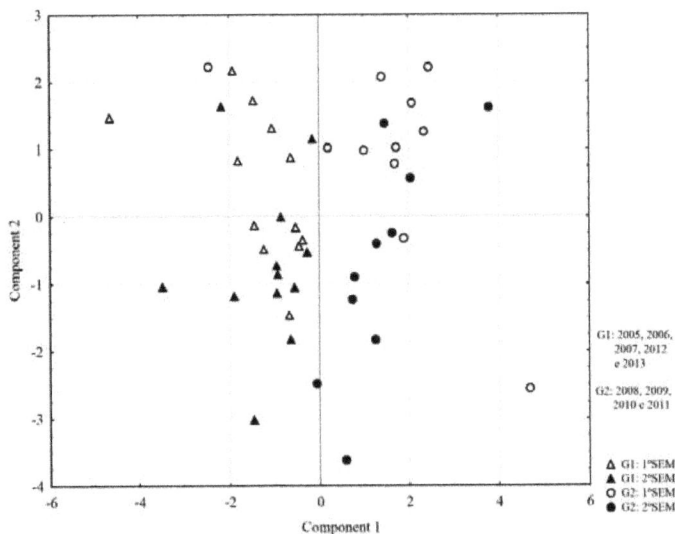

Figure 4. Scores of components plotted on the PC1 × PC2 plane.

With the aid of the score graph (Figure 4), it is possible to verify the tendency of grouping for the explanation of PC2. The graph shows that most of the white circles and triangles, which represent the first semester of each year studied, are located above the zero value on the x-axis, so the data for the first semester contributed positively to PC2. Adversely, the black triangles and circles, which represent the second semester of each year studied, are located below the zero value on the x-axis, so it is possible to conclude that the data of second semesters contributed negatively.

In fact, the largest pluviometric indexes in the portion of the Capibaribe river basin controlled by the Jucazinho dam occur in the months of February to June. The lowest values of the monthly average rainfall occur in the period from August to January and are less than 40 mm [19].

The observed trend suggests that the intensification of the drought period associated with an increase of the dissolved solids concentration exhibits behavior that changes annually, and is explained by PC1. The intensification of the process of photosynthesis is performed by cyanobacteria, ranging within six months, and is explained by PC2.

4. Conclusions

The Jucazinho reservoir located in the semi-arid region of Northeastern Brazil suffers from recurrent water level reductions, which are aggravated for years when drought is severe. The studies contemplated data of nine years, and, after the application of statistical techniques and subsequent interpretation of the results. The main conclusions are summarized in the following points.

1. The employment of PCA promoted a reduction of 10 parameters of surface water quality in two components, which together explain 55% of total variance, illuminating the main problems that interfere in the temporal variation in water quality.
2. The results of PCA showed a tendency toward formation over years, and even over months, of similar water quality parameters, conditioned by rainfall, indicating, in general, the temporal variations of the parameters analyzed. The temporal pattern obtained by the analysis shows that two factors are responsible for the variation in water quality during periods of drought, one observed over many years and other observed every six months.
3. Wet years increase the inflow to the reservoir, so more domestic and industrial sewage is contributed, which in turn increases the turbidity and the content of phosphorus in the spring. On the other hand, years with little or no precipitation provide low inflow to the reservoir. In these conditions, there is an increase in total solids (dissolved salts) and, consequently, in electrical conductivity in view of the increase of the evaporation of the repressed water volume. An increase photosynthetic activity can also be seen, as is evidenced by the increase in the oxygen content in the water and in pH.
4. The research shows the importance of water quality monitoring, where adequate statistical treatment can provide subsidies for better monitoring to preserve water quality for public water supply. In extreme dry periods, the identification of annual and semi-annual variation behavior can assist managers in making decisions regarding reservoir operation, monitoring the most important parameters and actions that minimize the impact of this extreme drought.
5. The results of the study were relevant in the conduction of new methodologies for monitoring and management of the reservoir, since the applied statistical treatment shed light on the most recurrent problems of extreme drought, and these findings can be extended to other reservoirs located in the semi-arid of Northeastern Brazil.

Author Contributions: Rafael Roney Camara de Melo and Iona Maria Beltrão Rameh Barbosa conceived the idea of the paper, planned the statistical methodology and wrote the paper. Aida Araújo Ferreira, Alessandra Lee Barbosa Firmo, Simone Rosa da Silva, José Almir Cirilo and Ronaldo Ribeiro Barbosa de Aquino contributed writing and reviewing of papper.

Conflicts of Interest: The authors declare no conflict of interest.

References

1. Barbosa, I.M.B.R.; Cirilo, J.A. Evolução do estado trófico do reservatório Jucazinho, Pernambuco. In *Simpósio de Recursos Hidrícos do Nordeste*, 1st ed.; [CD-ROM]; 11. 2012, Anais; ABRH: João Pessoa, Brazil, 2012.
2. Bezerra, L.A.V.; Paulino, W.D.; Garcez, D.S.; Becker, H.; Sanchez-Botero, J.I. Limnological characteristics of a reservoir in semiarid Northeastern Brazil subjective to intensive tilapia farming (*Orechromisniloticus* Linnaeus, 1758). *Acta Limnol. Bras.* **2014**, *26*, 47–59. [CrossRef]

3. Gunkel, G.; Lima, D.; Selge, F.; Sobral, M.; Calado, S. Aquatic ecosstem services of reservoirs in semi-arid áreas: Sustainability and reservoir management. *River Basin Manag.* **2015**, *197*, 187–200.

4. Santos, A.C. Noções de Hidroquímica. In *Hidrologia: Conceitos e Aplicações*; Cprm/Labhid-Ufpe: Fortaleza, Brazil, 1997.

5. Silva, A.P.C.; Costa, I.A.S. Biomonitoring ecological status of two reservoirs of the Brazilian semi-arid using phytoplankton assemblages (Q index). *Acta Limnol. Bras.* **2015**, *27*, 1–14. [CrossRef]

6. Chapra, S.C. *Surface Water-Quality Modeling*; Waveland Press, Inc.: Long Grove, IL, USA, 2008; p. 844.

7. Esteves, F.A. *Fundamentos da Limnologia*, 2nd ed.; Interciência: Rio de Janeiro, Brazil, 2011; 602p.

8. Friese, K.; Schmidt, G.; Lena, J.C.; Nalini, H.A., Jr.; Zachmann, D.W. Anthropogenic influence on the degradation of na urban lake—The Pampulha reservoir in Belo Horizonte, Minas Gerais, Brazil. *Limnologica* **2010**, *40*, 114–125. [CrossRef]

9. Reartes, S.B.R.; Estrada, V.; Bazan, R.; Larossa, N.; Cossavella, A.; Lopez, A.; Busso, F.; Diaz, M.S. Evaluation of ecological effects of anthropogenic nutrient loading scenarios in Los Molinos reservoir through a mathematical model. *Ecol. Model.* **2016**, *320*, 393–406. [CrossRef]

10. Mcdonald, R.I.; Weber, K.F.; Padowski, J.; Boucher, T.; Shemie, D. Estimating watershed degradation over the last centur and its impacto n water-treatment costs for the world's large cities. *Proc. Natl. Acad. Sci. USA* **2016**, *113*, 9117–9122. [CrossRef] [PubMed]

11. Chow, C.W.K.; Drikas, M.; House, J.; Burch, M.D.; Velzeboer, R.M.A. The impact of conventional water treatment process on cells of the cyanobacteriumMicrocystisaeruginosa. *Water Res.* **1999**, *33*, 3253–3262. [CrossRef]

12. Sawyer, C.N.; Maccarty, P.L.; Parkin, G.F. *Chemistry for Environmental Engineering*; McGraw-Hill Higher Education: New York, NY, USA, 2003; 752p.

13. Bernardi, J.V.E.; Lacerda, L.D.; Dórea, J.G.; Landim, P.M.B.; Gomes, J.P.O.; Almeida, R.; Manzatto, A.G.; Bastos, W.R. Aplicação da análise das componentes principais na ordenação dos parâmetros físico-químicos no alto Rio Madeira e afluentes, Amazônia Ocidental. *Geochim. Bras.* **2009**, *23*, 1–158.

14. Norusis, M.J. *SPSS Base System User's Guide*; SPSS Inc.: Chicago, IL, USA, 1990; 520p.

15. Hair, J.F., Jr.; Anderson, R.E.; Tatham, R.L. *Análise Multivariada de Dados*, 5th ed.; Bookman: Porto Alegre, Brazil, 2005.

16. Parinet, B.; Lhote, A.; Legube, B. Principal componentanalysis: An appropriate tool for water quality evaluation andmanagement—Application a tropical lake system. *Ecol. Model.* **2004**, *178*, 295–311. [CrossRef]

17. Liao, S.W.; Gau, H.S.; Lai, W.L.; Chen, J.J.; Lee, C.G. Identification of pollution of Tapeng Lagoon from neighbouring rivers using multivariate statistical method. *J. Environ. Manag.* **2008**, *88*, 286–292. [CrossRef] [PubMed]

18. Singh, K.P.; Malik, A.; Mohan, D.; Sinha, S. Multivariate statistical techniques for the evaluation of spatial and temporal variations in water quality of Gomti River (India)—A case study. *Water Res.* **2004**, *38*, 3980–3992. [CrossRef] [PubMed]

19. Departamento Nacional de Obras Contra as Secas-DNOCS, Ministério da Integração Nacional. *Hidrologia do Reservatório de Jucazinho*; Águasolos: Recife, Brazil, 1995; p. 59.

20. Secretaria de Recursos Hídricos e Energéticos. *ProjetecBRLi. Plano Hidroambiental da Bacia Hidrográfica do Rio do Capibaribe*; TOMO I (vol 1, 2 e 3) e TOMO IV; SRHE: Recife, Brazil, 2010.

21. Barbosa, I.B.R.; Cirilo, J.A. Contribuição média de fósforo em reservatório de abastecimento de água—Parte 1. *Engenharia Sanitária e Ambiental* **2015**, *20*, 39–46. [CrossRef]

22. Frieder, C.A.; Nam, S.H.; Martz, T.R.; Levin, L.A. High temporal and spatial variability of dissolved oxygen and pH inanearshore California kelp forest. *Biogeosciences* **2012**, *9*, 3917–3930. [CrossRef]

23. Andrade, E.M.D.; Araújo, L.D.F.; Rosa, M.F.; Disney, W.; Alves, A.B. Surface water quality indicators in low acaraú basin, Ceará, Brazil, using multivariable analysis. *Engenharia Agrícola* **2007**, *27*, 683–690. [CrossRef]

24. Palácio, H.A.; Araújo Neto, J.R.; Meireles, A.; Andrade, E.M.; Santos, J.C.; Chaves, L.C. Similaridade e fatores determinantes na salinidade das águas superficiais do Ceará, por técnicas multivariadas. *Revista Brasileira de Engenharia Agrícola e Ambiental* **2011**, *15*, 395–402. [CrossRef]

25. Helena, B.; Pardo, R.; Vega, M.; Barrado, E.; Fernandez, J.M.; Fernandez, L. Temporal evolution of groundwater composition in an alluvial aquifer (Pisuergariver, Spain) by principal component analysis. *Water Res.* **2000**, *34*, 807–816. [CrossRef]

26. Ministério da Saúde. *Dispõe Sobre os Procedimentos de Controle e de Vigilância da Qualidade da Água Para Consumo Humano e Seu Padrão de Potabilidade*; Portaria No. 2914; Ministério da Saúde: Brasília, Brazil, 2011. Available online: http://bvsms.saude.gov.br/bvs/saudelegis/gm/2011/prt2914_12_12_2011.html (accessed on 12 July 2017).

27. Luíz, Â.M.E.; Pinto, M.L.C.; Scheffer, E.W. Parâmetros de cor e turbidez como indicadores de impactos resultantes do uso do solo, na bacia hidrográfica do rio Taquaral, São Mateus do Sul-PR. *Revista o Espaço Geográfico em Análise* **2012**, 290–310. [CrossRef]

28. Zhang, X.; Wang, Q.; Liu, Y.; Wu, J.; Yu, M. Application of multivariate statistical techniques in the assessment of water quality in the Southwest New Territories and Kowloon, Hong Kong. *Environ. Monit. Assess.* **2010**, *137*, 17–27. [CrossRef] [PubMed]

29. Girão, E.G.; De Andrade, E.M.; de Freitas Rosa, M.; Pereira de Araújo, L.D.F.; Maia Meireles, A.C. Seleção dos indicadores da qualidade de água no Rio Jaibaras pelo emprego da análise da componente principal. *Revista Ciência Agronômica* **2007**, *38*, 17–24.

30. Palácio, H.A.Q. Índice de Qualidade das Águas na Parte Baixa da Bacia Hidrográfica do Rio Trussu, Ceará. Master's Thesis, Universidade Federal do Ceará, Ceará, Brazil, 2004.

31. Ministério de Minas e Energia. *Diagnóstico do Município de Santa Cruz do Capibaribe*; Ministério de Minas e Energia: Recife, Brazil, 2005. Available online: http://rigeo.cprm.gov.br/xmlui/bitstream/handle/doc/16692/Rel_Santa%20Cruz%20do%20Capibaribe.pdf?sequence=1 (accessed on 12 July 2017).

32. Aragão, N.K.C.V. Taxonomia, Distribuição e Quantificação de Populações de Cianobactérias em Reservatórios do Estado de Pernambuco (Nordeste do Brasil). Master's Thesis, Departamento deBiologia, Universidade Federal Rural de Pernambuco, Recife, Brazil, 2011.

33. Fernandes, V.O.; Cavati, B.; de Oliveira, L.B.; de Souza, B.D.Â. Ecologia de cianobactérias: Fatores promotores e conseqüências das florações. *Oecol. Bras.* **2009**, *13*, 247–258. [CrossRef]

34. Santos, J.C.N.; Andrade, E.M.; Araujo Neto, J.R.; Meireles, A.C.M.; Palacio, H.A.Q. Land use and trophic state dynamics in a tropical semi-arid reservoir. *Revista Ciência Agronômica* **2014**, *45*, 35–44. [CrossRef]

35. Ferraz, H.D.A. Associação da Ocorrência de Cianobactérias às Variações de Parâmetros de Qualidade da Água em Quatro Bacias Hidrográficas de Minas Gerais. Ph.D. Thesis, Mestrado em Saneamento, Meio Ambiente e Recursos Hídricos, Universidade Federal de Minas Gerais-Escola de Engenharia, Minas Gerais, Brazil, 2012.

36. Umetsu, C.A.; Umetsu, R.K.; Munhoz, K.C.A.; Dalmagro, H.J.; Krusche, A.V. Aspectos físico-químicos de dois rios da bacia do Alto Tapajós—Teles Pires e Cristalino—MT, durante período de estiagem e cheia. *Revista de Ciências Agro-Ambientais* **2007**, *5*, 59–70.

water

MDPI

Article

A Comparison of Preference Handling Techniques in Multi-Objective Optimisation for Water Distribution Systems

Gilberto Reynoso-Meza [1,*,†], Victor Henrique Alves Ribeiro [1,†] and Elizabeth Pauline Carreño-Alvarado [2,†]

1   Industrial and Systems Engineering Graduate Program (PPGEPS), Pontifical Catholic University of Parana (PUCPR), Curitiba 80215-901, Brazil; victor.henrique@pucpr.edu.br
2   Department of Hydraulic and Environmental Engineering, Universitat Politènica de Valècia, 46022 Valencia, Spain; elcaral@posgrado.upv.es
*   Correspondence: g.reynosomeza@pucpr.br; Tel.: +55-(41)-3271-2473
†   These authors contributed equally to this work.

Received: 31 October 2017; Accepted: 24 November 2017; Published: 19 December 2017

Abstract: Dealing with real world engineering problems, often comes with facing multiple and conflicting objectives and requirements. Water distributions systems (WDS) are not exempt from this: while cost and hydraulic performance are usually conflicting objectives, several requirements related with environmental issues in water sources might be in conflict as well. Commonly, optimisation statements are defined in order to address the WDS design, management and/or control. Multi-objective optimisation can handle such conflicting objectives, by means of a simultaneous optimisation of the design objectives, in order to approximate the so-called Pareto front. In such algorithms it is possible to embed preference handling mechanisms, with the aim of improving the pertinency of the approximation. In this paper we propose two mechanisms to handle such preferences based on the TOPSIS (Technique for Order of Preference by Similarity to Ideal Solution) and PROMETHEE (Preference Ranking Organisation METHod for Enrichment of Evaluations) methods. Performance evaluation on two benchmarks validates the usefulness of such approaches according to the degree of flexibility to capture designers' preferences.

Keywords: water distribution systems; multi-objective optimisation; evolutionary multi-objective optimisation

1. Introduction

Dealing with real world engineering problems often comes with facing multiple and conflicting objectives and requirements. Water distributions systems (WDS) are not exempt from this: while cost and hydraulic performance are usually conflicting objectives, several requirements related with environmental issues in water sources might be in conflict as well. Commonly, optimisation statements are defined in order to address the WDS design, management and/or control. Nevertheless, such problems become difficult since, besides their multi-objective conflicting nature, the optimisation problem might be non-linear (due to head-loss relationships for example) and/or discrete combinatorial (due to standardisation of pipe parameters) [1,2].

Multi-objective optimisation [3] can handle such an issue, by means of a simultaneous optimisation of the design objectives. At the end of this process, a potential set of solutions, the Pareto front, is approximated. In this set of solutions, there is not a best solution, but a preferable solution. This means that several solutions are calculated, with different trade-offs between conflicting objectives and the engineer will select among them the most preferable for the problem at hand.

Given that several solutions are calculated, the designer must perform a decision making stage. In this stage, it is required to express somehow the preferences according to the trade-offs, in order to select the most suitable (preferable) solution for the problem at hand. This might not be a trivial task, since most of the times to interpret such trade-offs is not easy, given the multidimensional structure of the problem. Therefore, visualisation techniques [4] and multi-criteria decision making methodologies are valuable and helpful for designers.

It is possible to use different decision making methodologies, such as TOPSIS [5] (Technique for Order of Preference by Similarity to Ideal Solution), Physical Programming [6], PROMETHEE [7] (Preference Ranking Organisation METHod for Enrichment of Evaluations), among others. While it is usual to apply such methodologies in the decision making step, it is also possible to embed them into the optimisation process. For example, the Physical Programming method has been used before in order to evolve the population of a multi-objective evolutionary algorithm (MOEA) towards the pertinent region of the objective space [8,9]. With such approach, it is possible to use preference-information actively in the optimisation, improving the usability of the approximated Pareto front, as well as dealing with more than 3 design objectives effectively. In this paper some modifications are proposed, incorporating the TOPSIS and the PROMETHEE mechanism for the same purpose in a MOEA.

The remainder of this paper is as follows: in Section 2 a review on multi-objective optimisation techniques for WDS is presented, identifying the necessity of preference handling techniques. In Section 3 the TOPSIS and PROMETHEE methods are incorporated into a MOEA for preference handling and they are evaluated in Section 4 with two MOPs. Finally, conclusions of this work are commented.

2. Review

A literature review on the optimisation of WDS is presented in [2], where the authors bring together over two hundred journal publications from the past three decades. From those publications, the authors create a table with substantial information from over one hundred of them, from which seventeen papers focus on the use of a MOO approach. The first papers on MOO for WDS focused solely on the optimisation of operation and maintenance costs. Next, the optimisation of water quality became the main interest by some researchers. Nowadays, research on the subject focuses on finding the trade-off between cost and water quality. A review of each paper is presented below, followed by Table 1, which resumes the MOO design characteristics of each publication. Background and definitions of the MOO process are presented in the Appendix for interested readers.

Table 1. Summary of MOOD procedures for WDS design concept. $J(\theta)$ refers to the number of objectives; θ to the number of sets of decision variables and $g(\theta)$, $h(\theta)$ to the number of sets of inequality and equality constraints respectively.

Reference	MOP			MOO		MCDM	
	$J(\theta)$	θ	$(g(\theta), h(\theta))$	Algorithm	Features	Plot	Insights
Savic et al. [10]	2	1	(2, 1)	Hybrid GA	Local search	-	-
Sotelo and Baran [11]	4	1	(4, 0)	SPEA	-	Scatter plot	-
Kelner and Leonard [12]	2	3	(2, 2)	GAPS	Penalised tournament selection scheme	Scatter plot	-

Table 1. *Cont.*

Reference	MOP			MOO		MCDM	
	$J(\theta)$	θ	$(g(\theta), h(\theta))$	Algorithm	Features	Plot	Insights
Baran et al. [13]	4	1	(4, 0)	SPEA NSGA NSGA-II CNSGA NPGA MOGA	Comparison between algorithms	-	-
Lopez-Ibanez et al. [14]	2	1	(3, 1)	SPEA2	Comparison between initial population generation methods	Scatter Plot	Attainment surfaces
Odan et al. [15]	2	1	(3, 1)	AMALGAM	Real-time	Scatter Plot	-
Stokes et al. [16]	2	1	(2, 0)	NSGA-II	-	Scatter Plot	Minimum objective values
Prasad et al. [17]	2	2	(3, 0)	NSGA-II	-	Scatter plot	-
Kurek and Brdys [18]	3	2	(5, 0)	NSGA-II	Problem specific modification	Scatter Plot	-
Ewald et al. [19]	3	2	(4, 0)	Distributed MOGA	Distributed application using grid computing	Scatter Plot	-
Alfonso et al. [20]	3	1	(2, 1)	NSGA-II	-	Scatter Plot	-
Giustolisi et al. [21]	2	1	(4, 1)	OPTIMOGA	-	Table	-
Kougias and Theodossiou [22]	3	1	(4, 0)	MO-HSA	-	Scatter Plot	-
Kurek and Ostfeld [23]	3	3	(3, 1)	SPEA2	-	Scatter Plot	Utopian solution mechanism
Kurek and Ostfeld [24]	2	2	(3, 1)	SPEA2	-	Scatter Plot	Most "balanced" solution
Mala-Jetmarova et al. [25]	2	1	(4, 0)	NSGA-II	-	Scatter Plot	-
Mala-Jetmarova et al. [26]	3	1	(4, 0)	NSGA-II	-	Scatter Plot	-

A multi-objective hybrid approach of the Genetic Algorithm (GA) is introduced by [10] to find the trade-off between the minimisation of: (a) energy; and (b) maintenance costs on a net with four pumps and one reservoir. One set of binary decision variables is used for this problem, which indicates the pump statuses for each hour on a twenty-four hours period. The recovery of the initial reservoir water level at the end of the simulation period is used as the equality constraint, while the minimum and maximum reservoir levels are set as inequality constraints.

A simplified system, composed of one source, five pumps and one elevated reservoir is the object of study by [11], where strength Pareto evolutionary algorithm (SPEA), using one set of binary decision variables for the pump statuses, finds the trade-off between the minimisation of the: (a) pump operating costs; (b) number of pump switches; (c) difference between initial and final levels in the reservoir; and (d) maximum daily power peak. The problem contains four inequality constraints: (a) minimum reservoir water levels; (b) maximum reservoir water levels; (c) minimum pipeline pressure; and (d) maximum pipeline pressure. At the end of the paper, a two-dimensional Pareto front is presented.

A WDS from Belgium is optimised by [12] using a multi-objective genetic algorithm (MOGA) with penalised tournament selection scheme, where two objectives are minimised: (a) the pump operating costs; and (b) the number of pump switches. Three sets of decision variables are defined: (a) the binary pump statuses; (b) the rotating speed of the pumps; and (c) the pressure loss coefficient for the control valve. The problem is composed of two sets of equality constraints: (a) the initial reservoir water level must be reached by the end of the optimisation; and (b) the consumer demands must be satisfied at any period of time. In addition, two sets of inequality constraints must be met: (a) the maximum; and (b) the minimum water levels for each reservoir. The authors plot a two-dimensional Pareto front, but do not choose a preferred solution.

A comparison of six MOO algorithms is performed by [13] using the same WDS from [11]. The compared algorithms are: SPEA, non-dominated sorting algorithm (NSGA), NSGA-II, controlled elitist non-dominated sorting genetic algorithm (CNSGA), niched Pareto genetic algorithm (NPGA), and MOGA. The MOP is designed with four objectives, the minimisation of: (a) pump operating costs; (b) number of pump switches; (c) difference between initial and final water levels in the reservoir; and (d) maximum daily power peak. Only one set of decision variables is used, the binary pump statuses, while four inequality constraints are used: (a) the minimum reservoir water levels; (b) the maximum reservoir water levels; (c) the minimum pipeline pressure; and (d) the maximum pipeline pressure. The algorithms are compared by six different metrics: (a) overall non-dominated vector generation; (b) overall non-dominated vector generation ratio; (c) error ratio; (d) generational distance; (e) maximum Pareto front error; and (f) spacing.

The simulation of a small WDS is optimised by [14] using the second version of SPEA, the SPEA2. One set of decision variables, the binary pump status, is used to find the trade-off between two objectives, the minimisation of: (a) pump operating costs; and (b) number of pump switches. One equality constraint is used, the pressure at demand nodes, while three inequality constraints are used: (a) maximum tank water levels; (b) minimum tank water levels; and (c) tank volume deficit at the end of the simulation. The authors compare the Pareto fronts of four different methods for the initial population generation, using scatter plots and the attainment surfaces as the metric.

A real-time pump scheduling framework is proposed by [15], where optimisation is performed using a multialgorithm genetically adaptive method (AMALGAM). The framework is applied to a WDS from Brazil, and the MOP is composed of two objectives: (a) the minimisation of pump operating costs; and (b) maximisation of operational reliability. One set of decision variables, the binary pump status, is considered. Three sets of inequality constraints are used: (a) the minimum pressure at any network node; (b) the tank water levels at the end of the scheduling period; and (c) the maximum number of pump switches. In addition, one equality constraint, where the occurrence of simulation errors must be equal to zero, is considered. The authors present the resulting Pareto front on a two-dimensional scatter plot.

Seven different scenarios are optimised by [16] using NSGA-II. Each scenario is composed of different emission factors and time horizons. The MOP is composed of one set of decision variables, the pump schedules, and two objectives, the minimisation of: (a) pump operating costs; and (b) greenhouse gas emissions associated with the use of electricity from fossil fuel sources. Two sets of inequality constraints are considered: (a) the minimum pressure at network nodes; and (b) minimum total volume of water pumped into each district metered area. The authors present the Pareto fronts on two-dimensional scatter plots, and the solutions selected for analysis are based on minimum values for both objectives.

The simulation of a real water utility network is subject to MOO by [17] using NSGA-II, where the trade-off between two objectives: (a) the minimisation of the total disinfectant dose; and (b) the maximisation of the volumetric percentage of water supplied with disinfectant residuals, is found. Two decision variables are used in this optimisation: (a) the locations of booster disinfection stations; and (b) the disinfection injections schedules. In addition, three inequality constraints are used: (a) non-negative disinfectant doses; (b) lower bound on the value of the Objective (b); and (c) upper bound on disinfectant concentrations at monitoring nodes. An analysis of the Pareto front is made by the authors using scatter plots.

The Pareto front of a WDS is found by [18] using NSGA-II. The MOP is elaborated with three objectives, the minimisation of: (a) the number of chlorine booster stations; (b) mean value of chlorine concentrations; and (c) mean value of instances not meeting quality requirements. In addition, two decision variables are used: (a) the presence of a booster stations at a network node; and (b) the chlorine concentrations at booster stations and treatment plants. In total, five inequality constraints are used in this problem: (a) the maximum number of booster stations; and (b) the minimum number of booster stations; (c) the maximum shlorine concentration and (d) the minimum chlorine concentrations; and (e) the minimum chlorine concentration at treatment plants. By the end of the paper, the authors present a scatter plot of the Objectives (b) and (c), which are grouped by the values of Objective (a).

A WDS from Poland is optimised by [19] using a distributed MOGA, based on the island GA. The problem is composed of the same objectives and decision variables presented by [18], but only uses the first four inequality constraints. The resulting Pareto front is presented by a two-dimension scatter plot of objectives (b) and (c) grouped by values of objective (a).

Two case studies are optimised by [20] using NSGA-II, one hypothetical and one simulation of a WDS from Colombia. For both cases, the objectives are defined as the minimisation of: (a) the number of polluted nodes; and (b) the number of the operational interventions needed. Only one set of decision variables is used for this problem, the operational interventions on pumps, valves and switches. In total, three sets of constraints are used on this publication, two inequality constraints, where: (a) nodes pressures must be positive; and (b) technical operational capacity for interventions must be met, and one equality constraint, where network connectivity must be ensured. The resulting Pareto front is presented as a two-dimensional scatter plot.

The optimisation of a WDS is performed by [21] using the optimised MOGA (OPTIMOGA). The MOP is composed of four inequality constraints: (a) the minimum pressure for sufficient pressure, expressed by the number of times which it is not satisfied; (b) the tank volume deficit at the end of the simulation; (c) the minimum tank levels, expressed as the times which it is not satisfied; and (d) the maximum tank levels. It is also composed of one equality constraint, the global mass balance in each tank (there is only one tank in the case study). One set of decision variables, the binary status of the pumps and gates, is used to minimise two objective functions. The first objective is an aggregate function of: (a) pump operating costs; and (b) water losses cost; and the second is the function of Inequality Constraints (a), (b) and (c). The authors present a table with the resulting Pareto front of the problem, and one solution is selected for being the only feasible solution.

Two multi-objective optimisation algorithms, based on the harmonic search algorithm (HSA), are developed by [22] in order to solve a pump scheduling problem, the multi-objective HSA (MO-HSA) and the polyphonic HSA (Poly-HSA). The MO-HSA is used to optimise an operational pumping field from Paraguay, and the problem is composed by the minimisation of four objectives: (a) pump operating costs; (b) quantity of pumped water; (c) electric energy peak consumption; and (d) number of pump switches. The optimisation is executed two times, Once with Objectives (a), (b) and (c), and again with Objectives (a), (b), and (d). The problem is also composed of one set of decision variables, the binary pump status, and four sets of inequality constraints, used within a penalty mechanism. Such constraints are: (a) the minimum water levels in storage tanks; and (b) the maximum water level in storage tanks; and the (c) the minimum volume deficit in the storage tanks at the end of the scheduling period; and (d) the maximum volume deficit in the storage tanks at the end of the scheduling period. The authors present, for both runs, a three-dimensional scatter plot with the Pareto front.

Two optimisation problems, one related to chlorine concentrations, and another related to water age, are solved by [23] using SPEA2. In total, four objectives are defined for the MOP, the minimisation of: (a) pump operating costs; (b) disinfectant concentrations at monitoring nodes; (c) water age for demand nodes; and (d) cost of tanks. The first optimisation model includes Objectives (a), (b) and (c), while the second model includes Objectives (a), (c) and (d). Three sets of decision variables are used for this problem: (a) the pump speeds; (b) the disinfectant concentrations at treatment plants; and (c) tank diameters. One equality constraint, the pressure at nodes, and three inequality constraints: (a) the minimum volume deficit at the end of the simulation; (b) the maximum volume surplus at the end of the simulation; and (c) the minimum amount of stored water at any time, are used in both optimisation models. The resulting Pareto fronts are presented on three-dimensional scatter plots, and three solutions are selected for each Pareto, one "balanced" solution, which is the closest an Utopian solution, and the other two are related to minimal values for Objectives (a) and (b).

The optimisation of two systems is performed in [24] using SPEA2 and the MOP defined in [23], but only the first two objectives and decision variables are used. The authors present a two-dimensional scatter plot for both examples, and a single "balanced" solution is selected on the MCDM stage, but no metrics were specified.

A total of fourteen different scenarios are optimised by [25] using NSGA-II. All scenarios are defined as a MOP with two objectives, the minimisation of: (a) pump operating costs; and (b) deviations of the actual constituent concentrations from the required values. For such problems, one set of decision variables, the binary pump status, are defined. Furthermore four sets of inequality constraints are considered: (a) the minimum pressure at customer nodes; (b) the minimum water level in the storage tanks; (c) the maximum water level in the storage tanks; and (d) the volume deficit in the storage tanks at the end of the scheduling period. The resulting Pareto front for all scenarios are presented on a two dimensional scatter plot and, for comparison with results from the literature, one "balanced" solution is selected for analysis, but no metrics were specified.

Six different scenarios based on a network with ninety-four nodes are optimised by [26] using NSGA-II. All scenarios are defined as a MOP with three objectives, the minimisation of: (a) pump operating costs; (b) the turbidity deviations from the allowed values; and (c) deviations of the actual constituent concentrations from the required values. Decision variables and constraints are defined as in [25]. The resulting Pareto fronts for all scenarios are presented on a three-dimensional scatter plot and three different two-dimensional scatter plots, for each two objectives combination, and two solutions were selected from two different scenarios for comparison purposes. No specific metrics were indicated for such selections.

In Table 1 a summary of such papers is shown. It is interesting to note that the vast majority of papers are focusing on MOPs with two or three design objectives. A possible reason for this might be the difficulties to perform a MCDM and to visualise the Pareto front approximation. For this reason, the idea of stating more than three design objectives for WDS is exploited in this paper, using different preference mechanisms to evolve towards the pertinent region of the Pareto front.

3. Proposal and Experiment Description

As it has been noticed before, tendencies regarding the number of design objectives for MOPs is to state two or three. One of the circumstances leading to this might be that any MOP with more than three design objectives is said to be a many-objectives MOP problem. In such an instance, mechanisms for diversity and convergence are in conflict. Therefore, additional mechanisms are often required in order to guarantee a suitable performance of the Pareto front approximation process. One of such mechanisms is the inclusion of preferences [27].

As it has been commented in the introduction, a MOEA using Physical Programming as preference handling mechanism, the sp-MODEII (Multi-Objective Differential Evolution with Spherical Pruning, version II) (Toolbox available at https://www.mathworks.com/matlabcentral/fileexchange/47035) has been proposed before [8,9]. The sp-MODEII is an evolutionary algorithm for multi-objective optimisation. Its main characteristics are:

- It uses Differential Evolution (DE) algorithm [28–30] to produce its offspring at each generation. It is used given its convergence properties and simplicity for MOO [31].
- It uses spherical pruning [32] in order to promote diversity in the approximated Pareto front. Basically, the objective space is partitioned using spherical coordinates, and one solution is selected in each spherical sector, avoiding overcrowding areas.
- It uses physical programming (PP) [6] for pertinency improvement and as a mechanism for many-objectives optimisation. It states such preferences in aspiration levels in a matrix M as depicted in Table 2. This PP index is used as an additional mechanism to prune solutions, according to the preference index, in order to get a manageable size of the Pareto front approximation.

Nevertheless, in spite of its usefulness, different mechanisms might substitute the Physical Programming approach requiring less information from the designer. For this reason, we modify such an algorithm with two additional mechanisms for pertinency improvement (Available at: https://www.mathworks.com/matlabcentral/fileexchange/65145). The first of them the TOPSIS mechanism [5], and the second the PROMETHEE II [7] method. That is, the original pruning mechanism using Physical Programming is modified. While the TOPSIS mechanism just require as input the Pareto front approximation, the PROMETHEE II method require information about (in) significant differences for each design objective (See Table 3). The idea is to evaluate and compare different preference information methods working actively in the MOO process. Among the TOPSIS, PROMETHEE II and PP methods, the former requires the less information, while the latter the most. Comparison of input required for each mechanism is depicted in Table 4.

Table 2. Preference matrix M. Five preference ranges have been defined: highly desirable (HD), desirable (D), tolerable (T) undesirable (U) and highly undesirable (HU).

				Preference Matrix							
	$\leftarrow$	HD	$\rightarrow \leftarrow$	D	$\rightarrow \leftarrow$	T	$\rightarrow \leftarrow$	U	$\rightarrow \leftarrow$	HU	$\rightarrow$
Objective	J_i^0		J_i^1		J_i^2		J_i^3		J_i^4		J_i^5
$J_1(x)$ [-]	.		.		.		.		.		.
...											
$J_n(x)$ [-]	.		.		.		.		.		.

Table 3. Matrix with (in)significant differences. Significant (S) and Insignificant (I) differences for each design objectives are defined.

I/S Differences Matrix		
Objective	I	S
$J_1(x)$ [-]	.	.
...		
$J_n(x)$ [-]	.	.

Table 4. Input required for each preference mechanism.

Method	Input
TOPSIS	Pareto front approximation.
PROMETHEE II	Pareto front approximation. Significant/Insignificant differences.
Physical Programming	Solution from the Pareto front approximation. Information about the desirability limits.

For visualisation, Level Diagrams (Toolbox available at https://www.mathworks.com/matlabcentral/fileexchange/62224) are used [33–35], due to their capabilities to depict m-dimensional information [4]. Its main characteristics are:

- It uses as many subplots as design objectives to depict trade-off information.
- Solutions are synchronised by the vertical axis, while the horizontal axis keeps their original units. That is, no normalisation deforming the units scale is used.
- Trade-off relationships might be propagated to design variables by synchronising the same vertical axis.

4. Test Cases

In order to evaluate the impact of substituting the original pruning mechanism in the sp-MODEII algorithm, two MOPs with 5 and 6 design objectives are stated.

4.1. Case Study 1: Dissolved Oxygen Control in a Waste-Water Treatment Process

The first MOP is a dissolved oxygen control problem for an activated sludge waste-water treatment process [36]. This case study is proposed in order to evaluate the MCDM tools at the end of the MOOD procedure. The process is modelled as a continuous state-space model as:

$$\frac{dx}{dt} = Ax + Bu \tag{1}$$

$$y = Cx + Du \tag{2}$$

where x is the state vector, u and y are the input and output vectors and A, B, C and D are the state-space matrices with the following values:

$$A = \begin{bmatrix} -100.03 & 115.00 \\ 167.77 & -211.47 \end{bmatrix} \tag{3}$$

$$B = \begin{bmatrix} 0.87 \\ -1.55 \end{bmatrix} \tag{4}$$

$$C = \begin{bmatrix} 7.55 & 0.32 \end{bmatrix} \tag{5}$$

$$D = 0 \tag{6}$$

The control problem consists of tuning a proportional-integral (PI) controller in order to keep the dissolved oxygen concentration within desired specifications, by manipulating the oxygen mass transfer coefficient in the treatment process. A total of five design objectives are stated:

$J_1(x)$: Settling time (day) for a setpoint reference change (minimise).
$J_2(x)$: Settling time (day) for an input disturbance in the sludge process (minimise).
$J_3(x)$: Maximum deviation from setpoint ($gCOD/m^3$) due to an input disturbance in the sludge process (minimise).

$J_4(x)$:   Total variation of oxygen mass transfer coefficient (day $^{-1}$) due to the setpoint reference change and the input disturbance (minimise).

$J_5(x)$:   Aeration energy cost (kWh/day) due to the setpoint reference change and the input disturbance (minimise). Given a value of the control action for a given instant u_i, the instant aeration energy cost AE_i is calculated as:

$$AE_i = 0.4032u_i^2 + 7.8408u_i \tag{7}$$

A PI controller $C(s)$ has 2 design variables: proportional gain $k_p = x_1$ and integral gain $k_i = x_2$. The Laplace expression of a PI controller is as follows:

$$C(s) = k_p + k_i \frac{1}{s} \tag{8}$$

Therefore, the MOP for the optimisation process is:

$$\min_x J(x) \quad = \quad [J_1(x), J_2(x), J_3(x), J_4(x), J_5(x)] \tag{9}$$

subject to

$$0.00 \le x_1 \le 1000 \tag{10}$$
$$0.00 \le x_2 \le 10000 \tag{11}$$

Results from optimisation process are depicted in Figure 1. Inflection point in design objective $J_5(x)$ is used for further interpretability, in order to identify objective vectors with $J_5(x) > 2.2 \times 10^4$ kWh/day and $J_5(x) \le 2.2 \times 10^4$ kWh/day. With such information is possible to track tendencies across subplots. For example, the lower $J_5(x)$ the bigger $J_1(x)$, $J_2(x)$ and $J_3(x)$. This means that a reduction on the aeration energy tends to worsen settling times and the load deviation capacity. In addition, it means that aeration energy and total variation of control action are correlated.

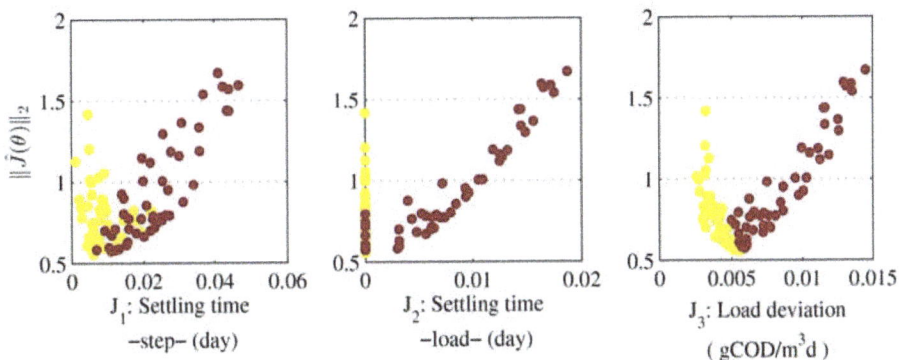

Figure 1. *Cont.*

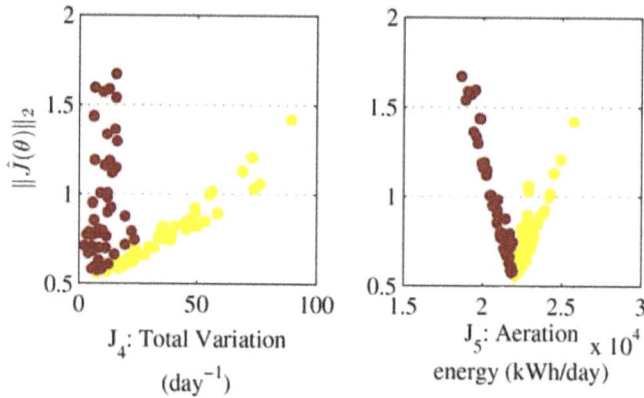

Figure 1. Pareto front approximation for MOP I.

Using preferences stated in Tables 5 and 6 and the TOPSIS method, three solutions have been selected for further control tests. Their time responses are depicted in Figure 2 when facing a setpoint change and a disturbance. As expected, none of such solutions is better than the others in an overall sense: each one has a unique trade-off. This approach might be used by a decision maker, in order to focus in a subset of approximated Pareto optimal solutions, to perform a final decision regarding the controller to implement. Next, we will actively use such methodologies and we will compare Pareto optimal solutions approximated in each case.

Table 5. Preference matrix m for MOP statement I. Five preference ranges have been defined: highly desirable (HD), desirable (D), tolerable (T) undesirable (U) and highly undesirable (HU).

		Preference Matrix					
Objective	$\overset{\leftarrow}{J_i^0}$	**HD** $\overset{\rightarrow\leftarrow}{J_i^1}$	**D** $\overset{\rightarrow\leftarrow}{J_i^2}$	**T** $\overset{\rightarrow\leftarrow}{J_i^3}$	**U** $\overset{\rightarrow\leftarrow}{J_i^4}$	**HU** $\overset{\rightarrow}{J_i^5}$
$J_1(x)$ (day)	0.03	0.04	0.05	0.06	0.07	0.08
$J_2(x)$ (day)	0.005	0.010	0.015	0.020	0.025	0.030
$J_3(x)$ ($gCOD/m^3day$)	0.000	0.005	0.010	0.0150	0.020	0.025
$J_4(x)$ (day^{-1})	20.00	30.00	40.00	50.00	60.00	70.00
$J_5(x)$ (kWh/day)	1.00	2.00	3.00	4.00	5.00	6.00

Table 6. Matrix with (in) significant differences for MOP statement I. Significant (S) and Insignificant (I) differences for each design objectives are defined.

I/S Differences Matrix		
Objective	**I**	**S**
$J_1(x)$ (day)	0.01	0.03
$J_2(x)$ (day)	0.05	0.05
$J_3(x)$ ($gCOD/m^3day$)	0.025	0.05
$J_4(x)$ (day^{-1})	5.0	10.00
$J_5(x)$ (kWh/day)	0.5	1.00

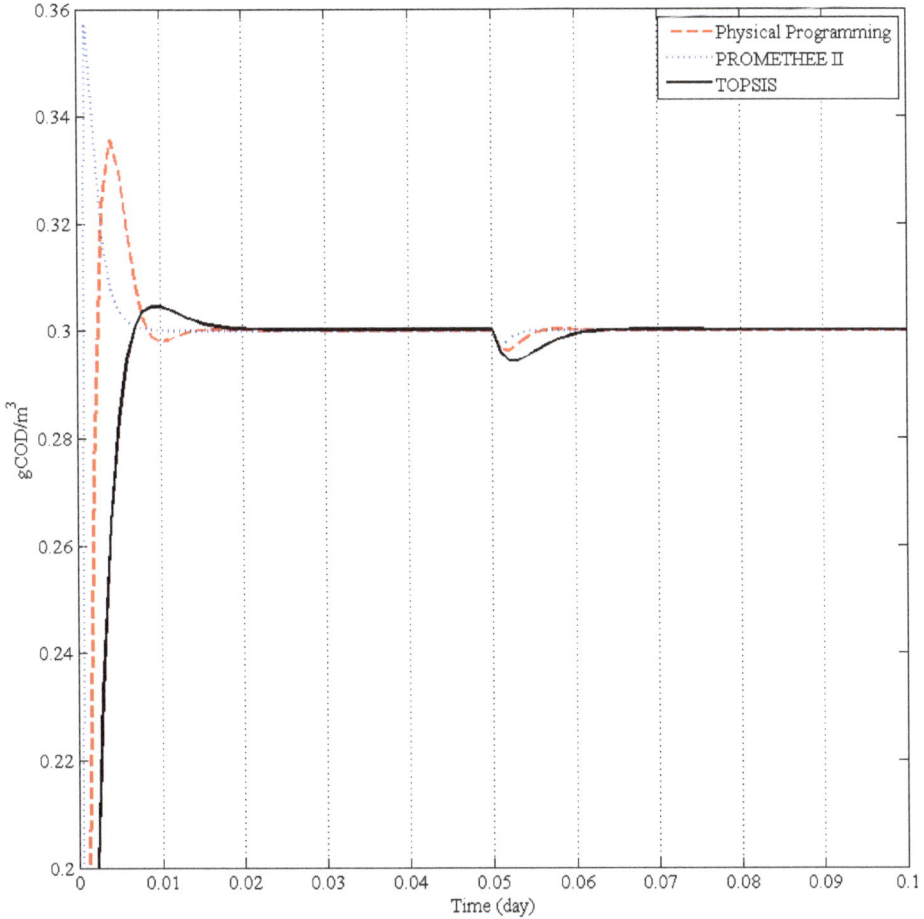

Figure 2. Time response comparison among selected controllers for MOP I.

4.2. Case Study 2: Pollution Management in Water Distribution Systems

This case study is based on the hypothetical condensed example of the Bow River Valley, as presented in [37], and it follows the preliminary results that were presented in [38]. It deals with the pollution problem due to a cannery industry (Pierce-Hall Cannery), to two sources of municipal waste (Bowville and Plymton), with a park in the middle (Robin State Park). Water quality in the river is evaluated via dissolved oxygen concentration (DO). Quality of the effluent from the three treatment plants is measured with the biochemical oxygen demanding material (BOD) which is separated into carbonaceous and nitrogenous material (BODc and BODn respectively). The major aim of this example is to evaluate structural differences in the approximated Pareto fronts when using different policies in the pruning mechanism of the sp-MODEII.

The MOP under consideration has six design objectives:

$J_1(x)$: DO level at Bowville (mg/L) (maximise).
$J_2(x)$: DO level at Robin State Park (mg/L) (maximise).
$J_3(x)$: DO level at Plymton (mg/L) (maximise).

$J_4(x)$: Return on equity (%) (maximise).
$J_5(x)$: Tax increment (Bowville) (minimise).
$J_6(x)$: Tax increment (Plymton) (minimise).

Additionally, the DO at the state line $G_1(x)$ (mg/L) is considered. Decision variables $x = [x_1, x_2, x_3]$ are the treatment levels of water discharge at the Pierce-Hall cannery, at Bowville and a Plymton, respectively. The constrained MOP for optimisation is as follows:

$$\min_x J(x) \quad = \quad [-J_1(x), -J_2(x), -J_3(x), -J_4(x), J_5(x), J_6(x)] \qquad (12)$$

subject to

$$G_1(x) \quad \geq \quad 3.5 \qquad (13)$$
$$0.3 \leq \quad x_1 \quad \leq 1.0 \qquad (14)$$
$$0.3 \leq \quad x_2 \quad \leq 1.0 \qquad (15)$$
$$0.3 \leq \quad x_3 \quad \leq 1.0 \qquad (16)$$

where

$$
\begin{aligned}
J_1(x) &= \quad 4.75 + 2.27(x_1 - 0.3) & (17) \\
J_2(x) &= \quad 2.0 + 0.524(x_1 - 0.3) + 2.79(x_2 - 0.3) + 0.882(w_1 - 0.3) + 2.65(w_2 - 0.3) & (18) \\
J_3(x) &= \quad 5.1 + 0.177(x_1 - 0.3) + 0.978(x_2 - 0.3) + 0.216(w_1 - 0.3) + 0.768(w_2 - 0.3) & (19)
\end{aligned}
$$

$$J_4(x) \quad = \quad 7.5 - 0.012\left(\frac{59}{1.09 - x_1^2} - 59\right) \qquad (20)$$

$$J_5(x) \quad = \quad 0.0018\left(\frac{532}{1.09 - x_2^2} - 532\right) \qquad (21)$$

$$J_6(x) \quad = \quad 0.0025\left(\frac{450}{1.09 - x_3^2} - 450\right) \qquad (22)$$

$$
\begin{aligned}
G_1(x) \quad = \quad & 1.0 + 0.0332(x_1 - 0.3) + 0.0186(x_2 - 0.3) + 3.34(x_3 - 0.3) + \\
& 0.0204(w_1 - 0.3) + 0.778(w_2 - 0.3) + 2.62(w_3 - 0.3) & (23)
\end{aligned}
$$

$$w_i = \frac{0.39}{1.39 - x_i^2}, i \in [1, 2, 3] \qquad (24)$$

In Tables 7 and 8 preferences stated are reported. Please note that congruence has been sought between them. That is, insignificant differences coincide with the tolerable interval, whilst significant differences coincide with the range from tolerability to highly desirability. This provides a guide in order to link both methods.

Pareto front and set approximations are depicted in Figures 3 and 4, respectively. It is possible to appreciate that solutions are clustering towards different regions in the Pareto front, according to the information provided. TOPSIS solutions (black x) describe a cluster at the bottom of Level Diagrams; this is expected, given that the TOPSIS method seeks for a similarity with the utopian solution, which is also used to normalise the Pareto front approximation in Level Diagrams and calculate the selected norm. The next cluster correspond to the PROMETHEE II pruning (blue *) and the one on the upper region to Physical Programming (red +).

Table 7. Matrix with (in)significant differences for MOP II statement. Significant (S) and Insignificant (I) differences for each design objectives are defined.

I/S Differences Matrix		
Objective	I	S
$J_1(x)$ (mg/L)	2.0	5.0
$J_2(x)$ (mg/L)	2.0	5.0
$J_3(x)$ (mg/L)	2.0	5.0
$J_4(x)$ ($)	1.0	3.0
$J_5(x)$ (%)	1.0	3.0
$J_6(x)$ (%)	1.0	3.0
$G_1(x)$ (mg/L)	0.5	1.5

Table 8. Preference matrix m for MOP II statement. Five preference ranges have been defined: highly desirable (HD), desirable (D), tolerable (T) undesirable (U) and highly undesirable (HU).

Preference Matrix					
Objective	$\leftarrow$ HD $\rightarrow$ J_i^0	$\leftarrow$ D $\rightarrow$ J_i^1	$\leftarrow$ T $\rightarrow$ J_i^2	$\leftarrow$ U $\rightarrow$ J_i^3	$\leftarrow$ HU $\rightarrow$ J_i^4	$\rightarrow$ J_i^5
$-J_1(x)$ (mg/L)	-9	-8	-6	-4	-2	0.0
$-J_2(x)$ (mg/L)	-9	-8	-6	-4	-2	0.0
$-J_3(x)$ (mg/L)	-9	-8	-6	-4	-2	0.0
$-J_4(x)$ ($)	-8	-7	-6	-5	-4	-3
$J_5(x)$ (%)	0	1	2	3	4	5
$J_6(x)$ (%)	0	1	2	3	4	5
$-G_1(x)$ (mg/L)	-5	-4.5	-4	-3.5	-3.0	-2.5

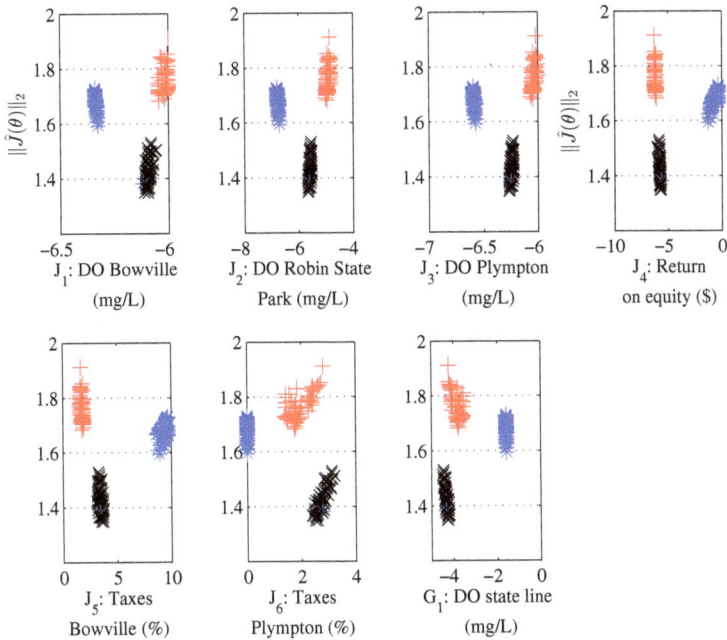

Figure 3. Pareto front approximation for MOP II with preference pruning: PP (red +); PROMETHEE II (blue *) and TOPSIS (black x).

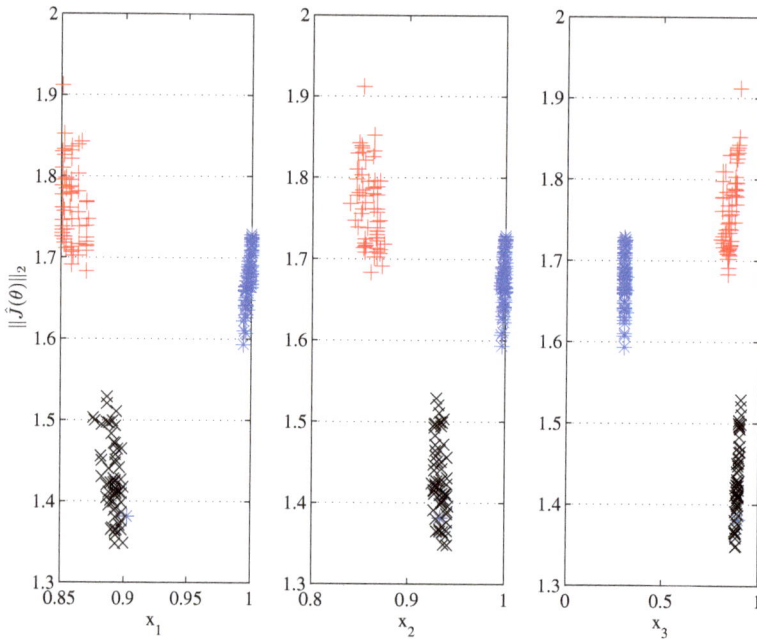

Figure 4. Pareto set approximation for MOP II with preference pruning: PP (red +), PROMETHEE II (blue ∗) and TOPSIS (black x).

Main difference between TOPSIS and PP, is regarding design objective $J_6(x)$: the preference matrix states that a solution with $J_6(x) > 3$ is undesirable; that is the reason because the pruning mechanism tends to worsen the remainder design objectives, with the aim of improving $J_6(x)$. This does not mean that the TOPSIS mechanism gives worst results; it is important to remember that such mechanisms did not have any information about such undesirability. The same apply with the PROMETHEE II mechanism: provided information about (in) significant differences was helpful to evolve towards a desirable region in several design objectives, but fails in some of them.

In any case, the fact that, the bigger the norm, the more the information used by the pruning mechanism, reveals the philosophy behind multi-objective optimisation: it might be not enough to minimise a given norm, but to analyse/incorporate the trade-off analysis in a different way. In conclusion, the most the information provided by the designer, the most the accurate the algorithm to approximate a pertinent region in the objective space. Obviously this is in exchange of investing more time in stating the preferences a priori.

5. Conclusions

In this paper, we incorporated two additional mechanisms to handle designers' preferences in multi-objective optimisation in the sp-MODEII algorithm. Such mechanisms are based on the TOPSIS and PROMETHEE methods, usually employed for multi-criteria analysis and decision making.

A comparison to handle designers' preferences on two benchmarks dealing with water distribution systems was performed. On the one hand, it was shown that an analysis in such *m*-dimensional spaces with more than three design objectives is possible via specialised visualisation tool (Level Diagrams). On the other hand, the structural differences between different approaches to approximate a pertinent region of the Pareto front was also analysed.

In the latter case, the capacity to approximate a compact set focusing in the region of interest of the decision maker is evaluated. It was shown that the main structural difference among approaches (TOPSIS, PROMETHEE II, and Physical Programming) is the closeness of their clusters to an ideal solution, defined by the Pareto front approximation itself.

Further work will focus on using additional mechanisms to handle such preferences actively in the optimisation stage. Besides, merging two or more mechanisms might be an interesting idea to explore, in order to exploit synergies between different approaches.

Acknowledgments: This work is under the research initiative Multi-objective optimisation design (MOOD) procedures for engineering systems: Industrial applications, unmanned aerial systems and mechatronic devices, supported by the National Council of Scientific and Technological Development of Brazil (CNPq) through the grant PQ-2/304066/2016-8. This work is also supported by the Brazilian Federal Agency for Support and Evaluation of Graduate Education (CAPES) through the grant PROSUC/159063/2017-0.

Author Contributions: G.R.-M. and E.P.C.-A. conceived benchmark experiments; G.R.-M. and E.P.C.-A. performed benchmarks experiments; G.R.-M. and V.H.A.R. analysed the data; and G.R.-M. and V.H.A.R. wrote the paper. All authors read and approved the manuscript.

Conflicts of Interest: The authors declare no conflict of interest.

Abbreviations

The following abbreviations are used in this manuscript:

EMO	Evolutionary Multi-Objective Optimisation
LD	Level Diagrams
MCDM	Multi-criteria Decision Making
MOP	Multi-Objective Problem
MOO	Multi-Objective Optimisation
PP	Physical Programming
WDS	Water Distribution System

Appendix A. Background

A multi-objective problem (MOP) with m objectives, can be stated as follows [3]:

$$\min_{x} J(x) \quad = \quad [J_1(x), \ldots, J_m(x)] \tag{A1}$$

subject to:

$$K(x) \quad \leq \quad 0 \tag{A2}$$
$$L(x) \quad = \quad 0 \tag{A3}$$
$$\underline{x_i} \leq x_i \quad \leq \quad \overline{x_i}, i = [1, \ldots, n] \tag{A4}$$

where $x = [x_1, x_2, \ldots, x_n]$ is defined as the decision vector with $\dim(x) = n$; $J(x)$ as the objective vector and $K(x)$, $L(x)$ as the inequality and equality constraint vectors respectively; $\underline{x_i}, \overline{x_i}$ are the lower and the upper bounds in the decision space.

It has been noticed that there is not a single solution in MOPs, because there is not generally a better solution in all the objectives. Therefore, a set of solutions, the Pareto set, is defined. Each solution in the Pareto set defines an objective vector in the Pareto front (see Figure A1). All the solutions in the Pareto front are a set of Pareto optimal and non-dominated solutions:

- Pareto optimality [3]: An objective vector $J(x^1)$ is Pareto optimal if there is not another objective vector $J(x^2)$ such that $J_i(x^2) \leq J_i(x^1)$ for all $i \in [1, 2, \ldots, m]$ and $J_j(x^2) < J_j(x^1)$ for at least one $j, j \in [1, 2, \ldots, m]$.

- Dominance: An objective vector $J(x^1)$ is dominated by another objective vector $J(x^2)$ if $J_i(x^2) \leq J_i(x^1)$ for all $i \in [1,2,\ldots,m]$ and $J_j(x^2) < J_j(x^1)$ for at least one j, $j \in [1,2,\ldots,m]$. This is denoted as $J(x_2) \preceq J(x_1)$.

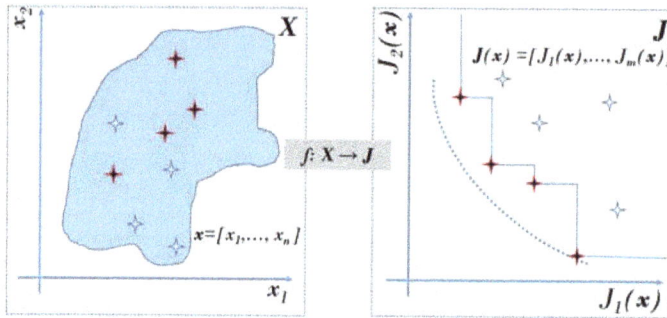

Figure A1. Pareto optimality and dominance concepts for a min-min problem. Dark solutions is the subset of non-dominated solutions which approximates a Pareto front (**right**) and a Pareto set (**left**). Remainder solutios are dominated solutions, because it is possible to find at least one solution with better values in all design objectives (Source: [39]).

To successfully implement the multi-objective optimisation approach, three fundamental steps are required: the MOP definition, the multi-objective optimisation (MOO) process and the multi-criteria decision making (MCDM) stage. This integral and holistic process will be denoted hereafter as a multi-objective optimisation design (MOOD) procedure [40]. In the MOP statement, design objectives are defined, as well as decision variables (with their bounds) and constraints for feasibility or suitability; in the MOO the major aim is to calculate a useful Pareto front approximation via an algorithm; in the MCDM stage, an analysis of the approximated Pareto front and trade-offs is carried out according to a set of preferences, in order to select the final solution to implement.

There are different methodologies for MCDM and visualisation approaches [4,41]. In the case of the MOO process, special (or particular) circumstances might require additional mechanisms to deal successfully with a given MOP [40,42]. Some of them are listed below:

- Constrained optimisation. Results from the optimisation problem are not always feasible in a practical sense; therefore constraints must be incorporated in order to assure their feasibility.
- Many-objectives optimisation. If a MOP has more than 3 design objectives, it is considered a many-objectives optimisation problem. It is important to consider such a sub-classification, given that converge and diversity mechanisms might be in conflict.
- Computational expensive optimisation. Extensive or exhaustive simulations might be required in order to compute one or more design objectives requires.
- Multi-modal optimisation. It might happen that two or more decision vector points to the same objective vector.

For the MOO process, multi-objective evolutionary algorithms (MOEAs) have shown to be a useful tool for a wide range of engineering problems [43].

References

1. Savic, D. Single-objective vs. multiobjective optimisation for integrated decision support. In Proceedings of the International Congress on Environmental Modelling and Software, Lugano, Switzerland, 24–27 June 2002.

2. Mala-Jetmarova, H.; Sultanova, N.; Savic, D. Lost in optimisation of water distribution systems? A literature review of system operation. *Environ. Model. Softw.* **2017**, *93*, 209–254.

3. Miettinen, K. Nonlinear Multiobjective Optimization. In *International Series in Operations Research and Management Science*; Springer Science & Business Media: Berlkin, Germany, 1999; Volume 12.

4. Tušar, T.; Filipič, B. Visualization of Pareto front approximations in evolutionary multiobjective optimization: A critical review and the prosection method. *IEEE Trans. Evol. Comput.* **2015**, *19*, 225–245.

5. Hwang, C.L.; Lai, Y.J.; Liu, T.Y. A new approach for multiple objective decision making. *Comput. Oper. Res.* **1993**, *20*, 889–899.

6. Messac, A. Physical programming: Effective optimization for computational design. *AIAA J.* **1996**, *34*, 149–158.

7. Behzadian, M.; Kazemzadeh, R.B.; Albadvi, A.; Aghdasi, M. PROMETHEE: A comprehensive literature review on methodologies and applications. *Eur. J. Oper. Res.* **2010**, *200*, 198–215.

8. Reynoso-Meza, G.; Sanchis, J.; Blasco, X.; García-Nieto, S. Multiobjective evolutionary algorithms for multivariable PI controller tuning. *Appl. Soft Comput.* **2014**, *24*, 341 – 362.

9. Reynoso-Meza, G. Controller Tuning by Means of Evolutionary Multiobjective Optimization: A Holistic Multiobjective Optimization Design Procedure. Ph.D. Thesis, Universitat Politècnica de València, València, Spain, 2014.

10. Savic, D.A.; Walters, G.A.; Schwab, M. Multiobjective genetic algorithms for pump scheduling in water supply. *AISB International Workshop on Evolutionary Computing*; Springer: Berlin, Germany, 1997; pp. 227–235.

11. Sotelo, A.; Baran, B. Optimizacion de los costos de bombeo en sistemas de suministro de agua mediante un algoritmo evolutivo multiobjetivo combinado (Pumping cost optimization in water supply systems using a multi-objective evolutionary combined algorithm). In Proceedings of the XV Chilean Conference on Hydraulic Engineering, Concepción, Chile, 7–9 November 2001; pp. 337–347.

12. Kelner, V.; Léonard, O. Optimal pump scheduling for water supply using genetic algorithms. In Proceedings of the Evolutionary Methods for Design, Optimization and Control with Applications to Industrial Problems (Eurogen 2003), Barcelona, Spain, 15–17 September 2003.

13. Barán, B.; von Lücken, C.; Sotelo, A. Multi-objective pump scheduling optimisation using evolutionary strategies. *Adv. Eng. Softw.* **2005**, *36*, 39–47.

14. Lopez-Ibanez, M.; Prasad, T.D.; Paechter, B. Multi-objective optimisation of the pump scheduling problem using SPEA2. In Proceedings of the 2005 IEEE Congress on Evolutionary Computation, Edinburgh, UK, 2–5 September 2005; Volume 1, pp. 435–442.

15. Odan, F.K.; Ribeiro Reis, L.F.; Kapelan, Z. Real-time multiobjective optimization of operation of water supply systems. *J. Water Resour. Plan. Manag.* **2015**, *141*, 04015011.

16. Stokes, C.S.; Maier, H.R.; Simpson, A.R. Water distribution system pumping operational greenhouse gas emissions minimization by considering time-dependent emissions factors. *J. Water Resour. Plan. Manag.* **2014**, *141*, 04014088.

17. Prasad, T.D.; Walters, G.; Savic, D. Booster disinfection of water supply networks: Multiobjective approach. *J. Water Resour. Plan. Manag.* **2004**, *130*, 367–376.

18. Kurek, W.; Brdys, M.A. Optimised allocation of chlorination stations by multi-objective genetic optimisation for quality control in drinking water distribution systems. *IFAC Proc. Vol.* **2006**, *39*, 232–237.

19. Ewald, G.; Kurek, W.; Brdys, M.A. Grid implementation of a parallel multiobjective genetic algorithm for optimized allocation of chlorination stations in drinking water distribution systems: Chojnice case study. *IEEE Trans. Syst. Man Cybern. C Appl. Rev.* **2008**, *38*, 497–509.

20. Alfonso, L.; Jonoski, A.; Solomatine, D. Multiobjective optimization of operational responses for contaminant flushing in water distribution networks. *J. Water Resour. Plan. Manag.* **2009**, *136*, 48–58.

21. Giustolisi, O.; Laucelli, D.; Berardi, L. Operational optimization: Water losses versus energy costs. *J. Hydraul. Eng.* **2012**, *139*, 410–423.

22. Kougias, I.P.; Theodossiou, N.P. Multiobjective pump scheduling optimization using harmony search algorithm (HSA) and polyphonic HSA. *Water Resour. Manag.* **2013**, *27*, 1249–1261.

23. Kurek, W.; Ostfeld, A. Multi-objective optimization of water quality, pumps operation, and storage sizing of water distribution systems. *J. Environ. Manag.* **2013**, *115*, 189–197.

24. Kurek, W.; Ostfeld, A. Multiobjective water distribution systems control of pumping cost, water quality, and storage-reliability constraints. *J. Water Resour. Plan. Manag.* **2012**, *140*, 184–193.

25. Mala-Jetmarova, H.; Barton, A.; Bagirov, A. Exploration of the trade-offs between water quality and pumping costs in optimal operation of regional multiquality water distribution systems. *J. Water Resour. Plan. Manag.* **2014**, *141*, 04014077.

26. Mala-Jetmarova, H.; Barton, A.; Bagirov, A. Impact of water-quality conditions in source reservoirs on the optimal operation of a regional multiquality water-distribution system. *J. Water Resour. Plan. Manag.* **2015**, *141*, 04015013.

27. Ishibuchi, H.; Tsukamoto, N.; Nojima, Y. Evolutionary many-objective optimization: A short review. In Proceedings of the IEEE Congress on Evolutionary Computation (CEC 2008), IEEE World Congress on Computational Intelligence, Hong Kong, China, 1–6 June 2008; pp. 2419–2426.

28. Storn, R.; Price, K. Differential evolution–a simple and efficient heuristic for global optimization over continuous spaces. *J. Glob. Optim.* **1997**, *11*, 341–359.

29. Das, S.; Suganthan, P.N. Differential evolution: A survey of the state-of-the-art. *IEEE Trans. Evol. Comput.* **2011**, *15*, 4–31.

30. Das, S.; Mullick, S.S.; Suganthan, P.N. Recent advances in differential evolution–an updated survey. *Swarm Evol. Comput.* **2016**, *27*, 1–30.

31. Zhou, A.; Qu, B.Y.; Li, H.; Zhao, S.Z.; Suganthan, P.N.; Zhang, Q. Multiobjective evolutionary algorithms: A survey of the state of the art. *Swarm Evol. Comput.* **2011**, *1*, 32–49.

32. Reynoso-Meza, G.; Sanchis, J.; Blasco, X.; Martínez, M. Design of continuous controllers using a multiobjective differential evolution algorithm with spherical pruning. *Appl. Evol. Comput.* **2010**, *6024*, 532–541.

33. Blasco, X.; Herrero, J.; Sanchis, J.; Martínez, M. A new graphical visualization of n-dimensional Pareto front for decision-making in multiobjective optimization. *Inf. Sci.* **2008**, *178*, 3908–3924.

34. Reynoso-Meza, G.; Blasco, X.; Sanchis, J.; Herrero, J.M. Comparison of design concepts in multi-criteria decision-making using level diagrams. *Inf. Sci.* **2013**, *221*, 124–141.

35. Blasco, X.; Reynoso-Meza, G.; Pérez, E.A.S.; Pérez, J.V.S. Asymmetric distances to improve n-dimensional Pareto fronts graphical analysis. *Inf. Sci.* **2016**, *340*, 228–249.

36. Holenda, B.; Domokos, E.; Redey, A.; Fazakas, J. Dissolved oxygen control of the activated sludge wastewater treatment process using model predictive control. *Comput. Chem. Eng.* **2008**, *32*, 1270–1278.

37. Monarchi, D.E.; Kisiel, C.C.; Duckstein, L. Interactive multiobjective programing in water resources: A case study. *Water Resour. Res.* **1973**, *9*, 837–850.

38. Reynoso-Meza, G.; Carreno-Alvarado, E.P.; Montalvo, I.; Izquierdo, J. Water pollution management with evolutionary multi-objective optimisation and preferences. In Proceedings of the Congress on Numerical Methods in Engineering CMN, Valencia, Spain, 3–5 July 2017.

39. Reynoso-Meza, G.; Sanchis, J.; Blasco, X.; Martínez, M. Preference driven multi-objective optimization design procedure for industrial controller tuning. *Inf. Sci.* **2016**, *339*, 108–131.

40. Reynoso-Meza, G.; Sanchis, J.; Blasco, X.; Martínez, M. Controller tuning using evolutionary multi-objective optimisation: current trends and applications. *Control Eng. Pract.* **2014**, *28*, 58–73.

41. Lotov, A.V.; Miettinen, K. Visualizing the Pareto frontier. In *Multiobjective Optimization*; Springer: Berlin, Germany, 2008; pp. 213–243.

42. Meza, G.R.; Ferragud, X.B.; Saez, J.S.; Durá, J.M.H. Tools for the Multiobjective Optimization Design Procedure. In *Controller Tuning with Evolutionary Multiobjective Optimization*; Springer: Berlin, Germany, 2017; pp. 59–88.

43. Coello, C.A.C.; Lamont, G.B. *Applications of Multi-Objective Evolutionary Algorithms*; World Science: Singapore, 2004; Volume 1.

water

MDPI

Article

Estimation of Water Demand in Water Distribution Systems Using Particle Swarm Optimization

Lawrence K. Letting [1,*], Yskandar Hamam [1] and Adnan M. Abu-Mahfouz [1,2]

[1] Department of Electrical Engineering, Tshwane University of Technology, Pretoria 0001, South Africa; HamamA@tut.ac.za (Y.H.); A.Abumahfouz@ieee.org (A.M.A.-M.)
[2] CSIR Meraka Institute, Pretoria 0081, South Africa
* Correspondence: LettingLK@gmail.com; Tel.: +27-12-382-4824

Received: 1 July 2017; Accepted: 3 August 2017; Published: 21 August 2017

Abstract: Demand estimation in a water distribution network provides crucial data for monitoring and controlling systems. Because of budgetary and physical constraints, there is a need to estimate water demand from a limited number of sensor measurements. The demand estimation problem is underdetermined because of the limited sensor data and the implicit relationships between nodal demands and pressure heads. A simulation optimization technique using the water distribution network hydraulic model and an evolutionary algorithm is a potential solution to the demand estimation problem. This paper presents a detailed process simulation model for water demand estimation using the particle swarm optimization (PSO) algorithm. Nodal water demands and pipe flows are estimated when the number of estimated parameters is more than the number of measured values. The water demand at each node is determined by using the PSO algorithm to identify a corresponding demand multiplier. The demand multipliers are encoded with varying step sizes and the optimization algorithm particles are also discretized in order to improve the computation time. The sensitivity of the estimated water demand to uncertainty in demand multiplier discrete values and uncertainty in measured parameters is investigated. The sensor placement locations are selected using an analysis of the sensitivity of measured nodal heads and pipe flows to the change in the water demand. The results show that nodal demands and pipe flows can be accurately determined from a limited number of sensors.

Keywords: water demand estimation; demand multipliers; underdetermined model; uncertain measurements; particle swarm optimization

1. Introduction

A water distribution system (WDS) performs the crucial role of supplying safe drinking water to the public. The main goal in WDS operation and control is to meet the desired water demand while ensuring the appropriate water quality and pressure is met in all the nodes. Water flow from the sources to the demand nodes is a continuous event with time-varying flow rates determined by water demand schedules. Nodal demands are state variables of interest in WDS operation because they are the driving inputs that determine the nodal pressures and pipe flow rates measured in the field. State estimation is the process of computing unknown network conditions from the knowledge of available measurements and other known network parameters [1]. A water demand estimation algorithm coupled with a WDS model forms a critical component for supervisory control and data acquisition (SCADA) systems.

In water demand estimation, the nodal demands are the unknown state variables, while pressure heads and pipe flows are determined from field measurements. Because of the complexity of pipe connections in large water distribution networks and the associated cost, only a few locations are selected for sensor placement. The accuracy of state estimation relies on the available field measurements, which are constrained by the number of installed pressure and flow sensors. The problem of sensor placement has been addressed in literature by considering budget constraints and the quality of measured data [2], location impact metrics [3,4], and the time taken to detect contamination events [5,6]. The number of sensor locations may also be reduced by clustering water demand nodes into groups [7], grouping nodes according to water quality characteristics [8], and using reduced network models [9]. A ranking technique is used to select nodes for locating pressure sensors in [10]. Loop flows are estimated in [11] using known nodal demands in order to address the problem of a low measurement redundancy. The installation of sensors in a WDS is also subject to site accessibility and consent from the landowner. With the observed constraints in sensor placement, the estimation of nodal demands and unmeasured pressure heads or pipe flows with data obtained from a limited number of locations is a current research problem. Because of the large number of unknown parameters compared to the available measurement data, the estimation problem is underdetermined. The contribution of this work is the formulation of a simulation optimization technique that estimates both nodal demands and unmeasured pressure heads when the number of measurements is less than the number of parameters to be estimated. The objective function minimizes the absolute error of the measured parameters using the particle swarm optimization (PSO) algorithm. The sensor placement locations are selected using an analysis of the sensitivity of measured nodal heads and pipe flows to the change in the water demand. The sensitivity of the estimated parameters to the number of sensors, the uncertainty in demand multiplier discrete values, and measurement errors is also investigated.

The next section presents the literature review on WDS state estimation, and Section 3 presents the formulation of the methodology for water demand estimation via simulation optimization. The simulation results and discussion are presented in Section 4, and Section 5 concludes the study.

2. Literature Review

Knowledge of the status of a water distribution network is realized by means of state estimation techniques that yield a set of variables that fully describe the status of the system. State estimation in WDSs is expressed using a nonlinear measurement model [12]:

$$z = g(x) + \epsilon \tag{1}$$

where $z \in \Re^T$ is the vector of available measurements (pressure heads, flows, and demands), $x \in \Re^T$ is the state vector formed by nodal heads and fixed-head node flows, and ϵ is a zero-mean random vector that models the measurement errors. The measured flows and pressure heads, combined with the hydraulic energy and mass conservation laws, provide the necessary system of equations used to implement the state estimation model defined in Equation (1).

Loop flows are estimated using weighted least squares in [11,13,14] from known nodal demands. Water demands are estimated using a Monte Carlo simulation with Kalman filtering in [15,16]. Genetic algorithms are used to calibrate predicted water demand data in [17,18]. The underdetermined nodal demand problem is solved in [19] using singular value decomposition (SVD) and in [20], by combining SCADA data and demand estimates. Compared to classical methods such as the weighted least squares and Langrangian multipliers, the SVD technique and evolutionary algorithms have the potential of estimating nodal demands when the number of measurements is less than the number of parameters to be estimated. In [11], known demand estimates are used in conjunction with measurements in order to ensure that the estimation problem is not underdetermined.

In this study, a simulation optimization technique based on the discrete event simulation paradigm was used to implement the PSO algorithm. The optimization algorithm was used to estimate nodal demands and unmeasured nodal heads and pipe flows. EPANET software [21] was used to implement and solve the WDS hydraulic model. The optimization algorithm was used for estimation while results from EPANET were used to simulate measured values one time step ahead. The results are compared with recent related works.

3. Methodology

3.1. Objective Function

The water demand estimation problem, which estimates the variation in nodal demand using information from measurements taken by pressure/flow sensors, is defined as a minimization of the absolute error:

$$Min \ J = \sum_{i=1}^{N_H} |H_{i,M} - H_{i,S}| + \sum_{j=1}^{N_Q} |Q_{j,M} - Q_{j,S}| \tag{2}$$

where $H_{i,M}$ and $Q_{j,M}$ are the measured nodal head and pipe flow rate at node i and pipe j, respectively; $H_{i,S}$ and $Q_{j,S}$ are the simulated head and flow rate at node i and pipe j, respectively; and N_H and N_Q are the number of head and flow sensors installed in the network. The constraints are the law of conservation of mass at each node, the law of conservation of energy in each loop, and minimum pressure head requirements. The water demands at each node are the decision variables of the optimization problem.

3.2. Modeling of Water Demand Multipliers

The node demand data used in a water distribution network analysis is normally based on estimates derived from monthly water meter readings. The nodal demand profiles are composed of static distribution factors, which vary with the consumer type. The water distribution system is classified into supply areas according to land use, such as residential, commercial, and industrial areas. The general demand pattern for the different consumer types in a 24 h period is shown in Figure 1. The demand pattern data is obtained from [22].

In order to solve the water demand estimation, the problem of how to set nodal demands used as the driving inputs for generating measured parameters in Equation (2) has to be addressed first. The nodal demand for node i at time step k is given by

$$D_{i,k} = m_{i,k} \times D_{0,i} \tag{3}$$

where $D_{i,k}$ is the nodal demand, $m_{i,k}$ is the demand multiplier for node i, and $D_{0,i}$ is the base demand for node i. The decision variables of the optimization problem are therefore conveniently represented as the set of demand multipliers $m = \{m_{1,k}, m_{2,k}, \dots, m_{N_T,k}\}$ at each time step k, where N_T is the total number of nodes in the water distribution network.

3.3. Process Simulation Model

The demand estimation problem is solved using the discrete event simulation optimization scheme shown in Figure 2. A discrete event simulation scheme eliminates the need for an explicit time loop. The estimation starts with the hydraulic simulator running with data from a defined demand–user pattern in order to generate the measured values of pressure head ($H_{i,m}$) and pipe flow ($Q_{j,m}$). The measured values are obtained from the selected sensor locations. The optimization algorithm uses randomly generated nodal demand multipliers to run the hydraulic simulator in order to generate the simulated values of the pressure head ($H_{i,s}$) and pipe flow ($Q_{j,s}$). At the end of each iteration of the optimization algorithm, the simulated data is passed to the optimization model. The objective Equation (2) function is used to evaluate the fitness of nodal demand multipliers. The optimization algorithm calibrates the demand multipliers on the basis of the fitness obtained. The hydraulic simulator runs in the steady-state mode.

The PSO algorithm [23], the optimization model, and the simulation optimization scheme of Figure 2 were implemented in C/C++. The EPANET software package source code in C and the programming toolkit [24] were utilized. The constraints of the optimization problem were solved by EPANET. The optimization algorithm was interfaced with EPANET using the interface module developed in [25].

The simulation was carried out by defining discrete event times that defined the transition of states in the system. An evolutionary algorithm such as PSO generates a population of potential solutions that are used to run the system, and the fitness of each solution is evaluated using Equation (2). A discrete event time T_0 is chosen as the time each potential solution of the optimization algorithm takes to run the system. The parameters of each potential solution are a set of demand multipliers that are updated at the end of each iteration. The parameters of each potential solution are passed to the hydraulic solver during the initialization phase in order to evaluate the fitness. Proper initialization ensures that the fitness of each potential solution is evaluated with the same initial conditions. After a simulation time equal to $T_1 = NT_0$, where is N is the population size, the fitness of each set of demand multipliers is evaluated and the optimization algorithm re-calibrates the demand multipliers. The hydraulic solver evaluates the reference values (measured values) for the optimization algorithm one time step ahead, that is, at $t = (k - 1)T_1$. In this paper, a discrete event time of $T_0 = 0.01$ s was used. The nodal water demands for each time step during the 24 h consumption period are estimated using demand multipliers, which are used to compute the actual demand using Equation (3). The demand multipliers are defined in the continuous range as shown in Figure 1. In order to increase the convergence speed of the estimation process, the demand multiplier range is converted into the discrete range with step increments. Step sizes of 0.01 and 0.05 were used and the results were compared. An integer coding scheme was used, for which the PSO particles were used to search for the optimal index in the defined range. Because the motion of the PSO particles during the search process is continuous, the position of the particle was rounded to the nearest integer. The particle position corresponds to an array index where the actual demand multiplier is stored.

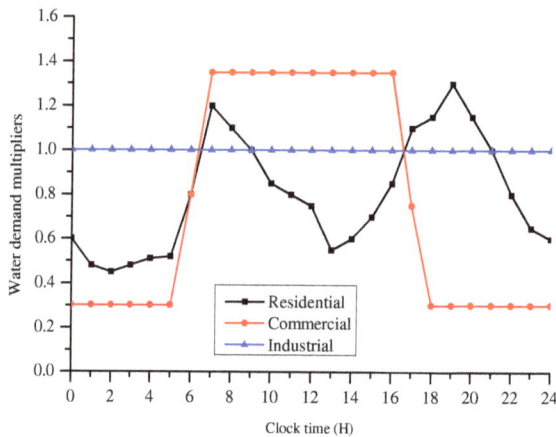

Figure 1. Demand curves for different user types in a period of 24 h.

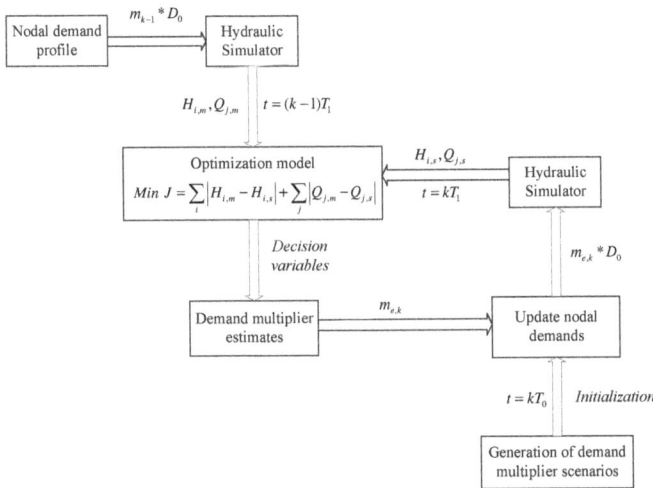

Figure 2. Methodology for estimation of water demand using an optimization algorithm.

3.4. Selection of Sensor Placement Locations

The sensor placement locations are selected on the basis of the sensitivity to the change in the water demand. The sensitivity matrix [26] is expressed as

$$S = \begin{bmatrix} \frac{\partial m_1}{\partial f_1} & \cdots & \frac{\partial m_1}{\partial f_n} \\ \cdot & \cdots & \cdot \\ \frac{\partial m_n}{\partial f_1} & \cdots & \frac{\partial m_n}{\partial f_n} \end{bmatrix} \tag{4}$$

where each element s_{ij} measures the effect of change in the nodal demand f_j on the measurement m_i at a selected pipe link or node. The sensitivity matrix has been used to select sensor placement locations for leakage detection in [27,28].

A simulated change in the water demand in all the nodes is used to generate a pipe flow sensitivity matrix and a nodal head sensitivity matrix. A change in the nodal demand of 1.5 liters per second (LPS) at the two nodes of each pipe link is considered at each step. The measurement locations are ranked according to the variance of the measurements at each location. The best locations for placing sensors are those that experience the highest variability in measurements as a result of change in the nodal demand. This simple approach was used in this study to select the measurement locations in a real network with a large number of pipes and nodes.

4. Simulation Results and Discussion

Simulation experiments were performed using three case studies. A small water network consisting of 9 nodes, 12 pipes, 1 pump and 1 tank was used to perform the simulation experiments in case studies I and II. The network is available in EPANET as example "Net1.net" [21]. The water network was as is shown in Figure 3. Case study I considers the water demand estimation using measurements from pressure sensors only, while Case study II considers measurements from both pressure sensors and flow sensors. All the nodes in the Net1 network share the same hourly demand multipliers, but each node has an independently defined base demand. Case study III considers a medium-sized network that represents the operation of a real water distribution network. The case studies utilized the optimization algorithm to estimate hourly nodal demands for each hour in a 24 h period.

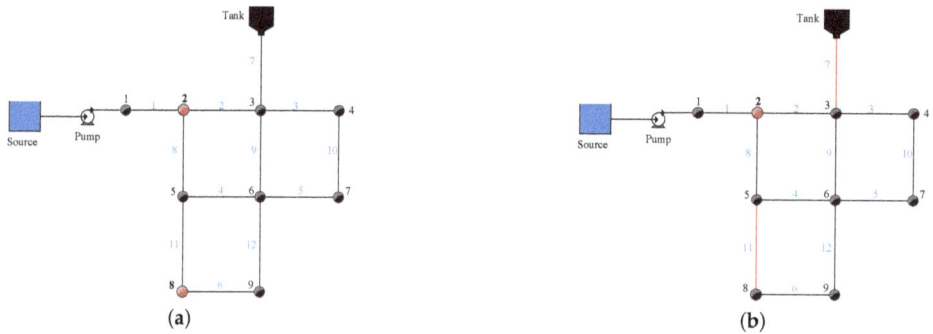

Figure 3. Water distribution network for case studies: (**a**) case study I, and (**b**) case study II.

4.1. Case Study I: Estimation of Nodal Demands with Measurements from Pressure Head Sensors

The sensor placement locations were identified using an exhaustive search. Tests were carried out starting with an ideal case where pressure sensors were placed in all the nodes. The sensors were then reduced sequentially while the performance was observed. Initial results showed that the water consumption at node 2 could not be accurately estimated without placing a pressure sensor at node 2. Node 2 was the source node that supplied the entire network. Node 1 was not selected for sensor placement as there was no water consumption at the node. One sensor was therefore permanently placed at node 2, while the sensors at the other locations were sequentially removed. Good results were obtained with only two sensors placed at nodes 2 and 8.

To assess the suitability of the identified locations, the nodal head sensitivity matrix of the network was generated and the variance of each row was used to rank the nodes. The results of the nodal head sensitivity ranking are shown in Figure 4. The pressure head at node 9 was the most sensitive, while that of node 3 was the least sensitive. The results showed that further engineering judgement was required in order to select the locations, as both nodes 8 and 9 were connected to the same pipe link. Nodes 2 and 5 also shared a common pipe link. It was also observed that node 1 had zero demand. Nodes 2 and 8 were therefore good locations to place pressure head sensors.

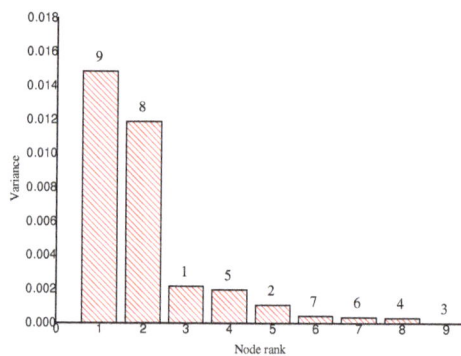

Figure 4. Nodal head sensitivity for Net1 network.

The PSO initialization parameters for the case study are given in Table 1. The first simulation experiment was carried out using step increments of 0.05 to encode the demand multipliers. The demand multipliers lay in the range from 0 to 2.0. Because of the stochastic nature of the optimization process, nodal demand estimates from 30 simulation runs were averaged and used for comparison. The estimated nodal demands for nodes 2, 4, 6 and 9 using a demand multiplier step increment of 0.05 are shown Figure 5. A summary of the average estimated and actual values of the water demands for the 24 h period is given in Table 2. The results in Table 3 were converted to gallons per minute (GPM) for easy comparison. The results were compared with those reported in [17] using a genetic algorithm (GA) model. The base demands for nodes 7, 8 and 9 in the GA model were different to those in the model used in this paper. Estimation was carried out in [17] using three pressure sensors placed at nodes 4, 6, and 8 and with a demand multiplier step increment of 0.02. The results in Table 2 show that the estimated demands for nodes 3, 6 and 7 were comparable, while the PSO model gave improved results at all the other nodes. The results obtained in this paper show that placing a pressure sensor at the source node contributes to an improvement in the estimated water demand. The results obtained from 30 (independent optimization trial) runs in this study compare with those obtained in 100 runs of the GA model [17]. There is therefore a need for further analyses of the sensitivity of the estimation error on the number of optimization runs used to obtain the average demand. A large number of optimization runs increases the computation time. A low computation time is necessary for the near-real-time on-line monitoring and control of water distribution systems.

Table 1. Case study I & II: Initialization of particle swarm optimization (PSO) parameters.

Parameter	Value
Population (P)	50
Number of iterations (N)	100
Inertia factor (w_{max}, w_{min})	0.5, 0.05
Social rate (C_1)	0.9
Cognitive rate (C_2)	2.5

Table 2. Case I: Estimated nodal demands (liters per second—LPS).

Node	Actual	Estimated	% Error
2	9.46	9.44	0.2
3	9.46	8.24	12.9
4	6.31	6.49	2.7
5	9.46	9.55	0.1
6	12.62	13.40	6.2
7	9.46	8.58	9.3
8	6.31	6.35	0.6
9	6.31	6.17	2.2

Table 3. Case I: Comparison of estimated nodal demands (gallons per minute—GPM) using particle swarm optimization (PSO) and the genetic algorithm (GA).

Node	Actual	Estimated (PSO)	Estimated (GA)	% Error (PSO)	% Error (GA)
2	150	149.60	118.67	0.2	20.9
3	150	130.59	131.45	12.9	12.4
4	100	102.85	93.15	2.7	6.9
5	150	151.35	164.15	0.1	9.43
6	200/(150)	212.36	140.58	6.2	6.3
7	150/(300)	135.97	(327.15)	9.3	9.1
8	100/(50)	100.63	(50.34)	0.6	0.7
9	100/(50)	97.78	(35.90)	2.2	28.2

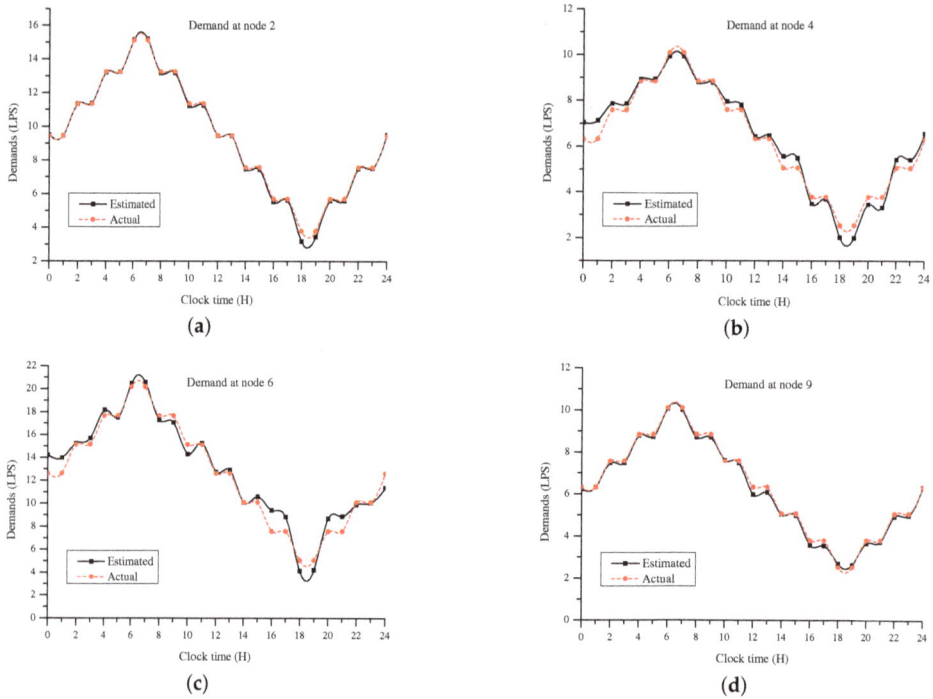

Figure 5. Case I: Demand estimation at nodes (**a**) 2, (**b**) 4, (**c**) 6, and (**d**) 9.

The second simulation experiment was carried out by adding noise to the nodal demands used to generate the measured values. White Gaussian noise with a zero mean and variance of 0.1 was added to the hourly nodal demand multipliers. The demand multipliers with noise scenario is shown in Figure 6. The purpose of adding noise was to investigate the sensitivity of estimating the water demand to uncertainty in the step increment of the demand multipliers. The results obtained using step increments of 0.05 and 0.01 were compared. A resolution of 0.05 resulted in a single nodal demand multiplier search space of 40, while that of 0.01 resulted in a search space of 200. The nodal demand estimates from 30 simulation runs were averaged and used for comparison. The estimated nodal demands for nodes 2, 4, 6, and 9 are shown in Figure 7. A summary of the estimated and actual values of the water demands for the 24 h period is given in Table 4. The values are highly correlated with $R = 0.981$ and $R = 0.991$ for $\Delta = 0.05$ and $\Delta = 0.01$, respectively. There was therefore no significant difference in the water demand estimates determined using a step increment of 0.05 or 0.01 when the discrete values of the demand multipliers had an uncertainty of ± 0.1.

Table 4. Case I: Estimated nodal demands (liters per second—LPS).

Node	Actual	Estimated ($\Delta = 0.05$)	% Error	Estimated ($\Delta = 0.01$)	% Error
2	9.39	9.33	0.3	9.37	0.2
3	9.39	8.58	8.6	8.75	6.8
4	6.26	6.39	2.1	6.43	2.7
5	9.39	9.75	3.8	9.57	1.9
6	12.53	12.87	2.7	13.10	4.5
7	9.39	8.65	7.9	8.57	8.7
8	6.26	6.22	0.6	6.30	0.6
9	6.26	6.17	1.4	6.15	1.7

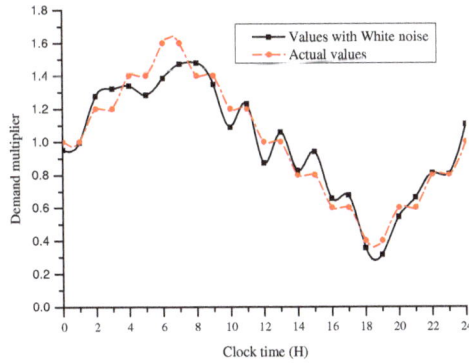

Figure 6. Nodal demand multipliers with added white noise.

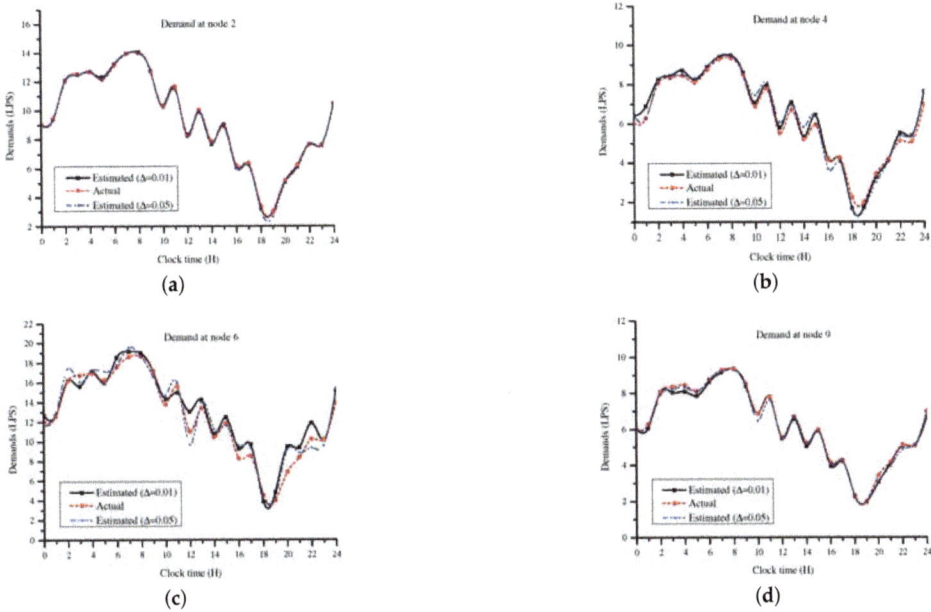

Figure 7. Case I: Estimation of water demands with demand multiplier resolutions of Δ = 0.05 and Δ = 0.01 at nodes (**a**) 2, (**b**) 4, (**c**) 6, and (**d**) 9.

4.2. Case Study II: Estimation of Nodal Demands with Measurements from Flow Sensors and Pressure Sensors

The simulation experiments in this case study were carried out by placing a pressure sensor at node 2 and flow sensors at pipes 7 and 11. Pipe 7 was selected as it supplied node 3, which had the highest error margin, as shown in Table 4.

To assess the suitability of the identified locations, the pipe flow rate sensitivity matrix of the network was generated, and the variance of each row was used to rank the pipe links. The results of the pipe flow rate sensitivity ranking are as shown in Figure 8. The pipe flow rate at pipe 9 was the most sensitive, while that of pipe 6 was the least sensitive. The results also show that further engineering judgement was required in order to select the locations, as pipe links 2, 7 and 9 were connected to node 3.

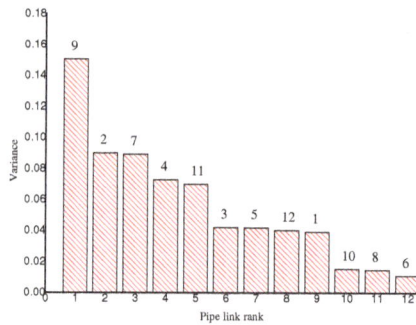

Figure 8. Pipe flow rate sensitivity for Net1 network.

Gaussian white noise was used to add a measurement uncertainty equal to ±10% of the measured value. A demand multiplier step increment of 0.05 was used in the simulations.

A summary of the averaged nodal demand estimates and pipe flow estimates are given in Tables 5 and 6. The averaged estimated nodal pressure heads are shown in Figure 9. The pipe flow rates were more sensitive to variation in nodal demands, compared to the pressure heads. The estimated flows for pipes 2, 5, 6, and 7 are shown Figure 10 and the estimated flows for pipes 3, 8, 10, and 11 are shown Figure 11. It can be observed from the results in Table 6 and Figures 10 and 11 that the pipes with low flow rates, for example, pipes 3, 5 and 6, had the highest error margin during estimation. The results show that these were the pipes with the highest sensitivity to changes in the water demand. The 24 h average pipe flow rates of Table 6 show that pipe 5 had an estimated average flow rate of 7.57 L/s with an error of 0.5%. However, the hourly flows in Figure 10b show that the estimated flow rate of pipe 5 had a correlation of $R = 0.567$ with the actual values. The flow rate of pipe 6 had an average estimation error of 8.3% and a correlation of $R = 0.721$ between the hourly estimated values and the actual values. It is therefore concluded that the correlation coefficient between actual and estimated values is a better measure of the quality of estimation, especially in pipes with low flow rates.

Table 5. Case II: Estimated nodal demands (liters per second—LPS) with uncertain measurements.

Node	Actual	Estimated	% Error
2	9.39	9.59	2.1
3	9.39	9.42	0.3
4	6.26	5.77	7.9
5	9.39	8.78	6.5
6	12.53	13.08	4.4
7	9.39	9.75	3.8
8	6.26	6.50	3.7
9	6.26	5.99	4.4

Table 6. Case II: Estimated pipe flows (liters per second—LPS) with uncertain measurements.

Pipe	Actual	Estimated	% Error
1	117.75	117.73	0.0
2	77.62	77.74	0.2
3	8.22	7.95	3.2
4	12.11	12.55	3.6
5	7.54	7.57	0.5
6	2.81	2.58	8.3
7	−48.44	−48.85	0.9
8	30.68	30.40	0.9
9	11.52	11.52	0.0
10	1.91	2.18	13.9
11	9.12	9.07	0.5
12	3.49	3.41	2.2

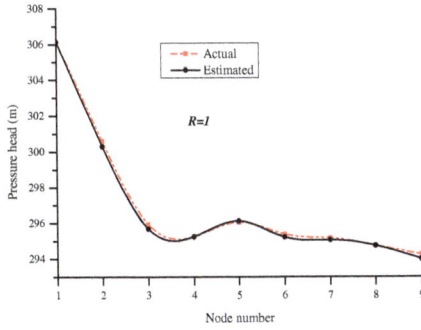

Figure 9. Case II: Estimated nodal pressure heads using uncertain measurements.

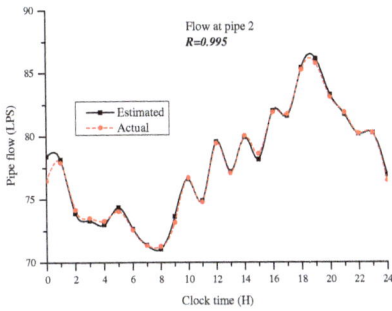

(**a**)

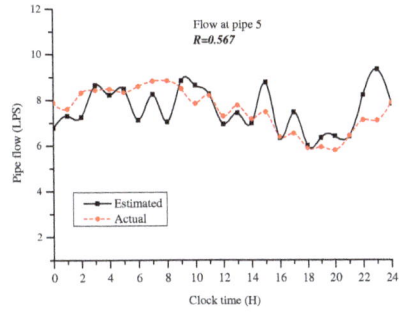

(**b**)

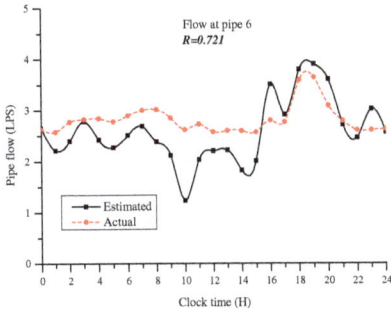

(**c**)

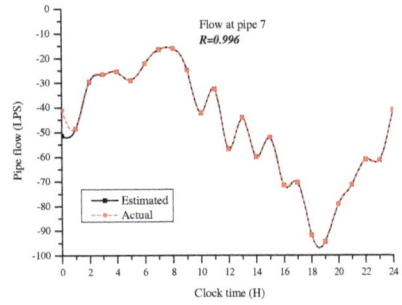

(**d**)

Figure 10. Case II: Estimation of pipe flow rates in pipes (**a**) 2, (**b**) 5, (**c**) 6, and (**d**) 7 using uncertain measurements.

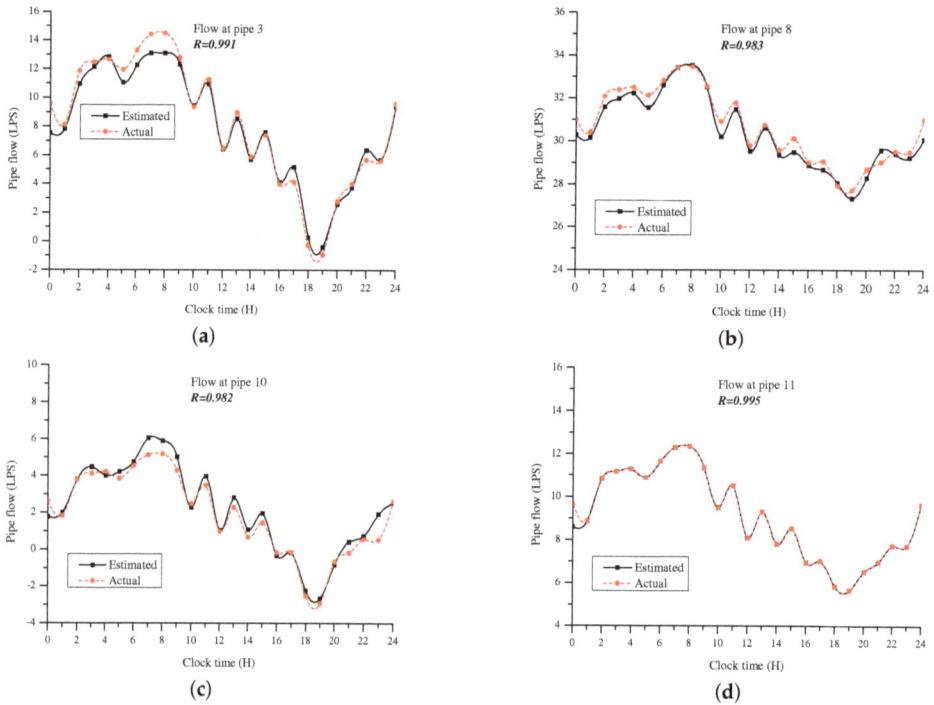

Figure 11. Case II: Estimation of pipe flow rates in pipes (**a**) 3, (**b**) 8, (**c**) 10, and (**d**) 11 using uncertain measurements.

The nodal demands estimated in case II had a maximum error margin of 7.9% at node 4, while those estimated in case I had a maximum error of 12.9% at node 3. The estimation error at node 3 reduced to 1.3% in case II. This could be attributed to the flow sensor placed at pipe 7 in case II. It is therefore concluded that the utilization of both pressure and flow sensors improves the overall results obtained during estimation.

4.3. Case Study III: Estimation of Water Demand in a Larger Water Network

The third case study considers a medium water network that consists of 40 pipes, 35 nodes, 1 pump station, and 1 water tank. The network is available in EPANET as example "Net2.net" [21]. The pump station was modeled as a node with a negative demand using the provided demand profile. The network data modeled the operation of a real network. The water demand of each node was independently defined and the water consumption represented a combination of different user types. The network layout is shown in Figure 12.

In this case study, the PSO algorithm was initialized to run with a population of 100 and 500 iterations. A demand multiplier range of 0 to 4.0 was used with a step increment of 0.02. The upper limit of the demand multiplier range was chosen by analyzing the nodal base demands and the provided demand profile data.

The sensitivity matrix was utilized to select the locations for placing flow and pressure sensors. The results of the nodal head and pipe flow rate sensitivity ranking are as shown in Figures 13 and 14, respectively. The results show that the pressure of nodal heads nearer to the pump station was more sensitive to the change in the water demand. The flow rates of pipe links were also generally more sensitive to the change in the water demand, compared to the nodal heads. Considering a case in which three pressure sensors and six flow sensors were available for placement, nodes 3, 6 and 10 and pipe links 13, 14, 15, 27, 28, and 30 were selected.

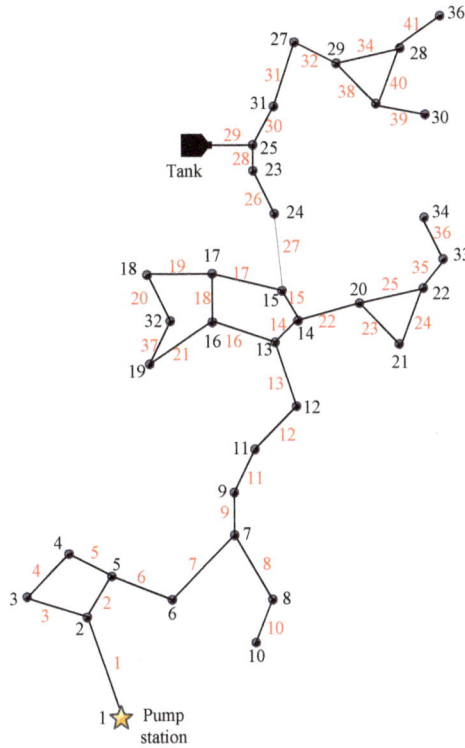

Figure 12. Case III: Water distribution network.

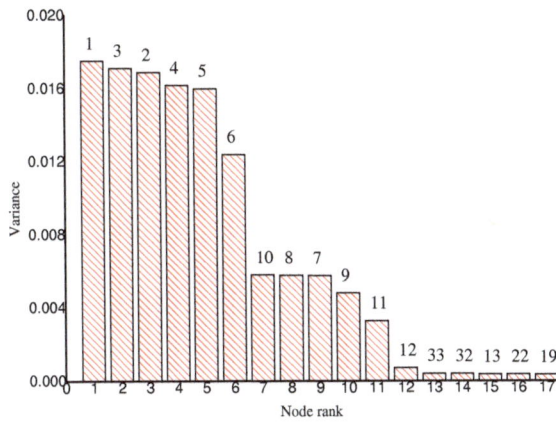

Figure 13. Nodal head sensitivity for Net2 network.

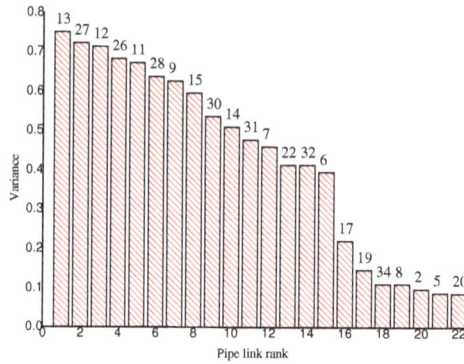

Figure 14. Pipe flow rate sensitivity for Net2 network.

The results for the estimated average water demand for a 24 h period are as shown in Figures 15 and 16. The results show that the nodes with a water demand of less than 0.3 LPS, for example, nodes 7, 32, 33 and 35, had estimation errors of more than 40%. The average of the estimation error for the actual values was 15.0%, while the average estimation error of the absolute values was 22.4%. The estimation error of the actual values was lower because the negative and positive errors offset each other. The average demand estimation error in [17] was 14.7% for the same network. The largest demand estimation error in this study was 56.2% at node 32. The average demand of node 32 was 0.1 LPS. However, Figure 15 shows that the absolute estimation error at node 32 was not significant. The results show that the nodes with low water demands had a high contribution to the overall water demand estimation error.

The estimation process of 500 iteration runs for a real-time of 62 s on a 2.6 GHz computer with 4 GB of memory. Further research is therefore required in order to reduce the estimation errors and improve the computation time. The proposed process simulation model is suitable for implementing and comparing the performance of evolutionary algorithms in water demand estimation. The future work for this study entails embedding a sensor placement algorithm in the developed process simulation model in order to assess the sensitivity of sensor placement locations on the water demand estimation errors.

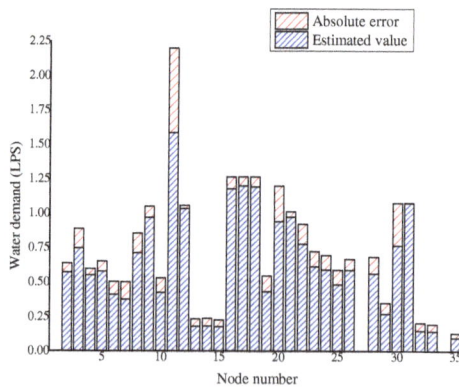

Figure 15. Case III: Estimated nodal water demands and absolute error.

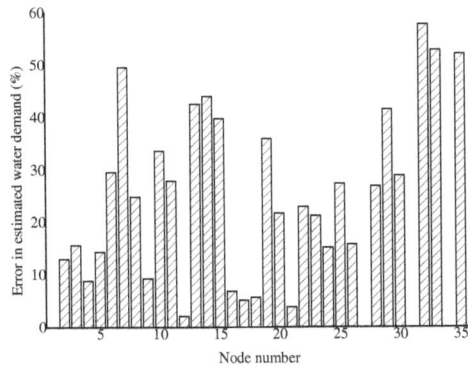

Figure 16. Case III: Absolute error (%) in estimated nodal water demands.

5. Conclusions

As a result of budgetary and other physical constraints, there is a need to estimate water demands in water distribution networks from a limited number of sensors. A detailed process simulation model for water demand estimation using the PSO algorithm was formulated and implemented. Nodal water demands and pipe flows were estimated when the number of estimated parameters was more than the number of measured values. The water demand at each node was determined by using the PSO algorithm to identify a corresponding nodal demand multiplier. The nodal demand multipliers were encoded with varying step sizes. The sensor placement locations were selected using an analysis of the sensitivity of measured nodal heads and pipe flows to the change in the water demand. The results show that accurate results can be determined using sensor measurements from a limited number of locations. The results also show that the estimated water demand is not sensitive to an uncertainty of ± 0.1 in demand multiplier discrete values or a $\pm 10\%$ uncertainty in measured parameters. Further research is required in order to address the problem of embedding a sensor placement algorithm in the developed process simulation model in order to assess the sensitivity of sensor placement locations on the water demand estimation errors.

Acknowledgments: The authors would like to thank The Council for Scientific and Industrial Research (CSIR), South Africa, for financial support.

Author Contributions: Lawrence K. Letting and Yskandar Hamam conceived and designed the experiments; Lawrence K. Letting developed the algorithm code in C++ and performed the simulation experiments; Adnan M. Abu-Mahfouz contributed to the analysis and interpretation of results. The manuscript was written by Lawrence K. Letting with contribution from all co-authors.

Conflicts of Interest: The authors declare no conflict of interest.

References

1. Della Giustina, D.; Pau, M.; Pegoraro, P.A.; Ponci, F.; Sulis, S. Electrical distribution system state estimation: Measurement issues and challenges. *IEEE Instrum. Meas. Mag.* **2014**, *17*, 36–42.
2. Sarrate, R.; Blesa, J.; Nejjari, F.; Quevedo, J. Sensor placement for leak detection and location in water distribution networks. *Water Sci. Technol. Water Supply* **2014**, *14*, 795–803.
3. Eliades, D.; Kyriakou, M.; Polycarpou, M. Sensor Placement in Water Distribution Systems Using the S-PLACE Toolkit. *Procedia Eng.* **2014**, *70*, 602–611.
4. Pourali, M.; Mosleh, A. A functional sensor placement optimization method for power systems health monitoring. In Proceedings of the IEEE 2012 Industry Applications Society Annual Meeting (IAS), Las Vegas, NV, USA, 7–11 October 2012; pp. 1–10.
5. Antunes, C.H.; Dolores, M. Sensor location in water distribution networks to detect contamination events—A multiobjective approach based on NSGA-II. In Proceedings of the 2016 IEEE Congress on Evolutionary Computation (CEC), Vancouver, BC, Canada, 24–29 July 2016; pp. 1093–1099.

6.   Schal, S.; Bryson, L.S.; Ormsbee, L.E. A simplified procedure for sensor placement guidance for small utilities. *Int. J. Crit. Infrastruct.* **2016**, *12*, 195–212.

7.   Jung, D.; Choi, Y.H.; Kim, J.H. Optimal Node Grouping for Water Distribution System Demand Estimation. *Water* **2016**, *8*, 160.

8.   Qin, T.; Boccelli, D.L. Grouping Water-Demand Nodes by Similarity among Flow Paths in Water-Distribution Systems. *J. Water Resour. Plan. Manag.* **2017**, *143*, 04017033.

9.   Preis, A.; Allen, M.; Whittle, A.J. On-line hydraulic modeling of a Water Distribution System in Singapore. In Proceedings of the Water Distribution Systems Analysis 2010, Tucson, AZ, USA, 12–15 September 2010; pp. 1336–1348.

10.   Ribeiro, L.; Sousa, J.; Marques, A.S.; Simões, N.E. Locating leaks with trustrank algorithm support. *Water* **2015**, *7*, 1378–1401.

11.   Andersen, J.H.; Powell, R.S. Implicit state-estimation technique for water network monitoring. *Urban Water* **2000**, *2*, 123–130.

12.   Bargiela, A. On-Line Monitoring of Water Distribution Networks. Ph.D. Thesis, Durham University, Durham, UK, 1984.

13.   Arsene, C.T.; Gabrys, B. Mixed simulation-state estimation of water distribution systems based on a least squares loop flows state estimator. *Appl. Math. Model.* **2014**, *38*, 599–619.

14.   Jung, D.; Lansey, K. Water distribution system burst detection using a nonlinear Kalman filter. *J. Water Resour. Plan. Manag.* **2014**, *141*, 04014070.

15.   Xie, X.; Zhang, H.; Hou, D. Bayesian Approach for Joint Estimation of Demand and Roughness in Water Distribution Systems. *J. Water Resour. Plan. Manag.* **2017**, *143*, 04017034.

16.   Hutton, C.J.; Kapelan, Z.; Vamvakeridou-Lyroudia, L.; Savic, D.A. Real-time demand estimation in water distrubtion systems under uncertainty. In Proceedings of the WDSA 2012: 14th Water Distribution Systems Analysis Conference, Adelaide, Australia, 24–27 September 2012; p. 1374.

17.   Do, N.C.; Simpson, A.R.; Deuerlein, J.W.; Piller, O. Calibration of Water Demand Multipliers in Water Distribution Systems Using Genetic Algorithms. *J. Water Resour. Plan. Manag.* **2016**, *142*, 04016044.

18.   Preis, A.; Whittle, A.J.; Ostfeld, A.; Perelman, L. Efficient hydraulic state estimation technique using reduced models of urban water networks. *J. Water Resour. Plan. Manag.* **2010**, *137*, 343–351.

19.   Sanz, G.; Pérez, R. Sensitivity analysis for sampling design and demand calibration in water distribution networks using the singular value decomposition. *J. Water Resour. Plan. Manag.* **2015**, *141*, 04015020.

20.   Davidson, J.; Bouchart, F.C. Adjusting nodal demands in SCADA constrained real-time water distribution network models. *J. Hydraul. Eng.* **2006**, *132*, 102–110.

21.   Rossman, L. *Epanet 2: Users Manual*; U.S. Environmental Protection Agency, Office of Research and Development, National Risk Management Research Laboratory: Washington, DC, USA, 2000.

22.   Kang, D.; Lansey, K. Real-time demand estimation and confidence limit analysis for water distribution systems. *J. Hydraul. Eng.* **2009**, *135*, 825–837.

23.   Kennedy, J.; Eberhart, R. Particle swarm optimization. In Proceedings of the IEEE International Conference on Neural Networks, Perth, Australia, 27 November–1 December 1995; Volume 4, pp. 1942–1948.

24.   Rossman, L.A. The EPANET programmer's toolkit for analysis of water distribution systems. In *WRPMD'99: Preparing for the 21st Century*; American Society of Civil Engineers: New York, NY, USA, 1999; pp. 1–10.

25.   Letting, L.; Hamam, Y.; Adnan, A.M. An Interface for Coupling Optimization Algorithms With EPANET in Discrete Event Simulation Platforms. In Proceedings of the IEEE 15th International Conference of Industrial Informatics, Emden, Germany, 24–26 July 2017.

26.   Pudar, R.S.; Liggett, J.A. Leaks in pipe networks. *J. Hydraul. Eng.* **1992**, *118*, 1031–1046.

27.   Pérez, R.; Puig, V.; Pascual, J.; Quevedo, J.; Landeros, E.; Peralta, A. Methodology for leakage isolation using pressure sensitivity analysis in water distribution networks. *Control Eng. Pract.* **2011**, *19*, 1157–1167.

28.   Casillas, M.V.; Puig, V.; Garza-Castañón, L.E.; Rosich, A. Optimal sensor placement for leak location in water distribution networks using genetic algorithms. *Sensors* **2013**, *13*, 14984–15005.

water

MDPI

Article

Water End Use Disaggregation Based on Soft Computing Techniques

L. Pastor-Jabaloyes (iD), **F. J. Arregui** * (iD) **and R. Cobacho**

ITA-Grupo de Ingeniería y Tecnología del Agua, Dpto. de Ingeniería del Agua y Medio Ambiente,
Universitat Politècnica de València, Camino de Vera s/n, 46022 València, Spain;
laupasja@ita.upv.es (L.P.-J.); rcobacho@ita.upv.es (R.C.)
* Correspondence: farregui@ita.upv.es

Received: 22 November 2017; Accepted: 5 January 2018; Published: 9 January 2018

Abstract: Disaggregating residential water end use events through the available commercial tools needs a great investment in time to manually process smart metering data. Therefore, it is extremely difficult to achieve a homogenous and sufficiently large corpus of classified *single-use* events capable of accurately describe residential water consumption. The main goal of the present paper is to develop an automatic tool that facilitates the disaggregation of the individual water consumptions events from the raw flow trace. The proposed disaggregation methodology is conducted through two actions that are iteratively performed: first, the use of an advanced two-step filter, whose calibration is automatically conducted by the Elitist Non-Dominated Sorting Genetic Algorithm NSGA-II; and second, a cropping algorithm based on the filtered water consumption flow traces. As a secondary goal, yet complementary to the main one, a semiautomatic massive classification process has been developed, so that the resulting single-use events can be easily categorized in the different water end uses in a household. This methodology was tested using water consumption data from two different case studies. The characteristics of the households taken as reference and their occupants were unequivocally dissimilar from each other. In addition, the monitoring equipment used to obtain the consumption flow traces had completely different technical specifications. The results obtained from the processing of the two studies show that the automatic disaggregation is both robust and accurate, and produces significant time saving compared to the standard manual analysis.

Keywords: water end uses; water microcomponents; high frequency smart metering data; residential water flow trace disaggregation; water flow trace filtering

1. Introduction

Since the Brundtland Report [1] was presented, sustainability in the use of water resources has been a steady concern in designing water policies [2–4]. This is a problem with many different faces, from the source (surface or ground water, desalination, reclamation, etc.) to the use (agriculture, residential, industrial, environmental preservation, etc.). All of them are relevant, but bearing in mind that most of the human population lives in cities, urban water management becomes an issue of paramount importance. Therefore, accounting urban water consumption and knowing about end-uses at each customer's household is not only key because of the amount of water resource that is used and/or can be saved, but also because of many other considerations. In this regard, there are well-founded reasons for the research currently being conducted on residential end-uses, such as reduction in treatment costs linked to water consumption; improvement and better effectiveness of conservation measures in the urban environment; conservation of energy linked to water consumption; design optimization of indoor piping systems; improved demand forecasting models; etc.

As an essential tool for enhancing urban water management, the new technologies being implemented today in smart meters are making possible a significant leap forward in recording and characterizing domestic water consumption. Further than the traditional monthly volume read, new meters may provide hourly consumption time patterns or a volume-flow pattern. They may also send alarms when a leak, a forgotten open faucet or a continuous back flow is detected, and all that information may be immediately sent through an AMR (automatic meter reading) system.

However, the above-mentioned capabilities are only the first tier when considering all the real possibilities current smart meters may yield. Though feasible today, a second, more advanced tier is not fully developed because of its notable complexity. It consists of high frequency monitoring—duration, volume and flow rate—of domestic water consumption, so that every single use in a household (hh) is accurately registered. Then, and after a detailed analysis, all consumption events can be categorized into the different end uses present in the household [5].

As soon as this will be soundly achieved on a large scale, new improvements in efficient water management strategies will be within reach. To name a few, from the consumer's perspective, water conservation measures could be tailored to each individual consumer [6], thus maximizing the saving potential in each case, or the variable term in the tariff could be designed according to consumer's characteristics to guarantee the balance between equity and income [7]. Furthermore, from the utility's view, water demand prediction models could be reliably produced from a more accurate bottom-up approach [8,9].

Nowadays, few commercial tools allow for this water end use analysis exercise—Trace Wizard® [10], Identiflow® [11] and BuntBrainForEndUses® [12]. However, any of these tools involves a great investment in time and human resources, as a significant part of the data processing work requires human intervention. Furthermore, the results from the analysis are unavoidably affected by arbitrary and constantly changing human criteria.

Alternatively, an automatic prototype based on machine learning algorithms was proposed by Nguyen et al. [13–16] to disaggregate and classify water consumption events. Unfortunately, the universal usability and compatibility of the tool is limited by the fact that the algorithms were trained with data originated from a specific water meter/data logger combination. In addition, all data were collected in the same geographical area from consumers sharing very similar water consumption habits and water appliances. Furthermore, the set of data employed for the training of the proposed machine learning tool has been obtained using Trace Wizard® software, which has limited capabilities for disaggregating overlapped consumption events [13]. Following a similar approach, Piga et al. [17] proposed an automated water and energy end use disaggregation, which has only been tested against electric energy data. Also, several start-ups claim to have developed software to automatically classify residential water consumption events into various uses [18–20]. Unfortunately, in this case there is not official or public information available about the processing tools and algorithms used, and the real performance achieved in the water end use classification for various types of households.

This paper presents a novel methodology to substitute manual water end use disaggregation and to produce more accurate sets of classified single end use events that can be employed as training sets for automatic recognition algorithms. Figure 1 depicts the general structure of the methodology, comprising two main processes: disaggregation, which is fully automatic and the main objective of this paper, and classification, which is semiautomatic at its current stage.

The proposed disaggregation process focuses on an advanced two-stage filtering based on the algorithm described in Pastor et al. [21], which is calibrated using the Elitist Non-Dominated Sorting Genetic Algorithm NSGA-II [22], and a new cropping algorithm having as input the filtered water consumption flow traces.

The contribution of this methodology can be summarized in two main aspects. The first one is the integration of a universal two-stage filtering algorithm that can be used to simplify, with a minimum loss of information, the flow traces originated in most commercial metering and logging equipment available in the market. The second one is the reduction of human intervention by automatically disaggregating *overlapped* water consumption events (as the one used as an example in Figure 1) into *single-use* events (examples in Appendix A), which are associated with individual uses of water through different appliances. Both features facilitate and improve the processing of flow traces generated during a long-term metering campaign.

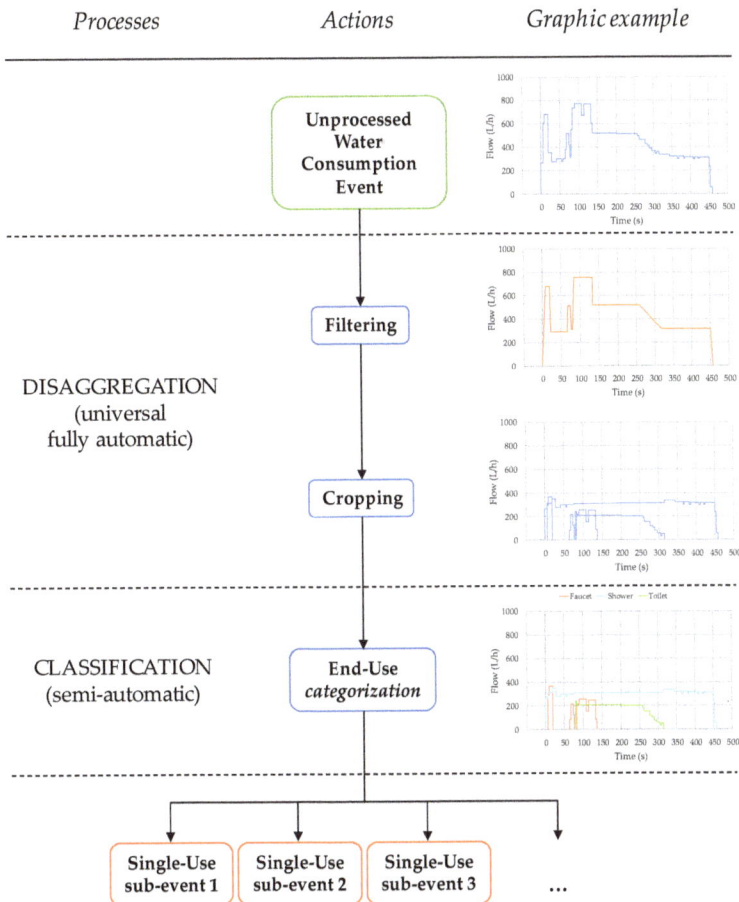

Figure 1. General structure of the proposed methodology.

The classification process corresponds to the use of unsupervised techniques to solve the classification of *single-use* events into the various water end use categories. This step is complementary to the disaggregation process and makes unnecessary the intervention of human analysts, except for validating the results, to classify thousands of automatically obtained *single-use* events. The paper presents a basic fully operational, yet semiautomatic, version to show the capabilities of these techniques for massive classification of consumption events.

As a case study, data originated from two independent water end uses studies conducted in distinct geographical areas were used to test the proposed filtering and disaggregation methodologies. The characteristics of the households and their occupants were unequivocally dissimilar from each other. Moreover, and in order to test the universal applicability of the methodology, the water consumption flow traces analyzed were obtained from monitoring equipment having completely different technical specifications.

BuntBrainForEndUses® was the commercial online software employed for the water end use analysis. This software offers the possibility of exporting raw flow traces and importing them back after some manipulation has been carried out by the user. This feature is particularly useful for the study conducted as the filters and disaggregating algorithms can be developed with a specialized external analysis software, completely independent from BuntBrainForEndUses®, and then have the results displayed and corrected in the online application. For the methodology presented, the filtering and disaggregation algorithms were programed in R statistics [23].

2. Materials and Methods

The methodology proposed is divided into two processes (Figure 1). The first one, disaggregation, works on the original, and generally *overlapped*, consumption events in an iterative way until all the resulting subevents are either *single-use* (the most) or *uncertain* (a few) events (examples in Appendix A). By means of this methodology, the resulting subevents are more homogenous than the ones obtained by manual processing of flow traces through, for example, BuntBrainForEndUses®. This software application follows the same analysis procedure and allows human analysts to graphically crop water consumption flow traces into its various individual components.

The second process of the methodology, classification, assigns a specific water end use category to each *single-use* event. In the case study developed, this classification is done by identifying homogeneous subsets of events by means of a non-supervised learning technique and assigning a water end use to each one of them. Whatever the classification technique used, its effectiveness is increased by the fact that the subevents generated by the disaggregating algorithms consistently create subevents using homogeneous and well-defined criteria.

2.1. Disaggregation Process

Figure 2 shows the flow chart of the proposed disaggregation process, which breaks down the unprocessed events defined by the raw flow trace into simpler consumption events, and classifies the resulting subevents as *single-use* or *uncertain*.

The reliability of the process strongly relies on the first analysis stage: filtering of the original flow trace. The filter is controlled by 10 parameters [21], and their calibration is automatically solved per consumption event by the *Elitist Non-Dominated Sorting Genetic Algorithm NSGA-II* [22] (R package *mco*). There are three objective functions to be minimized in this calibration: (a) number of points that describe the filtered flow trace; (b) total accumulated volume difference between raw and filtered flow trace (Figure 3(a2,b2), *Input* and *Output*, respectively); (c) maximum on the curve of accumulated volume difference. The first objective function (FO_1) leads NSGA-II algorithm to solutions that simplify the filtered flow trace, whereas the other two (FO_2 and FO_3) focus on improving its fitting quality respect the original raw flow trace.

The next steps are followed to calculate the curve of accumulated volume difference: (1) given a raw water flow trace, demonstrated as vector $qr = (qr_1, qr_2, \ldots qr_i \ldots, qr_m)$, and its corresponding filtered water flow trace $qf = (qf_1, qf_2, \ldots, qf_i, \ldots, qf_m)$, both expressed in litres per hour (L/h) and recorded at time t_i in seconds (s), two new synchronized time series are generated by linear interpolation, qrs and qfs, for the set of unique t_i that belong to qr and qf; (2) vector of time window tw and vector of reference flow $qref$ are defined as:

$$\begin{cases} tw_i = 0, & i = 1 \\ tw_i = t_i - t_{i-1}, & 2 \leq i < n \end{cases} \tag{1}$$

$$\begin{cases} qref_i = 0, & i = 1 \\ qref_i = \max(qrs_{i-1}, qrs_i), & 2 \leq i < n \end{cases} ; \tag{2}$$

(3) the curve of accumulated volume difference is defined for those components of tw_i greater than 0.001 s (all flow rate jumps that take place in the raw water flow trace have this duration, as it can be seen in Figure 3(a1), *Input*) as follows:

$$\begin{aligned} & \text{if}\left(\left(qref_j > q_{th}\right)\right) \left|V_{qr}\left[t \le (t_j + tw_j)\right] - V_{qf}\left[t \le (t_j + tw_j)\right]\right| \\ & \text{else } \left|V_{qr}\left[t \le (t_j + tw_j)\right] - V_{qf}\left[t \le (t_j + tw_j)\right]\right|/tw_i \end{aligned} \quad, \quad 2 \le j < m < n \quad (3)$$

where V_{qr} and V_{qf} are the accumulated volume along the raw and the filtered water flow traces, respectively, until $t = t_j + tw_j$. On the other hand, q_{th} is a user-defined threshold that has been established to properly process the events with a continuous low-flow water leak. In these cases, small differences in leakage flow rate between the raw and the filtered water flow traces can be maintained over a long period of time, resulting in large accumulated volume differences that decrease the representativeness of FO$_2$ and FO$_3$. To avoid this, the accumulated volume difference is divided by tw_i when $qref_j$ is less than the maximum level of leakage flow rate q_{th} defined by the user. In the same way, the curve of accumulated volume difference is defined for those components of tw_i greater than 0.001 s to reduce the appearance of noise when the flow rate is below q_{th}.

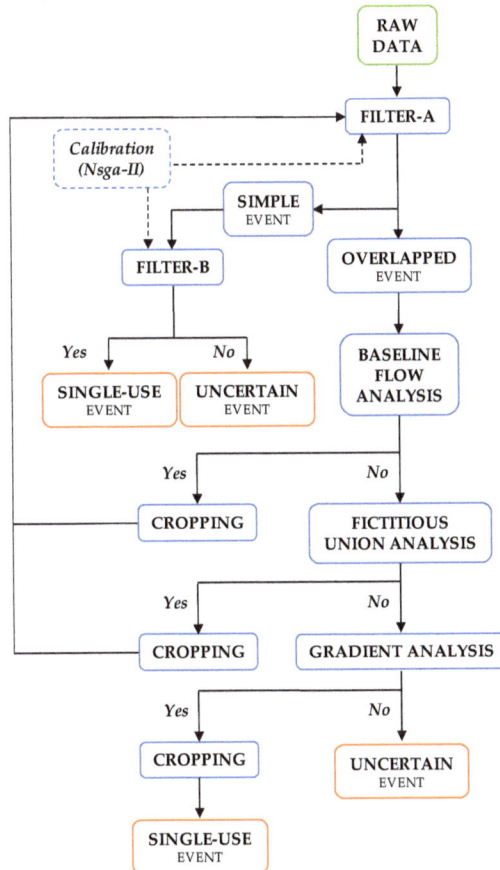

Figure 2. Flow chart of the disaggregation process and sorting of events as *single-use* or *uncertain*.

Regarding NSGA-II parameters, processing time was the most constraining factor to select the population size and the number of generations. The genetic algorithm achieves good results for the most intricate cases—long duration events with a great degree of overlapping, which typically come from households with leaks and high average daily consumption—with 24 individuals and 10 generations in a reasonable computing time. In relation to crossover and mutation probabilities, the default values taken were 0.7 and 0.2, respectively.

The result of this calibration process is a Pareto Front, and the chosen solution is the one for which the following expression is minimized:

$$FO_{select.sol} = w_1 * \frac{FO_1}{\max(FO_{1-PF})} + w_2 * \frac{FO_2}{\max(FO_{2-PF})} + w_3 * \frac{FO_3}{\max(FO_{3-PF})} ; \qquad (4)$$

where the maximum value reached in each objective function within the Pareto Front ($\max(FO_{1-PF})$, $\max(FO_{2-PF})$ y $\max(FO_{3-PF})$) was taken to standardize the corresponding term. A conservative criterion for *Filter-A*, which prioritizes the simplification of the flow traces, establishes the weights for each objective function ($w1 = 0.8$; $w2 = 0.1$; $w3 = 0.1$). This is necessary for subsequent disaggregation processes, which can only be applied if certain requirements are satisfied based on a strong filtering of the raw flow trace. Additionally, the calibration time can be limited by a user-defined threshold. In case a solution is not found within the established time limits (only happening in less than 0.01% of the sample events in the case studies below), the default values for the filter parameters will be used ($p1 = 150$ (h*ms)/L; $p2 = 0.16$ L; $p3 = 80$ L/h; $p4 = 40$ degrees; $p5 = 5\%$; $p6 = 100$ L/h; $p7 = 10,000$ ms; $p8 = 6000$ ms, $p9 = 5\%$, $p10 = 10$ degrees). These default values are the result of the authors' experience while developing and applying this process in several projects around the world. In any case, a future sensitivity analysis is already planned to improve the filter performance and improve the speed of the calibration procedure.

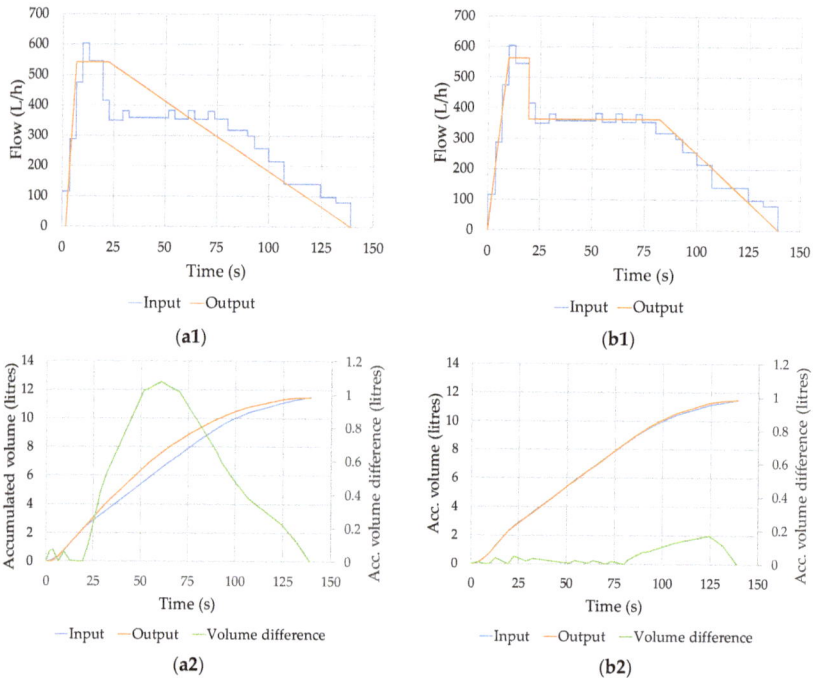

Figure 3. Comparison of raw flow traces vs filtered. (**a1,a2**) after *Filter-A* and (**b1,b2**) after *Filter-B*.

Once the consumption event defined by the raw flow trace has been filtered for the first time (*Filter-A* in Figure 2) the event can be classified as *simple*, constituted by only four vertexes, or *overlapped*. *Simple* events are analyzed by an additional filtering process (*Filter-B* in Figure 2). This process prioritises how accurately the filtered flow trace matches the original one. In this case, the solution selected from the Pareto Front should attain a value for the KGE index ([24]; R package *hydroGOF*) higher than 0.8 with a maximum $w1$ weight. Figure 3 compares the resulting filtered flow traces after going through the first and second filtering processes. Only if the final filtered flow trace (after *Filter-B*) is formed by only four vertexes, the event is finally classified as a *single-use* event. Otherwise, it will be classified as an *uncertain* event. The inclusion of a second filter significantly reduces the classification errors of *single-use* events compared to a one-stage filter approach.

An additional step of the disaggregation process is the analysis of the minimum/baseline flow (Q_{base}) in the filtered (*Filter-A*) events that have not been classified as *simple* events. The existence of a base event is considered when more than one horizontal section of the event satisfies the following two conditions: (i) the flow rate falls within a specified range $\{Q_{base} - tolerance, Q_{base} + tolerance\}$, where the tolerance is defined by the user; (ii) the volume associated with the horizontal section of the event is greater than a specified threshold. When these two conditions are met (Figure 4a), the events between horizontal sections are cropped from the base event (Figure 4b). If only the second condition is not satisfied (Figure 4a), the section is processed as a fictitious union, so that the events are separated and the union removed (Figure 4b). Fictitious unions appear when processing the raw flow traces. They do not actually correspond to any real water consumption (they are a distortion in the flow trace caused by the data-acquisition equipment). The end or starting times of the previous and following subevents are then recalculated to account for the volume removed from the fictitious union. Typically, this volume corresponds to one or two pulses from the pulse emitter of the water meter. The resulting subevents generated by these cropping operations are individually analyzed through the whole process again.

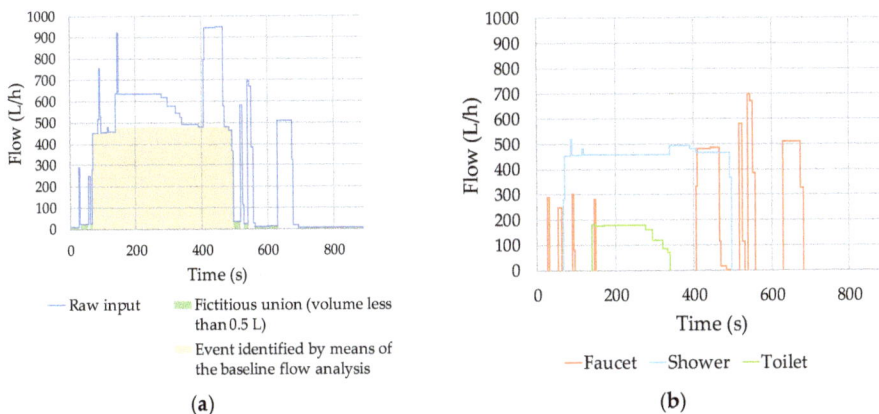

Figure 4. (**a**) Example of an *overlapped* event. (**b**) Resulting events obtained after disaggregating fictitious unions and conducting a baseline flow analysis.

Finally, if the event does not fall into any of the previous categories, a gradient analysis of the filtered flow trace is carried out (Figure 2). Only in case that the event has three major slopes, being the first one positive and other two negative or vice versa, it is considered that it is constituted by two or more different events that are *overlapped* in time, which begin or end within the same time range (Figure 5a). In this case, the events are cropped (Figure 5b), and the two new events are classified as *single-use* events. On the contrary, if the number of major slopes in the event is greater than three, it is directly categorized as an *uncertain* event. The key aspect of this separation process is to correctly identify the start or the end, depending on the case, of the second major slope on the raw flow trace (Figure 5a, point 2) and the flow rate at the analogous instant in the filtered flow trace (Figure 5a, point 1).

It should be highlighted as an important contribution of the proposed methodology that all previous separation processes are implemented on the raw flow trace. Thus, signal smoothing in the filtered flow trace do not generate any loss of information in the resulting subevents obtained. This particular feature can be clearly observed in Figure 5, where the separated consumption events (Figure 5b) maintain the details of the original flow trace (Figure 5a). The presence of these details can be used in a later stage to improve the effectiveness of the automatic classification tools that can be developed.

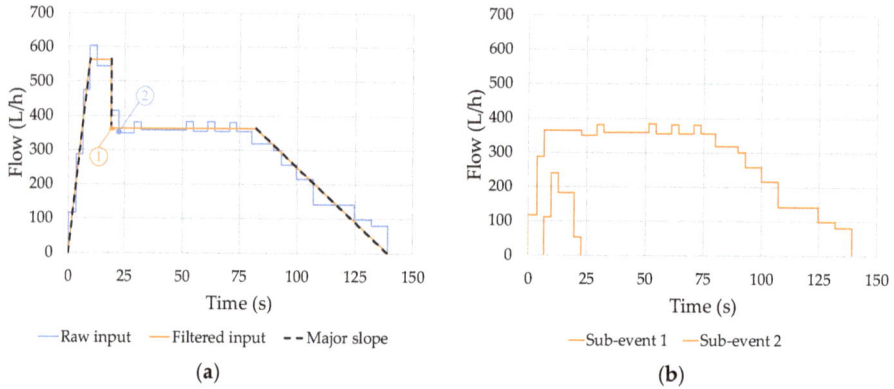

Figure 5. (**a**) *Overlapped* event in which the two subevents start at the same time. (**b**) Two resulting *single-use* events after the cropping operation.

2.2. Classification Process

At the end of the disaggregation process, all the events have been classified into two categories: *single-use* and *uncertain* events. As discussed below, the first group is the most numerous in any of the households analyzed, typically ranging from 75% to 92% depending on the amount of water consumption in the household. In addition, the amount of *single-use* events generated by the methodology is higher if the *uncertain* events having less than 3 L in volume are assumed to be *single-use* events (in this case the percentage of *single-use* events will range between 85% and 95%). The cropping and classification of the *uncertain* event group, corresponding to intricate events with high flow rate variability, will be the aim of future research. In this sense, it should be noted that high flow rate variability is not always associated to water consumption overlapping from different uses. Occasionally, the so-called *uncertain* events may be originated by pressure fluctuations, or the user changing the opening of a faucet for convenience or to adjust water temperature. Cropping and classification of *uncertain* events is not an easy task when accounting for the previous considerations and the fact that the overlapping of two uses may not produce a consumption flow rate that is the sum of the individual flow rates. The reason for this effect can be found in the pressure losses caused by the plumbing. Depending on the sizing of the pipes, the consumption flow rate of a given use may be significantly reduced by the appearance of additional water usages within the household.

The working hypothesis to categorizing *single-use* events is the following—those events with similar physical characteristics correspond to the same end use. In accordance with this, an initial clustering analysis is conducted by an unsupervised learning machine, Partition Around Mediods (PAM; [25]; R package *cluster*). One advantage of this partition clustering technique relies on the fact that it can work with different similarity measurements other than the Euclidean distance. In this work, the Gower distance has been chosen as the similarity measure, since it scales all the variables considered and allow for the definition of individual weights for the variables. On the other hand, the characteristics of the events taken into account as input data are the total volume and the average flow rate. Event duration was rejected due to the considerable noise generated by long water uses, which impeded clusters identification. This effect was observed in households with leakage, showing long *single-use* events with low consumption flow rates. However, in these cases the duration of the event is a variable used as a preliminary filter to allow the clustering analysis to solely focus on the bulk of *single-use* events. Once the clustering analysis is finished (Figure 6), the application allows the user to visualize random subsamples of events from each cluster and associate them with an end use according to their physical characteristics.

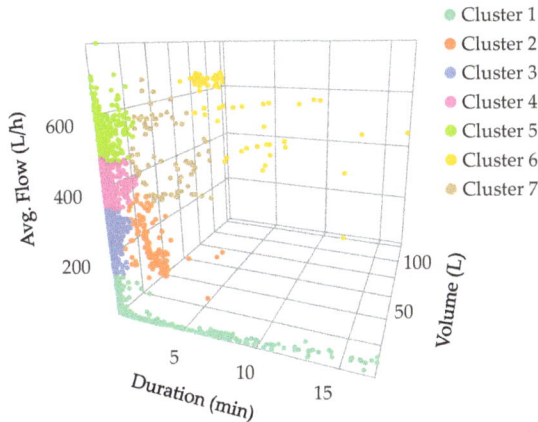

Figure 6. Result of Partition Around Mediods (PAM) algorithm with a number of clusters equal to 7 and similarity matrix based on Gower distance.

The subset of categorized *single-use* events and the subset of *uncertain* events are written in a flat CSV file at the end of the process. This file can be read by BuntBrainForEndUses®, a web application for manual processing and editing of water end uses. In this way, the user can correct misclassifications and further edit the *uncertain* events that have not been properly analyzed by the algorithms.

3. Case Study

The water consumption data utilized for testing the methodology was sourced from two different water end uses studies conducted in geographically distant regions. One of the main differences between the studies is the type of monitoring equipment employed. In the first study (R1), the smart meters installed for water consumption monitoring were ELSTER Y250 single-jet (maximum flow rate of 5 m^3/h) or ELSTER Y250M multi-jet (maximum flow rate of 7 m^3/s) depending on the type of residential household. These meters produce a pulse every 0.04 L or 0.06 L of water consumed, respectively. Specially designed data loggers calculated and recorded the average consumption flow rate at approximately 3-s intervals. This recording mode was chosen to optimize the file size while preserving the quality of the flow trace. Files were periodically sent (twice per day) to the server via GPRS/GSM. On the other hand, in the second study (R2) a piston type volumetric water meter was used (Aquadis+, ITRON (WA, United States)), which generates a pulse every 0.1 L. The data logger used, recorded the occurrence time of every pulse with a resolution of 0.02 s.

For this analysis, a selection of significant households of both studies was conducted according to the average daily consumption and the presence or not of continuous leakage. The final selected sample was composed by 20 households—10 from R1 and 10 from R2—for which two-week period of monitoring data were selected. For the first study, the data corresponds to consumption made during autumn 2015, while for the second study the data were collected during autumn 2016. In total

19,858 sampled events were analyzed during the period considered. Figure 7 shows the general characteristics of the selected households and events associated with them.

As shown in Figure 7, there is a considerable difference between the households and events characteristics of these two studies. In the first study, the average daily consumption of the sample was close to 1600 L/hh/day, while in the second it is less than 400 L/hh/day. The number of daily events are also completely different: 110 events/hh/day vs. 35 events/hh/day. The dissimilarities of the households considered in the analysis conducted in this paper emphasize the reliability of the methodology, which can be used independently of the data sources, as long as the quality of the flow traces allows for end use disaggregation.

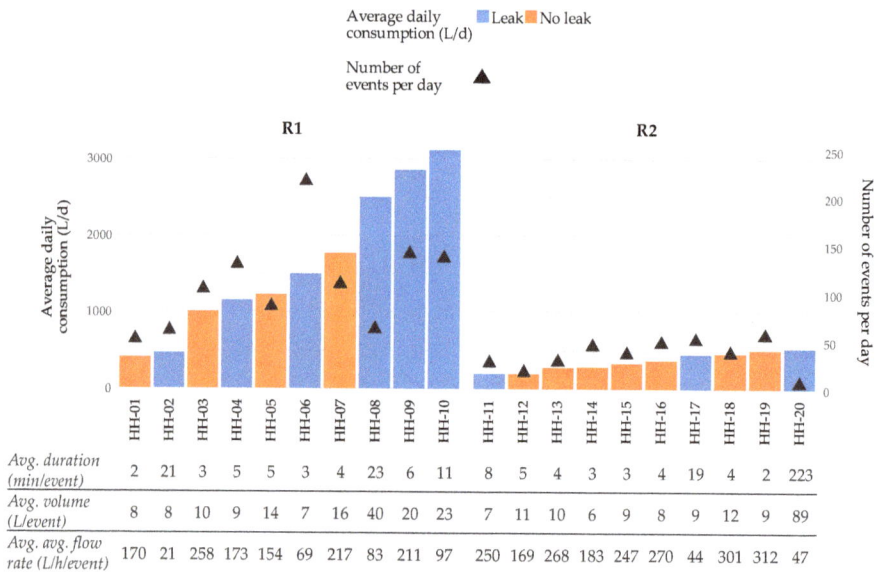

Avg. duration (min/event)	2	21	3	5	5	3	4	23	6	11	8	5	4	3	3	4	19	4	2	223
Avg. volume (L/event)	8	8	10	9	14	7	16	40	20	23	7	11	10	6	9	8	9	12	9	89
Avg. avg. flow rate (L/h/event)	170	21	258	173	154	69	217	83	211	97	250	169	268	183	247	270	44	301	312	47

Figure 7. General characteristics of the analyzed households and the events associated with them.

4. Results and Discussion

The proposed methodology has been applied to 19,858 unprocessed water consumption events. After applying the disaggregation process (filtering and cropping) to the flow signal, the total number of events increased to 46,721, being the average number of cropping operations per day equal to 121 and 58 for the studies R1 and R2, respectively (Table 1). The average processing time consumed per each one of these operations is 21.8 s, using an Intel Core i5-4440 processor. Per study, the average cropping time is equal to 18.6 s for R1 and 28.3 s for R2. The calibration of the filtering algorithm is the task requiring more processing time, which increases with the density per unit time of data points in the raw flow trace. For this reason, it takes longer to carry out a cropping operation in the case of an event belonging to the study R2, since in this study flow data were recorded with a lower temporal resolution (0.02 s vs. approx. 3 s). Currently, the research team is working in optimizing the calibration algorithms and reducing the required processing time. The strategies proposed for this optimization are: (1) developing a methodology to calibrate the filter per household rather than per event, without a considerable loss of filtering accuracy; (2) finding the filter parameters through an algorithm that combines heuristic and guided search methods.

Table 1. General statistics about performance of separation process.

	R1 Study	R2 Study	Total
Total number of unprocessed events (10 households)	14,648	5210	19,858
Total number of resulting events (10 households)	32,792	13,929	46,721
Total number of resulting events per household	3279	1393	2336
Average time consumed per cropping operation (s)	18.6	28.3	21.8
Average number of cropping operation per household and day	121	58	90

Analyzing in detail the consumption events selected from study R1 (Figure 8), the result of applying Filter-A to all 14,648 unprocessed events, was that 8768 events were classified as *simple*, whereas the remaining 5880 were classified as *overlapped*. As to the *simple* events, most of them were, as expected, *single-use* events (6394). The remaining 2374 *simple* events were not simple enough and were classified as *uncertain* events. None of these *uncertain* events originated from *simple* events could be cropped or further processed; however, most of them correspond to single-use events with unsteady consumption flows (for example, a faucet that is adjusted to the desired flow rate).

Overlapped events correspond to events that could not be simplified as four-nodes events after *Filter-A*. These events have a considerable degree of complexity due to the overlapping of water uses. The algorithm proposed is capable of separating into *single-use* events most of these *overlapped* events by accurately cropping the flow traces. The initially identified 5880 *overlapped* events were separated into 24,024 subevents after the disaggregation process. Most of these new subevents were classified as *single-use* events (17,908), and the rest (6116) as *uncertain*. In total, the analysis of the worst case scenario of study R1 produced 24,302 *single-use* events (74.1%) and 8490 *uncertain* events (25.9%). From the *uncertain* events, 4224 had a volume of less than 3 L. These, because of their low volume, could be also be added to the *single-use* group.

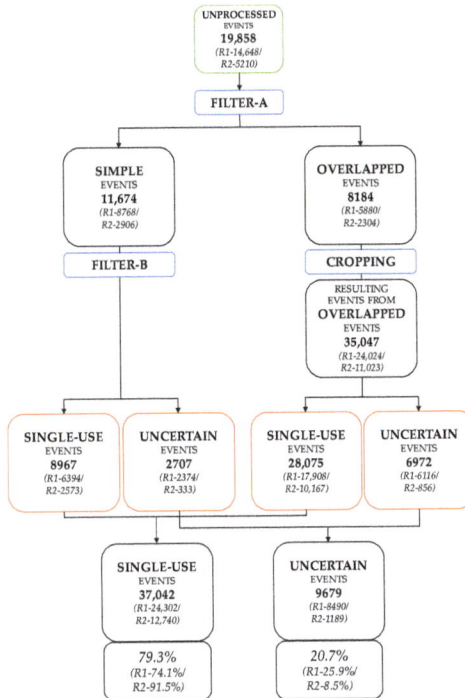

Figure 8. Classification of events after filtering and cropping (disaggregation process).

For study R2, showing simpler flow traces typical of water consumption profiles of a European family, the results are even more positive. As shown in Figure 8, 91.5% of 13,929 resulting events were considered to be *single-use* events and only 8.5% were catalogued as *uncertain* events. The difference between studies is mainly due to the characteristics of the households: the flow traces belonging to the sample R1 are notably more complex, obtained from households with high average daily consumption and frequent overlapping of water uses.

Overall, the methodology generated a number of *single-use* events equal to 37,042 events (79.3% of the 46,721 resulting events), of which 75.8% has been obtained through the proposed disaggregation process. Consequently, human intervention to crop and generate *single-use* events has been significantly minimized, with subsequently large human working time savings. In raw numbers, the total automatic disaggregation process has taken less than 4 days (96 computing hours); whereas, according to the authors' experience, the same work would have required about 45 human-working days (360 h).

The distribution of the physical characteristics of the *single-use* and *uncertain* events for both studies is presented in Figure 9. Some outliers, with a duration of more than 8 min, have been removed to improve the readability of the basic statistics (median, first and third quartile). As expected, the heterogeneity of the *uncertain* events is significantly larger than the one obtained for *single-use* events. In addition, the average duration, volume and flow rate of *uncertain* events are greater than those for *single-use* events. It should also be mentioned that some of the events considered here as *uncertain* correspond to single uses, and their flow rate variability can be caused by adjusting the faucet or variations in the input pressure.

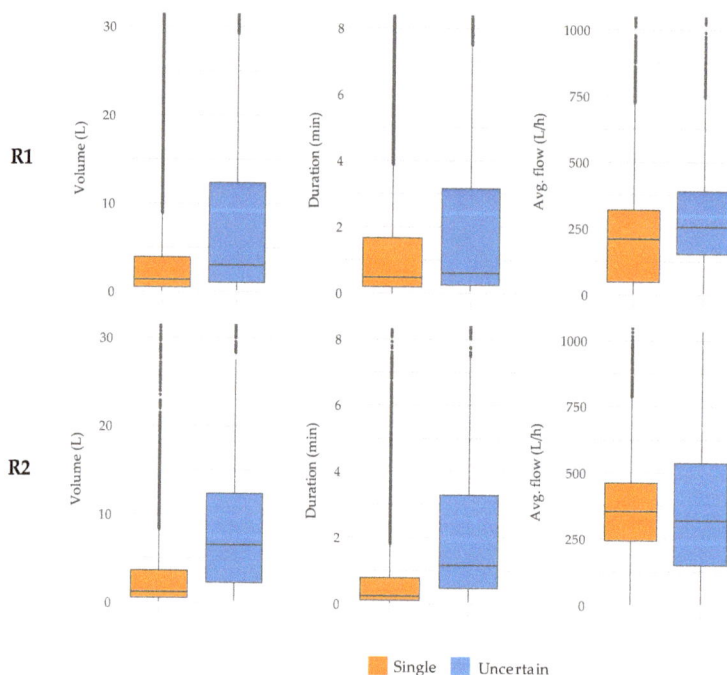

Figure 9. Distribution of the physical characteristics of the events classified as *single-use* and *uncertain* per study.

After the disaggregation process, *single-use* events could be categorized by clustering analysis or any other classification algorithm [15]. Similar methodologies have already been used in other fields, like non-intrusive electric load data disaggregation [26–29]. As an example of the results that can be achieved by these techniques, Figure 10 shows the findings for one of the most complex households, HH-06 of R1, showing an indoor leak and a high average daily consumption. The unsupervised learning technique used, allows to identify different types of water consumption uses: *Cluster 2* in Figure 10 mostly includes events corresponding to *toilets*, while *Cluster 6* is composed of *washing machine* and *shower* events.

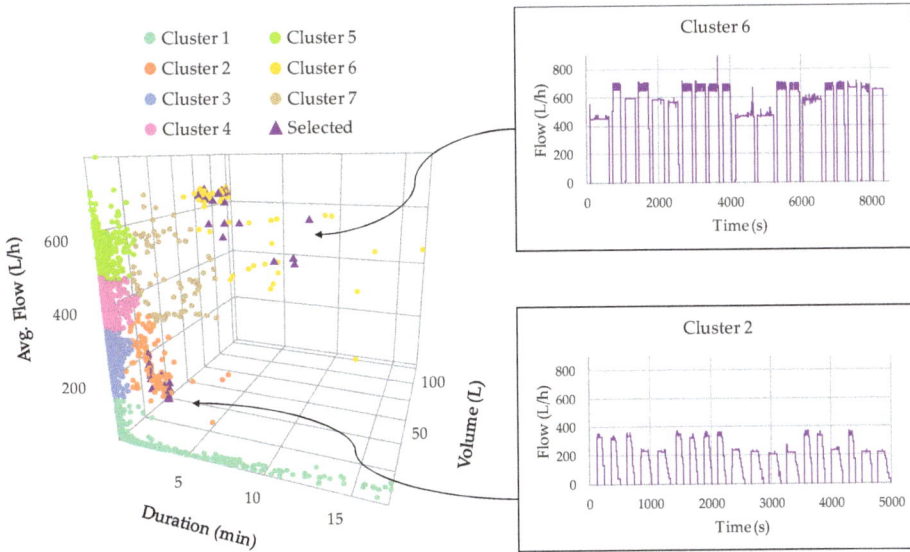

Figure 10. Display of 20 randomly selected individuals from clusters 2 and 6 for household HH-06 that belongs the R1 study.

Figure 11 shows the final results after assigning a water end use to each cluster for the household under study. Given the same monitoring period, the outcome of proposed methodology is compared with the one obtained manually. It can be observed that the physical characteristics of the events in each end use category tend to be similar. Nevertheless, there is a significant deviation, especially with respect to the mean flow rate and duration of the events: the average flow rate for each end use tends to be higher in the presented approach, while the average duration is generally shorter as more cropping operations are conducted through the automatic disaggregation process. In addition, both parameters—volume and duration of *single-use* events—are less dispersed when the flow traces are automatically cropped. This is directly related to the inherent defects of manual editing that can be seen in Figure 12 (additional examples in Appendix B): the automatic algorithms recognize the leakage event by means of a volume check (Figure 12b), whereas the analyst has subjectively decided in this specific case to ignore it (Figure 12a) and add the volume to the *toilet* use. When a human analyst processes the consumption data, the resulting average duration of *faucets* and *toilets* is larger and the average flow rate smaller. Additionally, for the same reason, a greater number of leakage events have been identified and separated from other consumption through the proposed methodology. These findings demonstrate that automatic disaggregation tools can generate a standardized corpus of processed data, which is more homogeneous and reliable because it is obtained as a result of cropping operations based on solid and well established criteria. Therefore, the *single-use* events obtained from the automatic disaggregation algorithms developed in this study are significantly more reliable, in terms of duration, average flow rate and shape than those resulting from a manual processing and cropping the water consumption flow traces. These results have direct implications in the probability functions used to characterize water consumption events frequency, duration and intensity [30–34].

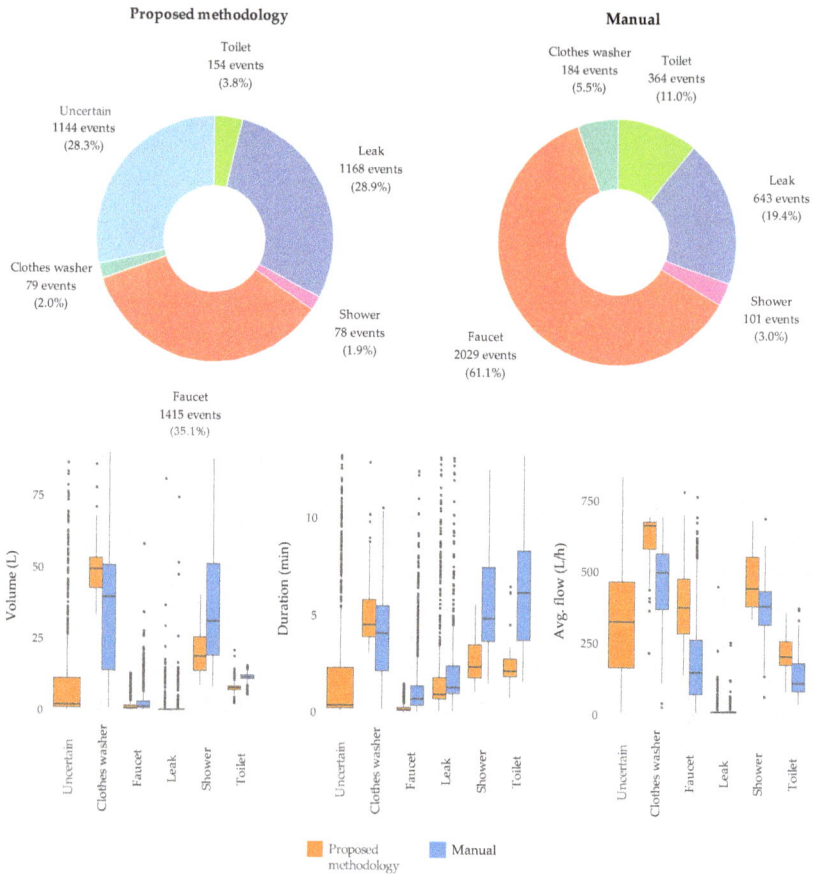

Figure 11. The manual vs proposed methodology final results of complete disaggregation processing for the household HH-06 of the R1 study.

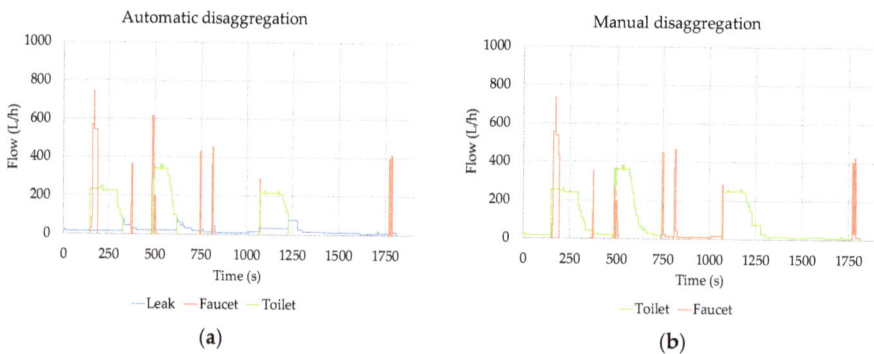

Figure 12. Manual vs automatic disaggregation (example from the household HH-06 of R1).

Obviously, more accurate classification techniques can be developed as processing experience is gained and larger and more reliable data sets are available for training the algorithms. The work presented should be considered as an important first stage to produce sets of individual events that are built consistent and accurately, which can be used to improve the training of automatic recognition algorithms. Therefore, the main contribution of the proposed methodology is mainly related to the quality of the *single-use* events obtained through an automatic separation technique that can be easily used for developing faster and better performing classification algorithms.

5. Conclusions

The work presented intends to be a step forward to the main objective of understanding in detail how water is consumed through end uses, and the reasons behind it. It proposes a new fully-automatic disaggregation process for water consumption events that is based on a two-step filtering and event cropping algorithm. An additional advantage achieved by the flexibility of the filter is that the whole process can be universally applied to different type of customers and monitoring equipment.

The disaggregation process presented is divided into two main stages:

(a) The raw water consumption events are filtered and categorized as *simple* or *overlapped* to facilitate subsequent operations. The filtering relies on an advanced algorithm that is automatically calibrated for each water consumption event by means of NSGA-II genetic algorithm. *Simple* events are then characterized as *single-use*, which correspond to actual individual water uses, or *uncertain* events.

(b) On the other hand, *overlapped* events, originated by simultaneous water uses, are cropped and separated into simpler *single-use* events. All cropping operations are implemented on the raw flow trace, and potential distortions in the filtered signal do not generate any loss of information in the resulting subevents. In other words, all the subevents created maintain the characteristics of the original flow trace. This particular feature increases the amount of information available for the classification algorithms that can be developed in the future, improving their effectiveness.

Finally, as a case study, an example of the way in which the events generated during the previous stages can be easily categorized into various end uses by a semiautomatic algorithm is added to the work presented. *Single-use* events are massively classified into various water end use categories by means of clustering analysis.

Regarding the performance analysis of the first and second stages for the case study presented, the following conclusions were raised: The original raw flow traces, of the 20 households belonging to the studies R1 and R2, covering a monitoring period of 15 days per household, contain 19,858 events. After the filtering and separation process, the number of subevents grew to 46,721, of which 79.3% (37,042 events) are single uses. In other words, the number of water consumption events increased by 130%, and 26,863 new events were created. Up to 75.8% of the *single-use* events that can be classified, have been obtained through the disaggregation process defined in this work. Therefore, the methodology proposed solves most of the cropping operations that need to be performed and reduces significantly the human intervention required to disaggregate the *overlapped* consumption events into *single-use* events, with significant time savings.

Finally, by comparing the manual and the automatically disaggregated events, it was observed that the characteristics of the events originated from the algorithms proposed are more homogeneous and consistent than the ones obtained by manual cropping. This result can be easily justified by the fact that the automatic separation algorithms always apply the same criteria, while a human analyst may change the cropping criteria while conducting the analysis. Furthermore, the inherent subjectivity of manual separation introduces dispersion in the physical characteristics of the events belonging to a specific water end use. More dispersion regarding the physical characteristics of the consumption events associated to an end use, unavoidably lead to poorer performance of whatever automatic classification technique that could be applied. This is why the *single-use* events obtained by the methodology proposed constitute a more reliable corpus for training and developing end use classifications algorithms.

Definitions

Classification: Process by which every *single-use* event is assigned to one of the potential water end uses in a household.

Cropping: Action by which one part of an *overlapped* event that has already been identified as a single water use (and still remains attached to the overlapped event) is effectively removed from it to become a new and independent *single-use* event.

Disaggregation: Process by which an *overlapped* event is effectively separated into all the *single-use* events integrating such event. In this work, the disaggregation process consists in two actions (filtering and cropping), and it is performed through a universal fully-automatic algorithm.

Fictitious union: When a water consumption starts, there is a significant time gap between the previous pulse and the initial pulse recorded. Therefore, the flow rate associated to the first pulse received (calculated as the ratio between the pulse volume and the time gap) is lower than the actual consumption flow rate. With this calculation it is also assumed that during the complete time gap between pulses, the consumption flow rate is constant and equal to the calculated value. Obviously, this calculation does not represent how water was really consumed in the time interval between the two pulses under analysis. Quite frequently, there will be part of the time (most) in which there is no consumption and another part (less) in which there is a consumption at a relative high flow.

Event: Every single water consumption, whatever its volume or duration. An event begins when the flow rate through the meter changes from zero to any positive value, and finishes when the flow rate becomes to zero again.

Unprocessed event: Initial event in the raw flow trace, before any kind of signal processing has been conducted.

Overlapped event: Complex event being the sum of more than one simultaneous single uses of water.

Pulse: Each of the signals sent by the pulse-emitter (attached to or embedded in the water meter) to the data-logger. Each signal corresponds to a fixed volume of water consumed. In the design stage of the monitoring project there are two crucial decisions related to pulse emitters: (i) the consumption volume associated to each pulse, which mostly depends on the water meter design, and (ii) the way pulses are recorded by the logger. Typically, the data loggers may store the number of pulses (volume) received at fixed intervals of time, or it may store the time at which each single pulse is received. Fictitious unions appear when the time of occurrence of the pulses are recorded in the data logger.

Simple event: Processed event, obtained after Filter-A, constituted by four vertexes. Most simple events will become single-use events at the end of the disaggregation process.

Single-use event: Final event obtained after the disaggregation process that corresponds to one specific end use.

Uncertain event: Final event obtained after the disaggregation process that cannot be further cropped into smaller single-use events nor classified as being a single use itself. Additional contextual information is needed to be able to split the event into smaller ones or to categorize it as one single-use event.

Acknowledgments: This study has received funding by the IMPADAPT project /CGL2013-48424-C2-1-R from the Spanish ministry MINECO with European FEDER funds and from the European Union's Seventh Framework Programme (FP7/2007e2013) under grant agreement No. 619172 (SmartH2O: an ICT Platform to leverage on Social Computing for the efficient management of Water Consumption).

Author Contributions: Methodology conception and definition: F.J.A., L.P.-J. and R.C. Data analysis tools and signal filtering algorithms: L.P.-J. End Use software design: F.J.A. Water consumption data gathering and field work: F.J.A. and R.C. All authors contributed to the preparation of the manuscript and approved it.

Conflicts of Interest: The authors declare no conflicts of interest.

Appendix A

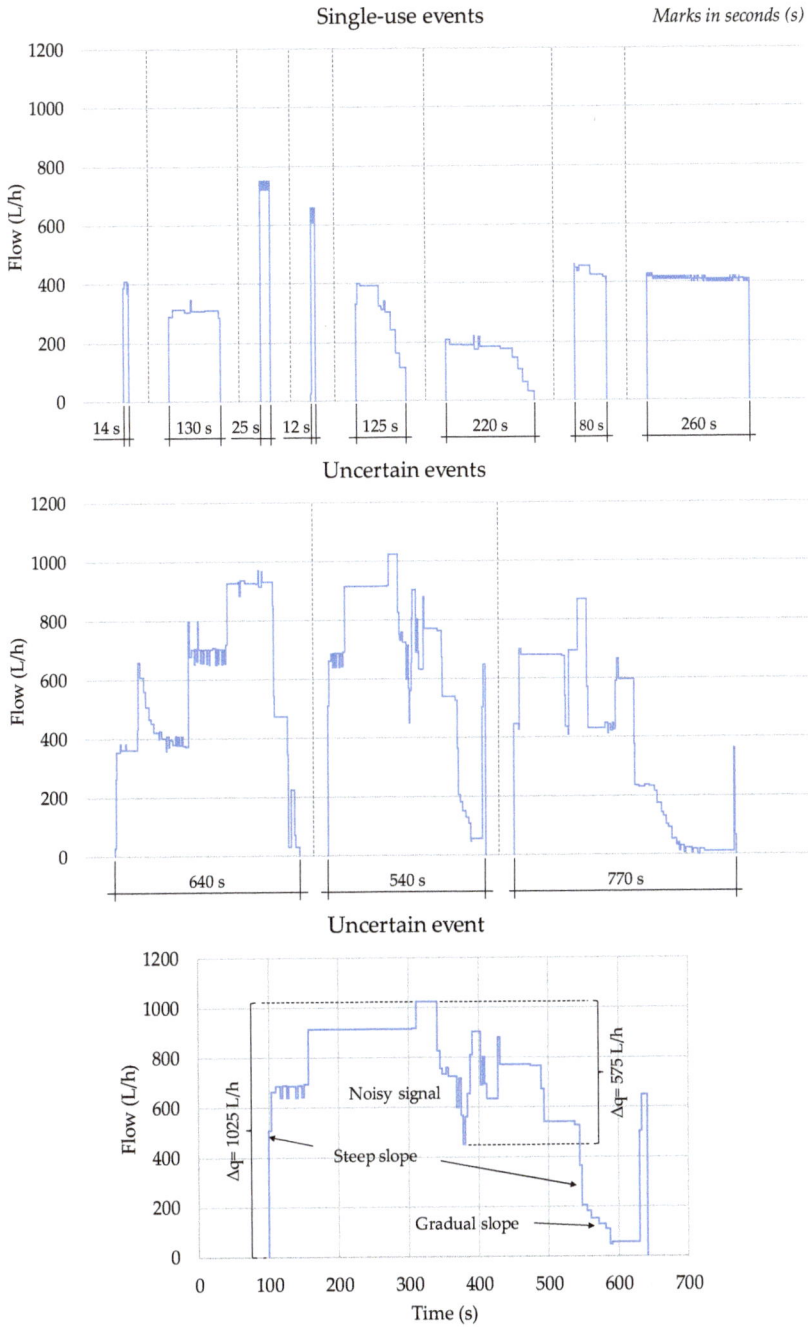

Figure A1. Examples of Single-Use and Uncertain Events.

Appendix B

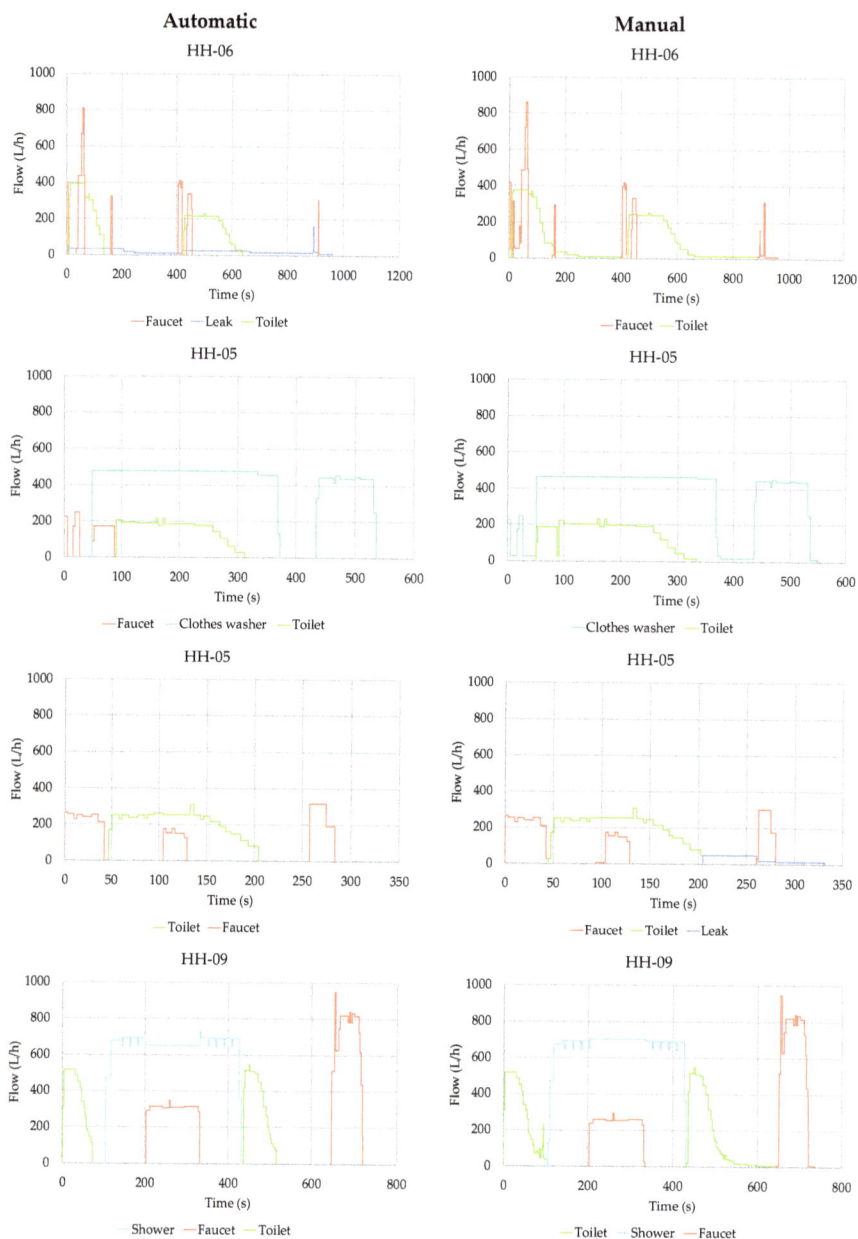

Figure A2. Examples of Manual vs. Automatic Disaggregation from R1 study.

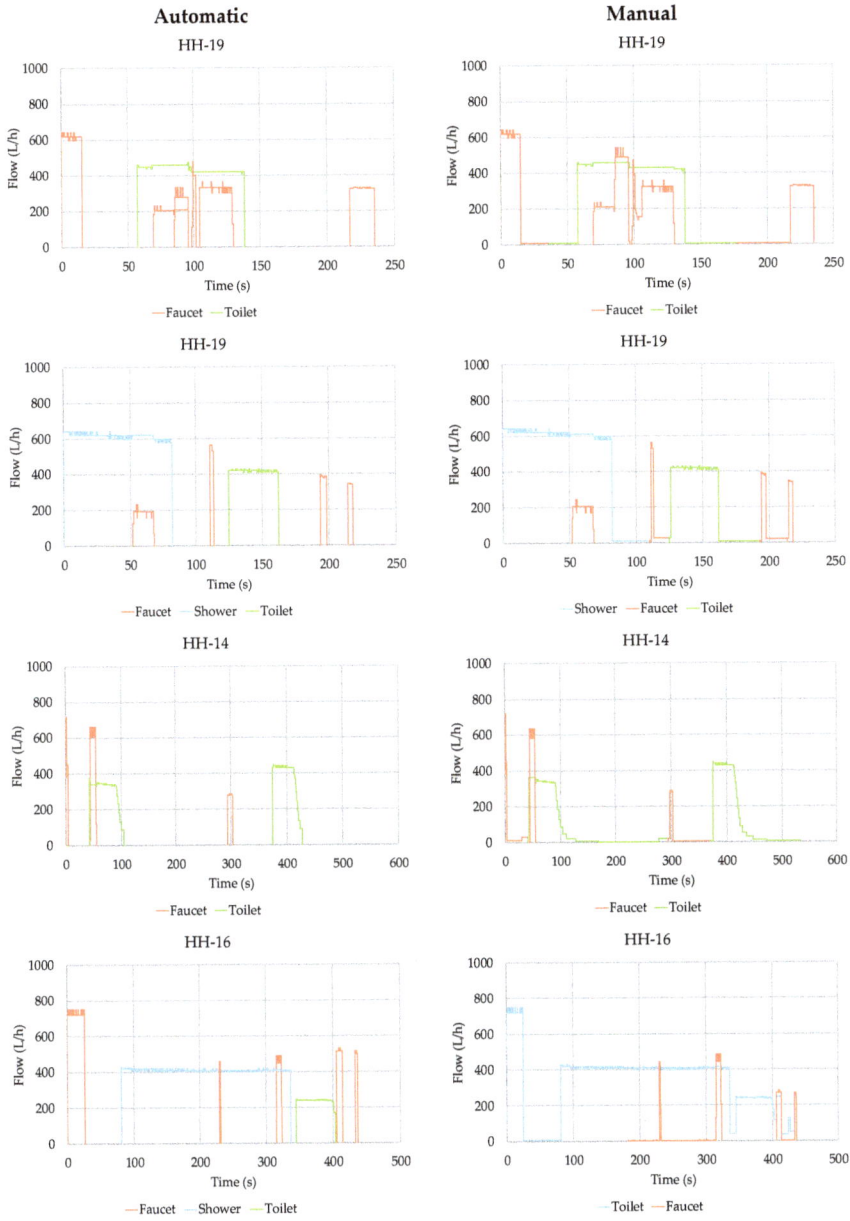

Figure A3. Examples of Manual vs. Automatic Disaggregation from R2 study.

References

1. World Commission on Environment and Development. *Our Common Future*; Oxford University Press: Oxford, UK, 1987; ISBN 019282080X.
2. Hoekstra, A.Y.; Mekonnen, M.M. The water footprint of humanity. *Proc. Natl. Acad. Sci. USA* **2012**, *109*, 3232–3237. [CrossRef] [PubMed]

3. Jaramillo, F.; Destouni, G. Local flow regulation and irrigation raise global human water consumption and footprint. *Science* **2015**, *350*, 1248–1251. [CrossRef] [PubMed]

4. Shiklomanov, I.A. Appraisal and Assessment of World Water Resources. *Water Int.* **2000**, *25*, 11–32. [CrossRef]

5. Cominola, A.; Giuliani, M.; Piga, D.; Castelletti, A.; Rizzoli, A.E. Benefits and challenges of using smart meters for advancing residential water demand modeling and management: A review. *Environ. Model. Softw.* **2015**, *72*, 198–214. [CrossRef]

6. Fielding, K.S.; Spinks, A.; Russell, S.; McCrea, R.; Stewart, R.A.; Gardner, J. An experimental test of voluntary strategies to promote urban water demand management. *J. Environ. Manag.* **2013**, *114*, 343–351. [CrossRef] [PubMed]

7. Sahin, O.; Bertone, E.; Beal, C.D. A system approach for assessing water conservation potential through demand-based water tariffs. *J. Clean. Prod.* **2017**, *148*, 773–774. [CrossRef]

8. Makki, A.A.; Stewart, R.A.; Beal, C.D.; Panuwatwanich, K. Novel bottom-up urban water demand forecasting model: Revealing the determinants, drivers and predictors of residential indoor end-use consumption. *Resour. Conserv. Recycl.* **2015**, *95*, 15–37. [CrossRef]

9. Bennet, C.; Stewart, R.A.; Beal, C.D. ANN-Based residential water end-use demand forecasting model. *Expert Syst. Appl.* **2013**, *40*, 1014–1023. [CrossRef]

10. DeOreo, W.B.; Heaney, J.P.; Mayer, P.W. Flow trace analysis to assess water use. *Am. Water Works Assoc.* **1996**, *88*, 79–90.

11. Kowalski, M.; Marshallsay, D. *A System for Improved Assessment of Domestic Water Use Components. II International Conference Efficient Use and Management of Urban Water Supply*; International Water Association: Tenerife, Spain, 2003.

12. Arregui, F. New software tool for water end-uses studies. Presented at the 8th IWA International Conference on Water Efficiency and Performance Assessment of Water Services, Cincinnati, OH, USA, 20–24 April 2015.

13. Nguyen, K.A.; Zhang, H.; Stewart, R.A. Development of an intelligent model to categorise residential water end use events. *J. Hydro-Environ. Res.* **2013**, *7*, 182–201. [CrossRef]

14. Nguyen, K.A.; Stewart, R.A.; Zhang, H. An intelligent pattern recognition model to automate the categorisation of residential water end-use events. *Environ. Model. Softw.* **2013**, *47*, 108–127. [CrossRef]

15. Nguyen, K.A.; Stewart, R.A.; Zhang, H. An autonomous and intelligent expert system for residential water end-use classification. *Expert Syst. Appl.* **2014**, *41*, 342–356. [CrossRef]

16. Nguyen, K.A.; Stewart, R.A.; Zhang, H.; Jones, C. Intelligent autonomous system for residential water end use classification: Autoflow. *Appl. Soft Comput.* **2015**, *31*, 118–131. [CrossRef]

17. Piga, D.; Cominola, A.; Giuliani, M.; Castelletti, A.; Rizzoli, A.E. A convex optimization approach for automated water and energy end use disaggregation. In Proceedings of the 36th IAHR World Congress, Hague, The Netherlands, 28 June–3 July 2015.

18. FLUID—The Learning Water Meter. Available online: http://www.fluidwatermeter.com (accessed on 17 November 2017).

19. WaterSmart—Platform Features. Take a Look Under the Hood. Available online: https://www.watersmart.com (accessed on 17 November 2017).

20. Aqubiq—Features. A Smart Path for a Greener Lifestyle. Available online: http://www.aqubiq.com (accessed on 17 November 2017).

21. Pastor, L.; Arregui, F.; Cobacho, R. Filtering smart metering data to improve detection of water end use events. In Proceedings of the 9th International Conference on Efficient Use and Management of Urban Water, Bath, UK, 18–20 July 2017.

22. Deb, K.; Pratap, A.; Agarwal, S.A. Fast and Elitist Multiobjective Genetic Algorithm: NSGAII. *IEEE Trans. Evolut. Comput.* **2002**, *6*, 182–197. [CrossRef]

23. R Core Team. *R: A Language and Environment for Statistical Computing*; R Foundation for Statistical Computing: Vienna, Austria, 2013.

24. Gupta, H.V.; Kling, H.; Yilmaz, K.K.; Martinez, G.F. Decomposition of the mean squared error and NSE performance criteria: Implications for improving hydrological modelling. *J. Hydrol.* **2009**, *377*, 80–91. [CrossRef]

25. Reynolds, A.; Richards, G.; de la Iglesia, B.; Rayward-Smith, V. Clustering rules: A comparison of partitioning and hierarchical clustering algorithms. *J. Math. Model. Algorithms* **1992**, *5*, 475–504. [CrossRef]

26. Amenta, V.; Tina, G.M. Load demand disaggregation based on simple load signature and user's feedback. *Energy Procedia* **2015**, *83*, 380–388. [CrossRef]

27. Elhamifar, E.; Sastry, S. Energy disaggregation via learning powerlets and sparse coding. In Proceedings of the National Conference on Artificial Intelligence, Austin, TX, USA, 25–30 January 2015; pp. 629–635.

28. Bonfigli, R.; Squartini, S.; Fagiani, M.; Piazza, F. Unsupervised algorithms for nonintrusive load monitoring: An up-to-date overview. In Proceedings of the 15th International Conference on Environment and Electrical Engineering (EEEIC), Rome, Italy, 10–13 June 2015; pp. 1175–1180.

29. Cominola, A.; Giuliani, M.; Piga, D.; Castelletti, A.; Rizzoli, A.E. A Hybrid Signature-based Iterative Disaggregation algorithm for Non-Intrusive Load Monitoring. *Appl. Energy* **2017**, *185*, 331–344. [CrossRef]

30. Guercio, R.; Magini, R.; Pallavicini, I. Instantaneous residential water demand as stochastic point process. *WIT Trans. Ecol. Environ.* **2001**, *48*, 129–138. [CrossRef]

31. Alvisi, S.; Franchini, M.; Marinelli, A. A stochastic model for representing drinking water demand at residential level. *Water Resour. Manag.* **2003**, *17*, 197–222. [CrossRef]

32. García, V.J.; García-Bartual, R.; Cabrera, E.; Arregui, F.; García-Serra, J. Stochastic model to evaluate residential water demands. *J. Water Resour. Plan. Manag.* **2004**, *130*, 386–394. [CrossRef]

33. Blokker, E.J.M.; Vreeburg, J.H.G.; van Dijk, J.C. Simulating residential water demand with a stochastic end-use model. *J. Water Resour. Plan. Manag.* **2010**, *136*, 375–382. [CrossRef]

34. Creaco, E.; Alvisi, S.; Farmani, R.; Vamvakeridou-Lyroudia, L.; Franchini, M.; Kapelan, Z.; Savic, D. Methods for preserving duration-intensity correlation on synthetically generated water-demand pulses. *J. Water Resour. Plan. Manag.* **2016**, *142*. [CrossRef]

Review

Overview, Comparative Assessment and Recommendations of Forecasting Models for Short-Term Water Demand Prediction

Amos O. Anele [1,*], Yskandar Hamam [1], Adnan M. Abu-Mahfouz [1,2] and **Ezio Todini [3]**

1   Department of Electrical Engineering, Tshwane University of Technology, Pretoria 0001, South Africa; hamama@tut.ac.za (Y.H.); AAbuMahfouz@csir.co.za (A.M.A.-M.)
2   Council for Scientific and Industrial Research, Pretoria 0081, South Africa
3   Department of Biological, Geological and Environmental Sciences, University of Bologna, Via Zamboni, 33-40126 Bologna, Italy; eziotodini@gmail.com
*   Correspondence: anelea@tut.ac.za or aneleamos@gmail.com; Tel.: +27-12-382-4191

Received: 8 September 2017; Accepted: 6 November 2017 ; Published: 13 November 2017

Abstract: The stochastic nature of water consumption patterns during the day and week varies. Therefore, to continually provide water to consumers with appropriate quality, quantity and pressure, water utilities require accurate and appropriate short-term water demand (STWD) forecasts. In view of this, an overview of forecasting methods for STWD prediction is presented. Based on that, a comparative assessment of the performance of alternative forecasting models from the different methods is studied. Times series models (i.e., autoregressive (AR), moving average (MA), autoregressive-moving average (ARMA), and ARMA with exogenous variable (ARMAX)) introduced by Box and Jenkins (1970), feed-forward back-propagation neural network (FFBP-NN), and hybrid model (i.e., combined forecasts from ARMA and FFBP-NN) are compared with each other for a common set of data. Akaike information criterion (AIC), originally proposed by Akaike (1974) is used to estimate the quality of each short-term forecasting model. Furthermore, Nash–Sutcliffe (NS) model efficiency coefficient proposed by Nash–Sutcliffe (1970), root mean square error (RMSE) and mean absolute percentage error (MAPE) are the forecasting statistical terms used to assess the predictive performance of the models. Lastly, as regards the selection of an accurate and appropriate STWD forecasting model, this paper provides recommendations and future work based on the forecasts generated by each of the predictive models considered.

Keywords: forecasting models; short-term; water demand simulation

---

1. Introduction

The most crucial factor in the planning, operation and management of water distribution systems (WDS) is the satisfaction of consumer demand. The stochastic nature of water demand during the day and week is influenced by several factors; namely, climatic and geographic conditions, commercial and social conditions of people, population growth, industrialisation, technical innovation, cost of supply, and condition of WDS [1–4]. Therefore, water utilities need accurate and appropriate short-term water demand (STWD) forecasts in order to continually satisfy consumers with quality water in adequate volumes, and at reasonable pressures [5–7]. STWD forecasting is an important component of the successful operation, management, and optimisation of any existing WDS. As a result, the selection of an accurate and appropriate STWD forecasting model is useful for [1,6,8–16]:

- explaining day-to-day demand variations
- minimising the operating cost of pumping stations
- pinpointing possible network failures (e.g., water leaks and pipe bursts)

- helping utilities plan and manage water demands for near-term events
- optimizing daily operations of the infrastructure (e.g., pump scheduling, control of reservoirs volume, pressure management, and water conservation program)

In the light of the above, the first objective of this paper is to present an overview of forecasting methods for STWD prediction. Based on that, the second objective is to conduct a comparative assessment of the performance of alternative forecasting models from the different methods. As regards the selection of an accurate and appropriate model, the third objective of the paper is to present recommendations and future work for the forecasts generated by the forecasting models considered.

2. Overview of STWD Forecasting Methods

In this section, the overview of univariate time series (UTS), time series regression (TSR), artificial neural network (ANN), and hybrid methods for STWD prediction is presented (see also Table 1).

2.1. UTS Forecasting Methods

UTS methods forecast future water demand based on past observations and associated error terms [17,18]. Exponential smoothing, autoregressive (AR), moving average (MA), autoregressive-moving average (ARMA), autoregressive integrated moving average (ARIMA) and seasonal ARIMA (SARIMA) are examples of UTS forecasting models. These models are useful for short-term operational forecasts. However, they may not be the most accurate alternative when weather changes are likely to occur in the underlying determinants of water demands [11,18]. Furthermore, it is discussed in [11] that stochastic process models (i.e., AR, MA, ARMA, and ARIMA) are used since exponential smoothing models sometimes cease to be adequate when time series data exhibit more complex profiles. Based on that, to achieve the second objective of this paper, the model processes of AR(p), MA(q), and ARMA(p, q) are respectively considered as given in Equations (1)–(3) [17,18].

$$Y_t = \mu + \sum_{k=1}^{p} \phi_k Y_{t-k} + \epsilon_t \tag{1}$$

$$Y_t = \mu + \epsilon_t + \sum_{k=1}^{q} \theta_k \varepsilon_{t-k} \tag{2}$$

$$Y_t = \mu + \sum_{k=1}^{p} \phi_k Y_{t-k} + \epsilon_t + \sum_{k=1}^{q} \theta_k \varepsilon_{t-k} \tag{3}$$

where p and q are the model orders, ϕ is the autoregressive parameter, θ is the moving average parameter, μ is the mean value of the process, and ϵ_t is the forecast error at time t. Y_t is the observed value of demand at time t, k is the number of historical periods, Y_{t-k} and ε_{t-k} are the observation at time $t-k$.

2.2. Time Series Regression (TSR) Forecasting Methods

Unlike the UTS models, TSR forecasting models consider the effects of exogenous variables. This is because they generate forecasts based on the relationship between water demand and its determinants [19–21]. TSR models include multiple linear regression (MLR), multiple and nonlinear regression (MNLR), ARMA with exogenous variable (ARMAX) and ARIMA with exogenous variable (ARIMAX). Among others, the ARMAX(p, q, b) model is considered to achieve the second objective of this paper. Equation (4) is useful in a case where the demand at time t is influenced by MA and AR terms, in addition to exogenous variables and their autoregressive terms [11].

$$Y_t = \mu + \sum_{k=1}^{p} \phi_k Y_{t-k} + \epsilon_t + \sum_{k=1}^{q} \theta_k \varepsilon_{t-k} + \sum_{k=0}^{b} \beta_k x_{t-k} \tag{4}$$

where b is a single exogenous variable considered for the ARMAX model. Additionally, β_k and x_{t-k} are respectively the coefficient and observed value of the kth independent variable.

2.3. Artificial Neural Network (ANN) Forecasting Methods

ANNs were introduced following Rosenblatt's concept of perceptron [22], and their application usually involves a comparative assessment of the performance with TSR models (e.g., feed-forward back-propagation neural network (FFBP-NN), generalized regression neural networks (GRNNs), radial basis neural networks (RBNNs), and MLR) [1,10,23–26], with UTS models (e.g., dynamic artificial neural network (DAN2), ARIMA and FFBP-NN) [27] or with both UTS and TSR models (e.g., simple linear regression (SLR), MLR, UTS models, and ANN models) [5,28,29]. Nonetheless, in order to achieve the second objective of this paper, FFBP-NN (see Equation (5)) is considered [1].

$$Y_t = \alpha_0 + \sum_{j=1}^{p} \alpha_j f(\sum_{i=1}^{h} \beta_{ij} Y_{t-j} + \beta_{0j}) + \epsilon_t \tag{5}$$

where p is the number of hidden nodes, h is the number of input nodes, f is a sigmoid transfer function, α_j is the vector of the weights from hidden to the output nodes, β_{ij} are the weights from the input to hidden nodes, and α_0 and β_{0j} are the weights of the arcs leaving from the bias terms.

2.4. Hybrid Forecasting Methods

Forecasting with hybrid models (i.e., combined forecasts from two or more predictive models) has found wide application [6,11,24,30–34], since it leads to better forecasting performance. For instance, Equation (6) is applied in a case where forecasts from different models are combined in order to obtain a hybrid forecast. As regards achieving the second objective of this paper, the combined forecast is obtained by using a UTS model (i.e., ARMA) and an ANN model (i.e., FFBP-NN).

$$\hat{Y}_t = \beta_0 + \sum_{i=1}^{n} \beta_i \hat{Y}_{i,t} \tag{6}$$

where $\hat{Y}_{i,t}$ is the predicted value of the time series at time t using the i_{th} model, β_0 is the regression intercept, β_i coefficients are determined by optimisation or least squares regression to minimise the mean square error (MSE) between the hybrid forecast $\hat{Y}_{i,t}$ and the actual data [11].

Table 1. Brief summary of short-term water demand (STWD) forecasting methods and models. UTS: univariate time series; MA: moving average; AR: autoregressive; ARIMA: autoregressive integrated moving average; ARMA: autoregressive-moving average; SARIMA: seasonal ARIMA; TSR: time series regression; MNLR: multiple and nonlinear regression; ARMAX: ARMA with exogenous variable; MLR: multiple linear regression; ARIMAX: ARIMA with exogenous variable; ANN: artificial neural network; FFBP-NN: feed-forward back-propagation neural network; GRNN: generalized regression neural network; RBNN: radial basis neural network; DAN2: dynamic artificial neural network; GARCH: generalized autoregressive conditional heteroskedasticity.

Forecasting Methods and Models	Quantitative Assessment of Forecast Accuracy	Forecast Purpose
UTS models [18,27,29]: MA, AR, ARIMA, exponential smoothing, ARMA, SARIMA	It can exhibit more complex profiles. However, it does not account for the effect of exogenous variables (e.g., weather data or price) [11].	Useful for short-term operational forecasts (i.e., to minimise the operating cost of pumping stations, etc.)
TSR models [1,25,26]: MNLR, ARMAX, MLR and ARIMAX	TSR models produce forecasts on the basis of the relationship between water demand and its determinants (e.g., weather data, income, demographics) [19].	Useful for better prediction of daily water demand [24]. Relevant for setting water rates, revenue forecasting, and financial planning exercises.
ANN models: FFBP-NN, GRNN, RBNN, DAN2 [5,35]	Used with TSR models [1,24–27], with UTS models [27], or with both UTS and TSR models [5,28,29]. According to [11], ANN outperforms UTS and TSR models. However, the results of [24,25] were inconclusive.	Useful for a better prediction of peak daily water demand. To inform optimal operating policy as well as pumping and maintenance scheduling.
Hybrid models: FFBP-NN and AR [36], Holt–Winters, ARIMA, and GARCH [37], Fuzzy logic and AR [38]	Different forecasting models are able to capture different aspects of the information available for prediction [11]. As a result, leading to better forecasting performance [33,36]	Useful for real-time, near-optimal control of water distribution systems (WDS) [6]. Necessary for operational purposes [33,36].

3. Presentation and Discussion of Results

In this paper, ARIMA-based models (i.e., AR, MA, ARMA, ARMAX) together with the widely used non-parametric forecasting model, FFBP-NN, have been compared with each other and against the hybrid model, a combination of two or more forecasting models (i.e., ARMA and FFBP-NN) for a common set of data (see Figure 1). Figure 1a shows the average water consumption for the 24 h of each day for a city in south-eastern Spain, obtained from all the available data provided in [1]. The predictive models considered in this paper were used to forecast hourly water demands. In addition, an average weekly data of 168 h was used, and based on that, the proportion of data used for the training and testing were 60% and 40% respectively.

Figure 1a shows a similar behaviour during the early morning (e.g., all curves grow from 6:00 a.m. until 10:00 a.m.). In addition, from 10:00 a.m. to 4:00 p.m., all the curves have decreasing and increasing trend (except on weekends). According to [1], temperature is said to be the main factor that influences multiple sources of water consumption (e.g., showers, water for garden, etc.). Hence, Figure 1b shows the single exogenous variable considered for ARMAX model.

The results shown in Figures 2–5 are obtained by computing Equations (1)–(6) in MATLAB. Figures 2–4 show the forecasts generated by AR and MA (see Figure 2), ARMA and ARMAX (see Figure 3), as well as FFBP-NN and hybrid model (see Figure 4). Figure 5a–c show the comparative assessment of the predictive performance of these models by using forecasting statistical terms such as root mean square error (RMSE), mean absolute percentage error (MAPE), and Nash–Sutcliffe (NS) [39]. This assessment was achieved by computing Equations (7)–(9). In addition, the estimate of the relative quality of AR, MA, ARMA, ARMAX, and FFBP-NN is shown in Figure 5d, and it was obtained by applying Akaike information criterion (AIC) [40], which is based on Equation (10). The forecasts presented in this paper were generated using the best model order, which is determined by the AIC.

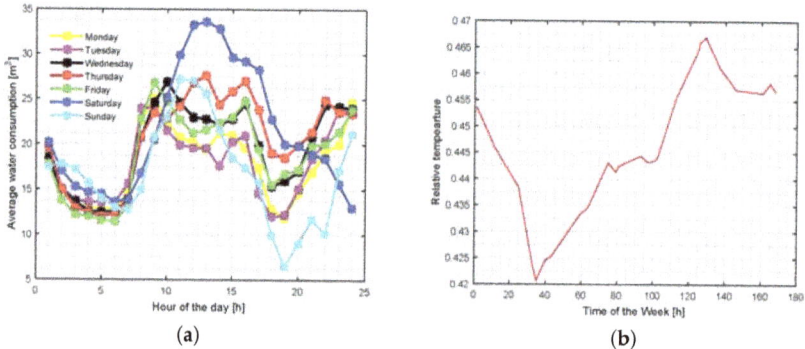

Figure 1. (a) Daily water demand profile and (b) Single exogenous variable "relative temperature".

Figures 2a,c, 3a,c, and 4a,c were obtained by using the training dataset, whereas the test dataset was used to obtain Figures 2b,d, 3b,d, and 4b,d. Based on the application of AIC [40], Figure 2a,b show that a model process of $AR(p = 2)$ was used to generate the forecasts for the training and test datasets. A model process of $MA(q = 3)$ was also used to obtain the forecasts presented in Figure 2c,d. The forecasts shown in Figure 3a,b were obtained based on a model process of $ARMA(p = 1, q = 1)$. The results of Figure 3c,d were obtained using a model process of $ARMAX(p = 1, q = 1, b = 1)$. A model order of three was used to obtain the results shown in Figure 4a,b. The configuration of the neural network was achieved with a feed-forward neural network of one hidden layer (10 hidden neurons) using a Levenberg–Marquardt optimisation-based backpropagation algorithm to train the neural network weights. The training was stopped using validation data (15% of the training datasets). This process was performed 10 times (i.e., 10 cross-validation) to select the feed-forward neural network with the best predictive accuracy to compensate for neural network training variations. The model orders mentioned in this paper are the best, and were obtained using AIC. Lastly, the optimal weighting of the hybrid forecast obtained using ARMA and FFBP-NN—as shown in Figure 4c,d—was achieved by using linear least square optimisation.

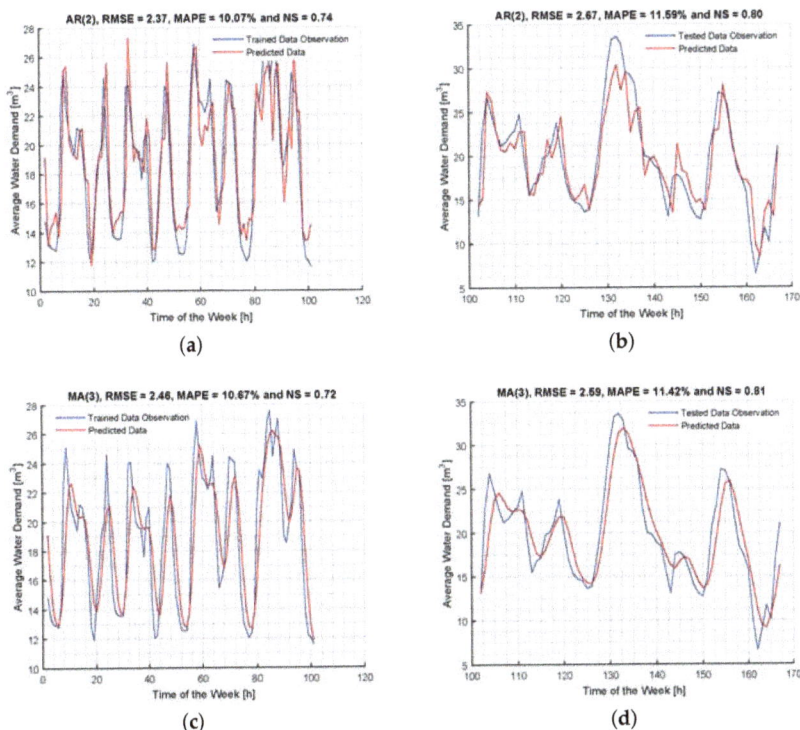

Figure 2. Forecasts generated using (**a**,**b**) AR model and (**c**,**d**) MA model. The best model orders, $AR(p = 2)$ and $MA(q = 3)$, were determined based on Akaike information criterion (AIC) computation [40]. MAPE: mean absolute percentage error; NS: Nash–Sutcliffe; RMSE: root mean square error.

The RMSE and MAPE were used to evaluate the forecasting accuracy of the predictive models. In addition, NS was used to estimate the forecasting power of the models. The results of Figures 2b,d, 3b,d, and 4b,d show that the hybrid model was the best forecasting model for STWD prediction (i.e., RMSE = 0.82, MAPE = 3.56%, NS = 0.98) followed by ARMAX (i.e., RMSE = 1.03, MAPE = 3.86%, NS = 0.95), ARMA (i.e., RMSE = 1.85, MAPE = 7.63%, NS = 0.91), MA (i.e., RMSE = 2.59, MAPE = 11.42%, NS = 0.81), AR (i.e., RMSE = 2.67, MAPE = 11.59%, NS = 0.8), and FFBP-NN (i.e., RMSE = 2.8, MAPE = 12.31%, NS = 0.78). In addition, the plots of RMSE, MAPE, and NS versus model order variation are also presented in Figure 5a–c. Compared to AR, MA, ARMA, and FFBP-NN, Figure 5d shows that the AIC value for ARMAX is the smallest. This implies that the quality of the ARMAX model compared to others (i.e., AR, MA, ARMA, and FFBP-NN) is estimated to be the best. The predictive accuracy of all models decreases as the model order increases. For instance, FFBP-NN model had a remarkable decrease in accuracy compared to other models. Due to the additional piece of information (i.e., relative temperature) as shown in Figure 1b, the results obtained in Figures 3 and 5 show that ARMAX(1,1,1) provided a better forecast than ARMA(1,1). Generally, based on the forecasting statistical terms considered in this paper, the comparative assessment shows that for STWD forecasting, the hybrid model (combined forecast from ARMA and FFBP-NN) was the best model, followed by ARMAX, ARMA, MA, AR, and FFBP-NN.

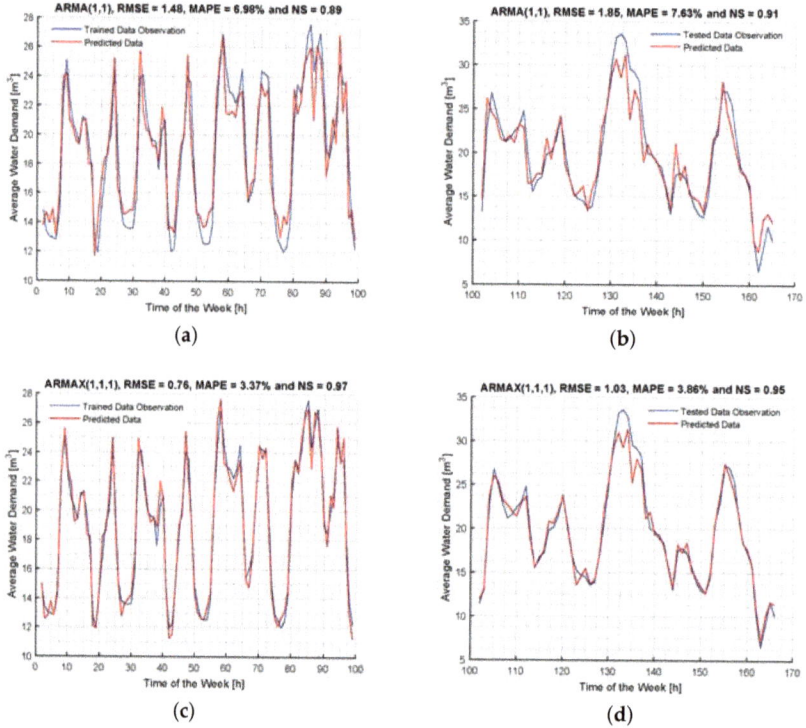

Figure 3. Forecasts generated using (**a,b**) ARMA model and (**c,d**) ARMAX model. Using AIC, the best model orders are $\mathrm{ARMA}(p=1, q=1)$ and $\mathrm{ARMAX}(p=1, q=1, b=1)$.

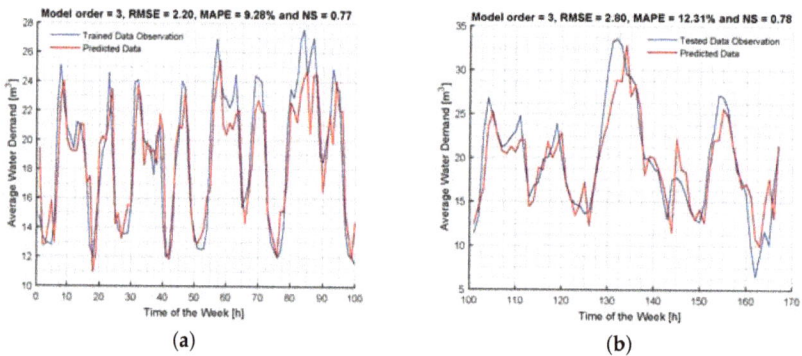

Figure 4. *Cont.*

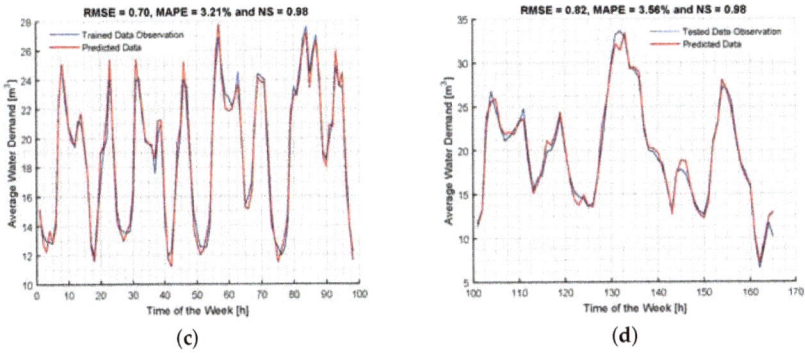

Figure 4. Forecasts generated using (**a,b**) FFBP-NN model and (**c,d**) Hybrid model. The hybrid forecast was obtained by the combined forecast from ARMA and FFBP-NN.

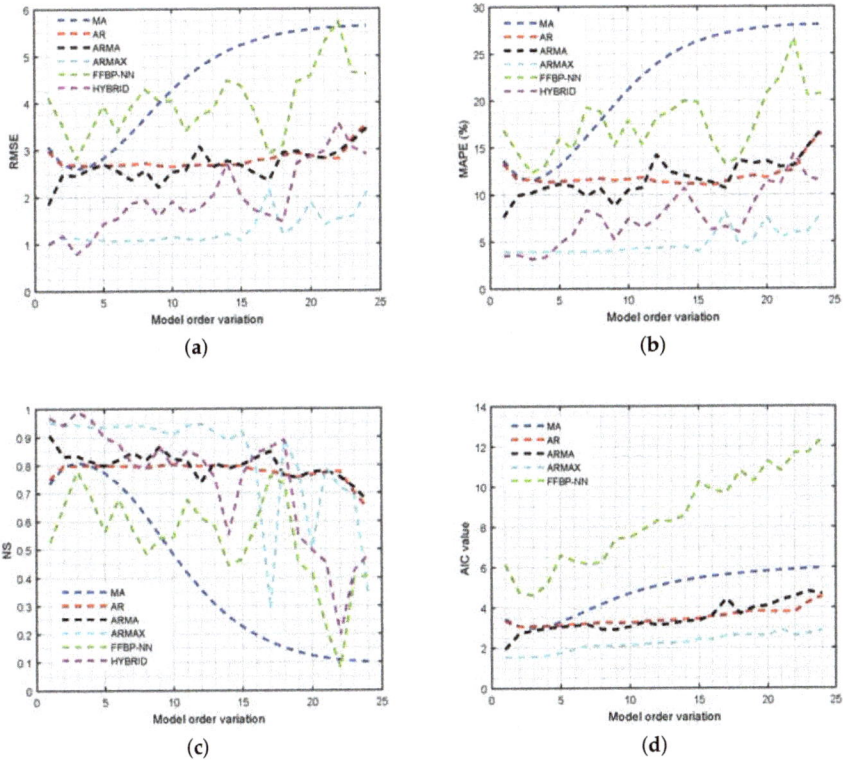

Figure 5. Comparative assessments of the STWD forecasting models using (**a**) RMSE; (**b**) MAPE; and (**c**) NS. (**d**) Estimated quality of AR, MA, ARMA, ARMAX, and FFBP-NN using AIC value.

$$MSE = \frac{1}{N}\sum_{t=1}^{N}(Y_t - \hat{Y}_t)^2$$
$$RMSE = \sqrt{MSE} \tag{7}$$

$$MAPE = \frac{100}{N}\sum_{t=1}^{N}\left|\frac{Y_t - \hat{Y}_t}{Y_t}\right| \tag{8}$$

$$NS = 1 - \frac{\sum\limits_{t=1}^{N}(Y_t - \hat{Y}_t)^2}{\sum\limits_{t=1}^{N}(Y_t - \mu_{Y_t})^2} \tag{9}$$

$$AIC = N\log(\frac{RSS}{N}) + 2k \tag{10}$$

where Y_t is the real observation, $\hat{Y}_t$ is the forecast value at time t, and μ_{Y_t} is the mean of real observation. RSS is the estimated residual of fitted model, and k is the number of estimated parameters in the model.

4. Recommendations of STWD Forecasting Models and Future Work

As regards the selection of accurate and appropriate forecasting models for STWD prediction, this section of the paper presents recommendations and future work based on the forecasts generated by AR, MA, ARMA, ARMAX, FFBP-NN, and hybrid models.

Concerning UTS forecasting models (i.e., AR, MA, and ARMA), the results obtained in Figures 2 and 3a,b show that ARMA is the best predictive model. It is useful for STWD operational forecasts to minimise the operating cost of pumping stations [1,6,15,16,18]. However, as regards influencing future water demand, a major criticism of UTS predictive models is their failure to account for the effects of changing exogenous variables [11,18]. In reference to UTS models, TSR models (i.e., ARMAX) is preferred since it offers a straightforward framework for quantifying the effects of exogenous variables (e.g., weather data, demographics) [11,19,24–26]. Figure 3d shows that the forecast generated by ARMAX is useful for better prediction of daily water demand and for setting water rates.

It is discussed in the scientific literature that ANN models (i.e., FFBP-NN) are designed to detect complex nonlinear relationships that may be harder to summarise. In addition, it is also discussed that it is useful for a better prediction of peak daily water demand to inform optimal operating policy as well as pumping and maintenance scheduling [1,5,24,26–29,35]. Nonetheless, it requires greater computational resources than most STWD forecasting methods [11]. Compared with AR, MA, ARMA, ARMAX, and hybrid model, the results obtained show that the forecasting performance of FFBP-NN was the least [24,25]. However, by combining the forecasts generated by FFBP-NN and ARMA, the result obtained in Figure 4d shows that the best forecasting performance was obtained. This shows that if ARMAX and FFBP-NN are used to generate a hybrid forecast, a better forecast compared to the combination of ARMA and FFBP-NN will be obtained. Hybrid forecasting is necessary for operational purposes because it is useful for real-time near-optimal control of WDS [11,33–38].

This study shows that UTS models (i.e., ARMA), TSR models (i.e., ARMAX), and hybrid model (combined forecast from two or more models such as ARMA and FFBP-NN) may be considered as the accurate and appropriate models for STWD prediction. However, these models are not applicable in more general decision problem frameworks, since they cannot be used to understand and analyse the overall level of uncertainty in future demand forecasts. Therefore, much more attention needs to be given to probabilistic forecasting methods for STWD prediction, since such best single valued forecasts obtained by hybrid model do not guarantee reliable and robust decisions, which can only be obtained via Bayesian Decision approaches requiring the estimation of the full predictive density [11,15,41–47]. Furthermore, given that the main objective of WDS management is to guarantee short-term user's demand, alternative approaches to predicting a future expected value as described in this paper will be analysed in the future. These approaches [15,42], based on the Bayesian maximisation of an "expected utility function", require forecasting the entire predictive density instead of the sole expected value, and can guarantee more reliable and robust decisions.

5. Conclusions

The main objective of WDS management is to guarantee short-term user demand, which implies making real-time rational decisions based on the best available information on future user demand. Deterministic forecasts such as the ones described in this paper are insufficient to provide the predictive probability distribution of future demand, conditional upon models' forecasts, which can be regarded as the maximum information to be used in any educated decision making process.

The selection of an accurate and appropriate STWD forecasting model is useful for the successive assessment of such predictive probability distribution. As a result, this paper overviews the forecasting methods and models for STWD prediction, assesses the the forecasting performances of AR, MA, ARMA, ARMAX, FFBP-NN, and hybrid model from the different methods overviewed, and provides recommendations and future work for the forecasts generated by these predictive models.

Furthermore, the forecasts generated by AR, MA, ARMA, ARMAX, FFBP-NN, and hybrid model (i.e., combined forecast using ARMA and FFBP-NN) have been compared with each other for a common set of data. AIC is used to estimate the quality of each model and forecasting statistical terms; namely, RMSE, MAPE, and NS model efficiency coefficient are used to assess the predictive performance of these models. The comparative assessment of the forecasting models show that ARMA, ARMAX, and the hybrid model may be considered as the best conditioning candidates for the assessment of the predictive probability distribution of future demands.

In a successive paper, we will show how to derive the above-mentioned predictive probability distribution conditional on one or more predictive models as the fundamental tool for estimating expected benefits (or expected losses) to be maximised (or minimised), within a Bayesian decision making framework.

Acknowledgments: The authors would like to thank the Council for Scientific and Industrial Research, South Africa and the Tshwane University of Technology, Pretoria, South Africa for their financial supports.

Author Contributions: Amos O. Anele, Yskandar Hamam and Adnan M. Abu-Mahfouz conceived and designed the experiments; Amos O. Anele developed the algorithm in MATLAB, performed the simulation experiments and analysed the results obtained; Ezio Todini contributed to the further analysis and interpretation of the results. The manuscript was written by Amos O. Anele with contribution from all co-authors.

Conflicts of Interest: The authors declare no conflict of interest.

References

1. Herrera, M.; Torgo, L.; Izquierdo, J.; Pérez-García, R. Predictive models for forecasting hourly urban water demand. *J. Hydrol.* **2010**, *387*, 141–150.
2. Gharun, M.; Azmi, M.; Adams, M.A. Short-term forecasting of water yield from forested catchments after Bushfire: A case study from southeast Australia. *Water* **2015**, *7*, 599–614.
3. Abu-Mahfouz, A.M.; Hamam, Y.; Page, P.R.; Djouani, K.; Kurien, A. Real-time dynamic hydraulic model for potable water loss reduction. *Procedia Eng.* **2016**, *154*, 99–106.
4. Pacchin, E.; Alvisi, S.; Franchini, M. A Short-Term Water Demand Forecasting Model Using a Moving Window on Previously Observed Data. *Water* **2017**, *9*, 172.

5. Jain, A.; Ormsbee, L.E. Short-term water demand forecast modeling techniques—Conventional methods versus AI. *J. Am. Water Works Assoc.* **2002**, *94*, 64–72.

6. Alvisi, S.; Franchini, M.; Marinelli, A. A short-term, pattern-based model for water-demand forecasting. *J. Hydroinf.* **2007**, *9*, 39–50.

7. Khatri, K.; Vairavamoorthy, K. Water Demand Forecasting for the City of the Future Against the Uncertainties and the Global Change Pressures: Case of Birmingham. In Proceedings of the World Environmental and Water Resources Congress, Kansas, MO, USA, 17–21 May 2009; pp. 17–21.

8. Hamam, Y.M.; Hindi, K.S. Optimised on-Line Leakage Minimisation in Water Piping Networks Using Neural Nets. In Proceedings of the IFIP Working Conference, Dagschul, Germany, 28 September–1 October 1992; pp. 57–64.

9. Hindi, K.; Hamam, Y. Locating pressure control elements for leakage minimization in water supply networks: An optimization model. *Eng. Opt.* **1991**, *17*, 281–291.

10. Hindi, K.; Hamam, Y. Pressure control for leakage minimization in water supply networks Part 1: Single period models. *Int. J. Syst. Sci.* **1991**, *22*, 1573–1585.

11. Donkor, E.A.; Mazzuchi, T.A.; Soyer, R.; Alan Roberson, J. Urban water demand forecasting: Review of methods and models. *J. Water Resour. Plan. Manag.* **2012**, *140*, 146–159.

12. Veiga, V.B.; Hassan, Q.K.; He, J. Development of Flow Forecasting Models in the Bow River at Calgary, Alberta, Canada. *Water* **2014**, *7*, 99–115.

13. Arampatzis, G.; Perdikeas, N.; Kampragou, E.; Scaloubakas, P.; Assimacopoulos, D. A Water Demand Forecasting Methodology for Supporting Day-to-Day Management of Water Distribution Systems. In Proceedings of the 12th International Conference "Protection & Restoration of the Environment", Skiathos, Greece, 29 June–3 July 2014.

14. Amponsah, S.; Otoo, D.; Todoko, C. Time series analysis of water consumption in the Hohoe municipality of the Volta region, Ghana. *Int. J. Appl. Math. Res.* **2015**, *4*, 393–403.

15. Alvisi, S.; Franchini, M. Assessment of predictive uncertainty within the framework of water demand forecasting using the Model Conditional Processor (MCP). *Urban Water J.* **2015**, *14*, 1–10.

16. Gagliardi, F.; Alvisi, S.; Kapelan, Z.; Franchini, M. A Probabilistic Short-Term Water Demand Forecasting Model Based on the Markov Chain. *Water* **2017**, *9*, 507.

17. Box, G.; Jenkins, G. *Time Series Analysis: Forecasting and Control*; Holden-Day: San Franciso, CA, USA, 1970.

18. Billings, R.B.; Jones, C.V. *Forecasting Urban Water Demand*; American Water Works Association: Denver, CO, USA, 2011.

19. Polebitski, A.S.; Palmer, R.N.; Waddell, P. Evaluating water demands under climate change and transitions in the urban environment. *J. Water Resour. Plan. Manag.* **2010**, *137*, 249–257.

20. Qi, G.; Hamam, Y.; Van Wyk, B.J.; Du, S. Model-free Prediction based on Tracking Theory and Newton Form of Polynomial. *World Acad. Sci. Eng. Technol.* **2011**, *5*, 882–889.

21. Candelieri, A. Clustering and Support Vector Regression for Water Demand Forecasting and Anomaly Detection. *Water* **2017**, *9*, 224.

22. Rosenblatt, F. The perceptron: A probabilistic model for information storage and organization in the brain. *Psychol. Rev.* **1958**, *65*, 386.

23. Hindi, K.; Hamam, Y. Locating pressure control elements for leakage minimisation in water supply networks by genetic algorithms. In *Artificial Neural Nets and Genetic Algorithms*, Springer: Vienna, Austria, 1993; pp. 583–587.

24. Jentgen, L.; Kidder, H.; Hill, R.; Conrad, S. Energy management strategies use short-term water consumption forecasting to minimize cost of pumping operations. *J. Am. Water Works Assoc.* **2007**, *99*, 86.

25. Cutore, P.; Campisano, A.; Kapelan, Z.; Modica, C.; Savic, D. Probabilistic prediction of urban water consumption using the SCEM-UA algorithm. *Urban Water J.* **2008**, *5*, 125–132.

26. Adamowski, J.; Karapataki, C. Comparison of multivariate regression and artificial neural networks for peak urban water-demand forecasting: evaluation of different ANN learning algorithms. *J. Hydrol. Eng.* **2010**, *15*, 729–743.

27. Ghiassi, M.; Zimbra, D.K.; Saidane, H. Urban water demand forecasting with a dynamic artificial neural network model. *J. Water Resour. Plan. Manag.* **2008**, *134*, 138–146.

28. Jain, A.; Varshney, A.K.; Joshi, U.C. Short-term water demand forecast modelling at IIT Kanpur using artificial neural networks. *Water Resour. Manag.* **2001**, *15*, 299–321.

29. Bougadis, J.; Adamowski, K.; Diduch, R. Short-term municipal water demand forecasting. *Hydrol. Process.* **2005**, *19*, 137–148.

30. Hamam, Y.; Brameller, A. Hybrid method for the solution of piping networks. *Proc. Inst. Electr. Eng. IET* **1971**, *118*, 1607–1612.

31. Zhou, S.L.; McMahon, T.A.; Walton, A.; Lewis, J. Forecasting daily urban water demand: A case study of Melbourne. *J. Hydrol.* **2000**, *236*, 153–164.

32. Gato, S.; Jayasuriya, N.; Roberts, P. Temperature and rainfall thresholds for base use urban water demand modelling. *J. Hydrol.* **2007**, *337*, 364–376.

33. Wang, X.; Sun, Y.; Song, L.; Mei, C. An eco-environmental water demand based model for optimising water resources using hybrid genetic simulated annealing algorithms. Part I. Model development. *J. Environ. Manag.* **2009**, *90*, 2628–2635.

34. Brentan, B.M.; Luvizotto, E., Jr.; Herrera, M.; Izquierdo, J.; Pérez-García, R. Hybrid regression model for near real-time urban water demand forecasting. *J. Comput. Appl. Math.* **2017**, *309*, 532–541.

35. Tiwari, M.; Adamowski, J.; Adamowski, K. Water demand forecasting using extreme learning machines. *J. Water Land Dev.* **2016**, *28*, 37–52.

36. Aly, A.H.; Wanakule, N. Short-term forecasting for urban water consumption. *J. Water Resour. Plan. Manag.* **2004**, *130*, 405–410.

37. Caiado, J. Performance of combined double seasonal univariate time series models for forecasting water demand. *J. Hydrol. Eng.* **2009**, *15*, 215–222.

38. Altunkaynak, A.; Özger, M.; Çakmakci, M. Water consumption prediction of Istanbul city by using fuzzy logic approach. *Water Resour. Manag.* **2005**, *19*, 641–654.

39. Nash, J.E.; Sutcliffe, J.V. River flow forecasting through conceptual models part I—A discussion of principles. *J. Hydrol.* **1970**, *10*, 282–290.

40. Akaike, H. A new look at the statistical model identification. *IEEE Trans. Autom. Control* **1974**, *19*, 716–723.

41. Todini, E. Using phase-state modelling for inferring forecasting uncertainty in nonlinear stochastic decision schemes. *J. Hydroinf.* **1999**, *1*, 75–82.

42. Todini, E. A model conditional processor to assess predictive uncertainty in flood forecasting. *Int. J. River Basin Manag.* **2008**, *6*, 123–137.

43. Anele, A.; Hamam, Y.; Todini, E.; Abu-Mahfouz, A. Predictive uncertainty estimation in water demand forecasting using the model conditional processor. *Water Resour. Manag.* **2017**, under review.

44. Coccia, G.; Todini, E. Recent developments in predictive uncertainty assessment based on the model conditional processor approach. *Hydrol. Earth Syst. Sci.* **2011**, *15*, 3253–3274.

45. Todini, E. The role of predictive uncertainty in the operational management of reservoirs. In Proceedings of the ICWRS2014—Evolving Water Resources Systems: Understanding, Predicting and Managing Water-Society Interactions, Bologna, Italy, 16 September 2014; pp. 4–6.

46. Barbetta, S.; Coccia, G.; Moramarco, T.; Todini, E. Case study: A real-time flood forecasting system with predictive uncertainty estimation for the Godavari River, India. *Water* **2016**, *8*, 463.

47. Reggiani, P.; Coccia, G.; Mukhopadhyay, B. Predictive Uncertainty Estimation on a Precipitation and Temperature Reanalysis Ensemble for Shigar Basin, Central Karakoram. *Water* **2016**, *8*, 263.

![water logo] *water*

MDPI

Article

A Probabilistic Short-Term Water Demand Forecasting Model Based on the Markov Chain

Francesca Gagliardi [1,*] [iD], Stefano Alvisi [1] [iD], Zoran Kapelan [2] and Marco Franchini [1] [iD]

[1] Department of Engineering, University of Ferrara, Via Saragat, Ferrara 44122, Italy;
     stefano.alvisi@unife.it (S.A.); marco.franchini@unife.it (M.F.)
[2] Centre for Water Systems, University of Exeter, Harrison Building, North Park Road, Exeter EX4 4QF, UK;
     Z.Kapelan@exeter.ac.uk
*   Correspondence: francesca.gagliardi@unife.it; Tel.: +39-053-297-4932

Received: 15 May 2017; Accepted: 7 July 2017; Published: 12 July 2017

Abstract: This paper proposes a short-term water demand forecasting method based on the use of the Markov chain. This method provides estimates of future demands by calculating probabilities that the future demand value will fall within pre-assigned intervals covering the expected total variability. More specifically, two models based on homogeneous and non-homogeneous Markov chains were developed and presented. These models, together with two benchmark models (based on artificial neural network and naïve methods), were applied to three real-life case studies for the purpose of forecasting the respective water demands from 1 to 24 h ahead. The results obtained show that the model based on a homogeneous Markov chain provides more accurate short-term forecasts than the one based on a non-homogeneous Markov chain, which is in line with the artificial neural network model. Both Markov chain models enable probabilistic information regarding the stochastic demand forecast to be easily obtained.

Keywords: water demand; forecasting; Markov chain; stochastic

1. Introduction

Water distribution systems fulfil the fundamental task of satisfying the water demand of users, and their long-term and short-term management can be supported by the use of water demand forecasting models. In fact, for the purpose of activities related to the design, maintenance and upgrading of water supply networks (e.g., expansion of distribution systems or replacement of parts of networks), which entail long-term planning, it is fundamental to have an accurate estimate of monthly or yearly demand in the years to come, that is, over the useful lifetime of the installation. Analogously, forecasts of daily or hourly demands over limited time horizons (for example a week or the next 24 h) can provide useful information for planning short-term or real-time management of the devices at the service of water distribution systems, such as the planning of pumping station operations, the control of valves, etc.

Depending on their features, water demand forecasting models can be divided into different categories. A first distinction can be made, in relation to the different practical objectives just mentioned, based on the forecasting frequency and the time horizon (i.e., the length of future time interval the forecast is intended to cover) (see [1]). In this work, the attention is focussed on short-term demand forecasting models, which generally predict hourly or sub-hourly demands over a time horizon normally ranging between 6 and 48 h [2–6].

A further distinction can be made in relation to the adopted modelling technique. Various water demand forecasting models based on data-driven techniques, in particular artificial neural networks (ANN), have recently been proposed [3,7–14]. A different approach to demand forecasting is based on the representation of periodic behaviours that typically characterise water demands. It is in fact possible to recognise trends or patterns, which are generally influenced by seasonality, climate conditions and the types and habits of the users served by the system. The structure of various water demand forecasting models is based precisely on the description of such behaviours [4], possibly in conjunction with techniques of time series analysis such as autoregressive processes [2], or Markov processes [15]. Most of the above techniques, especially the forecasting models based on ANN and similar methods, fall into the category of black box models (as opposed to white box models, such as physically-based models). Whilst accurate, black box models are simply not transparent in terms of mapping inputs (past demands and different explanatory factors) onto outputs (forecasted demands). The water demand forecasting model proposed here falls into the category of grey box-type models, i.e., with limited but better transparency when compared with black box models.

Finally, it is possible to formulate a further distinction between forecasting models based on the type of results they furnish. Water demand forecasting is characterised by a certain degree of uncertainty, due to the natural variability of the water consumption. Quantification/characterisation of this natural variability is very important in the planning or management of water distribution systems [16]. The models can therefore be classified into deterministic models, which provide a deterministic estimate of future water demands, and models that also provide an estimate of the stochastic behaviour of the demand forecast. Most models proposed in the scientific literature, including the ones mentioned thus far, belong to the former category, whereas the latter category embraces less recently proposed models, such as the Bayesian-based models [17–22], the approach proposed by Alvisi and Franchini [16] based on the use of the Model Conditional Processor [23], and the approach proposed by Cutore et al. [24], based on a neural network trained by means of the SCEM-UA algorithm [25]. All in all, stochastic models should be preferred to deterministic ones, due to additional information provided about the forecasted demands. However, existing stochastic forecasting models can be very computationally demanding, as they tend to use Monte Carlo simulations to assess the probability distribution of the future demand value and the related prediction bounds.

In this paper, a new approach for short-term water demand forecasting based on the statistical concept of the Markov chain (grey box modelling approach) is proposed. The model is capable of providing both a deterministic forecast of the future values of water demand, and a characterisation of the stochastic behaviour of the forecasted values (at reduced computational effort compared to most of the existing methods). The model, applied with the aim of estimating future hourly demands over a time horizon of 24 h relying solely on observed water demand data, is characterised by a simple structure, easy to comprehend and control.

The Markov chain technique is outlined in Section 2 below where, after giving the theoretical basis, the developed models are then illustrated. In Section 3, the case studies the models were applied to, and the benchmark models used by way of comparison, are described. After analysing and comparing the results obtained from the application of the Markov models and the benchmark models in Section 4, the final conclusions are presented in Section 5.

2. Markov Chain Based Demand Forecasting

2.1. Overview

The proposed approach is based on the statistical concept of the Markov Chain (MC) and aims to forecast short-term water demand while characterising the demand's stochastic behaviour. In the scientific literature, various examples of forecasting models applying this concept can be found, both in the context of water demands [15], and in other engineering contexts such as in the prediction of the performance of bridge decks [26], traffic flow forecasting [27], wind power forecasting [28] and streamflow forecasting for the prediction of flood events [29]. In particular, in the context of water demand, proposed by Shvarster et al. [15] propose an hourly water demand forecasting model in which each day is divided into three parts, referred to as "rising", "oscillating", and "falling" segments.

Within each segment, the water demand is modelled with a third-order autoregressive process. The transition from one segment to the next is modelled with a Markov process.

The approach proposed here, by contrast, exploits the Markov chain concept to characterise the probability that the demand in one or more successive times (for example, in the next hour or in the next 2, 3, ... hours) will fall into an assigned interval, where the interval at the current time step is known. In the sections that follow, the Markov chain concept is illustrated, and then its application to water demand forecasting is described.

2.2. The Markov Chain

A statistical process, i.e., the succession of a random variable $X(t)$, with $t \in T$, may be considered a *Markov process* if it exploits the *Markov theorem* [30]. Let us assume that the domain T of the variable t is the time, divided in discrete intervals Δt, and that the domain of existence of the variable $X(t)$ is known and divided into N intervals, each of which is defined as a *class* c_i (with $i = 1,.., N$) of the process (see Figure 1).

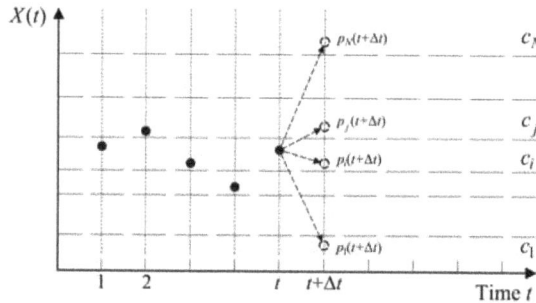

Figure 1. Reference diagram of a Markov process showing the N classes into which the domain of existence of the variable $X(t)$ is divided, and the probabilities referred to each class at the time $t + 1$.

The probability $p_i(t)$ that, at a generic time t, the process will be in a generic class c_i, can be defined as:

$$p_i(t) = \Pr[X(t) \in c_i],\tag{1}$$

The generic $p_i(t)$ represents the i-th component of the row-vector $\mathbf{p}(t) = [p_1(t), p_2(t), \dots, p_N(t)]$, which contains all the probabilities that the process is at the generic time t in each of the N classes. Of course, it results $\sum_{i=1}^{N} p_i(t) = 1$ for every t.

In passing from a class c_i at time t to a class c_j at the next time $t + \Delta t$, the process undergoes a *transition*, which is associated with a probability $\pi_{ij}(t)$, called *transition probability*. The probabilities associated with every possible transition from t to $t + \Delta t$ are the components making up the *transition matrix* $\mathbf{\Pi}(t) \in \square^{N \times N}$:

$$\mathbf{\Pi}(t) = \begin{bmatrix} \pi_{11}(t) & \cdots & \pi_{1i}(t) & \pi_{1j}(t) & \cdots & \pi_{1N}(t) \\ \cdots & \cdots & \cdots & \cdots & \cdots & \cdots \\ \pi_{i1}(t) & \cdots & \pi_{ii}(t) & \pi_{ij}(t) & \cdots & \pi_{iN}(t) \\ \pi_{j1}(t) & \cdots & \pi_{ji}(t) & \pi_{jj}(t) & \cdots & \pi_{jN}(t) \\ \cdots & \cdots & \cdots & \cdots & \cdots & \cdots \\ \pi_{N1}(t) & \cdots & \pi_{Ni}(t) & \pi_{Nj}(t) & \cdots & \pi_{NN}(t) \end{bmatrix},\tag{2}$$

where every row of the matrix $\mathbf{\Pi}(t)$ corresponds to the starting class of the process, and every column to the class of arrival: for instance, the probability $\pi_{ij}(t)$ of belonging in class c_j at time $t + \Delta t$ starting from class c_i at the preceding time t, is placed in the i-th row and j-th column of the matrix.

It should be noted that the transition matrix $\Pi(t)$ can vary at every step of the process, and this behaviour is generally indicated as a non-homogeneous Markov chain. A homogeneous Markov chain is based, by contrast, on the assumption that the transition probability is independent of t. This condition implies the existence of a single transition matrix, Π, which remains constant with variations in t and is characteristic of the entire process.

The transition matrix characterises the Markov process itself, since it quantifies the tendency of the process to move from one class to another in two successive times. With the transition matrix of a certain Markov process being known, the Markov chain theory allows the estimation of the probability that the process has to belong to each class one time ahead of the current one (for example in $t + \Delta t$ when t is the current time). In particular, in a real-time forecasting framework, with t being the current time, $\mathbf{p}(t)$ the corresponding probability vector of the process and $\Pi(t)$ the correspondent transition matrix, the probability vector of the process at time $t + \Delta t$, $\mathbf{p}^{for}(t+\Delta t)$, can be estimated as:

$$\mathbf{p}^{for}(t + \Delta t) = \mathbf{p}(t) \times \Pi(t), \tag{3}$$

This estimation is displayed in Figure 2a showing, at the time step centered in $t + \Delta t$, each class coloured with a shade of grey, correspondent to the calculated probability. More in general, by triggering a Markov chain and exploiting, for every time lag k following the first, the forecast made for the preceding times, it is possible to extend the time horizon to $k\Delta t$ ahead (with $k > 1$), i.e.,:

$$\mathbf{p}^{for}(t + k\Delta t) = \mathbf{p}^{for}(t + (k - 1)\Delta t) \times \Pi(t + (k - 1)\Delta t) \text{ with } k > 1, \tag{4}$$

What has been illustrated thus far can be used to obtain probability vectors at every time $t + k\Delta t$, one for every time lag k (with $k = 1, \ldots , K$) (see Figure 2b); these vectors characterise, for every time lag, *the probability that the process will fall into each of the N classes.*

The proposed Markov chain method exploits this concept in order to provide stochastic and deterministic forecasts of water demand, as detailed in the next section.

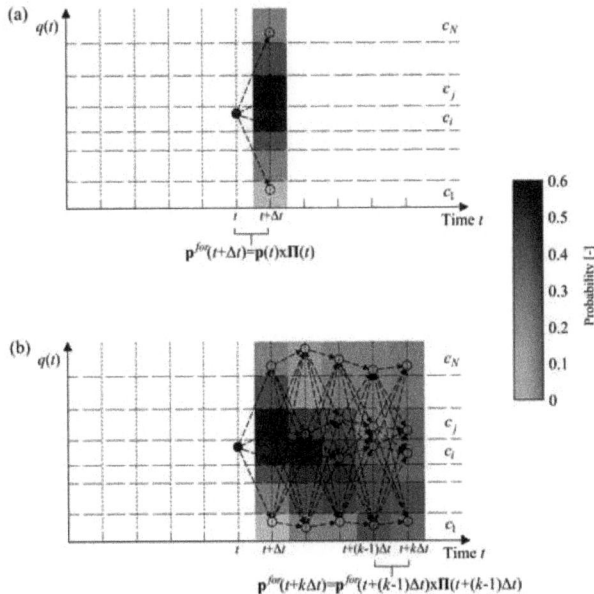

Figure 2. Estimated posterior probability vector $\mathbf{p}^{for}$ (**a**) for one step ahead and (**b**) for k steps ahead, highlighted using a shade of grey for each class.

2.3. Demand Forecasting Model

The proposed water demand forecasting model is based on the assumption that the trend in water demand can be defined as a Markov process. Generally speaking, it can be assumed that the random variable of the process at the time t can be identified with the mean water demand $q(t)$ in the generic time interval Δt (for example, $\Delta t = 1$ h). Moreover, it is possible to identify N classes $c_1, c_2, \ldots, c_N$ into which the entire range of variability of the water demand can be divided.

If the state, i.e., the class c_i, of the water demand at the current time t and the transition matrix referred to t are known, the Markov chain allows us to define what its state in the future Δt will be in probabilistic terms. In fact, Equation (3) can be used to estimate the probability vector $\mathbf{p}^{for}(t + \Delta t)$, where $\mathbf{p}(t)$ is referred to a real observed value, being in a context of real-time application of the model, thus composed of $N - 1$ null values and a value of 1 correspondent to the class the demand $q(t)$ belongs to. As regards the transition matrix, it represents a parameter of the model that can be estimated on the basis of the observed water demands used in the model calibration phase, as detailed below. Since it is a calibrated variable, it will be henceforth indicated as $\mathbf{\hat{\Pi}}(t)$. The forecast can be tied up to $k\Delta t$ ahead using Equation (4), thus estimating the probabilities of the demand to fall within each class at the time $t + k\Delta t$, $\mathbf{p}^{for}(t + k\Delta t)$, using the estimate made one time earlier and the correspondent transition matrix.

This information can also be used to obtain a deterministic forecast of future water demand $q^{for}(t + k\Delta t)$ at a generic time $t + k\Delta t$ in the following manner [28].

A weighted average of the N central values of the classes c_i (with $i = 1, \ldots, N$), represented in the vector $\mathbf{m} = [m_1, m_2, \ldots, m_N]$, is computed using the components $p_i^{for}(t + k\Delta t)$ (with $i = 1, .., N$) of the probability vector predicted for the time $t + k\Delta t$ as weights:

$$q^{for}(t + k\Delta t) = \sum_{i=1}^{N} m_i \cdot p_i^{for}(t + k\Delta t), \tag{5}$$

At this point it is important to set forth some considerations regarding the advisability of adopting a non-homogeneous or homogeneous Markov chain to predict the vector $\mathbf{p}^{for}(t + k\Delta t)$ in the case of short-term water demands (for example, to obtain hourly water demand forecasts for the next $K = 24$ h). Water demands are generally characterised by periodic patterns, present on different time scales. Considering, for example, the hourly water demands over the course of a day, it is possible to observe that they follow a trend or pattern that tends to reflect the type and the habits of the users served. In the case of residential users, the demand trends show morning and evening peaks, reduced demand during the night and variable demand in the afternoon hours; water use may also differ depending on whether it is a weekday or a holiday [2]. In the trends in demand over time, it is thus possible to distinguish different phases—rising, falling, etc.—characterised by a probability of demand transitioning from one class to another, which will clearly vary from phase to phase. As we are dealing with a time series characterised by periodic patterns, it would seem appropriate to use an approach based on a non-homogeneous Markov chain. On the other hand, prior to the application of the model, the demand time series could be suitably normalised (brought to a mean of zero and unit variance), thus creating the conditions for the use of an approach based on a homogeneous Markov chain.

Two different formulations of the model based respectively on the use of a non-homogeneous Markov chain (NHMC) and a homogeneous Markov chain (HMC), were developed on the basis of these considerations as detailed below.

2.3.1. Non-Homogeneous Markov Chain (NHMC) Model

In this first case, it is assumed to work directly on the water demand time series $q(t)$ and, given their periodic nature, to adopt a non-homogeneous Markov chain in order to take into account that the transition probability (and hence the transition matrix) varies as a function of time. In fact, given that, as mentioned earlier, the demand patterns generally show rising and falling phases during the day, each of them can be characterised by a different behaviour of the Markov process, in terms of transitions between classes. In the rising phases, for example, it should be more likely to see demands transition, in two consecutive instances, toward a higher class than the starting one, or at least remain in the same class. There is a reduced probability of observing demands transition to a lower class than the starting one. The opposite applies for the falling phases.

Assuming, moreover, that each of the phases making up the demand pattern over the course of the day is characterised by a single transition matrix, it is thus necessary to identify the F time ranges $f_1, f_2, \dots, f_F$, corresponding to the different rising and falling phases of the pattern. Incidentally, the assumption that each phase is characterised by a single transition matrix implies that the process is described through a sort of sequence of different homogeneous Markov processes.

From an operational viewpoint, the F time phases can be identified by making reference to the average trend in demand over the course of a day, possibly distinguishing between working and non-working days. Therefore, for the various phases $f_1, f_2, \dots, f_F$, the corresponding transition matrices $\hat{\Pi}_{f_1}, \hat{\Pi}_{f_2}, \dots, \hat{\Pi}_{f_F}$ can be estimated, using the observed calibration data. An estimation of the generic component $\hat{\pi}_{f_w, ij}$ (with $w = 1, \dots, F$ and $i, j = 1, \dots, N$) of the transition matrix $\hat{\Pi}_{f_w}$ is made, during the model calibration phase, by counting the transitions from c_i to c_j (with $i, j = 1, \dots, N$) between successive pairs of times, for which the starting time belongs to the phase f_w, and then dividing by the total transitions for which the starting time is inside the phase f_w, and which have the class c_i as the starting class, i.e.:

$$\hat{\pi}_{f_w, ij} = \frac{n_{f_w, ij}}{\sum\limits_{j=1}^{N} n_{f_w, ij}} \text{ with } i, j = 1, \dots, N, \ w = 1, \dots, F, \tag{6}$$

where $n_{f_w, ij}$ indicates the number of transitions from class c_i to class c_j in the consecutive times for which the starting time is inside the phase f_w. It is necessary to highlight that as the number of F phases increases, the accuracy of the estimate of the transition matrices will tend to decrease, because the number of data available for the purpose of the estimate decreases. Therefore, the number of the F phases adopted should not depend only on the trend in demand, but should also take into account the number of observed data available for calibrating the model.

During the actual application of the model, the forecast at a time lag 1, from the generic time t, to the time $t + \Delta t$ is made using the transition matrix $\hat{\Pi}_{f_w}$ associated with the phase f_w, in which the starting time t falls, i.e.:

$$\mathbf{p}^{for}(t + \Delta t) = \mathbf{p}(t) \times \hat{\Pi}_{f_w} \text{ with } t \in f_w, \tag{7}$$

while for time lags larger than one, the model is based on the following equation:

$$\mathbf{p}^{for}(t + k\Delta t) = \mathbf{p}^{for}(t + (k-1)\Delta t) \times \hat{\Pi}_{f_w} \text{ with } (t + (k-1)\Delta t) \in f_w, \tag{8}$$

In practical terms, therefore, in the event that the set of $k\Delta t$ data forecasted by the model at a generic time t straddles two different phases, the transition matrix used to estimate the vector $\mathbf{p}^{for}$ at the different time lags will change. The transition matrix used for each forecast will be "moving" with the forecast, instead of being fixed and equal to the one correspondent to t (i.e., the start time of the forecast). Thus, for every time lag k, the water demand of the generic time $t + k\Delta t$, which is based on the forecast made one time earlier $t + (k-1)\Delta t$, will be estimated basing on the transition matrix associated with the phase the time $t + (k-1)\Delta t$ belongs to (rather than being based on the one correspondent to the time t).

2.3.2. Homogeneous Markov Chain (HMC) Model

The second formulation of the Markov demand forecasting model is based on the application of a homogeneous Markov chain to a duly transformed demand time series. It may be noted, in fact, that in the case of a homogeneous Markov chain, the probability of a transition between classes in two successive instances is time-independent, and the estimate of the probability vector associated with the time $t + \Delta t$ made at the generic time t will always use the same transition matrix, irrespective of the time of the day at which the time t occurs. Thus, since, as previously observed, water demand time series are affected by significant periodicities, they are transformed through a normalisation process prior to the application of the HMC. Indeed, as shown in the numerical application, this transformation is likely to substantially reduce periodicities. More in detail, if we assume, for example, a time step $\Delta t = 1$ h, taking into account the daily periodicity of the data and distinguishing working days from non-working days, the original demand data are normalised in the following manner:

$$q^{norm}(t) = \frac{q(t) - \mu^h_{work/non\_work}}{\sigma^h_{work/non\_work}} \text{ with } h = 1, .., 24, \tag{9}$$

where $q(t)$ is the original generic water demand at time t, $q^{norm}(t)$ is the corresponding normalised value, and μ^h_{work/non_work} and σ^h_{work/non_work}, respectively, are the mean and the standard deviation of the data observed in the calibration phase in the h-th hour of the day, corresponding to the time t in which the original data $q(t)$ occurs, a distinction being made between the data related to working days (*work*) and non-working (*non_work*).

On the basis of the normalised data, the N normalised classes c_i^{norm} (with $i = 1, \ldots, N$) are then defined and the (only) transition matrix is estimated using the same approach as previously described for the NHMC model (see Equation (6)). However, the data are in no way dependent on time in this case, which requires counting the transitions n_{ij} from c_i^{norm} to c_j^{norm} (with $i,j = 1, \ldots, N$) between pairs of successive times within the entire calibration dataset, and dividing by the total number of transitions that have class c_i^{norm} as the starting class. The transition matrix $\hat{\Pi}$ thus estimated is used to estimate the probability that the normalised future water demand falls in each of the normalised classes by using the same approach as previously described for the NHMC model (see Equations (7) and (8)). However, in this case, the transition matrix does not change in time. Clearly, in this case, the vector $\mathbf{p}^{for}(t + k\Delta t)$ (with $k = 1, \ldots, 24$) provides an estimate of the probability that the normalised water demand will fall into each of the normalised classes c_i^{norm} (with $i = 1, \ldots, N$). This information must then be brought back to the original space by de-normalising the values at the ends of the classes using the mean μ^h_{work/non_work} and standard deviation σ^h_{work/non_work} previously defined at the time of normalisation, and relating to the h-th hour of the day (working or non-working) corresponding to the time $t + k\Delta t$ considered. For example, with cl_i^{norm} and cu_i^{norm} representing, respectively, the lower and upper ends of the i-th normalised class c_i^{norm}, the corresponding de-normalised lower and upper ends cl_i and cu_i are given by:

$$cl_i = \mu^h_{work/non\_work} + cl_i^{norm} \cdot \sigma^h_{work/non\_work} \text{ with } i = 1, \ldots, N, h = 1, \ldots, 24, \tag{10}$$

$$cu_i = \mu^h_{work/non\_work} + cu_i^{norm} \cdot \sigma^h_{work/non\_work} \text{ with } i = 1, \ldots, N, h = 1, \ldots, 24, \tag{11}$$

where h is the hour of the day corresponding to the time considered on each occasion and the type of day in which that time occurs is taken into account.

Incidentally, it is worth observing that the definition of classes in the normalised space, and the subsequent de-normalisation performed taking into account the hour and the type of day in which the data occurs, means that once de-normalised, the generic class c_i—which in the normalised space is the sole class and independent of the time considered—will have a width varying according to the hour and type of day in which the considered time occurs. In particular, the width will increase as the standard deviation σ^h_{work/non_work} increases. Therefore, for example, the classes corresponding to the hours of peak demand (for example, 7 in the morning), which are characterised by a high variability in water use, will be much wider than those corresponding to night-time hours, which are typically characterised by low variability.

3. Case Studies

The proposed models were applied to the observed water demand data relating to three district metering areas (DMAs) situated in Harrogate and Dales area of Yorkshire (UK), which have already been used in the past to assess water demand forecasting methods [17]. In all three cases, a time step $\Delta t = 1$ hour and a forecasting time horizon of $K = 24$ h were assumed. The periods of observation of water demands in the three DMAs were the following:

- DMA1: from 24 March 2011 to 19 December 2011 (270 days)
- DMA2: from 4 May 2011 a 31 October 2011 (181 days)
- DMA3: from 11 April 2011 a 20 November 2011 (224 days)

The number of users belonging to each metering area becomes progressively lower as we go from DMA1 to DMA3, as confirmed by the calculation of the mean flow rates: 26.7 L/s for DMA1, 24.6 L/s for DMA2 and 6.6 L/s for DMA3.

The two proposed models NHMC and HMC require calibration for the purpose of defining the classes and time phases (in the case of the non-homogeneous model) and estimating the transition matrix or matrices. Each data time series was thus divided into two sets: one for calibrating the parameters, made up of the first 80% of the data in the series, and one for validating the model, made up of the remaining 20% of the data.

The number N of classes of hourly flow rates was fixed equal to four for both the NHMC and HMC models. The classes were identified on the basis of the data in the calibration set (original or normalised, respectively, for models NHMC and HMC) in the following way. Once the minimum, mean and maximum values of the dataset were identified, two macro classes were defined; they were delimited, respectively, by the minimum and mean values of the data and by the mean and maximum values of the data. The mean values of the data belonging to each of the two macro classes were then calculated so as to identify two classes of values within each macro class and thereby obtain four classes altogether. Indeed, other approaches based on use of the median rather than mean or data-driven approaches, such as clustering techniques, could be used to identify the classes.

For the purpose of applying the NHMC model, it was necessary to identify the phases into which the generic day can be divided in order to take into account the periodic patterns of hourly demand over the course of the day. Since there was a clear difference between working and non-working days in terms of the hourly water demand patterns, a distinction was made accordingly, which resulted in the identification of different time phases for the two types of days. In particular, these phases, corresponding to the rising and falling phases of demand over the 24 h period, were defined on the basis of the average trend shown by the data relating to the three DMAs, as identified during calibration of the model (see Figure 3).

For the purpose of applying the HMC model, data are normalised by using Equation (9). In particular, subtraction at each original water demand $q(t)$ of the mean of the data observed in the calibration phase in the corresponding hour, a distinction being made between working and non-working days, substantially reduces periodicity. This is supported by Figure 4a,c, which show a comparison between observed and normalised hourly demands, respectively, for a one-week period in DMA1. Furthermore, a fast Fourier transform (FFT) algorithm [31] was applied to compute the amplitude of sinusoidal components, as a function of the frequency, characterising the original (Figure 4b) and normalised (Figure 4d) time series. In order to make the results of the analysis comparable, both the time series were preliminary scaled to belong to the [0 1] interval. It can be observed that the original data show strong dominant frequencies at 1.157×10^{-5} and 2.135×10^{-5} Hz, corresponding to 24 and 48 h. Even though periodicities are not completely removed by normalisation, the power spectral densities of the normalised series are much more smoothed, and the dominant frequencies at 24 and 48 h are less evident, leading to time series whereto the HMC can effectively be applied, as shown in the subsequent analysis of the numerical results. Similar considerations apply to DMA2 and DMA3.

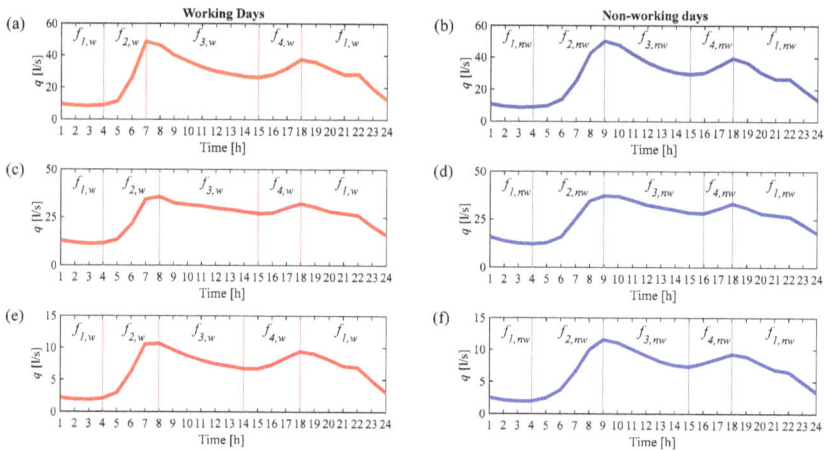

Figure 3. Average daily pattern of hourly demands during working (left-hand column, (**a,c,e**)) and non-working days (right-hand column, (**b,d,f**)) and initial and final ends of the time phases $f_1, f_2, f_3,$ and f_4 relating to DMA1 (row 1), DMA2 (row 2) and DMA3 (row 3).

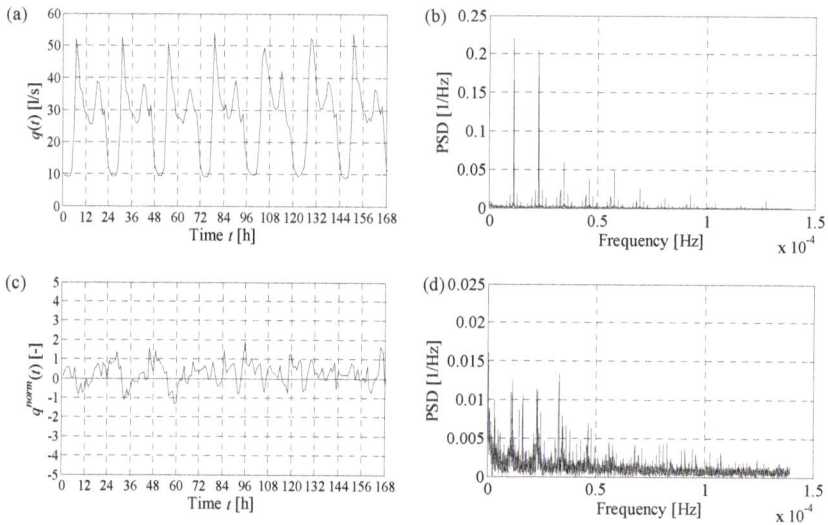

Figure 4. Comparison of (**a**) observed and (**c**) normalised hourly demands for a one-week period in DMA1 and the frequency analysis of the entire (**b**) observed and (**d**) normalised time series of DMA1.

Two benchmark models were applied by way of comparison to the same datasets, in order to assess the accuracy of the forecasts provided by the NHMC and HMC models. The first benchmark model adopted was a multi-layer perceptron artificial neural network (ANN) model structured in such a way as to provide water demand forecasts for the next $K = 24$ h. The network has the same structure of the one proposed by Alvisi and Franchini [16], whose inputs are the water demands observed in the past 24 h, normalised in the same manner seen for the HMC model, scaled in the [0 1] interval, and assigned a binary code identifying the type of the day (working or non-working). The second benchmark model used is of the naïve type. The model has a decidedly simpler structure than an artificial neural network, and the specific type used in this study is referred to in the scientific literature as 'mean' naïve [32], since it is based on the mean trend in demand during the day. The forecast water demand at each time is assumed to be equal to the mean value of the corresponding hour of the day computed by using the calibration data set.

4. Results and Discussion

The performances of the NHMC and HMC models were first evaluated by considering the deterministic forecasts obtained by these two models (see Equation (5)), and then comparing these with the corresponding forecasts obtained by the ANN and naïve deterministic models. This is followed by the analysis of the additional information provided by the HMC model with regard to the stochastic behaviour of forecasted demands.

For the purpose of evaluating the accuracy of the deterministic forecasts provided by the different models over different time horizons, use was made of the Nash–Sutcliffe index (*NS*) [33], computed for each forecasting time horizon k comprised between 1 and 24:

$$NS(k) = 1 - \frac{\sum\limits_{t=1}^{nd} \left(q^{obs}(t) - q^{for}(t|t - k\Delta t)\right)^2}{\sum\limits_{t=1}^{nd} \left(q^{obs}(t) - \mu_{q^{obs}}\right)^2}, \tag{12}$$

where $q^{obs}(t)$ is the observed water demands at the time instant t, $\mu_{q^{obs}}$ is the mean value of the observed demands, $q^{for}(t|t - k\Delta t)$ is the forecasted flow rate $k\Delta t$ instances before t and, finally, nd is the number of data of the forecasted time series.

Figure 5 shows the results obtained when the models were applied to the three datasets (i.e., DMAs) during calibration (left-hand column) and validation (right-hand column) phases.

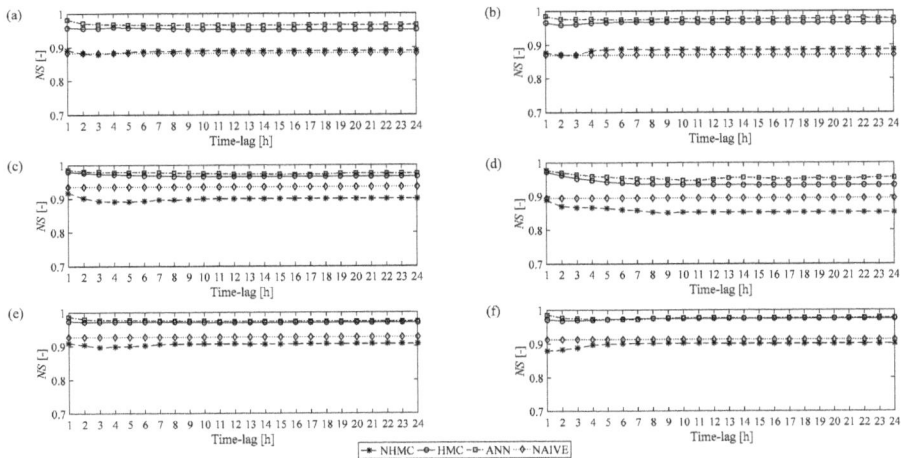

Figure 5. Nash–Sutcliffe index (*NS*) values obtained by the non-homogeneous Markov chain (NHMC), homogeneous Markov chain (HMC), artificial neural networks (ANN) and naïve models applied to the calibration (left-hand column, (**a**,**c**,**e**)) and validation (right-hand column, (**b**,**d**,**f**)) data for DMA1 (row 1), DMA2 (row 2) and DMA 3 (row 3).

With reference to DMA1, we can observe that in both the calibration (Figure 5a) and validation (Figure 5b) phases, the HMC and ANN models deliver higher predictive accuracy than the NHMC and naïve models. In the calibration phase, the *NS* indices calculated using the ANN model range from a maximum of 0.98 to a minimum of 0.97, and those calculated using the HMC model between a maximum of 0.96 and a minimum of 0.95. Moreover, both models appear to be stable, since the forecasting accuracy does not undergo appreciable decreases as the time horizon increases. The NHMC and naïve models, by contrast, provide a forecasting accuracy that is very similar, but distinctly worse than that of the HMC and ANN models, with values of NS ranging from 0.89 to 0.88 for NHMC and a value of 0.88 for the naïve model. Similar observations can be made with respect to the results obtained in the validation phase, again for DMA1 (Figure 5b): in the case of the HMC and ANN models, the values of the NS index remained very similar to the ones obtained in the calibration phase, whereas we observe a slight decline in the performance of the NHMC and naïve models. The performance of the models ANN and HMC was substantially the same also in the case of DMA2 and DMA3, as regards both calibration (Figure 5c,e) and validation (Figure 5d,f). For what concerns the NHMC model, it tends to lose effectiveness with respect to the naïve model. It can thus be observed that, insofar as deterministic water demand forecasting is concerned, the HMC model delivers a better predictive accuracy than the NHMC model, and an accuracy that is in line with that of the ANN model. It is worth highlighting that the outperformance of the HMC model with respect to the NHMC can mainly be due to the necessity to estimate several transition matrices for the non-homogeneous model, instead of a single transition matrix, such as for the homogeneous model. In fact, in the HMC model, all the calibration data are used for the estimation of the single transition matrix components, whereas in the NHMC, the calibration dataset is divided into eight sets (see Figure 3), one for each time phase and distinguishing working and non-working days. Thus, a lower number of data is used to calibrate each transition matrix, resulting in a less accurate estimate.

Given above, an analysis on the probabilistic results provided by the HMC model is performed in order to derive some considerations on its prediction capability, as the time horizon varies, to characterise the probability distribution of the water demand at time $t + k\Delta t$ when forecasted at time t. Figure 6 shows some examples of the results, in probabilistic terms, provided by the HMC model for DMA1. In particular, each graph shows the results provided by the model in relation to the forecasts made at a generic time for the next $K = 6$ h. The results are here shown with reference to a 6-h forecasting horizon, in order to make the graphs clearer and easier to comprehend and analyse. In fact, each graph contains a representation of the four classes (de-normalised) of water demand values for each of the 6 h of the forecasting time horizon; the background colour of each class, defined on a grey scale as shown in the legend, corresponds to the probability of the future value belonging to that class. The light colours, in particular, represent low probabilities that the future demand will belong to the class, whereas darker colours indicate high probabilities of them belonging to the class. Also shown are the observed and forecasted (deterministically) values for each hour of the time horizon.

For the purpose of interpreting the results, it is important to note first of all that the width of the classes does not depend on the time horizon k considered, but only on the hour of the day to which the forecast refers. In fact, in the case of the HMC model, as previously observed, the classes defined in the normalised space have a width, following de-normalisation, which depends on the standard deviation of the demand in different hours of the day. Let us consider, for example, Figure 6a, which shows the results related to the forecast made at 1 a.m. for the next 6 h, and thus the classes of de-normalised values for the hours between 2 and 7 in the morning. The classes associated with night-time hours (Figure 6a for $k = 1, \ldots, 4$) have a similar, reduced width, since water demand in these hours shows little variability relative to the mean. The width increases towards the first hours of light and the time the majority of users wake up, and culminates in the peak hour, 7 a.m. (Figure 6a $k = 6$), where the variability in demand is high. It is further important to observe that the width of the classes in Figure 6a for $k = 6$, i.e., 7 in the morning, and those in Figure 6b for $k = 1$ are the same, as Figure 6b refers to the forecast made at 6 a.m. for the next 6 h, and thus $k = 1$ likewise corresponds to 7 in the morning.

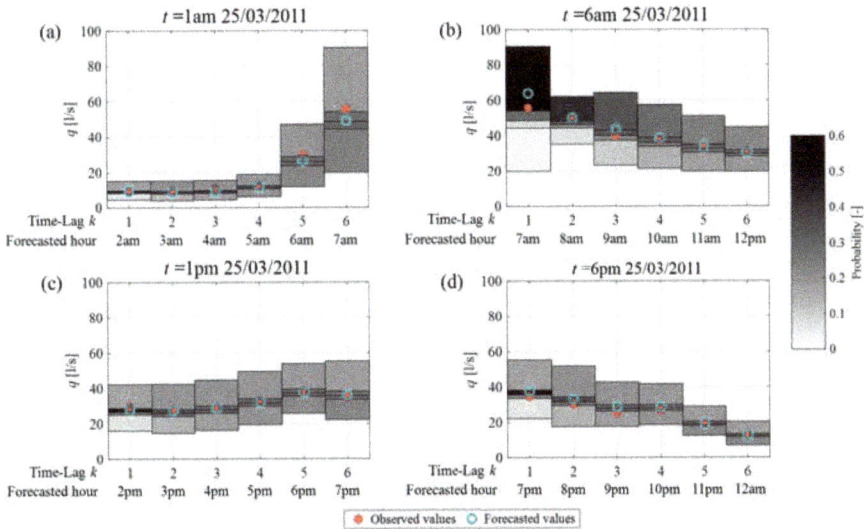

Figure 6. Probabilistic demand forecasts obtained by using model HMC model applied to DMA1 at following times: (**a**) 1 a.m. (with forecasts made from 2 a.m. to 7 a.m.); (**b**) 6 a.m. (with forecasts from 7 a.m. to 12 p.m.); (**c**) 1 p.m. (with forecasts 2 p.m. to 7 p.m.) and (**d**) 6 p.m. (with forecasts from 7 p.m. to 12 a.m.).

On the other hand, with reference to the Figure 6a, considering the shade of grey in the background of each class, which is proportional to the probability of the future value belonging to that same class, we can observe that for $k = 1$, the model indicates a high probability that the future water demand will fall in the middle classes 2 and 3 (and the observed data actually does fall in class 2), whereas the probabilities for classes 4 and 1 are much lower (very light grey). As the forecasting time horizon increases, the probability of the future value falling into each of the different classes tends to become uniform (as shown by increasingly similar shades of grey), which is indicative of higher uncertainty in defining the probability to be assigned to each class. This also emerges from a comparison of the probability distributions associated with forecasts of demand at 7 a.m. that are made one hour ahead (Figure 6b $k = 1$) and 6 h ahead (Figure 6a $k = 6$). This confirms that the uncertainty in defining the probabilities to be assigned to each class is, as expected, greater when the forecast is made several hours in advance. The same comment applies to other two metering areas, DMA 2 and DMA3.

It is worth remarking that the capability of the MC to define the probabilistic behaviour of future water demand is implicitly contained in its structure and, unlike other existing methods, it requires a minimum computational effort. In other words, due to its own structure, the MC can produce a deterministic forecast and, at the same time, a description of the expected probability distribution of the water demand at time $t + k\Delta t$, when the forecast is made at time t. As expected, as the lead time of forecast decreases, the probability estimates become more accurate.

It is also worth noting that we are not presenting, for the MC, uncertainty intervals limited by upper and lower bounds, but instead the entire, even though discretised, probability distribution of the future values as the time of forecast changes. This probability distribution overall characterises the uncertainty, or rather, the variability around the deterministic value forecasted by the model. Unlike the MC method, the ANN model used here, which is based on a standard multi-layer perceptron feedforward ANN, produces only deterministic demand forecasts. This does not mean that it is impossible to produce stochastic forecasts using the ANN—in fact, examples exist in the scientific literature where probabilistic forecasts have been presented e.g., [17,18,20–22]. However, this is possible only at the high computational cost due to Monte Carlo simulations typically used in a post-processing phase.

In summary, when compared to ANN-based and similar existing demand forecasting models, the advantage of the MC model presented here, is in its capacity to produce accurate deterministic forecasts and, at the same time, provide additional information on the probabilistic behaviour of future demands, thus characterising their dispersion around the forecasted value, all with minimum computational effort.

5. Conclusions

A new approach to short-term water demand forecasting based on the Markov chain is presented in this paper. The application of a homogeneous and non-homogeneous Markov chain gave rise to two models, which make it possible to estimate the probabilities of future water demand falling within pre-established classes/intervals and, based on these probabilities, provide a deterministic forecast of the future value.

The two models were applied to the water demand time series of three district metering areas, and the deterministic forecasts obtained were compared with the corresponding ones provided by the ANN and naïve forecasting models. The findings showed that the homogeneous Markov chain model (HMC) delivers better forecasting accuracy, matching the prediction accuracy of the ANN model and surpassing the Naïve model one. The homogeneous Markov chain model (HMC) proved to be distinctly more efficient than the non-homogeneous one (NHMC), and is hence preferred to the latter.

The application of the HMC model further demonstrated that probability estimates provided in relation to the state of future demands make it possible to derive, in a computationally efficient manner, useful considerations regarding the probability distribution of the forecasts. Note that this information is not readily available in either of the two benchmark models, and could be obtained only via post-processing analysis, by computationally using much more expensive Monte Carlo simulations. Note also that all this is achieved using a rather simple conceptual structure of the HMC model, which is easy to implement in a suitable software tool. The model also does not need a large amount of data for the calibration of its parameter, since just one transition matrix has to be estimated.

The future work will involve further comparison of the proposed model's performance to other short-term water demand forecasting models, both deterministic and probabilistic, on several different real life case studies.

Acknowledgments: This study was carried out as part of the PRIN 2012 project "Tools and procedures for an advanced and sustainable management of water distribution systems", No. 20127PKJ4X, funded by MIUR, and of the FIR2016 project "Metodologie gestionali innovative per le reti urbane di distribuzione idrica" funded by University of Ferrara, and under the framework of the Terra & Acqua Tech Laboratory.

Author Contributions: Each of the authors contributed to the design, analysis, and writing of the study.

Conflicts of Interest: The authors declare no conflict of interest. The founding sponsors had no role in the design of the study; in the collection, analyses, or interpretation of data; in the writing of the manuscript, and in the decision to publish the results.

References

1. Donkor, E.A.; Mazzucchi, T.A.; Soyer, R.; Roberson, A.J. Urban Water Demand Forecasting: A Review of Methods and Models. *J. Water Resour. Plan. Manag.* **2014**, *140*, 146–159. [CrossRef]
2. Alvisi, S.; Franchini, M.; Marinelli, A. A short-term, pattern-based model for water-demand forecasting. *J. Hydroinform.* **2007**, *9*, 39–50. [CrossRef]
3. Herrera, M.; Torgo, L.; Izquierdo, J.; Pérez-Garcìa, R. Predictive models for forecasting hourly urban water demand. *J. Hydrol.* **2010**, *387*, 141–150. [CrossRef]
4. Bakker, M.; Vreeburg, J.H.G.; van Schagen, K.M.; Rietveld, L.C. A fully adaptive forecasting model for short-term drinking water demand. *Environ. Model. Softw.* **2013**, *48*, 141–151. [CrossRef]

5. Arandia, E.; Ba, A.; Eck, B.; McKenna, S. Tailoring seasonal time series models to forecast short-term water demand. *J. Water Resour. Plan. Manag.* **2016**, *142*, 1–10. [CrossRef]

6. Tian, D.; Martinez, C.J.; Asce, A.M.; Asefa, T.; Asce, M. Improving Short-Term Urban Water Demand Forecasts with Reforecast Analog Ensembles. *J. Water Resour. Plan. Manag.* **2016**, *142*. [CrossRef]

7. Jain, A.; Varshney, A.K.; Joshi, U.C. Short-term water demand forecast modelling at IIT Kanpur using artificial neural networks. *Water Resour. Manag.* **2001**, *15*, 299–321. [CrossRef]

8. Babel, M.S.; Shinde, V.R. Identifying Prominent Explanatory Variables for Water Demand Prediction Using Artificial Neural Networks: A Case Study of Bangkok. *Water Resour. Manag.* **2011**, *25*, 1653–1676. [CrossRef]

9. Adamowski, J.; Fung Chan, H.; Prasher, S.O.; Ozga-Zielinski, B.; Sliusarieva, A. Comparison of multiple linear and nonlinear regression, autoregressive integrated moving average, artificial neural network, and wavelet artificial neural network methods for urban water demand forecasting in Montreal, Canada. *Water Resour. Res.* **2012**, *48*, W01528. [CrossRef]

10. Campisi-Pinto, S.; Adamowski, J.; Oron, G. Forecasting Urban Water Demand Via Wavelet-Denoising and Neural Network Models. Case Study: City of Syracuse, Italy. *Water Resour. Manag.* **2012**, *26*, 3539–3558. [CrossRef]

11. Odan, F.K.; Fernanda, L.; Reis, R. Hybrid Water Demand Forecasting Model Associating Artificial Neural Network with Fourier Series. *J. Water Resour. Plan. Manag.* **2012**, *138*, 245–256. [CrossRef]

12. Dos Santos, C.C.; Pereira Filho, A.J. Water Demand Forecasting Model for the Metropolitan Area of Sao Paulo, Brazil. *Water Resour. Manag.* **2014**, *28*, 4401–4414. [CrossRef]

13. Romano, M.; Kapelan, Z. Adaptive water demand forecasting for near real-time management of smart water distribution systems. *Environ. Model. Softw.* **2014**, *60*, 265–276. [CrossRef]

14. Al-Zahrani, M.A.; Abo-Monasar, A. Urban Residential Water Demand Prediction Based on Artificial Neural Networks and Time Series Models. *Water Resour. Manag.* **2015**, *29*, 3651–3662. [CrossRef]

15. Shvartser, L.; Shamir, U.; Feldman, M. Forecasting hourly water demands by pattern recognition approach. *J. Water Resour. Plan. Manag.* **1993**, *119*, 611–627. [CrossRef]

16. Alvisi, S.; Franchini, M. Assessment of the predictive uncertainty within the framework of water demand forecasting by using the model conditional processor (MCP). *Urban Water J.* **2017**, *14*, 1–10. [CrossRef]

17. Hutton, C.J.; Kapelan, Z. A probabilistic methodology for quantifying, diagnosing and reducing model structural and predictive errors in short term water demand forecasting. *Environ. Model. Softw.* **2015**, *66*, 87–97. [CrossRef]

18. Azadeh, A.; Neshat, N.; Hamidipour, H. Hybrid Fuzzy Regression-Artificial Neural Network for Improvement of Short-Term Water Consumption Estimation and Forecasting in Uncertain and Complex Environments: Case of a Large Metropolitan City. *J. Water Resour. Plan. Manag.* **2012**, *138*, 71–75. [CrossRef]

19. Bai, Y.; Wang, P.; Li, C.; Xie, J.; Wang, Y. A multi-scale relevance vector regression approach for daily urban water demand forecasting. *J. Hydrol.* **2014**, *517*, 236–245. [CrossRef]

20. Froelich, W. Forecasting Daily Urban Water Demand Using Dynamic Gaussian Bayesian Network. In Proceedings of the 11th International Conference on Beyond Databases, Architectures and Structures, Ustroń, Poland, 26–29 May 2015; Kozielski, S., Mrozek, D., Kasprowski, P., Małysiak-Mrozek, B., Kostrzewa, D., Eds.; Springer International Publishing: Cham, Germany, 2015; pp. 333–342.

21. Froelich, W.; Magiera, E. Forecasting Domestic Water Consumption Using Bayesian Model. In *8th KES International Conference on Intelligent Decision Technologies (KES-IDT 2016)—Part II*; Czarnowski, I., Caballero, A.M., Howlett, R.J., Jain, L.C., Eds.; Springer International Publishing: Cham, Germany, 2016; pp. 337–346.

22. Magiera, E.; Froelich, W. Application of Bayesian Networks to the Forecasting of Daily Water Demand. In Proceedings of the 7th KES International Conference on Intelligent Decision Technologies (KES-IDT 2015), Palace, Italy, 17–19 June 2015; Neves-Silva, R., Jain, L.C., Howlett, R.J., Eds.; Springer International Publishing: Cham, Germany, 2015; pp. 385–393.

23. Todini, E. A model conditional processor to assess predictivee uncertainty in flood forecasting. *Int. J. River Basin Manag.* **2008**, *6*, 123–137. [CrossRef]

24. Cutore, P.; Campisano, A.; Kapelan, Z.; Modica, C.; Savic, D. Probabilistic prediction of urban water consumption using the SCEM-UA algorithm. *Urban Water J.* **2008**, *5*, 125–132. [CrossRef]

25. Vrugt, J.A.; Gupta, H.V.; Bouten, W.; Sorooshian, S. A Shuffled Complex Evolution Metropolis algorithm for optimization and uncertainty assessment of hydrologic model parameters. *Water Resour. Res.* **2003**, *39*, 1201. [CrossRef]

26. Morcous, G. Performance Prediction of Bridge Deck Systems Using Markov Chains. *J. Perform. Constr. Facil.* **2006**, *20*, 146–155. [CrossRef]

27. Yu, G.; Hu, J.; Zhang, C.; Zhuang, L.; Song, J. Short-term traffic flow forecasting based on markov chain model. In Proceedings of the Intelligent Vehicles Symposium, Columbus, OH, USA, 9–11 June 2003; pp. 208–212.

28. Carpinone, A.; Giorgio, M.; Langella, R.; Testa, A. Markov chain modeling for very-short-term wind power forecasting. *Electr. Power Syst. Res.* **2015**, *122*, 152–158. [CrossRef]

29. Yapo, P.; Sorooshian, S.; Gupta, V. A Markov chain flow model with application to flood forecasting. *Water Resour. Res.* **1993**, *29*, 2427–2436. [CrossRef]

30. Benjamin, J.R.; Cornell, C.A. *Probability, Statistics, and Decision for Civil Engineers*; McGraw-Hill: New York, NY, USA, 1970; pp. 321–352.

31. Cooley, J.W.; Lewis, P.; Welch, P. The Fast Fourier Transform and its Applications. *IEEE Trans. Educ.* **1969**, *12*, 28–34. [CrossRef]

32. Gelažanskas, L.; Gamage, K. Forecasting Hot Water Consumption in Residential Houses. *Energies* **2015**, *8*, 12702–12717. [CrossRef]

33. Nash, J.E.; Sutcliffe, J.V. River Flow Forecasting Through Conceptual Models Part I—A Discussion of Principles. *J. Hydrol.* **1970**, *10*, 282–290. [CrossRef]

MDPI

St. Alban-Anlage 66

4052 Basel

Switzerland

Tel. +41 61 683 77 34

Fax +41 61 302 89 18

www.mdpi.com

Water Editorial Office

E-mail: water@mdpi.com

www.mdpi.com/journal/water

www.ingramcontent.com/pod-product-compliance
Lightning Source LLC
Chambersburg PA
CBHW051709210326
41597CB00032B/5419